Lecture Notes in Computer Science   10499

Commenced Publication in 1973
Founding and Former Series Editors:
Gerhard Goos, Juris Hartmanis, and Jan van Leeuwen

More information about this series at http://www.springer.com/series/7407

Mauricio Ayala-Rincón · César A. Muñoz (Eds.)

Interactive Theorem Proving

8th International Conference, ITP 2017
Brasília, Brazil, September 26–29, 2017
Proceedings

 Springer

Editors
Mauricio Ayala-Rincón ⓘD
University of Brasília
Brasília D.F.
Brazil

César A. Muñoz
NASA
Hampton, VA
USA

ISSN 0302-9743   ISSN 1611-3349   (electronic)
Lecture Notes in Computer Science
ISBN 978-3-319-66106-3   ISBN 978-3-319-66107-0   (eBook)
DOI 10.1007/978-3-319-66107-0

Library of Congress Control Number: 2017949523

LNCS Sublibrary: SL1 – Theoretical Computer Science and General Issues

Printed on acid-free paper

This Springer imprint is published by Springer Nature
The registered company is Springer International Publishing AG
The registered company address is: Gewerbestrasse 11, 6330 Cham, Switzerland

Preface

This volume contains the proceedings of the 8th Conference on Interactive Theorem Proving (ITP 2017) held in Brasília, Brazil, on September 26–29, 2017. The conference was organized by the departments of Computer Science and Mathematics of the Universidade de Brasília.

The ITP conference series is concerned with all topics related to interactive theorem proving, ranging from theoretical foundations to applications in program verification, security, and formalization of mathematics. ITP succeeded TPHOLs, which took place every year from 1988 until 2009. Since 2010, ITP has been held in Edinburgh (2010), Nijmegen (2011), Princeton (2012), Rennes (2013), Vienna (2014), Nanjing (2015), and Nancy (2016).

ITP 2017 was part of the Brasília Spring on Automated Reasoning and was co-located with the 26th International Conference on Automated Reasoning with Analytic Tableaux and Related Methods (Tableaux 2017) and the 11th International Symposium on Frontiers of Combining Systems (FroCoS 2017). In addition to the three main conferences, four workshops took place: 12th Logical and Semantic Frameworks with Applications (LSFA 2017), 5th Workshop for Proof eXchange for Theorem Proving (PxTP 2017), EPS - Encyclopedia of Proof Systems, and DaLí - Dynamic Logic: New Trends and Applications. The Brasília Spring on Automated Reasoning also included four tutorials: Proof Compressions and the Conjecture NP = PSPACE, General Methods in Proof Theory for Modal and Substructural Logics, From Proof Systems to Complexity Bounds, and PVS for Computer Scientists.

There were 65 submissions. Each submission was reviewed by at least 3 members of the Program Committee. The reviews were written by the 36 committee members and 69 external reviewers. An electronic PC meeting was held using the EasyChair system. The PC decided to accept 28 regular submissions and 2 rough diamond contributions. The program also included 3 invited talks by Moa Johansson on Automated Theory Exploration for Interactive Theorem Proving: An Introduction to the Hipster System, Cezary Kaliszyk on Automating Formalization by Statistical and Semantic Parsing of Mathematics, and Leonardo de Moura on Whitebox Automation. Cezary Kaliszyk, Katalin Bimbó, and Jasmin Blanchette presented joint TABLEAUX/FroCoS/ITP invited talks.

We would like to thank the PC members for their work, especially during the paper selection process, all the reviewers for writing high-quality reviews, the invited speakers for accepting our invitation and delivering insightful talks, and the authors who submitted their contributions to ITP 2017. Many people helped to make ITP 2017 a success. In particular, we are very grateful to Cláudia Nalon and Daniele Nantes-Sobrinho, who served as Local Organizers at the Universidade de Brasília. Claudia and Daniele worked hard and were highly instrumental in guaranteeing the success of the Brasília Spring on Automated Reasoning.

Last but not least, we are thankful to the sponsors of ITP 2017: Microsoft, the European Association for Artificial Intelligence (EurAI), the District Federal Research Support Foundation (FAPDF), the Coordination of Personnel Training in Higher Education of the Brazilian Education Ministry (CAPES), the Brazilian National Council for Scientific and Technological Development (CNPq), and the Departments of Computer Science and Mathematics of the Universidade de Brasília (UnB).

September 2017

Mauricio Ayala-Rincón
César Muñoz

Organization

Program Committee

Mauricio Ayala-Rincón	Universidade de Brasília (Co-chair), Brazil
César Muñoz	NASA (Co-chair), USA
María Alpuente	Universitat Politècnica de València, Spain
Vander Alves	Universidade de Brasília, Brazil
June Andronick	CSIRO—Data61 and UNSW, Australia
Jeremy Avigad	Carnegie Mellon University, USA
Sylvie Boldo	Inria, France
Ana Bove	Chalmers University of Technology, Sweden
Adam Chlipala	MIT, USA
Gilles Dowek	Inria and ENS Paris-Saclay, France
Aaron Dutle	NASA, USA
Amy Felty	University of Ottawa, Canada
Marcelo Frias	IT Buenos Aires, Argentina
Ruben Gamboa	University of Wyoming, USA
Herman Geuvers	Radboud University Nijmegen, The Netherlands
Elsa Gunter	University of Illinois at Urbana-Champaign, USA
John Harrison	Intel Corporation, USA
Nao Hirokawa	JAIST, Japan
Matt Kaufmann	University of Texas at Austin, USA
Mark Lawford	McMaster University, Canada
Andreas Lochbihler	Institute of Information Security, ETH Zurich, Switzerland
Assia Mahboubi	Inria, France
Panagiotis Manolios	Northeastern University, USA
Gopalan Nadathur	University of Minnesota, USA
Keiko Nakata	SAP Potsdam, Germany
Adam Naumowicz	Institute of Informatics, University of Bialystok, Poland
Tobias Nipkow	TU München, Germany
Scott Owens	University of Kent, UK
Sam Owre	SRI International, USA
Lawrence Paulson	University of Cambridge, UK
Leila Ribeiro	Universidade Federal do Rio Grande do Sul, Brazil
Claudio Sacerdoti Coen	University of Bologna, Italy
Augusto Sampaio	Universidade Federal de Pernambuco, Brazil
Monika Seisenberger	Swansea University, UK
Christian Sternagel	Universtity of Innsbruck, Austria
Sofiene Tahar	Concordia University, Canada
Christian Urban	King's College London, UK
Josef Urban	Czech Technical University in Prague, Czech Republic

ITP Steering Committee

Lawrence Paulson (Chair)	University of Cambridge, UK
David Basin	ETH Zurich, Switzerland
Yves Bertot	Inria, France
Amy Felty	University of Ottawa, Canada
Panagiotis Manolios	Northeastern University, USA
César Muñoz	NASA, USA
Michael Norrish	CSIRO—Data61 and ANU, Australia
Sofiène Tahar	Concordia University, Canada
Christian Urban	King's College London, UK
Jasmin Blanchette (Ex-officio)	Vrije Universiteit Amsterdam, The Netherlands

Organizing Committee

Cláudia Nalon	Universidade de Brasília, Brazil
Daniele Nantes-Sobrinho	Universidade de Brasília, Brazil
Elaine Pimentel	Universidade Federal do Rio Grande do Norte, Brazil
João Marcos	Universidade Federal do Rio Grande do Norte, Brazil

Additional Reviewers

Akbarpour, Behzad
Altenkirch, Thorsten
Asperti, Andrea
Azzi, Guilherme
Ballis, Demis
Bannister, Callum
Beckert, Bernhard
Berger, Ulrich
Besson, Frédéric
Brown, Chad
Castro, Thiago
Chau, Cuong
Claessen, Koen
Cohen, Cyril
Collins, Pieter
Daghar, Alaeddine
Danielsson, Nils Anders
Demeo, William

Escobar, Santiago
Faissole, Florian
Foster, Simon
Färber, Michael
Gacek, Andrew
Goel, Shilpi
Grabowski, Adam
Gutiérrez, Raúl
Helali, Ghassen
Herbelin, Hugo
Hunt, Warren A.
Iyoda, Juliano
Kaliszyk, Cezary
Keller, Chantal
Korniłowicz, Artur
Kozen, Dexter
Krebbers, Robbert
Kullmann, Oliver

Lammich, Peter
Larchey-Wendling, Dominique
Lawrence, Andrew
Lee, Holden
Magaud, Nicolas
Maggesi, Marco
Mahmoud, Mohamed Yousri
Maietti, Maria Emilia
Maric, Filip
Matichuk, Daniel
Melquiond, Guillaume
Miné, Antoine
Miquey, Étienne
Moscato, Mariano
Nakano, Keisuke
Narkawicz, Anthony

Nordvall Forsberg,
  Fredrik
Norrish, Michael
Popescu, Andrei
Rashid, Adnan
Setzer, Anton

Sewell, Thomas
Siddique, Umair
Sozeau, Matthieu
Sternagel, Thomas
Tan, Yong Kiam
Teixeira, Leopoldo

Théry, Laurent
Titolo, Laura
Van Oostrom, Vincent
Villanueva, Alicia
Wiedijk, Freek
Young, William D.

Local Sponsors

Coordination of Personnel Training
in Higher Education of the
Brazilian Education Ministry (CAPES)

District Federal Research Support
Foundation (FAPDF)

Brazilian National Council for Scientific
and Technological Development (CNPq)

Department of Computer Science
Universidade de Brasília - UnB

Department of Mathematics
Universidade de Brasília - UnB

Universidade de Brasília
Instituto de Ciências Exatas

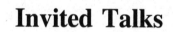

Invited Talks

Whitebox Automation

Leonardo de Moura[1], Jeremy Avigad[2], Gabriel Ebner[3],
Jared Roesch[4], and Sebastian Ullrich[5]

[1] Microsoft Research
leonardo@microsoft.com
[2] Carnegie Mellon University
avigad@andrew.cmu.edu
[3] Vienna University of Technology
gebner@gebner.org
[4] University of Washington
jroesch@cs.washington.edu
[5] Karlsruhe Institute of Technology
ullrich@kit.edu

Abstract. We describe the metaprogramming language currently in use in Lean, a new open source theorem prover that is designed to bridge the gap between interactive use and automation. Lean implements a version of the Calculus of Inductive Constructions. Its elaborator and unification algorithms are designed around the use of type classes, which support algebraic reasoning, programming abstractions, and other generally useful means of expression. Lean also has parallel compilation and checking of proofs, and provides a server mode that supports a continuous compilation and rich user interaction in editing environments such as Emacs, Vim, and Visual Studio Code. Lean currently has a conditional term rewriter, and several components commonly found in state-of-the-art Satisfiability Modulo Theories (SMT) solvers such as forward chaining, congruence closure, handling of associative and commutative operators, and E-matching. All these components are available in the metaprogramming framework, and can be combined and customized by users.

In this talk, we provide a short introduction to the Lean theorem prover and its metaprogramming framework. We also describe how this framework extends Lean's object language with an API to many of Lean's internal structures and procedures, and provides ways of reflecting object-level expressions into the metalanguage. We provide evidence to show that our implementation is performant, and that it provides a convenient and flexible way of writing not only small-scale interactive tactics, but also more substantial kinds of automation.

We view this as important progress towards our overarching goal of bridging the gap between interactive and automated reasoning. Users who develop libraries for interactive use can now more easily develop special-purpose automation to go with them thereby encoding procedural heuristics and expertise alongside factual knowledge. At the same time, users who want to use Lean as a back end to assist in complex verification tasks now have flexible means of adapting Lean's libraries and automation to their specific needs. As a result, our metaprogramming language opens up new opportunities, allowing for more natural and intuitive forms of interactive reasoning, as well as for more flexible and reliable forms of automation.

More information about Lean can be found at http://leanprover.github.io. The interactive book "Theorem Proving in Lean"[1] is the standard reference for Lean. The book is available in PDF and HTML formats. In the HTML version, all examples and exercises can be executed in the reader's web browser.

---

[1] https://leanprover.github.io/theorem_proving_in_lean.

Automated Theory Exploration for Interactive Theorem Proving

An Introduction to the Hipster System

Moa Johansson

Department of Computer Science and Engineering,
Chalmers University of Technology, Gothenburg, Sweden
moa.johansson@chalmers.se

Abstract. Theory exploration is a technique for automatically discovering new interesting lemmas in a mathematical theory development using testing. In this paper I will present the theory exploration system Hipster, which automatically discovers and proves lemmas about a given set of datatypes and functions in Isabelle/HOL. The development of Hipster was originally motivated by attempts to provide a higher level of automation for proofs by induction. Automating inductive proofs is tricky, not least because they often need auxiliary lemmas which themselves need to be proved by induction. We found that many such basic lemmas can be discovered automatically by theory exploration, and importantly, quickly enough for use in conjunction with an interactive theorem prover without boring the user.

Automating Formalization by Statistical and Semantic Parsing of Mathematics

Cezary Kaliszyk[1], Josef Urban[2], and Jiří Vyskočil[2]

[1] University of Innsbruck, Innsbruck, Austria
cezary.kaliszyk@uibk.ac.at
[2] Czech Technical University in Prague, Prague, Czech Republic

Abstract. We discuss the progress in our project which aims to automate formalization by combining natural language processing with deep semantic understanding of mathematical expressions. We introduce the overall motivation and ideas behind this project, and then propose a context-based parsing approach that combines efficient statistical learning of deep parse trees with their semantic pruning by type checking and large-theory automated theorem proving. We show that our learning method allows efficient use of large amount of contextual information, which in turn significantly boosts the precision of the statistical parsing and also makes it more efficient. This leads to a large improvement of our first results in parsing theorems from the Flyspeck corpus.

Contents

Automated Theory Exploration for Interactive Theorem Proving:
An Introduction to the Hipster System

Moa Johansson[✉]

Department of Computer Science and Engineering,
Chalmers University of Technology, Gothenburg, Sweden
moa.johansson@chalmers.se

Abstract. Theory exploration is a technique for automatically discovering new interesting lemmas in a mathematical theory development using testing. In this paper I will present the theory exploration system Hipster, which automatically discovers and proves lemmas about a given set of datatypes and functions in Isabelle/HOL. The development of Hipster was originally motivated by attempts to provide a higher level of automation for proofs by induction. Automating inductive proofs is tricky, not least because they often need auxiliary lemmas which themselves need to be proved by induction. We found that many such basic lemmas can be discovered automatically by theory exploration, and importantly, quickly enough for use in conjunction with an interactive theorem prover without boring the user.

1 Introduction

Theory exploration is a technique for discovering and proving new and interesting basic lemmas about given functions and datatypes. The concept of theory exploration was first introduced by Buchberger [3], to describe the workflow of a human mathematician: instead of proving theorems in isolation, like automated theorem provers do, mathematical software should support an exploratory workflow where basic lemmas relating new concepts to old ones are proved first, before proceeding to complex propositions. This is arguably the mode of usage supported in many modern proof assistants, including Buchberger's Theorema system [4] as well as Isabelle [13]. However, the discovery of new conjectures has mainly been the task for the human user. *Automated theory exploration systems,* [6,9,11,12], aims at addressing this by automatically both discover and prove basic lemmas. In the HipSpec system [6], automated theory exploration has been shown a successful technique for lemma discovery in inductive theorem proving solving several challenge problems where auxiliary lemmas were required. In this paper, we describe HipSpec's sister system Hipster, which is integrated with Isabelle/HOL and in addition produce certified proofs of lemmas and offer the user more flexibility and control over proof strategies.

Hipster consists of two main components: the *exploration component,* called QuickSpec [16], is implemented in Haskell and efficiently generates candidate

© Springer International Publishing AG 2017
M. Ayala-Rincón and C.A. Muñoz (Eds.): ITP 2017, LNCS 10499, pp. 1–11, 2017.
DOI: 10.1007/978-3-319-66107-0_1

conjectures using random testing and heuristics. The conjectures are then passed on to the *prover component* which is implemented in Isabelle. Hipster discards any conjectures with trivial proofs, and outputs snippets of proof scripts for each interesting lemma it discovers. The user can then easily paste the discovered lemmas and their proofs into the Isabelle theory file by a mouse-click, thus assisting and speeding up the development of new theories.

Example 1. As a first simple example consider the following small theory about binary trees with two functions, `mirror`, which recursively swaps the left and right subtrees and `tmap`, which applies a function to each element in the tree[1].

```
datatype 'a Tree =
  Leaf 'a
  | Node "'a Tree" 'a "'a Tree"

fun mirror :: "'a Tree =>'a Tree"
where
  "mirror (Leaf x) = Leaf x"
| "mirror (Node l x r) = Node (mirror r) x (mirror l)"

fun tmap :: "('a =>'b) =>'a Tree =>'b Tree"
where
  "tmap f (Leaf x) = Leaf (f x)"
| "tmap f (Node l x r) = Node (tmap f l) (f x) (tmap f r)"
```

We can ask Hipster to discover some properties about these two functions by issuing a command in the Isabelle theory file telling Hipster which functions it should explore:

`hipster tmap mirror`

Almost immediately, Hipster outputs the following two lemmas (and nothing else), which it has proved by structural induction followed by simplification, using the function definitions above:

```
lemma lemma_a [thy_expl]: "mirror (mirror y) = y"
  apply (induct y)
  apply simp
  apply simp
  done

lemma lemma_aa [thy_expl]: "mirror (tmap y z) = tmap y (mirror z)"
  apply (induct z)
  apply simp
  apply simp
  done
```

---

[1] This example can be found online: https://github.com/moajohansson/IsaHipster/blob/master/Examples/ITP2017/Tree.thy.

Here, Hipster was configured in such a way to consider lemmas requiring inductive proofs interesting, and other conjectures requiring only simplification trivial.

We believe our work on automated theory exploration can complement systems like Sledgehammer [14]. Sledgehammer is a popular tool allowing Isabelle users to call various external automated first order provers and SMT solvers. A key feature of Sledgehammer is its relevance filter which selects facts likely to be useful in proving a given conjecture from Isabelle's huge library, which otherwise would swap the external prover. However, if a crucial lemma is missing, Sledgehammer will fail, as might well be the case in a new theory development.

Current State of the Project

The first version of Hipster has been described in [10]. The Hipster project is ongoing and the system is under active development. The version described in this paper is a snapshot of forthcoming second version. It includes several improvements:

- Hipster now uses the recent QuickSpec 2 [16] as backend for conjecture generation, which is much more efficient than the previously used first version. QuickSpec 2 also has a generic interface via the TIP-language and tools [7,15] avoiding ad-hoc translation from Haskell to Isabelle. Figure 1 shows the new architecture of HipSpec.
- Hipster can use any Isabelle tactic as specified by the user, now also including Sledgehammer, which allows it to exploit knowledge from Isabelle's existing libraries more efficiently. The aim is to make it easy for the user to customise Hipster's proof strategies according to his/her needs.
- The proof output from Hipster has been improved, referring only to standard Isabelle tactics. Unlike the first version, which produced single-line proofs

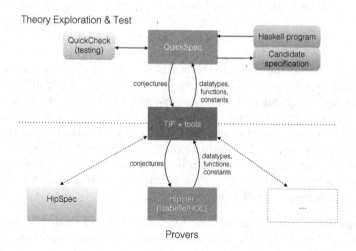

Fig. 1. Theory exploration architecture. Hipster and its sister-system HipSpec.

using a Hipster-specific tactic, the proofs now displays the variable on which Hipster did induction, as well as the tactics and lemmas it used to prove the base- and step cases. This also saves Isabelle re-doing a lot of search when the proof is replayed.

Hipster is open source with code available from GitHub: https://github.com/moajohansson/IsaHipster. We happily invite those interested in Hipster to try it out and welcome contributions to further development.

2   Architecture of a Theory Exploration System

A theory exploration system has two main tasks: First of all, it needs to generate candidate conjectures (of which at least the majority should be theorems) and secondly, it needs access to a sufficiently powerful automated theorem prover to (at least) prove most of the interesting conjectures, and dismiss uninteresting ones. Hipster's conjecture generation is outsourced to QuickSpec 2 [16] and proofs are performed by the tactics of Isabelle/HOL. In this section we describe both these parts.

2.1   Conjecture Generation

A trivial approach to conjecture generation would be to exhaustively generate all possible terms that could be constructed from the input functions and datatypes, but this would quickly become intractable, so some heuristics are necessary. Furthermore, we do not want to waste time trying to prove conjectures that are obviously false, so the conjecture generation should filter those out using testing, or if possible, avoid generating them in the first place.

Earlier theory exploration systems for Isabelle/HOL, IsaCoSy [9], and IsaScheme [12], took different approaches. IsaCoSy was restricted to generate only irreducible terms, starting from small term size, and interleaved inductive proofs with exploration before increasing the term size, so discovered equations could be used to further restrict the search space. IsaScheme generated conjectures by instantiating user provided term schemas (templates) and combined this rewriting and completion. To avoid false conjecture, both IsaCoSy and IsaScheme filtered the resulting conjectures through Isabelle's counter-example checker.

Hipster is considerably faster than both IsaCoSy and IsaScheme, much thanks to QuickSpec's clever conjecture generation. The key idea is that term generation is interleaved with testing and evaluation of terms, using Haskell's QuickCheck tool [5], which enables many terms to be tested at once, instead of one at the time (see Example 2 below). Put simply, the conjecture generation algorithm proceeds by iterating the following steps:

1. Generate new terms of the current term size and add them the the current universe of terms. The algorithm start from term size 1 and iterates up to user-specified max size.

2. Test and evaluate the terms generated so far using QuickCheck. Divide them into equivalence classes.
3. Extract equations from the equivalence classes. Using these equations, prune the search space for the next iteration of term generation when term size is increased.

Example 2 (Conjecture generation in QuickSpec). As a small example, suppose the universe of terms generated so far include the terms in the first column of Table 1 below. QuickSpec will generate many (by default 1000) random test cases and evaluate all terms on these. Initially, all terms are in one equivalence class, but as testing proceeds, terms are split according to which ones evaluate to the same value. Table 1 shows how our small set of terms are split into three equivalence classes using two random tests. Testing would then proceed on many more random values, but no more splits would occur. When the equivalence classes are stable, QuickSpec extracts two equations:

$$rev(rev\ xs) = xs \text{ and } sort(rev\ xs) = sort\ xs.$$

Table 1. How QuickSpec divides terms into equivalence classes based on their evaluation two random test cases. The first test case (top) splits the terms into two equivalence classes. The second test case (bottom) splits off a third equivalence class.

Test-case:	$xs \rightarrow [b, a]$, $ys \rightarrow [\ ]$	
Term	**Instance**	**Evaluation**
xs	[b,a]	[b,a]
rev(rev xs)	rev(rev [b,a])	[b,a]
sort xs	sort [b,a]	[a,b]
sort (rev xs)	sort (rev [b,a])	[a,b]
sort (xs @ ys)	sort([b,a] @ [ ])	[a,b]

Test-case:	$xs \rightarrow [b, a, c]$, $ys \rightarrow [c]$	
Term	**Instance**	**Evaluation**
xs	[b,a,c]	[b,a,c]
rev(rev xs)	rev(rev [b,a,c])	[b,a,c]
sort xs	sort [b,a,c]	[a,b,c]
sort (rev xs)	sort (rev [b,a,c])	[a,b,c]
sort (xs @ ys)	sort([b,a,c] @ [c])	[a,b,c,c]

Note that these conjectures have been tested many times, so they are likely to be true, but they have not yet been proved. QuickSpec's pruner will now use these two equations to restrict its search space. It will prune all terms of the shapes *rev(rev _)* and *sort(rev _)* on account of such terms being reducible by the two equations QuickSpec found. This stops generation of arguably less interesting equations, for example *rev(rev(xs @ ys)) = xs @ ys*, *rev(rev(xs @ ys @ zs)) = xs @ ys @ zs* and so on.

The new version of Hipster described here use QuickSpec 2 where the conjecture generation algorithm has been further refined compared to the simplified version described above, incorporating ideas from both IsaCoSy (avoiding generation of reducible terms) and IsaScheme (generation of schematic terms first). We refer to [16] for details of all heuristics in QuickSpec 2.

QuickSpec was originally designed to generate candidate specifications for Haskell programmes and can also be used as a stand alone light-weight verification tool for this purpose, producing a candidate specification consisting of equations that has been thoroughly tested, but not proved. With QuickSpec 2, an interface using the TIP-language [7], was added to facilitate communication with external systems such as Hipster and its sister system HipSpec [6]. Hipster translates its given input functions and datatypes into TIP before sending them to QuickSpec. Similarly, QuickSpec outputs the resulting conjectures in TIP format, and Hipster translates them back into Isabelle/HOL (see Fig. 1). The TIP-language is based on SMT-LIB [2], with extensions to accommodate recursive datatypes and functions. It was originally designed for creation of a shared benchmark repository for inductive theorem provers. TIP comes with a number of tools for translating between it and various other formats, such as standard SMT-LIB, TPTP [17] and Isabelle/HOL, as well as libraries for facilitating writing additional pretty printers and parsers for other prover languages [15]. QuickSpec should therefore be relatively easy to integrate with additional provers.

2.2 Proving Discovered Conjectures

When Hipster gets the set of candidate conjectures from QuickSpec, it enters its proof loop, where it tries to prove each conjecture in turn. The proof loop is shown in Fig. 2. Hipster is parametrised by two tactics, one for *easy reasoning* and one for *hard reasoning*, with the idea being that conjectures proved by the easy reasoning tactic are trivial, and not interesting enough to be presented to the user. The hard reasoning tactic is more powerful, and the conjectures requiring this tactic are considered interesting and are output to the user. Should the proof fail the first time around, Hipster retries the conjecture at the next iteration if any additional lemmas have been proved in between. Otherwise, the unproved conjectures are also presented to the user. As QuickSpec has tested each conjecture thoroughly it is likely to either be interesting as it is a theorem with a difficult proof, or, have a very subtle counter-example. So far, the combinations of hard and easy reasoning we have experimented with has been various combinations of simplification and/or first order reasoning for the easy reasoning tactic, and some form of induction for hard reasoning. We plan to do a more thorough experimentation with different tactic combinations to extract suitable heuristics.

As mentioned in the previous section, QuickSpec has its own heuristics for reducing the search space and removing seemingly trivial conjectures. However, QuickSpec does not know anything about Isabelle's libraries, nor does it assume

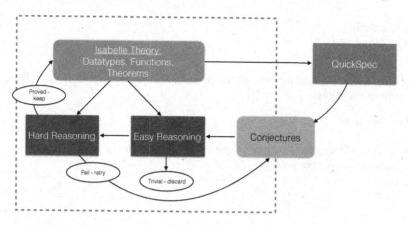

Fig. 2. Hipster's proof loop. Conjectures from QuickSpec are discarded if they can be proved by the *easy reasoning tactic* and presented to the user if they are proved by the *hard reasoning tactic*. Conjectures not proved at the first attempt might be retried once additional lemmas have been proved.

that it necessarily has access to the function definitions (when used as a stand-alone tool, it is designed to be able to explore properties also about Haskell programs which it can test, but not have the source code for). Hence, there will usually always be at least a few trivial conjectures from the Isabelle point of view. Priming QuickSpec with existing facts from Isabelle is future work.

Example 3 (Using Hipster for the maths homework). As an undergraduate at Chalmers, you might be faced with something like the following exercise in your discrete maths class: *Prove that the rev function and the tail-recursive qrev function produce the same result: rev xs = qrev xs [ ].* An Isabelle-savvy student formalises this as follows and tries to prove the exercise using induction and Isabelle's Sledgehammer tool[2]:

```
fun qrev :: "'a list =>'a list =>'a list"
where
  "qrev [] acc  = acc"
| "qrev (x#xs) acc = qrev xs (x#acc)"

theorem hardExercise: "rev xs = qrev xs []"
apply (induct xs)
apply auto
sledgehammer
```

Unfortunately, not even Sledgehammer succeeds in proving the step-case, here using the external provers CVC4 [1] and Z3 [8]:

---

[2] The source code is at: https://github.com/moajohansson/IsaHipster/blob/master/Examples/ITP2017/Qrev.thy.

```
Sledgehammering...
"cvc4": Timed out
"z3": Timed out
```

Something must be missing, so let us use Hipster to see what it can discover about these functions:

```
hipster rev qrev

lemma lemma_a [thy_expl]: "qrev (qrev z y) [] = qrev y z"
apply (induct z arbitrary: y)
apply simp
apply simp
done


lemma lemma_aa [thy_expl]: "rev y @ z = qrev y z"
apply (induct y arbitrary: z)
apply simp
apply simp
apply (metis append_eq_append_conv2 rev.simps(2) rev_append
            rev_singleton_conv rev_swap)
done
```

Hipster returns two interesting lemmas which it needed its *hard reasoning* tactic to prove. In this example, hard reasoning was set to structural induction followed by simplification and/or Sledgehammer[3], while the *easy reasoning* tactic was set to just simplification and/or Sledgehammer. Note that the second discovered lemma, lemma_aa, is a generalisation of our theorem. This is exactly what we need, as is confirmed by Sledgehammer:

```
theorem hardExercise: "rev xs = qrev xs []"
apply (induct xs)
apply auto
sledgehammer
by (metis lemma_aa) (*** This line is now found by Sledgehammer ***)
```

As a matter of fact, we could even prove the exercise without induction now, as it is a special case of lemma_aa.

Example 4 (Configuring Hipster's proof methods). If we were to study the intermediate output from Hipster while it is running on Example 3, we would notice that there are in fact 17 lemmas discovered by QuickSpec, most of which got discarded by Hipster. These include re-discovery of the function definitions (remember, QuickSpec does not assume it has direct access to the source code, only that it can test functions), a couple of lemmas about rev already present in Isabelle's library, and also theorem hardExercise from Example 3. Why did it get discarded?

---

[3] The proof command metis (followed by a list of required library facts) in the proof of lemma_aa is produced by Sledgehammer. Metis is Isabelle's built in first order prover used to reconstruct proofs from external provers.

The anser is simple: The conjectures returned from QuickSpec happens to come in an order so that Hipster tries to prove `hardExercise` before it has tried the essential `lemma_aa`. The first proof attempt therefore fails, and it is returned to the queue of open conjectures (see Fig. 2). In the next iteration of the proof-loop, Hipster has already proved `lemma_aa` and can prove `hardExercise` using just its easy reasoning tactic (here Sledgehammer). Suppose we consider Hipster a bit overzealous in its pruning, and want to see also proofs found by Sledgehammer. We can easily reconfigure it to use a different combination of tactics, for example an easy reasoning tactic which only use simplification with existing Isabelle facts, and a hard reasoning tactic which use Sledgehammer or induction[4]:

```
setup Tactic_Data.set_sledge_induct_sledge

hipster rev qrev
...


lemma lemma_ab [thy_expl]: "qrev (qrev (qrev x2 z) y) x3 =
                                  qrev y (qrev (qrev z x2) x3)"
apply (metis Qrev.lemma_aa append.assoc append.right_neutral lemma_a)
done


lemma lemma_ac [thy_expl]: "qrev y [] = rev y"
apply (metis Qrev.lemma_aa append.right_neutral)
done
```

Now, Hipster keeps two additional lemmas, which both follows from the previously discovered lemmas by first-order reasoning. `lemma_ac` is theorem `hardExercise` with the left- and right-hand sides flipped, while `lemma_ab` is a slightly exotic formulation of associativity for `qrev` and arguably not something a human would come up with.

3   Ongoing and Future Work

We plan to do a more comprehensive evaluation of various tactics in Hipster. As we saw in Example 4, the results of theory exploration are different depending on how we configure the hard- and easy reasoning tactics. Furthermore, there is a trade-off in run-time depending on how powerful we make the respective tactics. Experimental evaluation is needed to decide on some suitable heuristics and default combinations. In the examples shown here, we only used structural induction, but we would also like to compare it in detail to, for instance, recursion induction based on function definitions as default [18]. An extension to co-recursion and co-datatypes is also being developed as part of the MSc project of Sólrún Halla Einarsdottir at Chalmers.

---

[4] The interested reader may consult the file `Tactic_Data.ML` in the Hipster source code repository for details of several pre-defined combinations of easy/hard reasoning tactics, as well as how to define additional ones.

The version of Hipster described here is under active development, and not all features has yet been ported to the new version which uses QuickSpec 2. The first version of Hipster had some very basic support for discovery of conditional equations [18], where the user specified a predicate for the condition, which was passed to QuickSpec 1. Testing conditional equations is tricky, one need to generate test-cases where the condition holds which is a non-trivial task. In QuickSpec 1, the test-cases not satisfying the condition were just discarded, meaning that many extra test-cases had to be evaluated and testing become much slower and false conjectures are more likely to slip through. This has been improved in QuickSpec 2 [16], but at the time of writing not fully integrated in the new version of Hipster.

4   Summary

Hipster is a theory exploration system for Isabelle/HOL. It automatically conjectures and proves basic lemmas about given functions and datatypes, which can be particularly useful as part of an automated inductive theorem prover. Hipster is parametrised by two proof strategies which can be set by the user, one for *easy reasoning* and one for *hard reasoning*. Conjectures solved by easy reasoning (e.g. simplification) are considered trivial and uninteresting, while those requiring hard reasoning (e.g. induction) are considered worth presenting to the user.

Hipster use an external conjecture generation engine called QuickSpec. The systems are connected via an interface language called TIP, which is an extension of SMT-LIB, and related tools for parsing and pretty printing. We believe this interface has potential to be very useful for connecting additional provers wishing to benefit from theory exploration.

Hipster, QuickSpec and TIP are all under active development by our group at Chalmers. We invite anyone interested to test the tools and contribute to their development.

References

1. Barrett, C., Conway, C.L., Deters, M., Hadarean, L., Jovanović, D., King, T., Reynolds, A., Tinelli, C.: CVC4. In: Gopalakrishnan, G., Qadeer, S. (eds.) CAV 2011. LNCS, vol. 6806, pp. 171–177. Springer, Heidelberg (2011). doi:10.1007/978-3-642-22110-1_14
2. Barrett, C., Fontaine, P., Tinelli, C.: The SMT-LIB standard. http://smtlib.cs.uiowa.edu/standard.shtm
3. Buchberger, B.: Theory exploration with theorema. Analele Univ. Din Timis. ser. Mat.-Inform. **38**(2), 9–32 (2000)
4. Buchberger, B., Creciun, A., Jebelean, T., Kovacs, L., Kutsia, T., Nakagawa, K., Piroi, F., Popov, N., Robu, J., Rosenkranz, M., Windsteiger, W.: Theorema: towards computer-aided mathematical theory exploration. J. Appl. Log. **4**(4), 470–504 (2006). Towards Computer Aided Mathematics

5. Claessen, K., Hughes, J.: QuickCheck: a lightweight tool for random testing of Haskell programs. In: Proceedings of ICFP, pp. 268–279 (2000)
6. Claessen, K., Johansson, M., Rosén, D., Smallbone, N.: Automating inductive proofs using theory exploration. In: Bonacina, M.P. (ed.) CADE 2013. LNCS (LNAI), vol. 7898, pp. 392–406. Springer, Heidelberg (2013). doi:10.1007/978-3-642-38574-2_27
7. Claessen, K., Johansson, M., Rosén, D., Smallbone, N.: TIP: tons of inductive problems. In: Kerber, M., Carette, J., Kaliszyk, C., Rabe, F., Sorge, V. (eds.) CICM 2015. LNCS (LNAI), vol. 9150, pp. 333–337. Springer, Cham (2015). doi:10.1007/978-3-319-20615-8_23
8. De Moura, L., Bjørner, N.: Z3: an efficient SMT solver. In: Ramakrishnan, C.R., Rehof, J. (eds.) TACAS 2008. LNCS, vol. 4963, pp. 337–340. Springer, Heidelberg (2008). doi:10.1007/978-3-540-78800-3_24
9. Johansson, M., Dixon, L., Bundy, A.: Conjecture synthesis for inductive theories. J. Autom. Reason. **47**(3), 251–289 (2011)
10. Johansson, M., Rosén, D., Smallbone, N., Claessen, K.: Hipster: integrating theory exploration in a proof assistant. In: Watt, S.M., Davenport, J.H., Sexton, A.P., Sojka, P., Urban, J. (eds.) CICM 2014. LNCS (LNAI), vol. 8543, pp. 108–122. Springer, Cham (2014). doi:10.1007/978-3-319-08434-3_9
11. McCasland, R.L., Bundy, A., Smith, P.F.: Smith.: Ascertaining mathematical theorems. Electron. Notes Theor. Comput. Sci. **151**(1), 21–38 (2006)
12. Montano-Rivas, O., McCasland, R., Dixon, L., Bundy, A.: Scheme-based theorem discovery and concept invention. Expert Syst. Appl. **39**(2), 1637–1646 (2012)
13. Nipkow, T., Paulson, L.C., Wenzel, M.: Isabelle/HOL–A Proof Assistant for Higher-Order Logic. LNCS, vol. 2283. Springer, Heidelberg (2002)
14. Paulson, L.C., Blanchette, J.C.: Three years of experience with sledgehammer, a practical link between automatic and interactive theorem provers. In: IWIL-2010, (2010)
15. Rosén, D., Smallbone, N.: TIP: tools for inductive provers. In: Davis, M., Fehnker, A., McIver, A., Voronkov, A. (eds.) LPAR 2015. LNCS, vol. 9450, pp. 219–232. Springer, Heidelberg (2015). doi:10.1007/978-3-662-48899-7_16
16. Smallbone, N., Johansson, M., Koen, C., Algehed, M.: Quick specifications for the busy programmer. J. Funct. Program. **27**, e18 (2017)
17. Sutcliffe, G.: The TPTP problem library and associated infrastructure: the FOF and CNF parts, v3.5.0. J. Autom. Reason. **43**(4), 337–362 (2009)
18. Lobo Valbuena, I., Johansson, M.: Conditional lemma discovery and recursion induction in Hipster. In: ECEASST, vol. 72 (2015)

Automating Formalization by Statistical and Semantic Parsing of Mathematics

Cezary Kaliszyk[1]([⊠]), Josef Urban[2], and Jiří Vyskočil[2]

[1] University of Innsbruck, Innsbruck, Austria
cezary.kaliszyk@uibk.ac.at
[2] Czech Technical University in Prague, Prague, Czech Republic

Abstract. We discuss the progress in our project which aims to automate formalization by combining natural language processing with deep semantic understanding of mathematical expressions. We introduce the overall motivation and ideas behind this project, and then propose a context-based parsing approach that combines efficient statistical learning of deep parse trees with their semantic pruning by type checking and large-theory automated theorem proving. We show that our learning method allows efficient use of large amount of contextual information, which in turn significantly boosts the precision of the statistical parsing and also makes it more efficient. This leads to a large improvement of our first results in parsing theorems from the Flyspeck corpus.

1 Introduction: Learning Formal Understanding

Computer-understandable (formal) mathematics [17] is still far from taking over the mathematical mainstream. Despite recent impressive formalizations such as the Formal Proof of the Kepler conjecture (Flyspeck) [15], Feit-Thompson [9], seL4 [23], CompCert [26], and CCL [1], formalizing proofs is still largely unappealing to mathematicians. While research on AI and strong automation over large theories has taken off in the last decade [2], so far there has been little progress in automating the understanding of informal LATEX-written and ambiguous mathematical writings.

Automatic parsing of informal mathematical texts into formal ones has been for long time considered a hard or impossible task. Among the state-of-the-art Interactive Theorem Proving (ITP) systems such as HOL (Light) [16], Isabelle [31], Mizar [11] and Coq [4], none includes automated parsing, instead relying on sophisticated formal languages and mechanisms [7,10,13,28]. The past work in this direction – most notably by Zinn [33] – has often been cited as discouraging from such efforts.

C. Kaliszyk—Supported by the ERC Starting grant no. 714034 *SMART*.
J. Urban and J. Vyskočil—Supported by the ERC Consolidator grant no. 649043 *AI4REASON*. This work was supported by the European Regional Development Fund under the project AI&Reasoning (reg. no. CZ.02.1.01/0.0/0.0/15_003/0000466).

M. Ayala-Rincón and C.A. Muñoz (Eds.): ITP 2017, LNCS 10499, pp. 12–27, 2017.
DOI: 10.1007/978-3-319-66107-0_2

We have recently initiated [21,22] a *project to automatically learn formal understanding* of mathematics and exact sciences using a large corpus of alignments [8] between informal and formal statements. Such learning can additionally integrate strong semantic filtering methods such as typechecking combined with large-theory Automated Theorem Proving (ATP). In more detail, we believe that the current state of human-based formalization can be significantly helped by automatically learning how to formalize ("semanticize") informal texts, based on the knowledge available in existing large formal corpora. There are several justifications for this belief:

1. Statistical machine learning (data-driven algorithm design) has been responsible for a number of recent AI breakthroughs, such as web search, query answering (IBM Watson), machine translation (Google Translate), image recognition, autonomous car driving, etc. Given enough data to train on, data-driven algorithms can automatically learn complicated sets of rules that would be often hard to program and maintain manually.
2. The recent progress of formalization, provides reasonably large corpora such as the Flyspeck project [15]. These, together with additional annotation [14], can be used for experiments with machine learning of formalization. The growth of such corpora is only a matter of time, and automated formalization might gradually "bootstrap" this process, making it faster and faster.
3. Statistical machine learning methods have already turned out to be very useful in proof assistant automation in large theories [2], showing that data-driven techniques do apply also to mathematics.
4. Analogously, strong semantic *automated reasoning in large theories* [30] (ARLT) methods are likely to be useful in the formalization field also for complementing the statistical methods that learn formalization. This could lead to hybrid understanding/thinking AI methods that self-improve on large annotated corpora by cycling between (i) statistical prediction of the text disambiguation based on learning from existing annotations and knowledge, and (ii) improving such knowledge by confirming or rejecting the predictions by the semantic ARLT methods.

The last point (4) is quite unique to the domain of (informal/formal) mathematics, and a good independent reason to work on this AI research. There is hardly any other domain where natural language processing (NLP) could be related to such a firm and expressive semantics as mathematics has, which is additionally to a reasonable degree already checkable with existing ITP and ARLT systems. Gradually improving the computer understanding of how mathematicians (ab)use the normal imprecise vocabulary to convey ideas in the semantically well-grounded mathematical world, may even improve the semantic treatment of arbitrary natural language texts.

1.1   Contributions

This paper extends our previous short papers [21,22] on the informal-to-formal translation. We first introduce the informal-to-formal setting (Sect. 2), summarize our initial probabilistic context-free grammar (PCFG) approach of [21]

(Sect. 3), and extend this approach by fast context-aware parsing mechanisms that very significantly improve the performance.

- **Limits of the context-free approach.** We demonstrate on a minimal example, that the context-free setting is not strong enough to eventually learn correct parsing (Sect. 4) of relatively simple informal mathematical formulas.
- **Efficient context inclusion via discrimination trees.** We propose and efficiently implement modifications of the CYK algorithm that take into account larger parsing subtrees (context) and their probabilities (Sect. 5). This modification is motivated by an analogy with large-theory reasoning systems and its efficient implementation is based on a novel use of fast theorem-proving data structures that extend the probabilistic parser.
- **Significant improvement of the informal-to-formal translation performance.** The methods are evaluated, both by standard (non-semantic) machine-learning cross-validation, and by strong semantic methods available in formal mathematics such as typechecking combined with large-theory automated reasoning (Sect. 6).

2   Informalized Flyspeck and PCFG

The ultimate goal of the informal-to-formal traslation is to automatically learn parsing on informal LaTeX formulas that have been aligned with their formal counterparts, as for example done by Hales for his informal and formal Flyspeck texts [14,29]. Instead of starting with LaTeX, where only hundreds of aligned examples are so far available for Flyspeck, we reuse the first large informal/formal corpus introduced previously in [21], based on *informalized* (or *ambiguated*) formal statements created from the HOL Light theorems in Flyspeck. This provides about 22000 informal/formal pairs of Flyspeck theorems.

2.1   Informalized Flyspeck

We apply the following ambiguating transformations [21] to the HOL parse trees to obtain the aligned corpus:

- Merge the 72 overloaded instances defined in HOL Light/Flyspeck, such as (`"+"`, `"vector_add"`). The constant `vector_add` is replaced by + in the resulting sentence.
- Use the HOL Light infix operators to print them as infix in the informalized sentences. Since + is declared as infix, `vector_add u v`, would thus result in u + v.
- Obtain the "prefixed" symbols from the list of 1000 most frequent symbols by searching for: `real_`, `int_`, `vector_`, `nadd_`, `treal_`, `hreal_`, `matrix_`, `complex_` and make them ambiguous by forgetting the prefix.
- Overload various other symbols used to disambiguate expressions, for example the "c"-versions of functions such as `ccos cexp clog csin`, similarly for `vsum`, `rpow`, `nsum`, `list_sum`, etc.
- Remove parentheses, type annotations, and the 10 most frequent casting functors such as `Cx` and `real_of_num`.

2.2   The Informal-to-Formal Translation Task

The *informal-to-formal translation task* is to construct an AI system that will automatically produce the most probable formal (in this case HOL) parse trees for previously unseen informal sentences. For example, the informalized statement of the HOL theorem REAL_NEGNEG:

```
! AO -- -- AO = AO
```

has the formal HOL Light representation shown as a tree in Fig. 1.

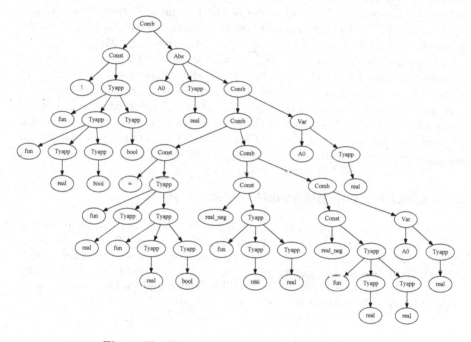

Fig. 1. The HOL Light parse tree of REAL_NEGNEG

Note that all overloaded symbols are disambiguated there, they are applied with the correct arity, and all terms are decorated with their result types. To solve the task, we allow (and assume) training on a sufficiently large corpus of such informal/formal pairs.

2.3   Probabilistic Context Free Grammars

Given a large corpus of corresponding informal/formal formulas, how can we train an AI system for parsing the next informal formula into a formal one? The informal-to-formal domain differs from natural-language domains, where millions of examples of paired (e.g., English/German) sentences are available for training machine translation. The natural languages also have many more words (concepts) than in mathematics, and the sentences to a large extent also lack the recursive structure that is frequently encountered in mathematics. Given that there are currently only thousands of informal/formal examples, purely statistical alignment methods based on n-grams seem inadequate. Instead, the methods have to learn how to compose larger parse trees from smaller ones based on those encountered in the limited number of examples.

A well-known approach ensuring such compositionality is the use of CFG (Context Free Grammar) parsers. This approach has been widely used, e.g., in word-sense disambiguation. A frequently used CFG algorithm is the CYK (Cocke–Younger–Kasami) chart-parser [32], based on bottom-up parsing. By default CYK requires the CFG to be in the Chomsky Normal Form (CNF). The transformation to CNF can cause an exponential blow-up of the grammar, however, an improved version of CYK gets around this issue [25].

In linguistic applications the input grammar for the CFG-based parsers is typically extracted from the *grammar trees* which correspond to the correct parses of natural-language sentences. Large annotated *treebanks* of such correct parses exist for natural languages. The grammar rules extracted from the treebanks are typically ambiguous: there are multiple possible parse trees for a particular sentence. This is why CFG is extended by adding a probability to each grammar rule, resulting in Probabilistic CFG (PCFG).

3   PCFG for the Informal-to-Formal Task

The most straightforward PCFG-based approach would be to directly use the native HOL Light parse trees (Fig. 1) for extracting the PCFG. However, terms and types are there annotated with only a few nonterminals such as: Comb (application), Abs (abstraction), Const (higher-order constant), Var (variable), Tyapp (type application), and Tyvar (type variable). This would lead to many possible parses in the context-free setting, because the learned rules are very universal, e.g.:

> Comb -> Const Var. Comb -> Const Const. Comb -> Comb Comb.

The type information does not help to constrain the applications, and the last rule allows a series of several constants to be given arbitrary application order, leading to uncontrolled explosion.

3.1   HOL Types as Nonterminals

The approach taken in [21] is to first re-order and simplify the HOL Light parse trees to propagate the type information at appropriate places. This gives the context-free rules a chance of providing meaningful pruning information. For example, consider again the raw HOL Light parse tree for REAL_NEGNEG (Fig. 1).

Instead of directly extracting very general rules such as Comb -> Const Abs, each type is first compressed into an opaque nonterminal. This turns the parse tree of REAL_NEGNEG into (see also Fig. 2):

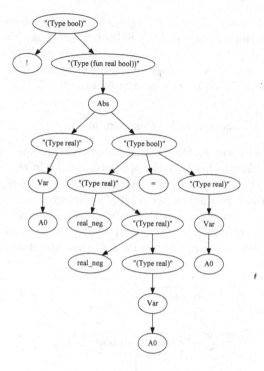

Fig. 2. Transformed tree of REAL_NEGNEG

```
("(Type bool)" ! ("(Type (fun real bool))" (Abs ("(Type real)" (Var A0)) ("(Type bool)"
("(Type real)" real_neg ("(Type real)" real_neg ("(Type real)" (Var A0)))) = ("(Type real)"
(Var A0))))))
```

The CFG rules extracted from this transformed tree thus become more targeted. For example, the two rules:

```
"(Type bool)" -> "(Type real)" = "(Type real)".
"(Type real)" -> real_neg "(Type real)".
```

say that equality of two reals has type bool, and negation applied to reals yields reals. Such learned *probabilistic typing rules* restrict the number of possible parses much more than the general "application" rules extracted from the original HOL Light tree. The rules still have a non-trivial generalization (learning) effect that is needed for the compositional behavior of the information extracted from the trees. For example, once we learn from the training data that the variable ''u'' is mostly parsed as a real number, i.e.:

```
"(Type real)" -> Var u.
```

we will be able to apply real_neg to u even if the subterm real_neg u has never yet been seen in the training examples, and the probability of this parse will be relatively high.

In other words, having the HOL types as *semantic categories* (corresponding e.g. to word senses when using PCFG for word-sense disambiguation) is a reasonable choice for the first experiments. It is however likely that even better semantic categories can be developed, based on more involved statistical and semantic analysis of the data such as *latent semantics* [5].

3.2 Semantic Concepts as Nonterminals

The last part of the original setting wraps ambiguous symbols, such as --, in their disambiguated *semantic/formal concept* nonterminals. In this case $#real_neg would be wrapped around -- in the training tree when -- is used as negation on reals. While the type annotation is often sufficient for disambiguation, such explicit disambiguation nonterminal is more precise and allows easier extraction of the HOL semantics from the constructed parse trees. The actual tree of REAL_NEGNEG used for training the grammar is thus as follows (see also Fig. 3):

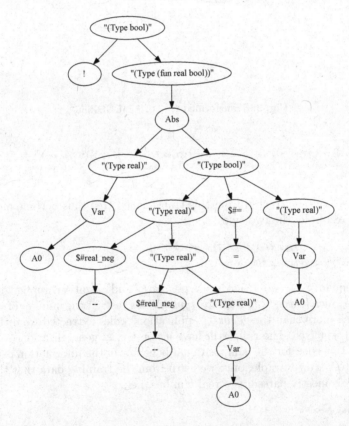

Fig. 3. The parse tree of REAL_NEGNEG used for the actual grammar training

```
("(Type bool)" ! ("(Type (fun real bool))" (Abs ("(Type real)" (Var A0)) ("(Type bool)"
("(Type real)" ($#real_neg --) ("(Type real)" ($#real_neg --) ("(Type real)" (Var A0))))
($#= =) ("(Type real)" (Var A0))))))
```

3.3 Modified CYK Parsing and Its Initial Performance

Once the PCFG is learned from such data, the CYK algorithm augmented with fast internal semantic checks is used to parse the informal sentences. The semantic checks are performed to require compatibility of the types of free variables in parsed subtrees. The most probable parse trees are then typechecked by HOL Light. This is followed by proof and disproof attempts by the HOL(y)Hammer system [18], using all the semantic knowledge available in the Flyspeck library (about 22000 theorems). The first large-scale disambiguation experiment conducted over "ambiguated" Flyspeck in [21] showed that about 40% of the ambiguous sentences have their correct parses among the best 20 parse trees produced by the trained parser. This is encouraging, but certainly invites further research in improving the statistical/semantic parsing methods.

4 Limits of the Context-Free Grammars

A major limiting issue when using PCFG-based parsing algorithms is the context-freeness of the grammar. This is most obvious when using just the low-level term constructors as nonterminals, however it shows often also in the more advanced setting described above. In some cases, no matter how good are the training data, there is no way how to set up the probabilities of the parsing rules so that the required parse tree will have the highest probability. We show this on the following simple example.

Example: Consider the following term t:

```
1 * x + 2 * x.
```

with the following simplified parse tree $T_0(t)$ (see also Fig. 4).

```
(S (Num (Num (Num 1) * (Num x)) + (Num (Num 2) * (Num x))) .)
```

When used as the training data (treebank), the grammar tree $T_0(t)$ results in the following set of CFG rules $G(T_0(t))$:

```
S -> Num .                          Num -> 1
Num ->   Num + Num                  Num -> 2
Num -> Num * Num                    Num -> x
```

This grammar allows exactly the following five parse trees $T_4(t), ..., T_0(t)$ when used on the original (non-bracketed) term t:

```
(S (Num (Num 1) * (Num (Num (Num x) + (Num 2)) * (Num x))) .)
(S (Num (Num 1) * (Num (Num x) + (Num (Num 2) * (Num x)))) .)
(S (Num (Num (Num 1) * (Num (Num x) + (Num 2))) * (Num x)) .)
(S (Num (Num (Num (Num 1) * (Num x)) + (Num 2)) * (Num x)) .)
(S (Num (Num (Num 1) * (Num x)) + (Num (Num 2) * (Num x))) .)
```

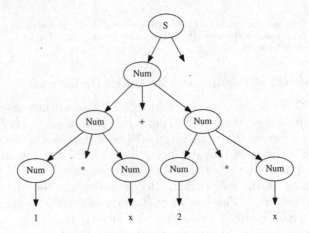

Fig. 4. The grammar tree $T_0(t)$.

Here only the last tree corresponds to the original training tree $T_0(t)$. No matter what probabilities $p(Rule_i)$ are assigned to the grammar rules $G(T_0(t))$, it is not possible to make the priority of + smaller than the priority of *. A context-free grammar forgets the context and cannot remember and apply complex mechanisms such as priorities. The probability of all parse trees is thus in this case always the same, and equal to:

$$p(T_4(t)) = ... = p(T_0(t)) = p(\texttt{S -> Num .}) \times p(\texttt{Num -> Num + Num})$$
$$\times p(\texttt{Num -> Num * Num}) \times p(\texttt{Num -> Num * Num})$$
$$\times p(\texttt{Num -> 1}) \times p(\texttt{Num -> 2}) \times p(\texttt{Num -> x}) \times p(\texttt{Num -> x})$$

While the example's correct parse does not strictly imply the priorities of + and * as we know them, it is clear that we would like the grammar to prefer parse trees that are in some sense *more similar* to the training data. One method that is frequently used for dealing with similar problems in the NLP domain is *grammar lexicalization* [3]. There an additional terminal can be appended to nonterminals and propagated from the subtrees, thus creating many more possible (more precise) nonterminals. This approach however does not solve the particular problem with operator priorities. We also believe that considering probabilities of larger subtrees in the data as we propose below is conceptually cleaner than lexicalization.

5   Using Probabilities of Deeper Subtrees

Our solution is motivated by an analogy with the n-gram statistical machine-translation models, and also with the large-theory premise selection systems. In such systems, characterizing formulas by all deeper subterms and subformulas is feasible and typically considerably improves the performance of the algorithms [20]. Considering subtrees of greater depth for updating the parsing probabilities may initially seem computationally involved. Below we however show that by using efficient ATP-style indexing datastructures such as discrimination trees, this approach becomes feasible, solving in a reasonably clean way some of the inherent problems of the context-free grammars mentioned above.

In more detail, our approach is as follows. We extract not just subtrees of depth 2 from the treebank (as is done by the standard PCFG), but all subtrees up to a certain depth. Other approaches – such as frequency-based rather than depth-based – are possible. During the (modified) CYK chart parsing, the probabilities of the parsed subtrees are adjusted by taking into account the statistics of such deeper subtrees extracted from the treebank. The extracted subtrees are technically treated as new "grammar rules" of the form:

root of the subtree – > list of the children of the subtree

Formally, for a treebank (set of trees) $\mathbb{T}$, we thus define $G^n(\mathbb{T})$ to be the grammar rules of depth n extracted from $\mathbb{T}$. The standard context-free grammar $G(\mathbb{T})$ then becomes $G^2(\mathbb{T})$, and we denote by $G^{n,m}(\mathbb{T})$ where $n \le m$ the union[1] $G^n(\mathbb{T}) \cup ... \cup G^m(\mathbb{T})$. The probabilities of these deeper grammar rules are again learned from the treebank. Our current solution treats the nonterminals on the left-hand sides as disjoint from the old (standard CFG) nonterminals when counting the probabilities (this can be made more complicated in the future). The right-hand sides of such new grammar rules thus contain larger subtrees, allowing to compute the parsing probabilities using more context/structural information than in the standard context-free case.

For the example term t from Sect. 4 this works as follows. After the extraction of all subtrees of depth 2 and 3 and the appropriate adjustment of their probabilities, we get a new extended set of probabilistic grammar rules $G^{2,3}(T_0(t)) \supset G(T_0(t))$. This grammar could again parse all the five different parse trees $T_4(t), ..., T_0(t)$ as in Sect. 4, but now the probabilities $p(T_4(t)), ..., p(T_0(t))$ would in general differ, and an implementation would be able to choose the training tree $T_0(t)$ as the most probable one. In the particular implementation that we use (see Sect. 5.1) its probability is:

$$
\begin{aligned}
p(T_0(t)) = {} & p(\texttt{Num -> (Num 1)}) \times p(\texttt{Num -> (Num x)}) \\
& \times \ p(\texttt{Num -> (Num\_2)}) \times p(\texttt{Num -> (Num x)}) \\
& \times \ p(\texttt{Num -> (Num Num * Num) + (Num Num * Num)}) \\
& \times \ p(\texttt{S -> Num .})
\end{aligned}
$$

Here the second line from the bottom stands for the probability of a subtree of depth 3. For the case of the one-element treebank $T_0(t)$, $p(T_0(t))$ would indeed be the highest probability. On the other hand, the probability of some of the other parses (e.g., $T_4(t)$ and $T_3(t)$ above) would remain unmodified, because in such parses there are no subtrees of depth 3 from the training tree $T_0(t)$.

---

[1] In general, a grammar could pick only some subtree depths instead of their contiguous intervals, but we do not use such grammars now.

5.1   Efficient Implementation of Deeper Subtrees

Discrimination trees [27], as first implemented by Greenbaum [12], index terms in a trie, which keeps single path-strings at each of the indexed terms. A discrimination tree can be constructed efficiently, by inserting terms in the traversal preorder. Since discrimination trees are based on path indexing, retrieval of matching subtrees during the parsing is straightforward.

We use a discrimination tree D to store all the subtrees $G^{n,m}(\mathbb{T})$ from the treebank $\mathbb{T}$ and to efficiently retrieve them together with their probabilities during the chart parsing. The efficiency of the implementation is important, as we need to index about half a million subtrees in D for the experiments over Flyspeck. On the other hand, such numbers have become quite common in large-theory reasoning recently and do not pose a significant problem. For memory efficiency we use OCaml maps (implemented as AVL trees) in the internal nodes of D. The lookup time thus grows logarithmically with the number of trees in D, which is the main reason why we so far only consider trees of depth 3.

When a particular cell in the CYK parsing chart is finished (i.e., all its possible parses are known), the subtree-based probability update is initiated. The algorithm thus consists of two phases: (i) the standard collecting of all possible parses of a particular cell, using the context-free rules $G^2(\mathbb{T})$ only, and (ii) the computation of probabilities, which involves also the deeper (contextual) subtrees $G^{3,m}(\mathbb{T})$.

In the second phase, every parse P of the particular cell is inspected, trying to find its top-level subtrees of depths $3, ..., m$ in the discrimination tree D. If a matching tree T is found in D, the probability of P is recomputed, using the probability of T. There are various ways how to combine the old context-free and the new contextual probabilities. The current method we use is to take the maximum of the probabilities, keeping them as competing methods. As mentioned above, the nonterminals in the new subtree-based rules are kept disjoint from the old context-free rules when computing the grammar rule probabilities. The usual effect is that a frequent deeper subtree that matches the parse P gives it more probability, because such a "deeper context parse" replaces the corresponding two shallow (old context-free) rules, whose probabilities would have to be multiplied.

Our speed measurement with depth 3 has shown that the new implementation is (surprisingly) faster. In particular, when training on all 21695 Flypeck trees and testing on 11911 of them with the limit of 10 best parses, the new version is 23% faster than the old one (10342.75 s vs. 13406.97 s total time). In this measurement the new version also failed to produce at least a single parse less often than the old version (631 vs 818). This likely means that the deeper subtrees help to promote the correct parse, which in the context-free version is considered at some point too improbable to make it into the top 10 parses and consequently discarded.

6   Experimental Evaluation

6.1   Machine Learning Evaluation

The main evaluation is done in the same cross-validation scenario as in [21]. We create the ambiguous sentences (Sect. 2) and the disambiguated grammar trees from all 21695 Flyspeck theorems,[2] permute them randomly and split into 100 equally sized chunks of about 217 trees and their corresponding sentences. The grammar trees serve for training and the ambiguous sentences for evaluation. For each testing chunk C_i ($i \in 1..100$) of 217 sentences we train the probabilistic grammar P_i on the union of the remaining 99 chunks of grammar trees (altogether about 21478 trees). Then we try to get the best 20 parse trees for all the 217 sentences in C_i using the grammar P_i. This is done for the simple context-free version (depth 2) of the algorithm (Sect. 3), as well as for the versions using deeper subtrees (Sect. 5). The numbers of correctly parsed formulas and their average ranks across the several 100-fold cross-validations are shown in Table 1.

Table 1. Numbers of correctly parsed Flyspeck theorems within first 20 parses and their average ranks for subtree depths 2 to 7 of the parsing algorithm (100-fold cross-validation).

Depth	Correct parse found (%)	Avg. rank of correct parse
2	8998 (41.5)	3.81
3	11003 (50.7)	2.66
4	13875 (64.0)	2.50
5	14614 (67.4)	2.34
6	14745 (68.0)	2.13
7	14379 (66.2)	2.17

It is clear that the introduction of deeper subtrees into the CYK algorithm has produced a significant improvement of the parsing precision. The number of correctly parsed formulas appearing among the top 20 parses has increased by 22% between the context-free (depth 2) version and the subtree-based version when using subtrees of depth 3, and it grows by 64% when using subtrees of depth 6.

The comparison of the average ranks is in general only a heuristic indicator, because the number of correct parses found differ so significantly between the methods.[3] However, since the number of parses is higher in the better-ranking methods, this improvement is also relevant. The average rank of the best subtree-based method (depth 6) is only about 56% of the context-free method. The

---

[2] About 1% of the longest Flyspeck formulas were removed from the evaluation to keep the parsing times manageable.

[3] If the context-free version parsed only a few terms, but with the best rank, its average rank would be 1, but the method would still be much worse in terms of the overall number of correctly parsed terms.

results of the best method say that for 68% of the theorems the correct parse of an ambiguous statement is among the best 20 parses, and its average rank among them is 2.13.

6.2   ATP Evaluation

In the ATP evaluation we measure how many of the correctly parsed formulas the HOL(y)Hammer system can prove, and thus help to confirm their validity. While the machine-learning evaluation is for simplicity done by randomization, regardless of the chronological order of the Flyspeck theorems, in the ATP evaluation we only allow facts that were already proved in Flyspeck before the currently parsed formula. Otherwise the theorem-proving task becomes too easy, because the premise-selection algorithm will likely select the theorem itself as the most relevant premise. Since this involves large amount of computation, we only compare the best new subtree-based method (depth 6) from Table 1 (*subtree-6*) with the old context-free method (*subtree-2*).

In the ATP evaluation, the number of the Flyspeck theorems is reduced from 21695 to 17018. This is due to omitting definitions and duplicities during the chronological processing and ATP problem generation. For actual theorem proving, we only use a single (strongest) HOL(y)Hammer method: the distance-weighted k-nearest neighbor (k-NN) [6] using the strongest combination of features [20] with 128 premises and running Vampire 4.0 [24]. Running the full portfolio of 14 AI/ATP HOL(y)Hammer strategies for hundreds of thousands problems would be too computationally expensive.

Table 2 shows the results. In this evaluation we also detect situations when an ambiguated Flyspeck theorem T_1 is parsed as a different known Flyspeck theorem T_2. We call the latter situation *other library theorem (OLT)*. The removal of definitions and duplicitites made the difference in the top-20 correctly parsed sentences even higher, going from 33.8% for subtree-2 to 63.1% in subtree-6. This is an improvement of 86.9%. A correspondingly high increase between subtree-2

Table 2. Statistics of the ATP evaluation for subtree-2 and subtree-6. The total number of theorems tried is 17018 and we require 20 best parses. OLT stands for other library theorem.

	Subtree-2 (%)	Subtree-6 (%)
At least one parse (limit 20)	14101 (82.9)	16049 (94.3)
At least one correct parse	5744 (33.8)	10735 (63.1)
At least one OLT parse	808 (4.7)	1584 (9.3)
At least one parse proved	5682 (33.3)	7538 (44.3)
Correct parse proved	1762 (10.4)	2616 (15.4)
At least one OLT parse proved	525 (3.1)	814 (4.8)
The first parse proved is correct	1168 (6.7)	2064 (12.1)
The first parse proved is OLT	332 (2.0)	713 (4.2)

and subtree-6 is also in the number of situations when the first parse is correct (or OLT) and HOL(y)Hammer can prove it using previous Flyspeck facts. The much greater easiness of proving existing library theorems than proving new theorems explains the high number of provable OLTs when compared to their total number of occurences. Such OLT proofs are however easy to filter out when using HOL(y)Hammer as a semantic filter for the informal-to-formal translation.

7   Conclusion and Future Work

In this paper, we have introduced our project aiming at automated learning of formal understanding of mathematics. In comparison to our first results [21], we have introduced efficient context-based learning and parsing, which significantly increases the success rate of the informal-to-formal translation task on the Flyspeck corpus. The overall improvement in the number of correct parses among the top 20 is 64%, and even higher (86.9%) when omitting duplicities and definitions. The average rank of the correct parse has decreased to about 56% of the previous approach. We believe that the contextual approach to enhancing CYK we took is rather natural (in particular more natural than lexicalization), the discrimination tree indexing scales to this task, and the performance increase is very impressive.

Future work includes adding further semantic checks and better probabilistic ranking subroutines directly into the parsing process. The chart-parsing algorithm is easy to extend with such checks and subroutines, and already the current semantic pruning of parse trees that have incompatible variable types is extremely important. While some semantic relations might eventually be learnable by less efficient learning methods such as recurrent neural networks (RNNs), we believe that the current approach allows more flexible experimenting and nontrivial integration and feedback loops between advanced deductive and learning components. A possible use of RNNs in such a setup is for better ranking of subtrees and for global focusing of the parsing process.

An example of a more sophisticated deductive algorithm that should be easy to integrate is congruence closure over provably equal (or equivalent) parsing subtrees. For example, ''a * b * c'' can be understood with different bracketing, different types of the variables and different interpretations of *. However, * is almost always associative across all types and interpretations. Human readers know this, and rather than considering differently bracketed parses, they focus on the real problem, i.e., which types to assign to the variables and how to interpret the operator in the current context. To be able to emulate this ability, we would cache directly in the chart parsing algorithm the results of large-theory ATP runs on many previously encountered equalities, and use them for fast congruence closure over the subtrees.

Similar ATP/logic-based components also seem necessary for dealing with more involved type systems and human-like parsing layers, such as the one used by the Mizar system. Our first experiments in combining the contextual parsing with ATPs to deal with phenomena like hidden variables and intersection types are described in [19].

References

1. Bancerek, G., Rudnicki, P.: A compendium of continuous lattices in MIZAR. J. Autom. Reason. **29**(3–4), 189–224 (2002)
2. Blanchette, J.C., Kaliszyk, C., Paulson, L.C., Urban, J.: Hammering towards QED. J. Formaliz. Reason. **9**(1), 101–148 (2016)
3. Collins, M.: Three generative, lexicalised models for statistical parsing. In: Cohen, P.R., Wahlster, W. (eds.) Proceedings of the 35th Annual Meeting of the Association for Computational Linguistics and 8th Conference of the European Chapter of the Association for Computational Linguistics, pp. 16–23. Morgan Kaufmann Publishers/ACL (1997)
4. The Coq Proof Assistant. http://coq.inria.fr
5. Deerwester, S.C., Dumais, S.T., Landauer, T.K., Furnas, G.W., Harshman, R.A.: Indexing by latent semantic analysis. JASIS **41**(6), 391–407 (1990)
6. Dudani, S.A.: The distance weighted K-nearest-neighbor rule. IEEE Trans. Syst. Man Cybern. **6**(4), 325–327 (1976)
7. Garillot, F., Gonthier, G., Mahboubi, A., Rideau, L.: Packaging mathematical structures. In: Berghofer, S., Nipkow, T., Urban, C., Wenzel, M. (eds.) TPHOLs 2009. LNCS, vol. 5674, pp. 327–342. Springer, Heidelberg (2009). doi:10.1007/978-3-642-03359-9_23
8. Gauthier, T., Kaliszyk, C.: Matching concepts across HOL libraries. In: Watt, S.M., Davenport, J.H., Sexton, A.P., Sojka, P., Urban, J. (eds.) CICM 2014. LNCS, vol. 8543, pp. 267–281. Springer, Cham (2014). doi:10.1007/978-3-319-08434-3_20
9. Gonthier, G., et al.: A machine-checked proof of the odd order theorem. In: Blazy, S., Paulin-Mohring, C., Pichardie, D. (eds.) ITP 2013. LNCS, vol. 7998, pp. 163–179. Springer, Heidelberg (2013). doi:10.1007/978-3-642-39634-2_14
10. Gonthier, G., Tassi, E.: A language of patterns for subterm selection. In: Beringer, L., Felty, A. (eds.) ITP 2012. LNCS, vol. 7406, pp. 361–376. Springer, Heidelberg (2012). doi:10.1007/978-3-642-32347-8_25
11. Grabowski, A., Korniłowicz, A., Naumowicz, A.: Mizar in a nutshell. J. Formaliz. Reason. **3**(2), 153–245 (2010)
12. Greenbaum, S.: Input transformations and resolution implementation techniques for theorem-proving in first-order logic. Ph.D. thesis, University of Illinois at Urbana-Champaign (1986)
13. Haftmann, F., Wenzel, M.: Constructive type classes in isabelle. In: Altenkirch, T., McBride, C. (eds.) TYPES 2006. LNCS, vol. 4502, pp. 160–174. Springer, Heidelberg (2007). doi:10.1007/978-3-540-74464-1_11
14. Hales, T.: Dense Sphere Packings a Blueprint for Formal Proofs, London Mathematical Society Lecture Note Series, vol. 400. Cambridge University Press, Cambridge (2012)
15. Hales, T.C., Adams, M., Bauer, G., Dang, D.T., Harrison, J., Hoang, T.L., Kaliszyk, C., Magron, V., McLaughlin, S., Nguyen, T.T., Nguyen, T.Q., Nipkow, T., Obua, S., Pleso, J., Rute, J., Solovyev, A., Ta, A.H.T., Tran, T.N., Trieu, D.T., Urban, J., Vu, K.K., Zumkeller, R.: A formal proof of the Kepler conjecture. CoRR, abs/1501.02155, 2015
16. Harrison, J.: HOL Light: a tutorial introduction. In: Srivas, M., Camilleri, A. (eds.) FMCAD 1996. LNCS, vol. 1166, pp. 265–269. Springer, Heidelberg (1996). doi:10.1007/BFb0031814
17. Harrison, J., Urban, J., Wiedijk, F.: History of interactive theorem proving. In: Siekmann, J.H. (ed.) Computational Logic. Handbook of the History of Logic, vol. 9. Elsevier, Amsterdam (2014)

18. Kaliszyk, C., Urban, J.: Learning-assisted automated reasoning with Flyspeck. J. Autom. Reason. **53**(2), 173–213 (2014)
19. Kaliszyk, C., Urban, J., Vyskocil, J.: System description: statistical parsing of informalized Mizar formulas. http://grid01.ciirc.cvut.cz/mptp/synasc17sd.pdf
20. Kaliszyk, C., Urban, J., Vyskočil, J.: Efficient semantic features for automated reasoning over large theories. In: Yang, Q., Wooldridge, M. (eds.) IJCAI 2015, pp. 3084–3090. AAAI Press, Menlo Park (2015)
21. Kaliszyk, C., Urban, J., Vyskočil, J.: Learning to parse on aligned corpora (rough diamond). In: Urban, C., Zhang, X. (eds.) ITP 2015. LNCS, vol. 9236, pp. 227–233. Springer, Cham (2015). doi:10.1007/978-3-319-22102-1_15
22. Kaliszyk, C., Urban, J., Vyskočil, J., Geuvers, H.: Developing corpus-based translation methods between informal and formal mathematics: project description. In: Watt, S.M., Davenport, J.H., Sexton, A.P., Sojka, P., Urban, J. (eds.) CICM 2014. LNCS, vol. 8543, pp. 435–439. Springer, Cham (2014). doi:10.1007/978-3-319-08434-3_34
23. Klein, G., Andronick, J., Elphinstone, K., Heiser, G., Cock, D., Derrin, P., Elkaduwe, D., Engelhardt, K., Kolanski, R., Norrish, M., Sewell, T., Tuch, H., Winwood, S.: seL4: formal verification of an operating-system kernel. Commun. ACM **53**(6), 107–115 (2010)
24. Kovács, L., Voronkov, A.: First-order theorem proving and VAMPIRE. In: Sharygina, N., Veith, H. (eds.) CAV 2013. LNCS, vol. 8044, pp. 1–35. Springer, Heidelberg (2013). doi:10.1007/978-3-642-39799-8_1
25. Lange, M., Leiß, H.: To CNF or not to CNF? an efficient yet presentable version of the CYK algorithm. Inform. Didact. **8**, 1–21 (2009). https://www.infor maticadidactica.de/uploads/Artikel/LangeLeiss2009/LangeLeiss2009.pdf
26. Leroy, X.: Formal verification of a realistic compiler. Commun. ACM **52**(7), 107–115 (2009)
27. Robinson, J.A., Voronkov, A. (eds.): Handbook of Automated Reasoning (in 2 Volumes). Elsevier and MIT Press, Cambridge (2001)
28. Rudnicki, P., Schwarzweller, C., Trybulec, A.: Commutative algebra in the Mizar system. J. Symb. Comput. **32**(1/2), 143–169 (2001)
29. Tankink, C., Kaliszyk, C., Urban, J., Geuvers, H.: Formal mathematics on display: a wiki for Flyspeck. In: Carette, J., Aspinall, D., Lange, C., Sojka, P., Windsteiger, W. (eds.) CICM 2013. LNCS, vol. 7961, pp. 152–167. Springer, Heidelberg (2013). doi:10.1007/978-3-642-39320-4_10
30. Urban, J., Vyskočil, J.: Theorem proving in large formal mathematics as an emerging AI field. In: Bonacina, M.P., Stickel, M.E. (eds.) Automated Reasoning and Mathematics. LNCS, vol. 7788, pp. 240–257. Springer, Heidelberg (2013). doi:10.1007/978-3-642-36675-8_13
31. Wenzel, M., Paulson, L.C., Nipkow, T.: The Isabelle framework. In: Mohamed, O.A., Muñoz, C., Tahar, S. (eds.) TPHOLs 2008. LNCS, vol. 5170, pp. 33–38. Springer, Heidelberg (2008). doi:10.1007/978-3-540-71067-7_7
32. Younger, D.H.: Recognition and parsing of context-free languages in time n^3. Inf. Control **10**(2), 189–208 (1967)
33. Zinn, C.: Understanding informal mathematical discourse. Ph.D. thesis, University of Erlangen-Nuremberg (2004)

A Formalization of Convex Polyhedra
Based on the Simplex Method

Xavier Allamigeon[1][(✉)] and Ricardo D. Katz[2]

[1] Inria and CMAP, Ecole Polytechnique, CNRS, Université Paris–Saclay,
Paris, France
xavier.allamigeon@inria.fr
[2] CIFASIS-CONICET, Rosario, Argentina
katz@cifasis-conicet.gov.ar

Abstract. We present a formalization of convex polyhedra in the proof assistant CoQ. The cornerstone of our work is a complete implementation of the simplex method, together with the proof of its correctness and termination. This allows us to define the basic predicates over polyhedra in an effective way (i.e. as programs), and relate them with the corresponding usual logical counterparts. To this end, we make an extensive use of the Boolean reflection methodology. The benefit of this approach is that we can easily derive the proof of several essential results on polyhedra, such as Farkas Lemma, duality theorem of linear programming, and Minkowski Theorem.

1   Introduction

Convex polyhedra play a major role in many different application areas of mathematics and computer science, including optimization and operations research, control theory, combinatorics, software verification, compilation and program optimization, constraint solving, etc. Their success mainly comes from the fact that they provide a convenient tradeoff between expressivity (conjunction of linear inequalities) and tractability. As an illustration of the latter aspect, linear programming, i.e., the class of convex optimization problems over linear inequality constraints, can be solved in polynomial time [14].

Among the aforementioned applications of polyhedra, there are some which are critical. For instance, in software verification or control theory, polyhedra are used to provide guarantees on the safety of programs [6] or the stability of dynamical systems [12]. On the mathematical side, polyhedra are still a very active research subject. Let us mention Steve Smale's 9th problem for the 21th century (whether linear programming can be solved in strongly polynomial complexity) [17], or the open questions on the diameter of polytopes following the

The authors were partially supported by the programme "Ingénierie Numérique & Sécurité" of ANR, project "MALTHY", number ANR-13-INSE-0003, by a public grant as part of the Investissement d'avenir project, reference ANR-11-LABX-0056-LMH, LabEx LMH and by the PGMO program of EDF and FMJH.

© Springer International Publishing AG 2017
M. Ayala-Rincón and C.A. Muñoz (Eds.): ITP 2017, LNCS 10499, pp. 28–45, 2017.
DOI: 10.1007/978-3-319-66107-0_3

disproof of the Hirsch conjecture [16]. In particular, (informal) mathematical software play an increasing role in testing or disproving conjectures (see e.g. [4]). All this strongly motivates the need to formalize convex polyhedra in a proof assistant, in order to increase the level of trust in their applications.

In this paper, we present the first steps of a formalization of the theory of convex polyhedra in the proof assistant CoQ. A motivation for using CoQ comes from the longer term objective of formally proving some mathematical results relying on large-scale computation (e.g., Santos' counterexample to the Hirsch conjecture [16]). The originality of our approach lies in the fact that our formalization is carried out in an effective way, in the sense that the basic predicates over polyhedra (emptiness, boundedness, membership, etc.) are defined by means of CoQ programs. All these predicates are then proven to correspond to the usual logical statements. The latter take the form of the existence of certificates: for instance, the emptiness of a polyhedron is shown to be equivalent to the existence of a certificate *a la* Farkas (see Corollary 1 for the precise statement). This equivalence between Boolean predicates and formulas living in the kind Prop is implemented by using the boolean reflection methodology, and the supporting tools provided by the Mathematical Components library and its tactic language [11]. The benefit of the effective nature of our approach is demonstrated by the fact that we easily arrive at the proof of important results on polyhedra, such as several versions of Farkas Lemma, duality theorem of linear programming, separation from convex hulls, Minkowski Theorem, etc.

Our effective approach is made possible by implementing the simplex method inside CoQ, and proving its correctness and termination. Recall that the simplex method is the first algorithm introduced to solve linear programming [7]. Two difficulties need to be overcome to formalize it. On the one hand, we need to deal with its termination. More precisely, the simplex method iterates over the so-called bases. Its termination depends on the specification of a pivoting rule, whose aim is to determine, at each iteration, the next basis. In this work, we have focused on proving that the lexicographic rule [8] ensures termination. On the other hand, the simplex method is actually composed of two parts. The part that we previously described, called Phase II, requires an initial basis to start with. Finding such a basis is the purpose of Phase I. It consists in building an extended problem (having a trivial initial basis), and applying to it Phase II. Both phases need to be formalized to obtain a fully functional algorithm.

We point out that our goal here is *not* to obtain a practically efficient implementation of the simplex method (e.g., via the code extraction facility of CoQ). Rather, we use the simplex method as a tool in our proofs and, in fact, it turns out to be the cornerstone of our approach, given the intuitionistic nature of the logic in CoQ. Thus, we adopt the opposite approach of most textbooks on linear programming where, firstly, theoretical results (like the ones mentioned above) are proven, and then the correctness of the simplex method is derived from them.

The formalization presented in this paper can be found in a library developed by the authors called Coq-Polyhedra.[1] As mentioned above, our formalization

---

[1] Available in a git repository at https://github.com/nhojem/Coq-Polyhedra.

is based on the Mathematical Components library (MathComp for short). On top of providing a convenient way to use Boolean reflection, this library contains most of the mathematical tools needed to formalize the simplex method (linear algebra, advanced manipulations of matrices, etc).

Related Work. Our approach has been strongly influenced by the formalization of abstract linear algebra in the Mathematical Components library, which is done in an effective way by exploiting a variant of Gaussian elimination [10].

As far as we know, this is the first formalization of the simplex method in the Calculus of Constructions. In this paradigm, the only work concerning convex polyhedra we are aware of is the implementation of Fourier–Motzkin elimination on linear inequalities in CoQ, leading to a proof of Farkas Lemma [15]. Our work follows a different approach, relying on the theory of linear programming, which has the advantage of providing certificates for the basic predicates over polyhedra. Concerning other families of logics, HOL Light provides a very complete formalization of convex polyhedra, including several important results (Farkas Lemma, Minkowski Theorem, Euler–Poincaré formula, etc) [13]. The classical nature of the logic implemented in HOL Light makes it difficult to compare this work with ours. In Isabelle, an implementation of a simplex-based satisfiability procedure for linear arithmetics has been carried out [18]. This is motivated by obtaining a practical and executable code for SMT solving purposes. Here, we are driven by using the simplex method for mathematical proving, which explains why we obtain a completely different kind of formalization.

Finally, the theory of convex polyhedra is widely used in the area of formal proving as an "informal backend" which helps to establish the validity of some linear inequalities. In more detail, such inequalities are proven by formally checking certificates which are built by untrusted oracles based on linear programming. As illustrations, this allows to automate the deduction of some linear inequalities in proof assistants (see e.g. [3]), or to certify the computations made by static analysis tools [9].

Organization of the Paper. In Sect. 2, we introduce basic concepts and results on polyhedra and linear programming. In Sect. 3, we describe the main components of the simplex method, and start its formalization. The lexicographic rule is dealt with in Sect. 4. The two phases of the simplex method are formalized in Sects. 5 and 6, along with some of the main mathematical results that can be derived from them. Finally, we discuss the outcome of our work in Sect. 7.

By convention, all CoQ definitions, functions, theorems, etc introduced in our work are highlighted in blue. This is to distinguish them from the existing material, in particular, the ones brought from the MathComp library. We inform the reader that the vast majority of the results described in this paper (especially the ones of Sects. 3 to 6) are gathered in the file `simplex.v` of Coq-Polyhedra.

2   Polyhedra, Linear Programming and Duality

A *(convex) polyhedron* is a set of the form $\mathcal{P}(A, b) := \{x \in \mathbb{R}^n \mid Ax \geq b\}$, where $A \in \mathbb{R}^{m \times n}$ and $b \in \mathbb{R}^m$. The notation $\geq$ stands for the partial ordering over vectors, meaning that $y \geq z$ when $y_i \geq z_i$ for all i. In geometric terms, a polyhedron corresponds to the intersection of finitely many halfspaces. A *(affine) halfspace* refers to a set of the form $\{x \in \mathbb{R}^n \mid \langle a, x \rangle \geq \beta\}$, where $a \in \mathbb{R}^n$, $\beta \in \mathbb{R}$, and $\langle \cdot, \cdot \rangle$ stands for the Euclidean scalar product, i.e., $\langle x, y \rangle := \sum_i x_i y_i$.

More generally, convex polyhedra can be defined over any ordered field. This is why our formalization relies on a variable R of the type `realFieldType` of MathComp, whose purpose is to represent an ordered field in which the inequality is decidable. Assume that m and n are variables of type `nat`. The types `'M[R]_(m,n)` and `'cV[R]_m` provided by MathComp respectively represent matrices of size m × n and column vectors of size m with entries of type R. In this paper, we usually omit R in the notation of these types, for the sake of readability. The polyhedron associated with the matrix A:`'M_(m,n)` and the vector b:`'cV_m` is then defined by means of a Boolean predicate, using the construction pred of MathComp:

`Definition polyhedron A b := [pred x:'cV_n | (A *m x) >=m b].`

Here, `*m` stands for the matrix product, and `>=m` for the entrywise ordering of vectors: `y <=m z` if and only if `y i 0 <= z i 0` for all i, where `y i 0` and `z i 0` are respectively the ith entry of the vectors y and z (see `vector_order.v`).

Linear programming consists in optimizing a linear map $x \in \mathbb{R}^n \mapsto \langle c, x \rangle$ over a polyhedron, such as:

$$\begin{array}{ll} \text{minimize} & \langle c, x \rangle \\ \text{subject to} & Ax \geq b, \; x \in \mathbb{R}^n \end{array} \tag{LP(A, b, c)}$$

Let us introduce a bit of terminology. A problem of the form LP(A, b, c) is referred to as a *linear program* (see Fig. 1 for an example). A vector $x \in \mathbb{R}^n$ satisfying the constraint $Ax \geq b$ is a *feasible point* of this linear program. The polyhedron $\mathcal{P}(A, b)$, which consists of the feasible points, is called the *feasible set*. The map $x \mapsto \langle c, x \rangle$ is the *objective* function. The *optimal value* is defined as the infimum of $\langle c, x \rangle$ for $x \in \mathcal{P}(A, b)$. A point $x \in \mathcal{P}(A, b)$ reaching this infimum is called *optimal solution*. When $\mathcal{P}(A, b)$ is not empty, the linear program LP(A, b, c) is said to be *feasible*, and its optimal value is either finite, or $-\infty$ (when the quantity $\langle c, x \rangle$ is not bounded from below over $\mathcal{P}(A, b)$). In the latter case, we say that the linear program is *unbounded (from below)*. Finally, when $\mathcal{P}(A, b)$ is empty, the linear program is *infeasible*, and its value is defined to be $+\infty$.

A fundamental result in linear programming relates the optimal value of LP(A, b, c) with the one of another linear program which is dual to it. In more detail, the *dual linear program* of LP(A, b, c) is the following linear program:

$$\begin{array}{ll} \text{maximize} & \langle b, u \rangle \\ \text{subject to} & A^T u = c, \; u \geq 0, \; u \in \mathbb{R}^m \end{array} \tag{DualLP(A, b, c)}$$

where A^T stands for the transpose of A. Notice that DualLP(A, b, c) is a linear program as well. Indeed, its constraints can be rewritten into a block system $\begin{pmatrix} A^T \\ -A^T \\ I_m \end{pmatrix} u \geq \begin{pmatrix} c \\ -c \\ 0 \end{pmatrix}$, where I_m is the $m \times m$ identity matrix. Besides, the maximization problem can be turned into a minimization problem with objective function $x \mapsto \langle -b, x \rangle$. We denote by $\mathcal{Q}(A, c)$ the feasible set of $\text{DualLP}(A, b, c)$, and we refer to it as the *dual polyhedron*. Assuming c is a variable of type 'cV_n (i.e., representing a vector in $\mathbb{R}^n$), we adopt a specific formalization for this polyhedron, as follows:

`Definition dual_polyhedron A c := [pred u:'cV_m | A^T *m u == c & (u >=m 0)].`

As opposed to the dual linear program, $\text{LP}(A, b, c)$ is referred to as the *primal linear program*. The interplay between the primal and dual linear programs is described by the following result:

Theorem 1 (Strong duality). *If one of the two linear programs* $\text{LP}(A, b, c)$ *or* $\text{DualLP}(A, b, c)$ *is feasible, then they have the same optimal value.*

In addition, when both are feasible, then the optimal value is attained by a primal feasible point $x^* \in \mathcal{P}(A, b)$ *and by a dual feasible point* $u^* \in \mathcal{Q}(A, c)$.

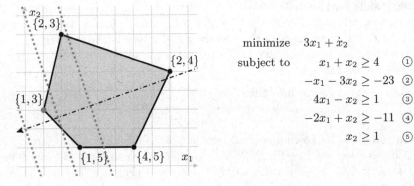

$$\text{minimize} \quad 3x_1 + \dot{x}_2$$
$$\text{subject to} \quad x_1 + x_2 \geq 4 \quad \text{①}$$
$$-x_1 - 3x_2 \geq -23 \quad \text{②}$$
$$4x_1 - x_2 \geq 1 \quad \text{③}$$
$$-2x_1 + x_2 \geq -11 \quad \text{④}$$
$$x_2 \geq 1 \quad \text{⑤}$$

Fig. 1. A linear program. The feasible set is depicted in gray. The direction in which the objective function decreases is represented by a dashdotted oriented line, and some level sets (i.e., sets of the form $\{x \in \mathbb{R}^2 \mid 3x_1 + x_2 = \alpha\}$, where $\alpha \in \mathbb{R}$) are drawn in blue dotted lines. The basic points are represented by dots, and are annotated with the corresponding bases. The optimal basic point is highlighted in green. (Color figure online)

In particular, when $\text{DualLP}(A, b, c)$ is feasible and its optimal value is $+\infty$, the primal linear program $\text{LP}(A, b, c)$ is necessarily infeasible. This holds for any choice of the vector c, including $c = 0$. Observe that $\text{DualLP}(A, b, 0)$ obviously admits $u = 0$ as a feasible point. Hence, we readily obtain a characterization of the emptiness of the polyhedron $\mathcal{P}(A, b)$:

Corollary 1 (Farkas Lemma). *The polyhedron* $\mathcal{P}(A, b)$ *is empty if, and only if, the optimal value of* $\text{DualLP}(A, b, 0)$ *is* $+\infty$, *or, equivalently, there exists* $u \in \mathbb{R}^m$ *such that* $u \geq 0$, $A^T u = 0$ *and* $\langle b, u \rangle > 0$.

The first part of Corollary 1 shows a way to formalize the emptiness property of polyhedra in an effective fashion, e.g., as a program computing the value of DualLP($A, b, 0$) inside the proof assistant and comparing it to $+\infty$. This is precisely the approach that we have adopted in this work. As we shall see in Sect. 7, it also applies to other properties over polyhedra.

3 The Three Ingredients of the Simplex Method

Bases and Basic Points. In order to solve the linear program LP(A, b, c), the simplex method iterates over the feasible bases, up to reaching one corresponding to an optimal solution or concluding that the optimal value is $-\infty$. A *basis* is a subset I of $\{1, \ldots, m\}$ with cardinality n such that the square matrix A_I, formed by the rows A_i of A indexed by $i \in I$, is invertible. With each basis I, it is associated a *basic point* defined as $x^I := (A_I)^{-1} b_I$. The basis I is said to be *feasible* when the point x^I is feasible. It is said to be *optimal* when x^I is an optimal solution of the linear program. We refer to Fig. 1 for an illustration.

In geometric terms, a basis corresponds to a set of n hyperplanes $A_i x = b_i$ which intersect in a single point. The basis is feasible when this point belongs to the feasible set $\mathcal{P}(A, b)$. It can be shown that feasible basic points precisely correspond to the vertices, i.e., the 0-dimensional faces, of the polyhedron $\mathcal{P}(A, b)$.

Formalization of bases and feasible bases is performed by introducing three layers of types. We start with a type corresponding to *prebases*, i.e., subsets of $\{1, \ldots, m\}$ with cardinality n.

```
Inductive prebasis := Prebasis (I: {set 'I_m}) of (#|I| == n).
```

Here, 'I_m stands for the finite subtype of i:nat such that i < m (cf. Interface finType of MathComp). A term I of type {set 'I_m} represents a finite set of elements of type 'I_m, and #|I| corresponds to its cardinality.

Defining bases then requires us to deal with submatrices of the form A_I. This is the purpose of the library row_submx.v of Coq-Polyhedra, where we define:

```
Definition row_submx (A:'M_(m,n)) (I:{set 'I_m}) :=
  (\matrix_(i < #|I|, j < n) A (enum_val i) j) : 'M_(#I, n).
```

In this definition, \matrix_(i < p,j < q) Expr(i,j) is the matrix (of type 'M_(p,q)) whose (i, j) entry is Expr(i,j). The function enum_val retrieves the ith element of the set I. Even when I has cardinality n, the submatrix row_submx A I does not have type 'M_n, i.e., that of square matrices of size n $\times$ n. Indeed, in MathComp, matrices are defined using dependent types (depending on the size). Thus the two types 'M_n and M_(#|I|,n) are distinct, and we use the function castmx to explicitly do the glueing job. The square matrix A_I is thus formalized as follows:

```
Definition matrix_of_prebasis (A:'M_(m,n)) (I:prebasis) :=
  castmx (prebasis_card I, erefl n) (row_submx A I) : 'M_n.
```

where prebasis_card I is a proof of the fact that #|I| = n and erefl n of the fact that n = n. Assuming the variables A:'M_(m,n) and b:'cV_m have been previously declared, the type representing bases is then defined by:

```
Inductive basis := Basis (I:prebasis) of (matrix_of_prebasis A I) \in unitmx.
```

where the type `unitmx` represents the set of invertible matrices. The basic point associated with a basis `I` is determined by a function called `point_of_basis`:

```
Definition point_of_basis (I:basis) :=
  (invmx (matrix_of_prebasis A I)) *m (matrix_of_prebasis b I).
```

where `invmx Q` returns the inverse of the matrix `Q`. From this, we can define the type of feasible bases:

```
Inductive feasible_basis :=
  FeasibleBasis (I:basis) of point_of_basis I \in polyhedron A b.
```

Reduced Costs. The simplex method stops when the current feasible basic point is an optimal solution of $\mathrm{LP}(A, b, c)$. This is determined thanks to the so-called *reduced cost vector*. The reduced cost vector associated with the basis I is defined as $u := A_I^{-T}c$, where A_I^{-T} denotes the inverse of the transpose matrix of A_I. On the Coq side, assuming `c` is a variable of type `'cV_n`, this leads to:

```
Definition reduced_cost_of_basis (I:basis) :=
  (invmx (matrix_of_prebasis A I)^T) *m c : 'cV_n.
```

where `Q^T` stands for the transpose of the matrix `Q`. When $u \geq 0$ and I is feasible, the associated basic point is optimal:

```
Lemma optimal_cert_on_basis (I:feasible_basis) :
  (reduced_cost_of_basis I) >=m 0 ->
    forall y, y \in polyhedron A b -> '[c, point_of_basis I] <= '[c,y].
```

Here, the notation `'[.,.]` corresponds to the scalar product $\langle \cdot, \cdot \rangle$ (see the file `inner_product.v` in Coq-Polyhedra).

Strong duality lies in the core of the simplex method. To see this, consider the *extended reduced cost vector* $\bar{u} \in \mathbb{R}^m$, which is defined by $\bar{u}_i := u_i$ if $i \in I$, and $\bar{u}_i := 0$ otherwise. On the Coq side, this extended vector is built by the function `ext_reduced_cost_of_basis: basis -> 'cV_m`. When $u \geq 0$, $\bar{u}$ is a feasible point of the dual linear program $\mathrm{DualLP}(A, b, c)$, and it has the same objective value as the basic point x^I:

```
Lemma ext_reduced_cost_dual_feasible (I:basis) :
  let: u := reduced_cost_of_basis I in
  u >=m 0 = (ext_reduced_cost_of_basis I \in dual_polyhedron A c).
Lemma eq_primal_dual_value (I:basis) :
  '[c, point_of_basis I] = '[b, ext_reduced_cost_of_basis I].
```

As a consequence, proving the termination of the simplex method is one of the possible ways to establish the duality theorem of linear programming.

Pivoting. *Pivoting* refers to the operation of moving from a feasible basis to a "better" one, chosen according to what is known as the *pivoting rule*. More precisely, when the reduced cost vector u associated with the current feasible basis I does not satisfy $u \geq 0$, the pivoting rule selects an index $i \in I$ such that $u_i < 0$, which is called the *leaving variable*, and builds the direction vector $d := (A_I)^{-1}e_i$ (where e_i is the ith vector of the canonical base of $\mathbb{R}^n$):

```
Definition direction (I:basis) (i:'I_n) :=
  let: ei := (delta_mx i 0):'cV_n in
  (invmx (matrix_of_prebasis A I)) *m ei.
```

along which the objective function $x \mapsto \langle c, x \rangle$ decreases:

```
Lemma direction_improvement c (I:basis) (i:'I_n) :
  let: u := reduced_cost_of_basis c I in
  u i 0 < 0 -> '[c, direction I i] < 0.
```

As a consequence, the simplex method moves along the halfline $\{x^I + \lambda d \mid \lambda \geq 0\}$ in order to decrease the value of the objective function. When d is a *feasible direction*, i.e., $Ad \geq 0$, this halfline is entirely contained in the polyhedron $\mathcal{P}(A, b)$. In this case, we can easily show that the linear program $\mathrm{LP}(A, b, c)$ is unbounded:

```
Definition feasible_dir A := [pred d | (A *m d) >=m 0].
Lemma unbounded_cert_on_basis (I:feasible_basis) (i:'I_n) :
  let: u := reduced_cost_of_basis c I in
  let: d := direction I i in
  feasible_dir A d -> u i 0 < 0 ->
  forall M, exists x, (x \in polyhedron A b) /\ ('[c,x] < M).
```

In contrast, if d is not a feasible direction, moving along the halfline $\{x^I + \lambda d \mid \lambda \geq 0\}$ makes the simplex method eventually hit the boundary of one of the halfspaces $\{x \in \mathbb{R}^n \mid A_j x \geq b_j\}$ delimiting $\mathcal{P}(A, b)$. This happens precisely when λ reaches the threshold value $\bar{\lambda}$ defined by:

$$\bar{\lambda} = \min_j \left\{ \frac{b_j - A_j x^I}{A_j d} \mid A_j d < 0 \right\}. \tag{1}$$

The indexes attaining the minimum in Eq. (1) correspond to the halfspaces which are hit. Then, the pivoting rule selects one of them, say j, which is called the *entering variable*, and the next basis is defined as $J := (I \setminus \{i\}) \cup \{j\}$. In this way, it can be shown that J is a feasible basis, and that $\langle c, x^J \rangle \leq \langle c, x^I \rangle$.

The major difficulty arising in this scheme is the possibility that $\bar{\lambda} = 0$, or, equivalently, that several bases correspond to the same basic point. Such bases are said to be *degenerate*, and constitute the only obstacle to the termination of the simplex method. In the presence of degenerate bases, the pivoting rule needs to choose carefully the entering and leaving variables in order to avoid cycling over them. Our formalization of the simplex method is based on a rule having this property, called the *lexicographic rule* [8], which is described in the next section.

4 Lexicographic Pivoting Rule

In informal terms, the lexicographic rule acts as if the vector b was replaced by a perturbed vector $\tilde{b}$ defined by $\tilde{b}_i := b_i - \varepsilon^i$, where ε is a small positive parameter (here ε^i is the usual exponentiation). The advantage of perturbing b in such a way is that there is no degenerate basis anymore. However, as we shall see, the feasible bases of the polyhedron $\mathcal{P}(A, \tilde{b})$ only form a subset of the feasible bases of $\mathcal{P}(A, b)$. The former are called *lex-feasible bases*, and they constitute the set of bases over which the simplex method with the lexicographic rule iterates.

In the formalization, which is carried out in Section Lexicographic_rule of simplex.v, we have chosen to use a symbolic perturbation scheme in order to avoid dealing with numerical values for ε.[2] In this symbolic perturbation scheme, a row vector $v = (v_0, \ldots, v_m) \in \mathbb{R}^{1 \times (1+m)}$ encodes the perturbed quantity $v_0 + \sum_{i=1}^{m} v_i \varepsilon^i$. The vector $\tilde{b}$ is then implemented as a row block matrix built from the column vector b and the opposite of the identity matrix -(1%:M) of order m:

Definition b_pert := (row_mx b -(1%:M)):'M_(m,1+m).

In this way, the matrix b_pert can be thought of as a column vector whose ith entry is the row vector $(b_i, 0, \ldots, 0, -1, 0, \ldots, 0)$, representing the quantity $b_i - \varepsilon^i$, as desired. Given a basis, the associated "perturbed" basic point is then:

Definition point_of_basis_pert (I:basis) :=
  (invmx (matrix_of_prebasis A I)) *m (matrix_of_prebasis b_pert I).

Now we can define the type of lex-feasible bases:

Definition is_lex_feasible (I:basis) :=
  [forall i, ((row i A) *m (point_of_basis_pert I)) >=lex (row i b_pert)].
Inductive lex_feasible_basis :=
  LexFeasibleBasis (I:basis) of is_lex_feasible I.

where >=lex is the lexicographic ordering over row vectors (see vector_order.v in Coq-Polyhedra). We first observe that any lex-feasible basis is feasible:

Lemma lex_feasible_basis_is_feasible (I:lex_feasible_basis): is_feasible I.

Following the description of the pivoting step in Sect. 3, we now assume that the variables I:lex_feasible_basis and i:'I_n have been declared, and we make the following assumptions:

Hypothesis leaving: (reduced_cost_vector_of_basis I) i 0 < 0.
Hypothesis infeas_dir: ~~(feasible_dir A (direction I i)).

where ~~b stands for the negation of the Boolean b. Our aim is to determine an entering variable j. In the symbolic perturbation scheme, every ratio appearing in Eq. (1) turns out to be a row vector encoding a perturbed quantity:

Definition lex_gap (d:'cV_n) (j:'I_m) :=
  let: x_pert := point_of_basis_pert I in
  ((A *m d) j 0)^-1 *: ((row j b_pert) - ((row j A) *m x_pert)) : 'rV_(1+m).

In order to obtain in the perturbed setting the analog of the threshold value $\bar{\lambda}$ defined in Eq. (1), we determine the minimum of these ratios in the lexicographic sense, using the function lex_min_seq S introduced in vector_order.v. The entering variable is then computed as follows:

Definition lex_ent_var_nat :=
  let: d := direction I i in
  let: J := [seq j <- (enum 'I_m) | (A *m d) j 0 < 0] in

---

[2] Finding how small ε must be chosen is tedious, and this would make proofs unnecessarily complicated.

```
let: min_gap := lex_min_seq [seq lex_gap d j | j <- J] in
  find (fun j => (j \in J) && (min_gap == lex_gap d j)) (enum 'I_m).
```

where the MathComp function `find p S` returns the index of the first item in the sequence `S` for which the predicate `p` holds, if any. Next, we prove that the result (of type `nat`) returned by `lex_ent_var_nat` is strictly less than `m`, which allows us to convert it into an element of type `'I_m` called `lex_ent_var`. We are finally ready to build the next basis:

Definition `lex_rule_set := lex_ent_var |: (I :\ (enum_val [...] i)))`.

where `k |: S` and `S :\ k` respectively adds and removes the element `k` from the set `S`. With this definition, we show that the lexicographic rule provides a lex-feasible basis called `lex_rule_lex_bas`, by proving the following successive results:

Lemma `lex_rule_card : #|lex_rule_set| == n`.
Lemma `lex_rule_is_basis : is_basis (Prebasis lex_rule_card)`.
Lemma `lex_rule_lex_feasibility : is_lex_feasible (Basis lex_rule_is_basis)`.
Definition `lex_rule_lex_bas := LexFeasibleBasis lex_rule_lex_feasibility`.

We finally prove that the analog of the objective function in the perturbed setting is *strictly* decreasing in the lexicographic sense:

Lemma `lex_rule_dec : let: J := lex_rule_lex_bas in`
  `(c^T *m point_of_basis_pert I) >lex (c^T *m point_of_basis_pert J)`.

As mentioned above, this comes from the fact that the analog of the threshold λ in this setting is nonzero, thanks to the absence of degenerate bases:

Lemma `eq_pert_point_imp_eq_bas (I I':basis)` :
  `point_of_basis_pert I = point_of_basis_pert I' -> I == I'`.

Let us sketch the proof of this key result. Recall that `point_of_basis_pert I` is a $n \times (1 + m)$-matrix. Given `j:'I_m`, we can show that the `(1+j)`th column of this matrix is nonzero if, and only if, `j` belongs to `I` (we refer to Lemma `col_point_of_basis_pert` in `simplex.v`). Indeed, since the matrix A_I is invertible, the `(1+j)`th column of `point_of_basis_pert I` is nonzero if, and only if, the `(1+j)`th column of `matrix_of_prebasis b_pert I` is. By construction of `b_pert`, the latter column vector has only zero entries, except in the case where `j \in I` (in this case, the entry corresponding to the index of `j` in `I` is -1).

5   Phase II of the Simplex Method, and Farkas Lemma

Phase II. In this section, we present our formalization of *Phase II* of the simplex method. We do it before the one of Phase I because as we will explain in Sect. 6, Phase II is used in Phase I. Phase II of the simplex method determines the optimal value of the linear program $LP(A, b, c)$, supposing that an initial feasible basis `bas0:feasible_basis` is known. *De facto*, this makes the underlying assumption that the linear program is feasible.

Our implementation of Phase II, which is developed in **Section Phase2** of `simplex.v`, consists in iterating the function `lex_rule_lex_bas` until finding an optimal basis (i.e. identifying that the associated reduced cost vector is non-negative), or determining that the linear program is unbounded (i.e. identifying that the direction vector is feasible). Termination is expected to be guaranteed by Lemma `lex_rule_dec` and the fact that the number of bases is finite. In addition, it looks reasonable to start the iteration of `lex_rule_lex_bas` from the basis `bas0`. However, albeit feasible, the basis `bas0` has no reason to be lex-feasible. Fortunately, it can be shown that, up to reordering the inequalities defining $\mathcal{P}(A, b)$, we can make `bas0` be lex-feasible. Instead of applying permutations on the rows of A and b, we choose to apply the inverse permutation on the symbolic perturbation components of `b_pert`, and leave the initial problem $LP(A, b, c)$ unchanged. As a consequence, we modify the previous definition of `b_pert` as follows:

```
Definition b_pert := (row_mx b (-(perm_mx s))).
```

where `s : 'S_m` represents a permutation of the set $\{1, \ldots, m\}$, and `perm_mx` builds the corresponding permutation matrix (see the libraries `perm` and `matrix` of MathComp). All the previous results remain valid under this change. The only difference is that they are now additionally parametrized by the permutation `s`, appearing as a global variable in Section `Lexicographic_rule`. For reason of space, we omit the description of the construction of the permutation `s0` associated with `bas0`. We only mention that it satisfies the expected result:

```
Lemma feasible_to_lex_feasible : is_lex_feasible s0 bas0.
```

The function performing one iteration of the Phase II algorithm with the lexicographic rule is built as follows:

```
Definition basic_step (bas: lex_feasible_basis) :=
let u := reduced_cost_of_basis c bas in
if [pick i | u i 0<0] is Some i (* picks i such that u i 0<0, if any *)
    then let d := direction bas i in
       if (@idPn (feasible_dir A d)) is ReflectT infeas_dir
       then Lex_next_basis (lex_rule_lex_bas infeas_dir)
       else Lex_final (Lex_res_unbounded (bas, i))
    else Lex_final (Lex_res_optimal_basis bas).
```

where `@idPn (feasible_dir A d)` returns a proof `infeas_dir` of the fact that the direction vector `d` is not feasible, when this is the case. As a consequence, the function `basic_step` returns either a next basis (constructor `Lex_next_basis`), or indicates that the method should stop (constructor `Lex_final`).

The recursive function which iterates the function `basic_step` is the following:

```
Function lex_phase2 bas {measure basis_height bas} :=
  match basic_step bas with
  | Lex_final final_res => final_res
  | Lex_next_basis bas' => lex_phase2 bas'
  end.
```

It is defined in the framework provided by the library `RecDef` of COQ, see [2]. More precisely, its termination (and subsequently, the fact that COQ accepts the definition) is established by identifying an integer quantity which is strictly decreased every time the function `basic_step` returns a next basis:

```
Definition basis_height bas :=  #| [ set bas':(lex_feasible_basis s0) |
  (c^T *m (point_of_basis_pert s0 bas')) <lex
      (c^T *m (point_of_basis_pert s0 bas)) ] |.
```

This quantity represents the number of lex-feasible bases for which the value of the "perturbed" objective function is (lexicographically) strictly less than the value of this function at the current lex-feasible basis. The fact that `basis_height` decreases at every iteration is a consequence of Lemma `lex_rule_dec`.

Gathering all these components, we finally arrive at the definition of the function implementing Phase II:

```
Definition phase2 :=
  let: lex_bas0 := LexFeasibleBasis feasible_to_lex_feasible in
  lex_to_phase2_final_result ((@lex_phase2 s0) c lex_bas0).
```

We present the correctness specification of this function by means of an adhoc inductive predicate. Such a presentation is idiomatic in the Mathematical Components library. The advantage is that it provides a convenient way to perform case analysis on the result of `phase2`.

```
Inductive phase2_spec : phase2_final_result -> Type :=
| Phase2_unbounded (p: feasible_basis * 'I_n) of
  (reduced_cost_of_basis c p.1) p.2 0 < 0 /\ feasible_dir A
    (direction p.1 p.2) : phase2_spec (Phase2_res_unbounded p)
| Phase2_optimal_basis (bas: feasible_basis) of
  (reduced_cost_of_basis c bas) >=m 0 :
    phase2_spec (Phase2_res_optimal_basis bas).skip
Lemma phase2P : phase2_spec phase2.
```

More precisely, Lemma `phase2P` states that when the function `phase2` returns a result of the form `Phase2_res_unbounded (bas, i)`, the pair `(bas, i)` satisfies `(reduced_cost_of_basis c bas) i 0 < 0` and `feasible_dir A (direction bas i)`. It precisely corresponds to the hypotheses of Lemma `unbounded_cert_on_basis`, and indicates that $LP(A, b, c)$ is unbounded. Similarly, if the result of `phase2` is of the form `Phase2_res_optimal_basis bas`, we have `(reduced_cost_of_basis c bas) >=m 0`, i.e., the basis `bas` is an optimal basis (see Lemma `optimal_cert_on_basis`).

Effective Definition of Feasibility, and Farkas Lemma. We can now formalize the notion of feasibility, i.e., the property that the polyhedron $\mathcal{P}(A, b)$ is empty or not, as a Boolean predicate.[3] We still assume that the variables A and b are declared. Following the discussion at the end of Sect. 2, the predicate is defined by means of the function `phase2` executed on the dual problem $DualLP(A, b, 0)$. To this end, we first build a feasible basis `dual_feasible_bas0` for this problem, whose associated basic point is the vector $0 \in \mathbb{R}^m$. Feasibility of the polyhedron $\mathcal{P}(A, b)$ is then defined as follows:

---

[3] We make a slight abuse of language, since feasibility usually applies to linear programs. By extension, we apply it to polyhedra: $\mathcal{P}(A, b)$ is *feasible* if it is nonempty.

```
Definition feasible :=
  if phase2 dual_feasible_bas0 (-b) is Phase2_res_optimal_basis _ then
    true else false.
```

Note that -b corresponds to the objective function of the dual linear program (when written as a minimization problem). The correctness of our definition is established by showing that the predicate feasible is equivalent to the existence of a point $x \in \mathcal{P}(A, b)$. This is presented by means of Boolean reflection, using the reflect relation of MathComp:

```
Lemma feasibleP : reflect (exists x, x \in polyhedron A b) feasible.
```

We point out that the feasibility certificate x is constructed from the extended reduced cost vector of the optimal basis of DualLP$(A, b, 0)$ returned by phase2.

In a similar way, we prove the following characterization of the emptiness of $\mathcal{P}(A, b)$, which precisely corresponds to Farkas Lemma:

```
Definition dual_feasible_dir := [pred d | (A^T *m d == 0) && (d >=m 0)].
Lemma infeasibleP :
  reflect (exists d, dual_feasible_dir A d /\ '[b,d] > 0) (~~feasible).
```

Indeed, ~~feasible amounts to the fact that phase2 returns an unboundedness certificate Phase2_res_unbounded (bas,i) for DualLP$(A, b, 0)$. The emptiness certificate d of $\mathcal{P}(A, b)$ is then obtained from the dual feasible direction direction bas i.

6   Complete Implementation of the Simplex Method

The Pointed Case. In order to obtain a full formalization of the simplex method, it remains to implement a *Phase I* algorithm. Its purpose is twofold: (i) determine whether the linear program LP(A, b, c) is feasible or not, (ii) in the former case, return an initial feasible basis for Phase II. There is one obstacle to the definition of such a Phase I algorithm: even if a linear program is feasible, it may not have any feasible basis. For instance, consider the linear program over the variables x_1, x_2 which aims at minimizing x_2 subject to $-1 \leq x_2 \leq 1$. The feasible set is a cylinder around the x_1-axis, and it does not have any vertex, or, equivalently, basic point. A necessary and sufficient condition for the existence of a feasible basis is that the rank of A is n. When this condition is fulfilled, the feasible set $\mathcal{P}(A, b)$ is said to *pointed*. We now describe the Phase I algorithm under this assumption. This is developed in Section Pointed_simplex of simplex.v.

From the hypothesis on the rank of A, we can extract an invertible square submatrix of A, which provides an initial basis bas0 of LP(A, b, c). Beware that this basis is not necessarily a feasible one. As a consequence, we split the inequalities in the system $Ax \geq b$ into two complementary groups, $A_K x \geq b_K$ and $A_L x \geq b_L$, where the K is the set of indexes $i \in \{1, \ldots, m\}$ for which the basic point point_of_basis bas0 does not satisfy the inequality $A_i x \geq b_i$, and $L := \{1, \ldots, m\} \setminus K$. We denote by p the cardinality of the set K. Phase I is based on applying Phase II algorithm to the following "extended" problem over the vector $z = (x, y) \in \mathbb{R}^{n+p}$:

$$\text{minimize} \quad \langle e, y - A_K x \rangle$$
$$\text{subject to} \quad A_K x \le b_K + y, \ A_L x \ge b_L, \ y \ge 0, \ (x, y) \in \mathbb{R}^{n+p} \qquad (\text{LP}_{\text{Phase I}})$$

where $e \in \mathbb{R}^p$ stands for the all-1-vector. The constraints defining $\text{LP}_{\text{Phase I}}$ are gathered into a single system $A_{\text{ext}} z \ge b_{\text{ext}}$. Similarly, the objective function of $\text{LP}_{\text{Phase I}}$ can be rewritten as a sole linear function $z = (x, y) \mapsto \langle c_{\text{ext}}, z \rangle$.

The linear program $\text{LP}_{\text{Phase I}}$ has two important properties. On the one hand, its optimal value can be bounded (from below) by the quantity $M_{\text{ext}} := \langle e, -b_K \rangle$:

```
Definition Mext := '[const_mx 1, - (row_submx b K)].
Lemma cext_min_value z : (z \in polyhedron Aext bext) -> '[cext, z] >=
    Mext.
```

On the other hand, the optimal value of $\text{LP}_{\text{Phase I}}$ is equal to M_{ext} if, and only if, the original problem $\text{LP}(A, b, c)$ is feasible. The "only if" implication follows from the following lemma, which also provides a feasibility witness of $\text{LP}(A, b, c)$:

```
Lemma feasible_cext_eq_min_active z :
  z \in polyhedron Aext bext -> '[cext,z] = Mext ->
  (usubmx z \in polyhedron A b).
```

where the MathComp function usubmx returns the upper subvector x of a block vector of the form $z = \binom{x}{y}$. Regarding the "if" implication, an infeasibility certificate of $\text{LP}(A, b, c)$ can be constructed by means of a feasible point $\bar{u} \in \mathbb{R}^{m+p}$ of the dual of $\text{LP}_{\text{Phase I}}$ whose objective value $\langle b_{\text{ext}}, \bar{u} \rangle$ is strictly greater than M_{ext}. This certificate is built by the following function:

```
Definition dual_from_ext (u:'cV[R]_(m+p)) :=
  \col_i (if i \in K then 1 - (usubmx u) i 0 else (usubmx u) i 0).
```

where \col_i Expr(i) is the column vector whose ith entry is Expr(i). As expected, this certificate satisfies:

```
Lemma dual_polyhedron_from_ext u :
  (u \in dual_polyhedron Aext cext) -> dual_feasible_dir A (dual_from_ext
  u).
Lemma dual_from_ext_obj u :'[bext, u] > Mext ->'[b, dual_from_ext u] > 0.
```

In this way, we readily obtain a proof that $\text{LP}(A, b, c)$ is infeasible, by using Lemma infeasibleP.

Finally, we can build an initial feasible basis **feasible_bas0_ext** for $\text{LP}_{\text{Phase I}}$ by considering the union of the basis bas0 with the set $\{m + 1, \ldots, m + p\}$ of the indexes of the last p constraints $y \ge 0$ of $\text{LP}_{\text{Phase I}}$.[4] As a consequence, we can apply phase2 to solve $\text{LP}_{\text{Phase I}}$, starting from the basis feasible_bas0_ext. In this way, we obtain an optimal basis bas of $\text{LP}_{\text{Phase I}}$. If the associated basic point z satisfies '[cext,z] > Mext, we build an infeasibility certificate of $\text{LP}(A, b, c)$ using the function dual_from_ext, as described above. Otherwise, we construct a feasible basis bas' of $\text{LP}(A, b, c)$. This is performed by the function extract_feasible_basis which we do not describe here for the sake of concision. Then, we use bas' to execute phase2 on $\text{LP}(A, b, c)$ and finally obtain its optimal value.

---

[4] We let the reader check that the associated basic point is $\binom{x}{0} \in \mathbb{R}^{n+p}$, where x is the basic point associated with the basis bas0, and that this point is feasible.

The previous discussion precisely describes the way we have implemented the function `pointed_simplex`, which completely solves the linear program $\mathrm{LP}(A, b, c)$ under the pointedness assumption.

The General Case. In general, we can always reduce to the pointed case by showing that $\mathrm{LP}(A, b, c)$ is equivalent to the following linear program in which the original variable $x \in \mathbb{R}^n$ is substituted by $v - w$ with $v, w \geq 0$:

$$
\begin{aligned}
\text{minimize} \quad & \langle c, (v - w) \rangle \\
\text{subject to} \quad & A(v - w) \geq b, \ v \geq 0, \ w \geq 0, \ (v, w) \in \mathbb{R}^{n+n}
\end{aligned}
\qquad (\mathrm{LP_{Pointed}})
$$

The feasible set of $\mathrm{LP_{Pointed}}$ is pointed because of the constraints $v, w \geq 0$. Thus, we can apply to it the function `pointed_simplex` of the previous section. In this way, we define the function `simplex`, which is able to solve any linear program $\mathrm{LP}(A, b, c)$. It is implemented in Section `General_simplex` of `simplex.v`. Its correctness proof is formalized by means of the following inductive type:

```
Inductive simplex_spec : simplex_final_result -> Type :=
| Infeasible d of (dual_feasible_dir A d /\ '[b, d] > 0):
    simplex_spec (Simplex_infeasible d)
| Unbounded p of [/\ (p.1 \in polyhedron A b), (feasible_dir A p.2) &
    ('[c,p.2] < 0)] : simplex_spec (Simplex_unbounded p)
| Optimal_point p of [/\ (p.1 \in polyhedron A b),
    (p.2 \in dual_polyhedron A c) & '[c,p.1] = '[b, p.2]] :
    simplex_spec (Simplex_optimal_point p).
Lemma simplexP: simplex_spec simplex.
```

In other words, when `simplex` returns a result of the form `Simplex_infeasible d`, then `d` is a certificate of infeasibility of $\mathrm{LP}(A, b, c)$, see Lemma `infeasibleP`.

Similarly, the unboundedness of the linear program $\mathrm{LP}(A, b, c)$ is characterized by the fact that `simplex` returns a result of the form `Simplex_unbounded (x,d)`. Equivalently, we can define a predicate corresponding to this situation, and prove that it is correct, as follows:

```
Definition unbounded :=
  if simplex is (Simplex_unbounded _) then true else false.
Lemma unboundedP :
  reflect (forall M, exists y, y \in polyhedron A b /\ '[c,y] < M)
    unbounded.
```

Given any M, the certificate `y` is built by taking a point of the form $x + \lambda\,d$, where $\lambda \geq 0$ is sufficiently large.

Finally, when `simplex` returns `Simplex_optimal_point (x,u)`, this means that `x` is an optimal solution of $\mathrm{LP}(A, b, c)$, and `u` is a dual feasible element which certifies its optimality (i.e., $\langle c, x \rangle = \langle b, u \rangle$). Thanks to this, we can define in an effective way the fact that $\mathrm{LP}(A, b, c)$ admits an optimal solution (we say that the linear program is *bounded*), and, in this case, deal with the optimal value:

```
Definition bounded :=
```

```
   if simplex is (Simplex_optimal_point _) then true else false.
Definition opt_value :=
   if simplex is (Simplex_optimal_point (x,_)) then '[c,x] else 0.
Lemma boundedP :
   reflect ((exists x, x \in polyhedron A b /\ '[c,x] = opt_value) /\ (
       forall y, y \in polyhedron A b -> opt_value <= '[c,y])) bounded.
```

7   Outcome of the Effective Approach

Duality results immediately follow from the correctness statements of the simplex method and the resulting predicates feasible, unbounded and bounded. For instance, when $LP(A, b, c)$ and $DualLP(A, b, c)$ are both feasible, we have:

```
Theorem strong_duality : feasible A b -> dual_feasible A c ->
   exists p, [/\ p.1 \in polyhedron A b, p.2 \in dual_polyhedron A c
         & '[c,p.1] = '[b,p.2]].
```

which corresponds to the second part of Theorem 1. The remaining cases of Theorem 1 (when one of the two linear programs is infeasible) are dealt with in the file duality.v. All these statements are obtained in a few lines of proof. We also obtain another well-known form of Farkas Lemma, characterizing the logical implication between linear inequalities (Lemma farkas_lemma_on_inequalities).

The membership to the convex hull of a finite set of points is another property which can be defined in an effective way in our framework. Recall that a point $x \in \mathbb{R}^n$ belongs to the convex hull of a (finite) set $V = \{v^i\}_{1 \leq i \leq p} \subset \mathbb{R}^n$ if there exists $\lambda \in \mathbb{R}^p$ such that $x = \sum_{i=1}^p \lambda_i v^i$, $\lambda \geq 0$ and $\sum_i \lambda_i = 1$. The latter constraints define a polyhedron over $\lambda \in \mathbb{R}^p$, and the membership of x amounts to fact that this polyhedron is feasible. This is how we arrive at the definition of a Boolean predicate is_in_convex_hull, see the file minkowski.v. The *separation result* states that if x does not belong to the convex hull of V, then there is a hyperplane *separating* x from V. This means that x is located on one side of the hyperplane, while the points of V are on the other side. Formalizing V as the matrix of size $n \times p$ with columns v^i, we establish this result as follows:

```
Theorem separation (x: 'cV_n) :
~~(is_in_convex_hull x) -> exists c, [forall i, '[c, col i V] > '[c, x]].
```

The certificate c can be built directly from the infeasibility certificate of the underlying polyhedron over $\lambda \in \mathbb{R}^p$. Our proof of the separation result reduces to the technical manipulations of block matrices performing this conversion.

Finally, Minkowski Theorem states that every bounded polyhedron equals the convex hull of its vertices. We recover this result as the extensional equality of the predicates polyhedron A b and is_in_convex_hull matrix_of_vertices, where matrix_of_vertices is the matrix whose columns are the basic points of $\mathcal{P}(A, b)$:

```
Theorem minkowski : bounded_polyhedron A b ->
   polyhedron A b =i is_in_convex_hull matrix_of_vertices.
```

The most difficult part of the statement is proven in a few lines: if $x \in \mathcal{P}(A, b)$ does not belong to the convex hull of the basic points, $\texttt{Lemma separation}$ exhibits a separating hyperplane c such that $\langle c, x \rangle < \langle c, x^I \rangle$ for all feasible bases I of $\mathcal{P}(A, b)$. However, the program $\texttt{pointed\_simplex}$ is able to provide an optimal feasible basis I^*, i.e., which satisfies $\langle c, x^{I^*} \rangle \leq \langle c, x \rangle$. This yields a contradiction.

8   Conclusion

We have presented a formalization of convex polyhedra in COQ. Its main feature is that it is based on an implementation of the simplex method, leading to an effective formalization of the basic predicates over polyhedra. We have illustrated the outcome of this approach with several results of the theory of convex polyhedra. As a future work, we plan to deal with faces, which are a central notion in the combinatorial theory of polyhedra (the early steps of an effective definition of faces are already available in the file $\texttt{face.v}$ of Coq-Polyhedra). The simplex method should also greatly help us to prove adjacency properties on faces, in particular, properties related with the connectivity of the (vertex-edge) graph of polyhedra. Another direction of work is to exploit our library to certify computational results on polyhedra, possibly on large-scale instances. A basic problem is to formally check that a certain polyhedron (defined by inequalities) is precisely the convex hull of a certain set of points. This is again a problem in which the simplex method plays an important role [1]. To cope with the computational aspects, we plan to investigate how to translate our formally proven statements to lower-level data structures, like in [5].

Acknowledgments. The authors are very grateful to A. Mahboubi for her help to improve the presentation of this paper, and to G. Gonthier, F. Hivert and P.-Y. Strub for fruitful discussions. The second author is also grateful to M. Cristiá for introducing him to the topic of automated theorem proving. The authors finally thank the anonymous reviewers for their suggestions and remarks.

References

1. Avis, D., Fukuda, K.: A pivoting algorithm for convex hulls and vertex enumeration of arrangements and polyhedra. Discrete Comput. Geom. **8**(3), 295–313 (1992)
2. Barthe, G., Forest, J., Pichardie, D., Rusu, V.: Defining and reasoning about recursive functions: a practical tool for the Coq proof assistant. In: Hagiya, M., Wadler, P. (eds.) FLOPS 2006. LNCS, vol. 3945, pp. 114–129. Springer, Heidelberg (2006). doi:10.1007/11737414_9
3. Besson, F.: Fast reflexive arithmetic tactics the linear case and beyond. In: Altenkirch, T., McBride, C. (eds.) TYPES 2006. LNCS, vol. 4502, pp. 48–62. Springer, Heidelberg (2007). doi:10.1007/978-3-540-74464-1_4
4. Bremner, D., Deza, A., Hua, W., Schewe, L.: More bounds on the diameters of convex polytopes. Optim. Methods Softw. **28**(3), 442–450 (2013)
5. Cohen, C., Dénès, M., Mörtberg, A.: Refinements for free! In: Gonthier, G., Norrish, M. (eds.) CPP 2013. LNCS, vol. 8307, pp. 147–162. Springer, Cham (2013). doi:10.1007/978-3-319-03545-1_10

6. Cousot, P., Halbwachs, N.: Automatic discovery of linear restraints among variables of a program. In: Proceedings of POPL 1978, Tucson, Arizona. ACM Press (1978)
7. Dantzig, G.B.: Maximization of a linear function of variables subject to linear inequalities. In: Activity Analysis of Production and Allocation. Wiley (1951)
8. Dantzig, G.B., Orden, A., Wolfe, P.: The generalized simplex method for minimizing a linear form under linear inequality restraints. Pac. J. Math. 5(2), 183–195 (1955)
9. Fouilhe, A., Boulmé, S.: A certifying frontend for (sub)polyhedral abstract domains. In: Giannakopoulou, D., Kroening, D. (eds.) VSTTE 2014. LNCS, vol. 8471, pp. 200–215. Springer, Cham (2014). doi:10.1007/978-3-319-12154-3_13
10. Gonthier, G.: Point-free, set-free concrete linear algebra. In: van Eekelen, M., Geuvers, H., Schmaltz, J., Wiedijk, F. (eds.) ITP 2011. LNCS, vol. 6898, pp. 103–118. Springer, Heidelberg (2011). doi:10.1007/978-3-642-22863-6_10
11. Gonthier, G., Mahboubi, A., Tassi, E.: A small scale reflection extension for the Coq system. Research Report RR-6455, Inria Saclay Ile de France (2016)
12. Guglielmi, N., Laglia, L., Protasov, V.: Polytope Lyapunov functions for stable and for stabilizable LSS. Found. Comput. Math. 17(2), 567–623 (2017)
13. Harrison, J.: The HOL light theory of Euclidean space. J. Autom. Reason. 50, 173–190 (2013)
14. Khachiyan, L.: Polynomial algorithms in linear programming. USSR Comput. Math. Math. Phys. 20(1), 53–72 (1980)
15. Sakaguchi, K.: VASS (2016). https://github.com/pi8027/vass
16. Santos, F.: A counterexample to the Hirsch conjecture. Ann. Math. 176(1), 383–412 (2012)
17. Smale, S.: Mathematical problems for the next century. Math. Intell. 20, 7–15 (1998)
18. Spasić, M., Marić, F.: Formalization of incremental simplex algorithm by stepwise refinement. In: Giannakopoulou, D., Méry, D. (eds.) FM 2012. LNCS, vol. 7436, pp. 434–449. Springer, Heidelberg (2012). doi:10.1007/978-3-642-32759-9_35

A Formal Proof of the Expressiveness of Deep Learning

Alexander Bentkamp[1,2(✉)], Jasmin Christian Blanchette[1,3], and Dietrich Klakow[2]

[1] Vrije Universiteit Amsterdam, Amsterdam, The Netherlands
{a.bentkamp,j.c.blanchette}@vu.nl
[2] Universität des Saarlandes, Saarland Informatics Campus, Saarbrücken, Germany
s8albent@stud.uni-saarland.de, dietrich.klakow@lsv.uni-saarland.de
[3] Max-Planck-Institut für Informatik,
Saarland Informatics Campus, Saarbrücken, Germany
jasmin.blanchette@mpi-inf.mpg.de

Abstract. Deep learning has had a profound impact on computer science in recent years, with applications to image recognition, language processing, bioinformatics, and more. Recently, Cohen et al. provided theoretical evidence for the superiority of deep learning over shallow learning. We formalized their mathematical proof using Isabelle/HOL. The Isabelle development simplifies and generalizes the original proof, while working around the limitations of the HOL type system. To support the formalization, we developed reusable libraries of formalized mathematics, including results about the matrix rank, the Borel measure, and multivariate polynomials as well as a library for tensor analysis.

1 Introduction

Deep learning algorithms enable computers to perform tasks that seem beyond what we can program them to do using traditional techniques. In recent years, we have seen the emergence of unbeatable computer go players, practical speech recognition systems, and self-driving cars. These algorithms also have applications to image recognition, bioinformatics, and many other domains. Yet, on the theoretical side, we are only starting to understand why deep learning works so well. Recently, Cohen et al. [14] used tensor theory to explain the superiority of deep learning over shallow learning for one specific learning architecture called convolutional arithmetic circuits (CACs).

Machine learning algorithms attempt to model abstractions of their input data. A typical application is image recognition—i.e., classifying a given image in one of several categories, depending on what the image depicts. The algorithms usually learn from a set of data points, each specifying an input (the image) and a desired output (the category). This learning process is called training. The algorithms generalize the sample data, allowing them to imitate the learned output on previously unseen input data.

M. Ayala-Rincón and C.A. Muñoz (Eds.): ITP 2017, LNCS 10499, pp. 46–64, 2017.
DOI: 10.1007/978-3-319-66107-0_4

CACs are based on sum–product networks (SPNs), also called arithmetic circuits [30]. An SPN is a rooted directed acyclic graph with input variables as leaf nodes and two types of inner nodes: sums and products. The incoming edges of sum nodes are labeled with real-valued weights, which are learned by training.

CACs impose the structure of the popular convolutional neural networks (CNNs) onto SPNs, using alternating convolutional and pooling layers, which are realized as collections of sum nodes and product nodes, respectively. These networks can be shallower or deeper—i.e., consist of few or many layers—and each layer can be arbitrarily small or large, with low- or high-arity sum nodes. CACs are equivalent to similarity networks, which have been demonstrated to perform as well as CNNs, if not better [13].

Cohen et al. prove two main theorems about CACs: the fundamental and the generalized theorem of network capacity (Sect. 3). The generalized theorem states that CAC networks enjoy complete depth efficiency: In general, to express a function captured by a deeper network using a shallower network, the shallower network must be exponentially larger than the deeper network. By "in general," we mean that the statement holds for all CACs except for a Lebesgue null set S in the weight space of the deeper network. The fundamental theorem is a special case of the generalized theorem, where the expressiveness of the deepest possible network is compared with the shallowest network. Cohen et al. present both theorems in a variant where weights are shared across the networks and a more flexible variant where they are not.

As an exercise in mechanizing modern research in machine learning, we developed a formal proof of the fundamental theorem for networks with nonshared weights using the Isabelle/HOL proof assistant [27]. To simplify our work, we recast the original proof into a more modular version (Sect. 4), which generalizes the result as follows: S is not only a Lebesgue null set, but also a subset of the zero set of a nonzero multivariate polynomial. This stronger theorem gives a clearer picture of the expressiveness of deep CACs.

The formal proof builds on general libraries that we either developed or enriched (Sect. 5). We created a library for tensors and their operations, including product, CP-rank, and matricization. We added the matrix rank and its properties to Thiemann and Yamada's matrix library [33], generalized the definition of the Borel measure by Hölzl and Himmelmann [19], and extended Lochbihler and Haftmann's polynomial library [17] with various lemmas, including the theorem stating that zero sets of nonzero multivariate polynomials are Lebesgue null sets. For matrices and the Lebesgue measure, an issue we faced was that the definitions in the standard Isabelle libraries have types that are too restrictive: The dimensionality of the matrices and of the measure space is parameterized by types that encode numbers, whereas we needed them to be terms.

Building on these libraries, we formalized the fundamental theorem for networks with nonshared weights (Sect. 6). CACs are represented using a datatype that is flexible enough to capture networks with and without concrete weights. We defined tensors and polynomials to describe these networks, and used the datatype's induction principle to show their properties and deduce the fundamental theorem.

Our formalization is part of the *Archive of Formal Proofs* [2] and is described in more detail in Bentkamp's M.Sc. thesis [3]. It comprises about 7000 lines of Isabelle proofs, mostly in the declarative Isar style [34], and relies only on the standard axioms of higher-order logic, including the axiom of choice.

2   Mathematical Preliminaries

Tensors. Tensors can be understood as multidimensional arrays, with vectors and matrices as the one- and two-dimensional cases. Each index corresponds to a *mode* of the tensor. For matrices, the modes are called "row" and "column." The number of modes is the *order* of the tensor. The number of values an index can take in a particular mode is the *dimension* in that mode. Thus, a real-valued tensor $\mathscr{A} \in \mathbb{R}^{M_1 \times \cdots \times M_N}$ of order N and dimension M_i in mode i contains values $\mathscr{A}_{d_1, \ldots, d_N} \in \mathbb{R}$ for $d_i \in \{1, \ldots, M_i\}$.

Like for vectors and matrices, addition $+$ is defined as componentwise addition for tensors of identical dimensions. The product $\otimes$ is a binary operation on two arbitrary tensors that generalizes the outer vector product. The canonical polyadic rank, or CP-rank, associates a natural number with a tensor, generalizing the matrix rank. The matricization $[\mathscr{A}]$ of a tensor $\mathscr{A}$ is a matrix obtained by rearranging $\mathscr{A}$'s entries using a bijection between the tensor and matrix entries. It has the following property:

Lemma 1. *Given a tensor $\mathscr{A}$, we have* $\operatorname{rank}[\mathscr{A}] \leq \operatorname{CP-rank} \mathscr{A}$.

Lebesgue Measure. The Lebesgue measure is a mathematical description of the intuitive concept of length, surface, or volume. It extends this concept from simple geometrical shapes to a large amount of subsets of $\mathbb{R}^n$, including all closed and open sets, although it is impossible to design a measure that caters for all subsets of $\mathbb{R}^n$ while maintaining intuitive properties. The sets to which the Lebesgue measure can assign a volume are called *measurable*. The volume that is assigned to a measurable set can be a nonnegative real number or ∞. A set of Lebesgue measure 0 is called a *null set*. If a property holds for all points in $\mathbb{R}^n$ except for a null set, the property is said to hold *almost everywhere*.

The following lemma [12] about polynomials will be useful for the proof of the fundamental theorem of network capacity.

Lemma 2. *If $p \not\equiv 0$ is a polynomial in d variables, the set of points $\mathbf{x} \in \mathbb{R}^d$ with $p(\mathbf{x}) = 0$ is a Lebesgue null set.*

3   The Theorems of Network Capacity

Figure 1 gives the formulas for evaluating a CAC and relates them to the network's hierarchical structure. The $*$ operator denotes componentwise multiplication. The inputs $\mathbf{x}_1, \ldots, \mathbf{x}_N$ of a CAC are N real vectors of length M, where N must be a power of 2. The output $\mathbf{y}$ is a vector of length Y. The network's depth d can be any number between 1 and $\log_2 N$. The first $d - 1$ pooling layers

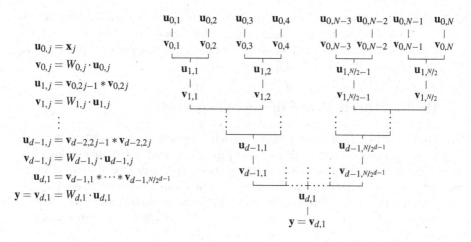

$$\mathbf{u}_{0,j} = \mathbf{x}_j$$
$$\mathbf{v}_{0,j} = W_{0,j} \cdot \mathbf{u}_{0,j}$$
$$\mathbf{u}_{1,j} = \mathbf{v}_{0,2j-1} * \mathbf{v}_{0,2j}$$
$$\mathbf{v}_{1,j} = W_{1,j} \cdot \mathbf{u}_{1,j}$$
$$\vdots$$
$$\mathbf{u}_{d-1,j} = \mathbf{v}_{d-2,2j-1} * \mathbf{v}_{d-2,2j}$$
$$\mathbf{v}_{d-1,j} = W_{d-1,j} \cdot \mathbf{u}_{d-1,j}$$
$$\mathbf{u}_{d,1} = \mathbf{v}_{d-1,1} * \cdots * \mathbf{v}_{d-1,N/2^{d-1}}$$
$$\mathbf{y} = \mathbf{v}_{d,1} = W_{d,1} \cdot \mathbf{u}_{d,1}$$

Fig. 1. Definition and hierarchical structure of a CAC with d layers

consist of binary nodes. The last pooling layer consists of a single node with an arity of $N/2^{d-1} \geq 2$.

The calculations depend on the learned weights, which are organized as entries of a collection of real matrices $W_{l,j}$, where l is the index of the layer and j is the position in that layer where the matrix is used. Matrix $W_{l,j}$ has dimensions $r_l \times r_{l-1}$ for natural numbers $r_{-1}, \ldots, r_d$ with $r_{-1} = M$ and $r_d = Y$. The *weight space* of a CAC is the space of all possible weight configurations. For a given weight configuration, the network expresses the function $(\mathbf{x}_1, \ldots, \mathbf{x}_N) \mapsto \mathbf{y}$.

The above definitions are all we need to state the main result proved by Cohen et al.:

Theorem 3 (Generalized Theorem of Network Capacity). *Consider two CACs with identical N, M, and Y parameters: a deeper network of depth d_1 with weight matrix dimensions $r_{1,l}$ and a shallower network of depth $d_2 < d_1$ with weight matrix dimensions $r_{2,l}$. Let $r = \min\{M, r_{1,0}, \ldots, r_{1,d_2-1}\}$ and assume*

$$r_{2,d_2-1} < r^{N/2^{d_2}}$$

Let S be the set of configurations in the weight space of the deeper network that express functions also expressible by the shallower network. Then S is a Lebesgue null set.

Intuitively, to express the same functions as the deeper network, almost everywhere in the weight space of the deeper network, r_{2,d_2-1} must be at least $r^{N/2^{d_2}}$, which means the shallower network needs exponentially larger weight matrices than the deeper network.

The special case of this theorem where $d_1 = \log_2 N$ and $d_2 = 1$ is called the fundamental theorem of network capacity. This is the theorem we formalized. Cohen et al. extended the result to CACs with an initial representational layer that applies a collection of nonlinearities to the inputs before the rest of the network is evaluated. Independently, they also showed that the fundamental and generalized theorems hold when the same weight matrix is applied within each layer l—i.e., $W_{l,1} = \cdots = W_{l,N/2^l}$.

4   Restructured Proof of the Theorems

The proof of either theorem of network capacity depends on a connection between CACs and measure theory, using tensors, matrices, and polynomials along the way. Briefly, the CACs and the functions they express can be described using tensors. Via matricization, these tensors can be analyzed as matrices. Polynomials bridge the gap between matrices and measure theory, since the matrix determinant is a polynomial, and zero sets of polynomials are Lebesgue null sets (Lemma 2).

The proof by Cohen et al. is structured as a monolithic induction over the deep network structure. It combines tensors, matrices, and polynomials in each induction step. Before launching Isabelle, we restructured the proof into a more modular version that cleanly separates the mathematical theories involved, resulting in the following sketch:

I. We describe the function expressed by a CAC for a fixed weight configuration using tensors. We focus on an arbitrary entry y_i of the output vector $\mathbf{y}$. If the shallower network cannot express the output component y_i, it cannot represent the entire output either. Let $\mathscr{A}_i(w)$ be the tensor that represents the function $(\mathbf{x}_1, \ldots, \mathbf{x}_N) \mapsto y_i$ expressed by the deeper network with a weight configuration w.

II. We define a function φ that reduces the order of a tensor. The CP-rank of $\varphi(\mathscr{A})$ indicates how large the shallower network must be to express a function represented by a tensor $\mathscr{A}$: If the function expressed by the shallower network is represented by $\mathscr{A}$, then $r_{2,d_2-1} \geq$ CP-rank $(\varphi(\mathscr{A}))$.

III. We construct a multivariate polynomial p, mapping the weights configurations w of the deeper network to a real number $p(w)$. It has the following properties:

  (a) If $p(w) \neq 0$, then rank $[\varphi(\mathscr{A}_i(w))] \geq r^{N/2^{d_2}}$. Hence CP-rank $(\varphi(\mathscr{A}_i(w))) \geq r^{N/2^{d_2}}$ by Lemma 1.

  (b) The polynomial p is not the zero polynomial. Hence its zero set is a Lebesgue null set by Lemma 2.

By properties IIIa and IIIb, the inequation CP-rank $(\varphi(\mathscr{A}_i(w))) \geq r^{N/2^{d_2}}$ holds almost everywhere. By step II, we need $r_{2,d_2-1} \geq r^{N/2^{d_2}}$ almost everywhere for the shallower network to express functions the deeper network expresses.

The restructuring helps us keep the induction simple, and we can avoid formalizing some lemmas of the original proof. Furthermore, the restructured proof allows us to state a stronger property than in the original proof, which Cohen et al. independently discovered later [16]: The set S from Theorem 3 is not only a

Lebesgue null set, but also a subset of the zero set of the polynomial p. This fact can be used to derive further properties of S. Zero sets of polynomials are well studied in algebraic geometry, where they are known as algebraic varieties. This generalization partially addresses an issue that arises when applying the theorem to actual implementations of CACs: Cohen et al. assume that the weight space of the deeper network is a Euclidean space, but in practice it will always be discrete. They also show that S is a closed null set, but since these can be arbitrarily dense, this gives no information about the discrete counterpart of S.

We can estimate the size of this discrete counterpart of S using our generalization in conjunction with a result from algebraic geometry [11, 24] that allows us to estimate the size of the ε-neighborhood of the zero set of a polynomial. The ε-neighborhood of S is a good approximation of the discrete counterpart of S if ε corresponds to the precision of computer arithmetic. Unfortunately, the estimate is trivial, unless we assume $\varepsilon < 2^{-170\,000}$, which largely exceeds the precision of modern computers. Thus, shallow CACs are perhaps more expressive than Theorem 3 suggests. On the other hand, our analysis is built upon inequalities, which only provide an upper bound. A mathematical result estimating the size of S with a lower bound would call for an entirely different approach.

5   Formal Libraries

Matrices. We had several options for the choice of a matrix library, of which the most relevant were Isabelle's analysis library and Thiemann and Yamada's matrix library [33]. The analysis library fixes the matrix dimensions using type parameters, a technique introduced by Harrison [18]. The advantage of this approach is that the dimensions are part of the type and need not be stated as conditions. Moreover, it makes it possible to instantiate type classes depending on the type arguments. However, this approach is not practical when the dimensions are specified by terms. Therefore, we chose Thiemann and Yamada's library, which uses a single type for matrices of all dimensions and includes a rich collection of lemmas.

We extended the library in a few ways. We contributed a definition of the matrix rank, as the dimension of the space spanned by the matrix columns:

definition (**in** *vec_space*) rank :: α *mat* $\Rightarrow$ *nat* **where**
rank A = vectorspace.dim F (span_vs (set (cols A)))

Moreover, we defined submatrices and proved that the rank of a matrix is larger than any submatrix with nonzero determinant, and that the rank is the maximum amount of linearly independent columns of the matrix.

Tensors. The *Tensor* entry [31] of the *Archive of Formal Proofs* might seem to be a good starting point for a formalization of tensors. However, despite its name, this library does not contain a type for tensors. It introduces the Kronecker product, which is equivalent to the tensor product but operates on the matricizations of tensors.

The *Groups, Rings and Modules* entry [22] of the *Archive of Formal Proofs* could have been another potential basis for our work. Unfortunately, it introduces the tensor product in a very abstract fashion and does not integrate well with other Isabelle libraries.

Instead, we introduced our own type for tensors, based on a list that specifies the dimension in each mode and a list containing all of its entries:

typedef α tensor $= \{(ds :: nat\ list,\ as :: \alpha\ list).$ length $as = \prod ds\}$

We formalized addition, multiplication by scalars, product, matricization, and the CP-rank. We instantiated addition as a semigroup (*semigroup_add*) and product as a monoid (*monoid_mult*). Stronger type classes cannot be instantiated: Their axioms do not hold collectively for tensors of all sizes, even though they hold for fixed tensor sizes. For example, it is impossible to define addition for tensors of different sizes while satisfying the cancellation property $a + c = b + c \Longrightarrow a = b$.

For proving properties of addition, scalar multiplication, and product, we devised a powerful induction principle on tensors that uses tensor slices. The induction step amounts to showing a property for a tensor $\mathscr{A} \in \mathbb{R}^{M_1 \times \cdots \times M_N}$ assuming it holds for all slices $\mathscr{A}_i \in \mathbb{R}^{M_2 \times \cdots \times M_N}$, which are obtained by fixing the first index $i \in \{1, \ldots, M_1\}$.

Matricization rearranges the entries of a tensor $\mathscr{A} \in \mathbb{R}^{M_1 \times \cdots \times M_N}$ into a matrix $[\mathscr{A}] \in \mathbb{R}^{I \times J}$. This rearrangement can be described as a bijection between $\{0, \ldots, M_1 - 1\} \times \cdots \times \{0, \ldots, M_N - 1\}$ and $\{0, \ldots, I - 1\} \times \{0, \ldots, J - 1\}$, assuming that indices start at 0. The operation is parameterized by a partition of the set of tensor indices into two sets $\{r_1 < \cdots < r_K\} \uplus \{c_1 < \cdots < c_L\} = \{1, \ldots, N\}$. The proof of Theorem 3 uses only standard matricization, which partitions the indices into odd and even numbers, but we formalized the more general formulation [1]. The matrix $[\mathscr{A}]$ has $I = \prod_{i=1}^{K} r_i$ rows and $J = \prod_{j=1}^{L} c_j$ columns. The rearrangement function is

$$(i_1, \ldots, i_N) \mapsto \left(\sum_{k=1}^{K} \left(i_{r_k} \cdot \prod_{k'=1}^{k-1} M_{r_{k'}}\right), \sum_{l=1}^{L} \left(i_{c_l} \cdot \prod_{l'=1}^{l-1} M_{c_{l'}}\right)\right)$$

The indices $i_{r_1}, \ldots, i_{r_K}$ and $i_{c_1}, \ldots, i_{c_L}$ serve as digits in a mixed-base numeral system to specify the row and the column in the matricization, respectively. This is perhaps more obvious if we expand the sum and product operators and factor out the bases M_i:

$$(i_1, \ldots, i_N) \mapsto \big(i_{r_1} + M_{r_1} \cdot (i_{r_2} + M_{r_2} \cdot \ldots \cdot (i_{r_{K-1}} + M_{r_{K-1}} \cdot i_{r_K}) \ldots)\big),$$
$$i_{c_1} + M_{c_1} \cdot (i_{c_2} + M_{c_2} \cdot \ldots \cdot (i_{c_{L-1}} + M_{c_{L-1}} \cdot i_{c_L}) \ldots)\big)\big)$$

To formalize the matricization operation, we defined a function calculating the digits of a number n in a given mixed-based numeral system:

```
fun encode :: nat list ⇒ nat ⇒ nat list where
  encode [] n = []
| encode (b # bs) n = (n mod b) # encode bs (n div b)
```

We then defined matricization as

> **definition** matricize :: *nat set* $\Rightarrow$ α *tensor* $\Rightarrow$ α *mat* **where**
> matricize R $\mathscr{A}$ = mat ($\prod$ sublist (dims $\mathscr{A}$) R) ($\prod$ sublist (dims $\mathscr{A}$) ($-R$))
> ($\lambda(r, c)$. lookup $\mathscr{A}$ (weave R
> (encode (sublist (dims $\mathscr{A}$) R) r)
> (encode (sublist (dims $\mathscr{A}$) ($-R$)) c)))

The matrix constructor mat takes as arguments the matrix dimensions and a function that returns each matrix entry given the indices r and c. Defining this function amounts to finding the corresponding indices of the tensor, which are essentially the mixed-base encoding of r and c, but the digits of these two encoded numbers must be interleaved in an order specified by the set $R = \{r_1, \ldots, r_K\}$.

To merge two lists of digits in the right way, we defined a function weave. This function is the counterpart of sublist from the standard library, which reduces a list to those entries whose indices belong to a set I:

> **lemma** *weave_sublists*: weave I (sublist *as* I) (sublist *as* ($-I$)) = *as*

The main concern when defining such a function is to determine how it should behave in corner cases—in our scenario, when $I = \{\}$ and the first list argument is nonempty. We settled on a definition such that the property length (weave I *xs* *ys*) = length *xs* + length *ys* holds unconditionally:

> **definition** weave :: *nat set* $\Rightarrow$ α *list* $\Rightarrow$ α *list* $\Rightarrow$ α *list* **where**
> weave I *xs* *ys* = map
> (λi. if $i \in I$ then *xs* ! $|\{a \in I.\ a < i\}|$ else *ys* ! $|\{a \in -I.\ a < i\}|$)
> $[0\ ..< $ length *xs* + length *ys*]

(The ! operator returns the list element at a given index.) This definition allows us to prove lemmas about weave I *xs* *ys* ! a and length (weave I *xs* *ys*) very easily. Other properties, such as the *weave_sublists* lemma above, are justified using an induction over the length of a list, with a case distinction in the induction step on whether the new list element is taken from *xs* or *ys*.

Another difficulty arises with the rule rank $[\mathscr{A} \otimes \mathscr{B}]$ = rank $[\mathscr{A}] \cdot$ rank $[\mathscr{B}]$ for standard matricization and tensors of even order, which seemed tedious to formalize. Restructuring the proof eliminates one of its two occurrences (Sect. 4). The remaining occurrence is used to show that rank $[\mathbf{a}_1 \otimes \cdots \otimes \mathbf{a}_N] = 1$, where $\mathbf{a}_1, \ldots, \mathbf{a}_N$ are vectors and N is even. A simpler proof relies on the observation that the entries of $[\mathbf{a}_1 \otimes \cdots \otimes \mathbf{a}_N]$ can be written as $f(i) \cdot g(j)$, where f depends only on the row index i, and g depends only on the column index j. Using this argument, rank $[\mathbf{a}_1 \otimes \cdots \otimes \mathbf{a}_N] = 1$ can be shown for generalized matricization and an arbitrary N, which we used to prove Lemma 1:

> **lemma** *matrix_rank_le_cp_rank*:
> **fixes** $A :: (\alpha :: field)$ *tensor*
> **shows** mrank (matricize R A) $\leq$ cprank A

Lebesgue Measure. Isabelle's analysis library defines the Borel measure on $\mathbb{R}^n$ but not the closely related Lebesgue measure. The Lebesgue measure is the completion of the Borel measure. The two measures are identical on all sets that are Borel measurable, but the Lebesgue measure has more measurable sets. The proof by Cohen et al. allows us to show that the set S defined in Theorem 3 is a subset of a Borel null set. It follows that S is a Lebesgue null set, but not necessarily a Borel null set.

To resolve this mismatch, we considered three options: (1) Prove that S is a Borel null set, which we believe is the case, although it does not follow trivially from S's being a subset of a Borel null set; (2) define the Lebesgue measure, using the already formalized Borel measure and measure completion; (3) formulate the theorem using the almost-everywhere quantifier ($\forall_{ae}$) instead of the null set predicate.

We chose the third approach, because it seemed simpler. Theorem 3, as expressed in Sect. 3, defines the set S as set of configurations in the weight space of the deeper network that express functions also expressible by the shallower network, and then states that S is a null set. In the formalization, we state it as follows: Almost everywhere in the weight space of the deeper network, the deeper network expresses functions not expressible by the shallower network. This formulation is equivalent to asserting that S is a subset of a null set, which we can easily prove for the Borel measure as well.

There is, however, another issue with the definition of the Borel measure from Isabelle's analysis library:

> **definition** lborel :: (α :: *euclidean_space*) *measure* **where**
>   lborel = distr ($\prod_{\mathsf{M}} b \in$ Basis. interval_measure ($\lambda x.\ x$)) borel
>   ($\lambda f.\ \sum b \in$ Basis. $f\ b *_{\mathsf{R}} b$)

The type α specifies the number of dimensions of the measure space. In our proof, the measure space is the weight space of the deeper network, and its dimension depends on the number N of inputs and the size r_l of the weight matrices. The number of dimensions is a term in our proof. We described a similar issue with Isabelle's matrix library already.

The solution is to provide a new definition of the Borel measure whose type does not fix the number of dimensions. The multidimensional Borel measure is the product measure ($\prod_{\mathsf{M}}$) of the one-dimensional Borel measure (lborel :: *real measure*) with itself:

> **definition** lborel$_f$ :: $nat \Rightarrow (nat \Rightarrow real)$ *measure* **where**
>   lborel$_f$ n = ($\prod_{\mathsf{M}} b \in \{.. < n\}$. lborel)

The argument n specifies the dimension of the measure space. Unlike with lborel, the measure space of lborel$_f$ n is not the entire universe of the type: Only functions that map to a default value for numbers $\geq n$ are contained in the measure space, which is available as space (lborel$_f$ n). With the above definition, we could prove the main lemmas about lborel$_f$ from the corresponding lemmas about lborel with little effort.

Multivariate Polynomials. Multivariate polynomial libraries have been developed to support other formalization projects in Isabelle. Sternagel and Thiemann [32] formalized multivariate polynomials designed for execution, but the equality of polynomials is a custom predicate, which means that we cannot use Isabelle's simplifier to rewrite polynomial expressions. Immler and Maletzky [20] formalized an axiomatic approach to multivariate polynomials using type classes, but their focus is not on the evaluation homomorphism, which we need. Instead, we chose to extend a previously unpublished multivariate polynomial library by Lochbihler and Haftmann [17]. We derived induction principles and properties of the evaluation homomorphism and of nested multivariate polynomials. These were useful to formalize Lemma 2:

> **lemma** *lebesgue_mpoly_zero_set*:
>   **fixes** $p :: real\ mpoly$
>   **assumes** $p \neq 0$ **and** vars $p \subseteq \{.. < n\}$
>   **shows** $\{x \in$ space (lborel$_f$ n). insertion x $p = 0\} \in$ null_sets (lborel$_f$ n)

6   Formalization of the Fundamental Theorem

With the necessary libraries in place, we undertook the formal proof of the fundamental theorem of network capacity, starting with the CACs. A recursive datatype is appropriate to capture the hierarchical structure of these networks:

> **datatype** α cac = | Input nat | | Conv α (α cac) | | Pool (α cac) (α cac) |

To simplify the proofs, Pool nodes are always binary. Pooling layers that merge more than two branches are represented by nesting Pool nodes to the right.

The type variable α can be used to store weights. For networks without weights, it is set to $nat \times nat$, which associates only the matrix dimension with each Conv node. For networks with weights, α is $real\ mat$, an actual matrix. These two network types are connected by insert_weights $:: (nat \times nat)cac \Rightarrow (nat \Rightarrow real) \Rightarrow real\ mat\ cac$, which inserts weights into a weightless network. The weights are specified by the second argument f, of which only the first values $f\ 0,\ f\ 1,\ \ldots, f\ (k-1)$ are used, until the necessary number of weights, k, is reached. Sets over $nat \Rightarrow real$ can be measured using lborel$_f$.

The following function describes how the networks are evaluated, where $\otimes_{mv}$ multiplies a matrix with a vector and component_mult multiplies vectors componentwise:

> **fun** evaluate_net $:: real\ mat\ cac \Rightarrow real\ vec\ list \Rightarrow real\ vec$ **where**
>   evaluate_net (Input M) is = hd is
> | evaluate_net (Conv A m) is = $A \otimes_{mv}$ evaluate_net m is
> | evaluate_net (Pool m_1 m_2) is = component_mult
>   (evaluate_net m_1 (take (length (input_sizes m_1)) is))
>   (evaluate_net m_2 (drop (length (input_sizes m_1)) is))

The *cac* type can represent networks with arbitrary nesting of Conv and Pool nodes, going beyond the definition of CACs. Moreover, since we focus on the fundamental theorem of network capacities, it suffices to consider a deep model with $d_1 = \log_2 N$ and a shallow model with $d_2 = 1$. These are specified by generating functions:

fun
    deep_model$_0$ :: $nat \Rightarrow nat\ list \Rightarrow (nat \times nat)\ cac$ **and**
    deep_model :: $nat \Rightarrow nat \Rightarrow nat\ list \Rightarrow (nat \times nat)\ cac$
where
    deep_model$_0$ Y [] = Input Y
    | deep_model$_0$ Y (r # rs) = Pool (deep_model Y r rs) (deep_model Y r rs)
    | deep_model Y r rs = Conv (Y, r) (deep_model$_0$ r rs)

fun shallow_model$_0$ :: $nat \Rightarrow nat \Rightarrow nat \Rightarrow (nat \times nat)\ cac$ **where**
    shallow_model$_0$ Z M 0 = Conv (Z, M) (Input M)
    | shallow_model$_0$ Z M (Suc N) =
    Pool (shallow_model$_0$ Z M 0) (shallow_model$_0$ Z M N)

definition shallow_model :: $nat \Rightarrow nat \Rightarrow nat \Rightarrow nat \Rightarrow (nat \times nat)\ cac$ **where**
    shallow_model Y Z M N = Conv (Y, Z) (shallow_model$_0$ Z M N)

Two examples are given in Fig. 2. For the deep model, the arguments Y # r # rs correspond to the weight matrix sizes $[r_{1,d}\ (= Y), r_{1,d-1}, \ldots, r_{1,0}, r_{1,-1}\ (= M)]$. For the shallow model, the arguments Y, Z, M correspond to the parameters $r_{2,1}\ (= Y)$, $r_{2,0}$, $r_{2,-1}\ (= M)$, and N gives the number of inputs minus 1.

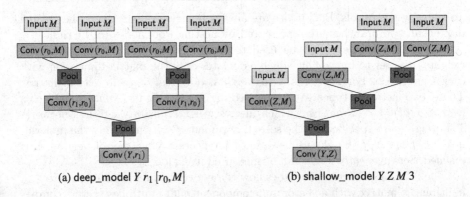

(a) deep_model Y r_1 $[r_0, M]$ (b) shallow_model Y Z M 3

Fig. 2. A deep and a shallow network represented using the *cac* datatype

The rest of the formalization follows the proof sketch presented in Sect. 4.

Step I. The following operation computes a list, or *vector*, of tensors representing a network's function, each tensor standing for one component of the output vector:

fun tensors_from_net :: *real mat cac* $\Rightarrow$ *real tensor vec* **where**
    tensors_from_net (Input M) = Matrix.vec M ($\lambda i.$ unit_vec M i)

| tensors_from_net (Conv A m) =
    mat_tensorlist_mult A (tensors_from_net m) (input_sizes m)
| tensors_from_net (Pool m_1 m_2) =
    component_mult (tensors_from_net m_1) (tensors_from_net m_2)

For an Input node, we return the list of unit vectors of length M. For a Conv node, we multiply the weight matrix A with the tensor list computed for the subnetwork m, using matrix–vector multiplication. For a Pool node, we compute, elementwise, the tensor products of the two tensor lists associated with the subnetworks m_1 and m_2. If two networks express the same function, the representing tensors are the same:

> **lemma** *tensors_from_net_eqI*:
>   **assumes** valid_net' m_1 **and** valid_net' m_2 **and** input_sizes m_1 = input_sizes m_2
>   **and** $\forall is$. input_correct is $\longrightarrow$ evaluate_net m_1 is = evaluate_net m_2 is
>   **shows** tensors_from_net m_1 = tensors_from_net m_2

The fundamental theorem fixes an arbitrary deep network. It is useful to fix the deep network parameters in a locale—a sectioning mechanism that fixes variables and assumptions on them across definitions and lemmas:

> **locale** deep_model_correct_params =
>   **fixes** rs :: *nat list*
>   **assumes** *deep*: length rs $\geq$ 3
>   **and** *no_zeros*: $\bigwedge r$. $r \in$ set rs $\implies r > 0$

The list rs completely specifies one specific deep network model:

> **abbreviation** deep_net = deep_model (rs ! 0) (rs ! 1) (tl (tl rs))

The other parameters of the deep network can be defined based on rs:

> **definition** r = min (last rs) (last (butlast rs))
> **definition** N_half = $2^{\text{length rs} - 3}$
> **definition** weight_space_dim = count_weights deep_net

The shallow network must have the same input and output sizes as the deep network, if it is to express the same function as the deep network. This leaves only the parameter $Z = r_{2,0}$, which specifies the weight matrix sizes in the Conv nodes and the size of the vectors multiplied in the Pool nodes of the shallow network:

> **abbreviation** shallow_net Z = shallow_model (rs ! 0) Z (last rs) (2*N_half-1)

Following the proof sketch, we consider a single output component y_i. We do so using a second locale that introduces a constant i for i.

> **locale** deep_model_correct_params_output_index =
>   deep_model_correct_params +
>   **fixes** i :: *nat*
>   **assumes** *output_index_valid*: i < rs ! 0

Then we can define the tensor $\mathscr{A}_i$, which describes the behavior of the function expressed by the deep network at the output component y_i, depending on the weight configuration w of the deep network:

definition $\mathscr{A}_i\ w =$ tensors_from_net (insert_weights deep_net w) ! i

We want to analyze for which w the shallow network can express the same function, and is hence represented by the same tensor.

Step II. We must show that if a tensor $\mathscr{A}$ represents the function expressed by the shallow network, then $r_{2,d_2-1} \geq$ CP-rank $(\varphi(\mathscr{A}))$. For the fundamental theorem of network capacity, φ is the identity and $d_2 = 1$. Hence, it suffices to prove that $Z = r_{2,0} \geq$ CP-rank $(\mathscr{A})$:

lemma *cprank_shallow_model*:
    cprank (tensors_from_net (insert_weights w (shallow_net Z)) ! i) $\leq Z$

This lemma can be proved easily from the definition of the CP-rank.

Step III. We define the polynomial p and prove that it has properties IIIa and IIIb. Defining p as a function is simple:

definition $p_{func}\ w =$ det (submatrix $[\mathscr{A}_i\ w]$ rows_with_1 rows_with_1)

where $[\mathscr{A}_i\ w]$ abbreviates the standard matricization matricize $\{n.\ \text{even}\ n\}$ $(\mathscr{A}_i\ w)$, and rows_with_1 is the set of row indices with 1s in the main diagonal for a specific weight configuration w that will be defined in Step IIIb. We try to make the submatrix as large as possible while maintaining the property that p is not the zero polynomial. The bound on Z in the statement of the final theorem is derived from the size of this submatrix.

The function p_{func} must be shown to be a polynomial function. We introduce a predicate polyfun, which is true if a function is a polynomial function:

definition polyfun $N\ f = (\exists p.\ \text{vars}\ p \subseteq N \wedge (\forall x.\ \text{insertion}\ x\ p = f\ x))$

This predicate is preserved from constant and linear functions through the tensor representation of the CAC, matricization, choice of submatrix, and determinant:

lemma *polyfun_p*:
    polyfun $\{..< \text{weight\_space\_dim}\}$ p_{func}

Step IIIa. We must show that if $p(w) = 0$, then CP-rank $(\mathscr{A}_i(w)) \geq r^{N/2}$. The Isar proof is sketched below:

lemma *if_polynomial_0_rank*:
    assumes $p_{func}\ w \neq 0$
    shows $r^{N\_half} \leq$ cprank $(\mathscr{A}_i\ w)$
    proof $-$
        have $r^{N\_half} =$ dim$_r$ (submatrix $[\mathscr{A}_i\ w]$ rows_with_1 rows_with_1)
            by calculating the size of the submatrix
        also have $\cdots \leq$ mrank $[\mathscr{A}_i\ w]$

using the assumption and the fact that the rank is larger than submatrices
with nonzero determinant
also have $\cdots \leq$ cprank $(\mathscr{A}_i\ w)$
using Lemma 1
finally show *?thesis* .
qed

Step IIIb. To prove that p is not the zero polynomial, we must exhibit a witness
weight configuration where p is nonzero. Since weights are arranged in matrices,
we define concrete matrix types: matrices with 1 s on their main diagonal and 0 s
elsewhere (eye_matrix), matrices with 1 s everywhere (all1_matrix), and matrices
with 1 s in the first column and 0 s elsewhere (copy_first_matrix). For example,
the last matrix type is defined as follows:

definition copy_first_matrix :: *nat* $\Rightarrow$ *nat* $\Rightarrow$ *real mat* **where**
copy_first_matrix *nr nc* = mat *nr nc* $(\lambda(r, c).$ if $c = 0$ then 1 else 0)

For each matrix type, we show how it behaves under multiplication with a
vector:

lemma *mult_copy_first_matrix*:
assumes $i < nr$ **and** dim$_v\ v > 0$
shows (copy_first_matrix *nr* (dim$_v\ v$) $\otimes_{mv}\ v$) ! $i = v$! 0

Using these matrices, we can define the deep network containing the witness
weights:

fun
witness$_0$ :: *nat* $\Rightarrow$ *nat list* $\Rightarrow$ *real mat cac* **and**
witness :: *nat* $\Rightarrow$ *nat* $\Rightarrow$ *nat list* $\Rightarrow$ *real mat cac*
where
witness$_0$ Y [] = Input Y
| witness$_0$ Y $(r \# rs)$ = Pool (witness Y r rs) (witness Y r rs)
| witness Y r rs = Conv ((if length rs = 0 then eye_matrix else
    if length rs = 1 then all1_matrix else copy_first_matrix) Y r) (witness$_0$ r rs)

The network's structure is identical to deep_model. For each Conv node, we
carefully choose one of the three matrix types we defined, such that the representing tensor of this network has as many 1 s as possible on the main diagonal
and 0 s elsewhere. This in turn ensures that its matricization has as many 1 s
as possible on its main diagonal and 0 s elsewhere. The rows_with_1 constant
specifies the row indices that contain the 1s.

The witness weights can be extracted from the witness network as follows:

definition witness_weights :: *nat* $\Rightarrow$ *real* **where**
witness_weights =
$(\varepsilon w.$ witness (rs ! 0) (rs ! 1) (tl (tl rs)) = insert_weights deep_net w)

This could also be achieved without using Hilbert's choice operator, by defining
a recursive function that extracts the weights from weighted networks.

We prove that the representing tensor of the witness network, which is equal to $\mathscr{A}_i$ witness_weights, has the desired form. This step is rather involved: We show how the defined matrices act in the network and perform a tedious induction over the witness network. Then we can show that the submatrix characterized by rows_with_1 of the matricization of this tensor is the identity matrix of size $r^{\text{N\_half}}$:

lemma *witness_submatrix*:
   submatrix $[\mathscr{A}_i$ witness_weights] rows_with_1 rows_with_1 $=$
   eye_matrix $r^{\text{N\_half}}\ r^{\text{N\_half}}$

As a consequence of this lemma, the determinant of this submatrix, which is the definition of p_{func}, is nonzero. Therefore, p is not the zero polynomial:

lemma *polynomial_not_zero*:
   p_{func} witness_weights $\neq 0$

Fundamental Theorem. The results of Steps II and III can be used to establish the fundamental theorem of network capacity:

theorem *fundamental_theorem_of_network_capacity*:
   $\forall_{\text{ae}}\ w_{\text{d}}$ w.r.t. lborel$_f$ weight_space_dim. $\nexists w_{\text{s}}\ Z$.
     $Z < r^{\text{N\_half}} \wedge$
     $\forall is$. input_correct $is \longrightarrow$
       evaluate_net (insert_weights deep_net w_{d}) $is =$
       evaluate_net (insert_weights (shallow_net Z) w_{s}) is

The $r^{\text{N\_half}}$ bound corresponds to the size of the identity matrix in *witness_submatrix*.

The theorem statement is independent of the tensor library, and is therefore correct regardless of whether the library faithfully captures tensor-related notions.

7   Discussion and Related Work

Extension with Shared Weights and the Generalized Theorem. We formalized the fundamental theorem for nonshared weights. The case of shared weights is so similar that Cohen et al. discharge it with a one-sentence proof by analogy. Using copy and paste, we could easily extend the formalization to cover this case, but to reduce code duplication we would need more abstract definitions of the involved networks.

The generalized theorem of network capacity is mostly a straightforward generalization. To formalize it, we would need to define CACs for arbitrary depths, which our datatype allows. Moreover, we would need to define the function φ and prove some of its properties. Then, we would generalize the existing lemmas. We focused on the fundamental theorem because it contains all the essential ideas.

Sledgehammer and SMT. To discharge proof obligations, we used Sledge-hammer [28] extensively. This Isabelle tool heuristically selects a few hundred lemmas from the thousands available (using machine learning [8]); translates the proof obligation and the selected lemmas to first-order logic; invokes exter-nal automatic theorem provers on the translated problem; and, in case of suc-cess, translates the derivations found by the external provers to Isar proof texts that can be inserted in the formalization. In the best-case scenario, Sledgeham-mer quickly produces a one-line proof text consisting of an invocation of the *metis* proof method [29], Isabelle's internal superposition prover. Unfortunately, Sledgehammer sometimes returns only cryptic structured Isar proofs [7] or, if all else fails, proofs that depend on the *smt* method [10].

The *smt* method relies on the SMT solver Z3 [25] to find a proof, which it then replays using Isabelle's inference kernel. Relying on a highly heuristic third-party prover is fragile; some proofs that are fast with a given version of the prover might time out with a different version, or be unreplayable due to some incompleteness in *smt*. As a result, entries in the *Archive of Formal Proofs* cannot use it. Sledgehammer often generates *smt* proofs, especially in proof obligations about sums and products of reals, existential quantifiers, and λ-expressions. We ended up with over 60 invocations of *smt*, which we later replaced one by one with structured Isar proofs, a tedious process. The following equation on reals is an example that can only be proved by *smt*, with suitable lemmas:

$$\sum_{i \in I} \sum_{j \in J} a \cdot b \cdot f(i) \cdot g(j) = \left(\sum_{i \in I} a \cdot f(i)\right) \cdot \left(\sum_{j \in J} b \cdot g(j)\right)$$

We could not solve it with other proof methods without engaging in a detailed proof involving multiple steps. This particular example relies on *smt*'s partial support for λ-expressions through λ-lifting, an instance of what we would call "easy higher-order."

Similar Theoretical Results about Other Deep Learning Architec-tures. CACs are relatively easy to analyze but little used in practice. In a follow-up paper [15], Cohen et al. connected their tensor analysis of CACs to the frequently used CNNs with rectified linear unit (ReLU) activation. Unlike CACs, ReLU CNNs with average pooling are not universal—that is, even shal-low networks of arbitrary size cannot express all functions a deeper network can express. Moreover, ReLU CNNs do not enjoy complete depth efficiency; the ana-logue of the set S for those networks has a Lebesgue measure greater than zero. This leads Cohen et al. to conjecture that CACs could become a leading app-roach for deep learning, once suitable training algorithms have been developed.

Related Formal Proofs. We are aware of a few other formalizations of machine learning algorithms, including hidden Markov models [23], perceptrons [26], expectation maximization, and support vector machines [6]. To our knowledge, our work is the first formalization about deep learning.

Some of the mathematical libraries underlying our formalizations have coun-terparts in other systems, notably Coq. For example, the Mathematical Compo-nents include comprehensive matrix theories [5], which are naturally expressed using dependent types. The tensor formalization by Boender [9] restricts itself to the Kronecker product on matrices. Bernard et al. [4] formalized multivariate polynomials and used them to show the transcendence of e and π. Kam formalized the Lebesgue measure as part of a formalization of the Lebesgue integral, which in turn was used to state and prove Markov's inequality [21].

8 Conclusion

We applied a proof assistant to formalize a recent result in a field where they have been little used before, namely machine learning. We found that the functionality and libraries of a modern proof assistant such as Isabelle/HOL were mostly up to the task. Beyond the formal proof of the fundamental theorem of network capacity, our main contribution is a general library of tensors.

Admittedly, even the formalization of fairly short pen-and-paper proofs can require a lot of work, partly because of the need to develop and extend libraries. On the other hand, not only does the process lead to a computer verification of the result, but it can also reveal new ideas and results. The generalization and simplifications we discovered illustrate how formal proof development can be beneficial to research outside the small world of interactive theorem proving.

Acknowledgment. We thank Lukas Bentkamp, Robert Lewis, Anders Schlichtkrull, Mark Summerfield, and the anonymous reviewers for suggesting many textual improvements. The work has received funding from the European Research Council under the European Union's Horizon 2020 research and innovation program (grant agreement No. 713999, Matryoshka).

References

1. Bader, B.W., Kolda, T.G.: Algorithm 862: MATLAB tensor classes for fast algorithm prototyping. ACM Trans. Math. Softw. **32**(4), 635–653 (2006)
2. Bentkamp, A.: Expressiveness of deep learning. Archive of Formal Proofs (2016). http://isa-afp.org/entries/Deep_Learning.shtml. Formal proof development
3. Bentkamp, A.: An Isabelle formalization of the expressiveness of deep learning. M.Sc. thesis, Universität des Saarlandes (2016). http://matryoshka.gforge.inria.fr/pubs/bentkamp_msc_thesis.pdf
4. Bernard, S., Bertot, Y., Rideau, L., Strub, P.: Formal proofs of transcendence for e and π as an application of multivariate and symmetric polynomials. In: Avigad, J., Chlipala, A. (eds.) CPP 2016, pp. 76–87. ACM (2016)
5. Bertot, Y., Gonthier, G., Ould Biha, S., Pasca, I.: Canonical big operators. In: Mohamed, O.A., Muñoz, C., Tahar, S. (eds.) TPHOLs 2008. LNCS, vol. 5170, pp. 86–101. Springer, Heidelberg (2008). doi:10.1007/978-3-540-71067-7_11
6. Bhat, S.: Syntactic foundations for machine learning. Ph.D. thesis, Georgia Institute of Technology (2013). https://smartech.gatech.edu/bitstream/handle/1853/47700/bhat_sooraj_b_201305_phd.pdf
7. Blanchette, J.C., Böhme, S., Fleury, M., Smolka, S.J., Steckermeier, A.: Semi-intelligible Isar proofs from machine-generated proofs. J. Autom. Reason. **56**(2), 155–200 (2016)

8. Blanchette, J.C., Greenaway, D., Kaliszyk, C., Kühlwein, D., Urban, J.: A learning-based fact selector for Isabelle/HOL. J. Autom. Reason. **57**(3), 219–244 (2016)
9. Boender, J., Kammüller, F., Nagarajan, R.: Formalization of quantum protocols using Coq. In: Heunen, C., Selinger, P., Vicary, J. (eds.) QPL 2015. EPTCS, vol. 195, pp. 71–83 (2015)
10. Böhme, S., Weber, T.: Fast LCF-style proof reconstruction for Z3. In: Kaufmann, M., Paulson, L.C. (eds.) ITP 2010. LNCS, vol. 6172, pp. 179–194. Springer, Heidelberg (2010). doi:10.1007/978-3-642-14052-5_14
11. Bürgisser, P., Cucker, F., Lotz, M.: The probability that a slightly perturbed numerical analysis problem is difficult. Math. Comput. **77**(263), 1559–1583 (2008)
12. Caron, R., Traynor, T.: The zero set of a polynomial. Technical report, University of Windsor (2005). http://www1.uwindsor.ca/math/sites/uwindsor.ca.math/files/05-03.pdf
13. Cohen, N., Sharir, O., Shashua, A.: Deep SimNets. In: CVPR 2016, pp. 4782–4791. IEEE Computer Society (2016)
14. Cohen, N., Sharir, O., Shashua, A.: On the expressive power of deep learning: a tensor analysis. In: Feldman, V., Rakhlin, A., Shamir, O. (eds.) COLT 2016. JMLR Workshop and Conference Proceedings, vol. 49, pp. 698–728. JMLR.org (2016)
15. Cohen, N., Shashua, A.: Convolutional rectifier networks as generalized tensor decompositions. In: Balcan, M., Weinberger, K.Q. (eds.) ICML 2016. JMLR Workshop and Conference Proceedings, vol. 48, pp. 955–963. JMLR.org (2016)
16. Cohen, N., Shashua, A.: Inductive bias of deep convolutional networks through pooling geometry. CoRR abs/1605.06743 (2016)
17. Haftmann, F., Lochbihler, A., Schreiner, W.: Towards abstract and executable multivariate polynomials in Isabelle. In: Nipkow, T., Paulson, L., Wenzel, M. (eds.) Isabelle Workshop 2014 (2014)
18. Harrison, J.: A HOL theory of Euclidean space. In: Hurd, J., Melham, T. (eds.) TPHOLs 2005. LNCS, vol. 3603, pp. 114–129. Springer, Heidelberg (2005). doi:10.1007/11541868_8
19. Hölzl, J., Heller, A.: Three chapters of measure theory in Isabelle/HOL. In: Eekelen, M., Geuvers, H., Schmaltz, J., Wiedijk, F. (eds.) ITP 2011. LNCS, vol. 6898, pp. 135–151. Springer, Heidelberg (2011). doi:10.1007/978-3-642-22863-6_12
20. Immler, F., Maletzky, A.: Gröbner bases theory. Archive of Formal Proofs (2016). http://isa-afp.org/entries/Groebner_Bases.shtml. Formal proof development
21. Kam, R.: Case studies in proof checking. Master's thesis, San Jose State University (2007). http://scholarworks.sjsu.edu/cgi/viewcontent.cgi?context=etd_projects&article=1149
22. Kobayashi, H., Chen, L., Murao, H.: Groups, rings and modules. Archive of Formal Proofs (2004). http://isa-afp.org/entries/Group-Ring-Module.shtml. Formal proof development
23. Liu, L., Aravantinos, V., Hasan, O., Tahar, S.: On the formal analysis of HMM using theorem proving. In: Merz, S., Pang, J. (eds.) ICFEM 2014. LNCS, vol. 8829, pp. 316–331. Springer, Cham (2014). doi:10.1007/978-3-319-11737-9_21
24. Lotz, M.: On the volume of tubular neighborhoods of real algebraic varieties. Proc. Amer. Math. Soc. **143**(5), 1875–1889 (2015)
25. Moura, L., Bjørner, N.: Z3: an efficient SMT solver. In: Ramakrishnan, C.R., Rehof, J. (eds.) TACAS 2008. LNCS, vol. 4963, pp. 337–340. Springer, Heidelberg (2008). doi:10.1007/978-3-540-78800-3_24
26. Murphy, T., Gray, P., Stewart, G.: Certified convergent perceptron learning (unpublished draft). http://oucsace.cs.ohiou.edu/~gstewart/papers/coqperceptron.pdf

27. Nipkow, T., Wenzel, M., Paulson, L.C. (eds.): Isabelle/HOL. LNCS, vol. 2283. Springer, Heidelberg (2002)
28. Paulson, L.C., Blanchette, J.C.: Three years of experience with Sledgehammer, a practical link between automatic and interactive theorem provers. In: Sutcliffe, G., Schulz, S., Ternovska, E. (eds.) IWIL-2010. EPiC, vol. 2, pp. 1–11. EasyChair (2012)
29. Paulson, L.C., Susanto, K.W.: Source-level proof reconstruction for interactive theorem proving. In: Schneider, K., Brandt, J. (eds.) TPHOLs 2007. LNCS, vol. 4732, pp. 232–245. Springer, Heidelberg (2007). doi:10.1007/978-3-540-74591-4_18
30. Poon, H., Domingos, P.M.: Sum-product networks: a new deep architecture. In: Cozman, F.G., Pfeffer, A. (eds.) UAI 2011, pp. 337–346. AUAI Press (2011)
31. Prathamesh, T.V.H.: Tensor product of matrices. Archive of Formal Proofs (2016). http://isa-afp.org/entries/Matrix_Tensor.shtml. Formal proof development
32. Sternagel, C., Thiemann, R.: Executable multivariate polynomials. Archive of Formal Proofs (2010). http://isa-afp.org/entries/Polynomials.shtml. Formal proof development
33. Thiemann, R., Yamada, A.: Matrices, Jordan normal forms, and spectral radius theory. Archive of Formal Proofs (2015). http://isa-afp.org/entries/Jordan_Normal_Form.shtml. Formal proof development
34. Wenzel, M.: Isar—a generic interpretative approach to readable formal proof documents. In: Bertot, Y., Dowek, G., Théry, L., Hirschowitz, A., Paulin, C. (eds.) TPHOLs 1999. LNCS, vol. 1690, pp. 167–183. Springer, Heidelberg (1999). doi:10.1007/3-540-48256-3_12

Formalization of the Lindemann-Weierstrass Theorem

Sophie Bernard[(✉)]

Université Côte d'Azur, Inria, Valbonne, France
Sophie.Bernard@inria.fr

Abstract. This article details a formalization in Coq of the Lindemann-Weierstrass theorem which gives a transcendence criterion for complex numbers: this theorem establishes a link between the linear independence of a set of algebraic numbers and the algebraic independence of the exponentials of these numbers. As we follow Baker's proof, we discuss the difficulties of its formalization and explain how we resolved them in Coq. Most of these difficulties revolve around multivariate polynomials and their relationship with the conjugates of a univariate polynomial. Their study ultimately leads to alternative forms of the fundamental theorem of symmetric polynomials. This formalization uses mainly the Mathcomp library for the part relying on algebra, and the Coquelicot library and the Coq standard library of real numbers for the calculus part.

Keywords: Coq · Formal proofs · Multivariate polynomials · Polynomial conjugates · Transcendance

1 Introduction

Natural, integer, rational, real, complex . . . We are so used to this classification of numbers that we tend to forget it is not the only one. After all, integers are only the solutions of simple additive equations in $\mathbb{N}$, and rationals are the solutions of the simple multiplicative ones in $\mathbb{Z}$. The next move would normally be to mix them both in $\mathbb{Q}$. We then obtain the *algebraic* numbers: the set of all the roots of polynomials whose coefficients lie in $\mathbb{Q}$.

The Lindemann-Weierstrass theorem gives a criterion to recognize *transcendental* numbers, that is non-algebraic numbers. More precisely, it explains that a set of exponentials of algebraic numbers which respect certain conditions can never verify some polynomial equation. The usual statement gives a result which links linear independence of these algebraic numbers and the algebraic independences of their exponentials, both over $\mathbb{Q}$.

This kind of criterion was pretty new at the time: only in 1844 has Liouville [14] shown that there exist transcendental numbers, by explicitly exhibiting one. In the next few years, Hermite [11] used a special function to show that e is also transcendental. This same function led to the transcendance of π by Lindemann,

© Springer International Publishing AG 2017
M. Ayala-Rincón and C.A. Muñoz (Eds.): ITP 2017, LNCS 10499, pp. 65–80, 2017.
DOI: 10.1007/978-3-319-66107-0_5

and its generalization, an earlier version of the theorem we formalized [13]. The work of Weierstrass eventually resulted in the Lindemann-Weierstrass theorem in its usual form [16].

Later on, Gelfond and Schneider worked independently of some generalizations, which gave results on linear forms of complex logarithms, opening the door for many applications: diophantine equations, elliptic curves, cryptography, ...

In this paper, we formalized, in Coq, one of the many proofs that exist. To our knowledge, it is the first formalization of this theorem in any formal proof assistant. We were given different choices for the proof, the main ones were the following:

- Hermite, Lindemann and Weierstrass: it has the advantage of explaining exactly why a certain function is introduced.
- Baker: it uses the same steps as the previous one, except that it goes straight for the result.
- Lang: a less elementary proof relying on field extensions and Galois theory [12].

Lang's proof could be interesting to formally prove but the goal here was to continue the previous work on multivariate polynomials [3] and give a small interface for the study of polynomial roots. That's why Baker's proof was chosen. It can be found in Baker's book [2].

In this paper, we will dedicate Sect. 2 to the first part of the proof: it will serve as an overview of the useful libraries and the proof, as well as introducing the theorem statement. Then, in Sect. 3, we will follow the different parts of Baker's proof. For each part, we will explain the ideas of the mathematical proof, its difficulties and how we resolved them by formalizing several notions such as the conjugates roots of a polynomial or a special kind of symmetric polynomials. Finally, we compare our solution to related work, and present what could be the next step (Sect. 4).

All the files of this formalization can be found on [1], they are relying on Coq 8.5 [8], coquelicot 2.1.2, and development versions of mathcomp 1.6.1, multinomials, and finmap [6].

2   Context

In this section, we will formally define the already introduced terms, motivate our choice of libraries in order to be able to give the Coq statement of the Lindemann-Weierstrass theorem.

2.1   Mathematical Context

Firstly, we say a finite set of numbers $A = \{a_1, \ldots, a_n\}$ is *linearly independent* over $\mathbb{Q}$ when a rational linear combination of A is null only if the coefficients are all zero.

$$\forall (q_1, \ldots, q_n) \in \mathbb{Q}^n, q_1 a_1 + \cdots + q_n a_n = 0 \implies \forall i \in \{1, \ldots, n\}, q_i = 0. \quad (1)$$

The same way a *polynomial* over a commutative ring $\mathbb{K}$ is the linear combination of coefficients in $\mathbb{K}$ and exponents of the indeterminate, a *multivariate polynomial* over a commutative ring $\mathbb{K}$ is a linear combination of coefficients in $\mathbb{K}$ and *monomials* which are products of indeterminates. Usually, there are no more than 3 indeterminates, they are written X, Y and Z, otherwise we call them X_k where k ranges from 1 to n if n indeterminates are needed.

A number is said to be *algebraic* if it is one of the roots of a polynomial over $\mathbb{Q}$. It is transcendental if it never is a root of a polynomial over $\mathbb{Q}$. A finite set of numbers are *algebraically independent* over $\mathbb{Q}$ if no multivariate polynomial over $\mathbb{Q}$ has exactly this set as his roots.

With these definitions, we can finally explicitly state the theorem.

Theorem 1 (Lindemann-Weierstrass). *For any non-zero natural number n and any algebraic numbers $a_1, \ldots, a_n$, if the set $\{a_1, \ldots, a_n\}$ is linearly independent over $\mathbb{Q}$, then $\{e^{a_1}, \ldots, e^{a_n}\}$ is algebraically independent over $\mathbb{Q}$.*

2.2 Stating the Lindemann-Weierstrass Theorem in Coq

In order to formally prove Theorem 1, the previous definitions need to be transfered in Coq, like the complex numbers. The library MathComp, originally developed to prove the four-color [9] and Feit-Thompson theorems [10], provides a nice frame to define algebraic structures. In fact, to define complex numbers in module Cstruct, we use the Coq standard library of real numbers, and a MathComp file to define complexes as pairs over a real-closed field.

```
CoInductive complex (R : Type) : Type := Complex { Re : R; Im : R }.
Definition complexR := (complex R).
```

Thanks to the real-closed field structure of the real numbers and the Math-Comp tools, the type complexR automatically inherits a closed field structure, equipped with a norm and a complete order on this norm. There exists a morphism QtoC from rat to complexR. This allows the use of the predicate algebraicOver QtoC x which states that there exists a polynomial over rat such that when its coefficients are embedded into complexR by QtoC, x is one of its roots.

```
Notation QtoC := (ratr : rat → complexR).
Notation "x'is_algebraic'" := (algebraicOver QtoC x).
```

We can then define the complex exponential from the embedding RtoC, and the Coq standard library functions, cos, sin and exp.

```
Definition Cexp (z : complexR) : complexR :=
  RtoC (exp(Re_R z)) * (RtoC (cos (Im_R z)) + 'i * RtoC (sin (Im_R z))).
```

In MathComp, as soon as a type has a monoid structure, an interface (bigop) is provided for the repeated use of the monoid operation. This usually translates in the possibility to easily write sums or products. For example, we call *exp-linear* combination of the tuples β and α the linear combination of the exponentials of α's with the β's as coefficients. For instance, the exp-linear combination of $(2,3)$ and $(4,5)$ is $2e^4 + 3e^5$.

```
Definition Cexp_span (n : nat) (a : complexR^n) (alpha : complexR^n) :=
  \sum_(i : 'I_n) a i * Cexp (alpha i).
```

Mathematical tuples are used a lot in the proof of the Lindemann-Weierstrass theorem. In MathComp, they can either be viewed as a sequence of fixed size (`tuple`) or as a function with finite domain (`finfun`), which are actually constructed above `tuple`. For instance, with functions, the type of a mathematical n-tuple of complexes is `complexR^ n`. In the context of multivariate polynomials, a monomial is also a tuple of natural numbers: to each ordinal i it associates the exponents of X_i.

Finally, concerning polynomials, whether univariate or multivariate, the predicate `\is a polyOver P` (resp. `mpolyOver`) states that all the coefficients of the polynomial respect a certain predicate P. In each case, we also have a way to evaluate them: `p.[x]` for univariate polynomials and `p.@[t]` for multivariate polynomials.

To prove the Lindemann-Weierstrass theorem, we have to recognize the rational numbers amongst the complex. In the case of algebraic numbers, there exists a boolean function that returns true if a number is rational, and false in the other case [7]. In our context, we already added an axiom (from the Coq standard library: Epsilon, only on equality) which makes it possible to transform the equality between two reals into a boolean function. This boolean equality with the fact that `complexR` verifies a modified version of the archimedean property implies that testing whether a complex is an integer or not is now a boolean function.

```
Definition archimedean_axiom (R : numDomainType) : Prop :=
  forall x : R, exists ub : nat, '|x| < ub%:R.
```

This means we have two boolean predicates `Cnat` and `Cint` that recognize if a complex number is a natural number or an integer. This also means that any `numDomainType` (approximately, an integral domain with a norm, a partial order and a boolean equality) which verifies this archimedean property can be automatically equipped with these two predicates, like `int`, `rat`, or `algC` (algebraic complex numbers).

Instead of adding yet another axiom to obtain a boolean predicate for the rationals, we preferred changing small parts of the proof to equivalent ones that don't rely on rational numbers. Moreover, as the MathComp library is designed around reflections between properties and booleans, this really improves both the feasibility and the ease of use for the proof.

The last missing piece concerns the linear and algebraic independence of a set of numbers. Their formalization follows almost exactly the mathematical definitions except that the set over which the independence is considered ($\mathbb{Q}$ until now, P in the following definition) is represented as a predicate.

```
Definition lin_indep_over (P : pred_class) {n : nat} (x : complexR^n) :=
  forall (lambda : complexR^n),
    lambda \in ffun_on P →
    lambda != 0 →
    \sum_(i < n) (lambda i * x i) != 0.
```

```
Definition alg_indep_over (P : pred_class) {n : nat} (x : complexR^n) :=
  forall (p : {mpoly complexR[n]}),
    p \is a mpolyOver _ P →
    p != 0 →
    p.@[x] != 0.
```

We can now state Theorem 1 in Coq, using the predicate `Cint`, as the linear (or algebraic) independence over $\mathbb{Q}$ is the same as over $\mathbb{Z}$.

```
Theorem Lindemann (n : nat) (a : complexR^n) :
  (n > 0)%N →
  (forall i : 'I_n, alpha i is_algebraic) →
  lin_indep_over Cint alpha →
  alg_indep_over Cint (finfun (Cexp \o alpha)).
```

2.3 Proof Context

The proof has a small part involving real analysis on functions with complex values, and more precisely, derivatives and integrals of such functions. For this part, we use the Coquelicot library [5] which gives us a way to get an upper bound for an integral (`RInt`).

```
RInt (fun x => norm ((T i)^+ p).[x *: alpha j]) 0 1 <= M ^+ p.
```

This means that our development is based on multiple libraries which were not meant to be used together. In Fig. 1, we show how the different libraries interact with each other. As the context for this proof is almost exactly the same as for the direct proof of e and π, you can refer to [3] for more details.

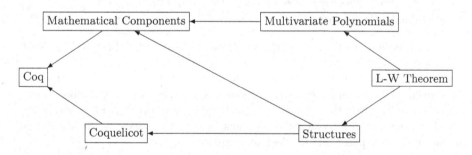

Fig. 1. Link between the different libraries

3 Following the Proof

Baker's proof doesn't actually prove Theorem 1 but rather a reformulation which we call Theorem 2. This reformulation provides the advantage of eliminating the study of linear and algebraic independence.

Theorem 2 (Baker's reformulation). *For any non-zero natural number l, any distinct algebraic numbers $\alpha_1, \ldots, \alpha_l$ and any non-zero algebraic numbers $\beta_1, \ldots, \beta_l$, we have:*

$$\beta_1 e^{\alpha_1} + \ldots + \beta_l e^{\alpha_l} \neq 0. \tag{2}$$

Its proof has 3 main steps. The first one focuses on the β's and strengthens their hypothesis, using the first lemma (Lemma 3). The second one deals with the α's and adds some conditions on them, showed in Lemma 4. The final part actually deals with the last lemma and proves it. We will present the ideas of these different steps in the following subsections.

In Fig. 2, the dashed arrows are the trivial implications that are not proved in this paper, as we focus on the proof of the Lindemann-Weierstrass theorem.

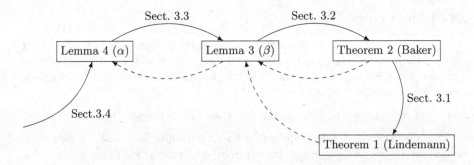

Fig. 2. Implications between the different theorems and lemmas

3.1 Baker's Reformulation

In this subsection, we prove that the Lindemann-Weierstrass theorem is at least as strong as Baker's reformulation (previously called Theorem 2).

As the notations in Sect. 2 cover our needs, we can immediately give the Coq translation of Theorem 2, and notice the change on l: to avoid any problem later on, we explicitly state that the tuple is not empty, as its size is l.+1. It should be noted that the conditions that the α's must be distinct is seen as the injectivity of the function alpha from 'I_l.+1 to complexR.

```
Theorem LindemannBaker :
  forall (l : nat) (alpha : complexR^l.+1) (a : complexR^l.+1),
  injective alpha →
  (forall i : 'I_l.+1, alpha i is_algebraic) →
  (forall i : 'I_l.+1, a i != 0) →
  (forall i : 'I_l.+1, a i is_algebraic) →
  (Cexp_span a alpha != 0).
```

In order to prove this reformulation is at least as strong as Theorem 1, we need to recall some definitions around multivariate polynomials. The *support* of a multivariate polynomial p is the set of all the monomials which have a non-zero coefficients, and is written `msupp p` in Coq. Finally, it is worth noticing that when one evaluates a monomial on a set of exponential of numbers, one obtains the exponential of a linear combination of these numbers.

As shown below, the proof of the implication of the Lindemann-Weierstrass theorem by Baker's reformulation is pretty straightforward.

Proof. Let us call `t` the tuple of the support of `p`, `alpha` (resp. `beta`) the tuple of the evaluation of the monomials of `t` on the exponentials of `a` (resp. the coefficients of the monomials of `t` in `p`). We can then recognize an exp-linear combination: we prove the following equality and rewrite it.

```
P.@[finfun (Cexp \o a)] = Cexp_span beta alpha.
```

By applying Theorem 2, we are left with all its hypothesis to prove. All of them are almost instantaneous except the injectivity of `alpha`. To prove this last bit, it suffices to unfold `alpha` so that we obtain an equality between two linear combinations of the `a`, which can we reduced to a single linear combination.

```
\sum_(k < n) (((t i) k)%:R - ((t j) k)%:R) * a k != 0
```

By linear independence, and unicity of the support of a polynomial, the proof is finished.   □

3.2   Simplifying the β's

The goal of the first part of Baker's proof is to show that we can assume, without loss of generality, that the β's are integers. Thus, this subsection gives some insights about the set of roots of a polynomial, the minimal polynomial, a total order on the complex and the formalization of the maximum of a tuple. With all these facts, we prove that Lemma 3 below implies Theorem 2.

Lemma 3. *For any non-zero natural number l, any distinct algebraic numbers $\alpha_1, \ldots, \alpha_l$ and any non-zero integers $\beta_1, \ldots, \beta_l$, we have:*

$$\beta_1 e^{\alpha_1} + \ldots + \beta_l e^{\alpha_l} \neq 0. \tag{3}$$

The main difference between the mathematical statement and the Coq one is that we replace the condition on l to be non-zero by the explicit value `1.+1`.

```
Lemma wlog1 :
  (forall (l : nat) (alpha : complexR^l.+1) (a : complexR^l.+1),
  injective alpha →
  (forall i : 'I_l.+1, alpha i is_algebraic) →
  (forall i : 'I_l.+1, a i != 0) →
  (forall i : 'I_l.+1, a i \is a Cint) →
  (Cexp_span a alpha != 0)).
```

To explain the proof, we need some definitions about symmetric polynomials and minimal polynomials. A multivariate polynomial is said to be *symmetric* if it is left unchanged by any permutation of its variables. For instance, the following polynomial is symmetric, but its first half isn't.

$$X^3Y^2Z + X^2YZ^3 + XY^3Z^2 + X^3YZ^2 + X^2Y^3Z + XY^2Z^3. \tag{4}$$

The *fundamental theorem of symmetric polynomials* states that any symmetric polynomial in a ring $\mathbb{K}$ can be obtained from a particular set of symmetric polynomials using only addition, multiplication, and multiplication by coefficients in the ring $\mathbb{K}$. As we shall develop more on this topic later, let us just remark that as a consequence of this theorem, we have that for any symmetric polynomial whose coefficients are in $\mathbb{K}$, its evaluation on the set of roots of a polynomial is in $\mathbb{K}$.

To more easily track the set of roots of a polynomial, we gave ourselves a small definition, which expresses that a set `f` is the set of roots of a polynomial. To be more general, we consider an archimedean closed field T with a norm and partial order, a predicate S and say that the polynomial whose roots are exactly `f` has all its coefficients in S when multiplied by a number `c`.

```
(f \is a set_roots S c) =
  ((c *: \prod_(x <- enum_fset f) ('X - x%:P)) \is a polyOver S).
```

In our case, the predicate will be `Cint` and the field `complexR`. Once more, as we can't recognize the rational numbers among the complex, we workaround this problem by specifying a multiple (`c`) of the expected least common multiple of the denominator of all the coefficients. The negative point of having one more parameter is once again largely compensated by the gain in ease of proof from a boolean predicate.

The *minimal polynomial* of an algebraic complex number x is the polynomial of smallest degree which has x as a root, whose coefficients are in $\mathbb{Q}$ and whose leading coefficient is 1 (we also say *monic*). We defined two related notions for the minimal polynomial: the one whose coefficients are in $\mathbb{Z}$, and the one already embedded in the complex.

The first one, `polyMinZ`, is the irreducible, separable polynomial whose coefficients are in `int`, whose `zcontents` (the greatest common divisor of all its coefficients with the same sign as the leading coefficient) is 1, and which divides exactly all the polynomials which have x as a root. It corresponds to the minimal polynomial multiplied by the least common multiple of the denominators of all its coefficients, and if needed, by -1 so that the leading coefficient is positive.

```
{p : {poly int} | [∧ (zcontents p = 1),
     irreducible_poly p, separable_poly (map_poly ZtoC p)
   & forall q : {poly int}, root (map_poly ZtoC q) x = (p %| q)]}.
```

To prove the existence of such a polynomial, we used the fact that any number of type `algC` has a minimal polynomial over `rat`. As shown in Fig. 3, from an algebraic complex number x, we obtain a polynomial p which has x as a root. We can then proceed by recurrence on the degree of p to prove the existence of the minimal polynomial of x: if it is below or equal to 2, we already have a candidate for the minimal polynomial. From this point, we extract a root xC in algC of the polynomial p embedded in algC, and call its minimal polynomial q. If x is a root of q, we found a candidate for its minimal polynomial. If not, then q divides p, and we obtain a new polynomial with coefficients in rat with a smaller size and which has x as a root. By recurrence, we obtain a candidate. Once we have a candidate, we have to transform it to be in int instead of rat, by multiplying it by the least common multiple of the denominators of all its coefficients.

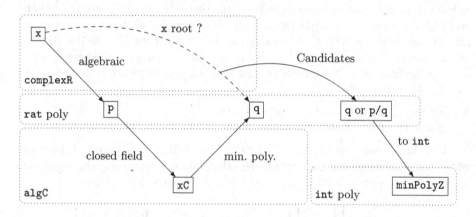

Fig. 3. Existence of a minimal polynomial

Proof (Lemma 3 $\implies$ Theorem 2). We proceed by contrapositive, so we assume we have some l, α's and β's that respect Theorem 2's assumptions but contradict its conclusion: the linear combination of the β's and the exponentials of α's is considered equal to 0. Every β is algebraic: for each k which ranges from 1 to l, we call B_k the set of roots of the minimal polynomial of β_k. Then, we multiply all the expressions obtained by letting the β_k's to run independently through the B_k's in the left-hand side of (2).

$$\prod_{\beta_1' \in B_1} \cdots \prod_{\beta_l' \in B_l} (\beta_1' e^{\alpha_1} + \ldots + \beta_l' e^{\alpha_l}) = 0. \tag{5}$$

Finally, (5) is symmetric with respect to each B_k so that the new coefficients of the linear combination are rationals. It then suffices to multiply by the least common multiple of the denominators to ensure the new β's are integers. $\quad\square$

In this proof, we can notice three main problems. In usual mathematical practice, we don't even bother checking if there is at least one new β and α to contradict Lemma 3. Secondly, we recognize an expression as symmetric and then directly state that the result is what we expected. Actually, we are supposed to begin by showing it is a symmetric polynomial evaluated on some numbers, then continue by applying the fundamental theorem of symmetric polynomials and end by verifying we obtain a $\mathbb{Q}$-exp-linear combination. Finally, even the application of the theorem is misleading: we have to apply it for each B_k.

The first point can be resolved by following the highest α according to the lexical ordering on complex numbers following the real part and then the imaginary part:

```
Definition letc x y :=
  ('Re x < 'Re y) || (('Re x == 'Re y) && ('Im x <= 'Im y)).
```

In order to follow the highest of the α's, we define the maximum of a tuple. This is the repeated use of the `maxc` operation on the tuple, which evokes the `bigop` module. The structure (`complexR, maxc`) is not equipped with a monoid structure and we cannot use directly the big operations provided by the MathComp library. In the case of a non-empty tuple f, we invoke the sequence case whose default value is the first value of the tuple (`f ord0`), and whose sequence is the sequence value of the tuple (`codom f`).

```
Definition bmaxf n (f : complexR^n.+1) :=
  bigmaxc (f ord0) (codom f).
```

The third point can be resolved by changing a little bit the proof. Instead of letting each β_k to run only within B_k, they can now run through the whole $B = \biguplus_{k=1}^{l} B_k$, seen as a multiset. Expression (5) is then changed into:

$$\prod_{\beta_1' \in B} \cdots \prod_{\beta_l' \in B} (\beta_1' e^{\alpha_1} + \ldots + \beta_l' e^{\alpha_l}) = 0. \tag{6}$$

It presents two immediate benefits. First, (6) is now completely symmetric in the β's, which means we now only have to use the fundamental theorem of symmetric polynomials once. Secondly, it makes the expression clearer, and thus, more convincing. Finally, we can notice that the products in (6) can be replaced by a single product sign over the l-tuples of elements of B.

For the second point, we can identify explicitly the left-hand side of (2) with a multivariate polynomial whose coefficients are also multivariate polynomials evaluated on the right points: the first set of indeterminates represents the β's, the second set is used to hide the exponentials and verify that, in the end, we still obtain a linear combination of rationals and exponentials. For instance, (6) is changed into the following, where `L.+1` is the cardinal of B.

```
((\prod_(f : 'I_L.+1 ^ 1.+1) \sum_(i : 'I_1.+1) 'X_i *: 'X_(f i))
  .@[finfun ((@mpolyC 1.+1 complexR_ringType) \o beta)])
  .@[finfun (Cexp \o alpha)] = 0.
```

We can also notice that Lemma 3 is a direct consequence of the Lindemann-Weierstrass theorem.

3.3 Simplifying the α's

The second and third steps in Baker's proof both revolve around the same new lemma which strengthens the conditions on the α's. But, we need some last definitions to be able to write it.

First of all, a *partition* of a set S is a set of sets such that none of its elements is the null set, all of its elements are disjoint, and the union of all its elements is exactly S. For instance, $\{\{1,3,6\},\{2\},\{4,5\}\}$ is a partition of $\{1,2,3,4,5,6\}$. In MathComp, the notion already existed and is based on the null set (set0), the union of all its parts (cover P) and trivIset which is a predicate that treats the disjoint condition by studying the cardinal of the parts of P.

```
Definition partition (T : finType) (P : {set {set T}}) (D : {set T}) :=
  [&& cover P == D, trivIset P & set0 \notin P].
```

Secondly, two complex numbers are called *conjugates* if they are roots of a same minimal polynomial. In particular, in order to find a minimal polynomial, we need at least one of those numbers to be algebraic. In this case, an algebraic number x and a complex number y are conjugates if y is a root of the minimal polynomial of x.

```
Lemma conjOfP (x y : complexR) (x_alg : x is_algebraic) :
  reflect (y \is a conjOf x_alg) (root (polyMin x_alg) y).
```

We also say a set of conjugates is *complete* if it is exactly the set of all the roots of a minimal polynomial. For instance, $\{\sqrt[3]{2}, e^{\frac{2i\pi}{3}}\sqrt[3]{2}, -e^{\frac{2i\pi}{3}}\sqrt[3]{2}\}$ is a complete set of conjugates. Formalizing this property with a boolean predicate leads to a strange definition which once again uses an additional parameter c with the same role. In particular, given a complex c and a set f of complex numbers, our definition checks if P := c *: \prod_(x <- enum fset f) ('X - x%:P) is non-zero, non-constant and only has integer coefficients, and from the proofs of these three statements, it constructs the minimal polynomial of a root of P. Then, it suffices to check if these two polynomials are equal up to an integer multiplication. In particular, any complex in a complete set of conjugates is algebraic.

```
Lemma setZconj_algebraic (c x : complexR) (f : {fset complexR}) :
  x \in f → f \is a setZconj c → x is_algebraic.
```

Lemma 4. *For any non-zero natural number l, any distinct algebraic numbers $\alpha_1, \ldots, \alpha_l$ and any non-zero integers $\beta_1, \ldots, \beta_l$, such that the α's can be grouped into a partition P, if for each part in P, the α's form a complete set of conjugates, and on each part in P, the β's are constant, then:*

$$\beta_1 e^{\alpha_1} + \ldots + \beta_l e^{\alpha_l} \neq 0. \tag{7}$$

With the previously introduced notation, we obtain the following lemma in Coq.

```
Lemma wlog2 :
  forall (l : nat) (c : complexR) (alpha : complexR^l.+1)
    (part : {set {set 'I_l.+1}}) (a : complexR^l.+1),
  c != 0 →
  c \is a Cint →
  injective alpha →
  partition part [set: 'I_l.+1] →
  {in part, forall P : {set 'I_l.+1},
    [fset (alpha i) | i in P]%fset \is a setZroots c} →
  (forall i : 'I_l.+1, a i != 0) →
  (forall i : 'I_l.+1, a i \is a Cint) →
  {in part, forall P : {set 'I_l.+1}, constant [seq a i | i in P]} →
  Cexp_span a alpha != 0.
```

To continue the proof of the Lindemann-Weierstrass theorem, we need to prove that Lemma 4 implies Lemma 3. Its proof follows the same steps as the proof in Sect. 3.2 so we won't give too much details. The main differences and sources of difficulties between the two proofs can be found in Fig. 4. They come from the fact that we need to be far more specific in each argument.

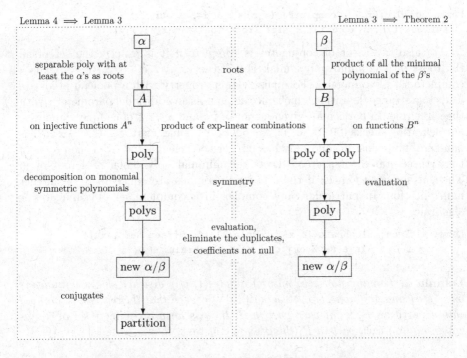

Fig. 4. Comparison of the proofs of Sect. 3.3 (left) and Sect. 3.2 (right)

For instance, in Sect. 3.2, we introduced the fundamental theorem of symmetric polynomials. Here, we need a more precise version that states that any symmetric polynomial in a ring $\mathbb{K}$ can be obtained as a $\mathbb{K}$-linear combination of a particular set of symmetric polynomials: *the monomial symmetric polynomials*. Each one is exactly the *smallest symmetric polynomial* of a monomial $X_1^{k_1} X_2^{k_2} \ldots X_n^{k_n}$, that is the sum of all the monomials $X_{\sigma(1)}^{k_1} X_{\sigma(2)}^{k_2} \ldots X_{\sigma(n)}^{k_n}$ that can be obtained from all the permutations σ of the indeterminates. In Coq, we called them `mmsym x` or `'m_x` if x is a monomial.

3.4   Proving the Final Lemma

The final part of the proof uses even more precise arguments around multivariate polynomials and symmetry.

Proof. To actually prove Lemma 4, which corresponds to the third part of Baker's proof, we proceed by contradiction. We then need to introduce many notations and a big enough prime number p. This p only depends on the values of the α's and β's, and must be an upper bound (UB) of a certain polynomial T on $[0;1]$. Once it is defined, we construct other polynomials and values I, J and K, as shown in Fig. 5 and study them in two ways. First, with algebra, we can show they have some properties of divisibility which lead to a lower bound for K. Then, as I can be recognized as an integral of a function of T, we can find an upper bound, and consequently, an upper bound for K, too. Finally, we combine both inequations involving K:

$$(p-1)!^n \le K < (p-1)!^n. \tag{8}$$

This is a logical contradiction which completes the proof.   □

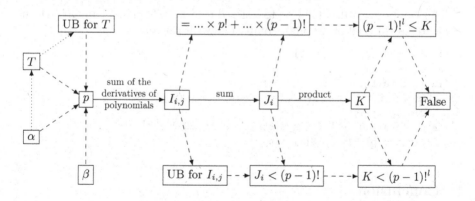

Fig. 5. Proof structure of Lemma 4

This proof follows the same steps as the common lemma of the transcendence of e and π up until the results on J. In the last few different steps which are only in the algebra part, Baker proposes to impose being algebraic integers on the α's and then speaks of divisibility of algebraic integers by natural numbers. Because of the path chosen in the common lemma proof, we can finish the proof without ever explicitly using algebraic integers.

The proofs of divisibility on K relies heavily on repeated uses of the fundamental theorem of symmetric polynomials. In fact, K is the evaluation of a multivariate polynomial K_m on the α's. For each part of the partition P, K_m is symmetric on the subset of indeterminates of the part but not fully symmetric. This means that the theorem can not be applied directly on the full set of indeterminates, but repeatedly on each subset of indeterminates corresponding to a part of the partition, that is on a subset of indeterminates which make K_m symmetric and corresponds to a complete set of conjugates. All the indeterminates must be considered once and only once, hence the need for a partition on complete set of conjugates.

We formalized a predicate `symmetric_for` indicating if a multivariate polynomial is symmetric on a given subset of its indeterminates in the same way as a fully symmetric polynomial: it must be equal to itself on any permutation of this subset of indeterminates. This was done with an already existing predicate `perm_on` on permutations that states that outside the subset, the permutation maps a value to itself.

Thanks to this predicate and extensions of other notations such as the smallest symmetric polynomial containing a monomial, we were able to obtain a new formalized corollary of the fundamental theorem of symmetric polynomials.

Lemma 5. *Given a natural number m, a m-tuple l of unique complex numbers, a m-variate polynomial p whose coefficients are rationals, and P a partition of $\{1, \ldots, m\}$, if p is symmetric on each part of P for each part of P, the corresponding l's form a complete set of conjugates, then $P(l_1, \ldots, l_m)$ is rational.*

This can be generalized in Coq to a closed field T (instead of $\mathbb{C}$) and a boolean predicate kS (instead of rational) compatible with ring operations.

```
partition P [set: 'I_m.+1] →
injective l →
{in P, forall Q : {set 'I_m.+1},
   [fset l i | i in Q]%fset \is a set_roots kS c} →
p \is a mpolyOver m.+1 kS →
{in P, forall Q, p \is (@symmetric_for _ _ Q)} →
c ^+ (msize p).-1 * p.@[l] \in kS.
```

4   Conclusion

4.1   Related Work

Concerning transcendance formalizations, this work is a follow-up of the proof of transcendence of e by Bingham [4] in HOL Light, and of the proofs of transcendance of e and π in Coq.

The work on multivariate polynomials is primarily based on a development made by Strub, and continued by Hivert who also formalized monomial symmetric polynomials but relied upon the definitions of partitions of an integer. We prefered to have a self-sufficient definition which introduces the notion of the smallest symmetric polynomial containing a monomial. Nevertheless, we added a lemma to ensure the compatibility of both definitions.

To our knowlegdge, this is the first formalization of multivariate polynomials that are symmetric on a subset of their indeterminates. This is not really surprising as the main current use of multivariate polynomials consists in studying the Bernstein polynomials [15] in order to approximate polynomial inequalities. Moreover, it would be easier to separate the indeterminates in two sets (symmetric or not) of indeterminates, rather than staying with a fixed number of indeterminates.

4.2 Future Work

In the future, the multivariate polynomial library will evolve to pick the indeterminates differently, allowing the change of the number of indeterminates to be far easier. This will probably change the statement of the corollary of the fundamental theorem of symmetric polynomials, and make its proof way easier.

It could be interesting to try to formalize the same theorem using Galois theory as it is already formalized in Mathcomp, to compare the proofs. In the same spirit, proving the existence of an embedding from the algebraic complex numbers (type `algC`) to the complex numbers (type `complexR`) could offer a different view on the proof, with different definitions or different proofs. Nevertheless, the construction of the complex numbers would still be necessary for the analysis part.

The next step could probably be to prove Baker's theorem on the linear form of complex logarithms, as it is the basis for several developments derived from transcendence.

References

1. Formalization of the Lindemann-Weierstrass theorem in Coq. http://www-sop.inria.fr/marelle/lindemann/
2. Baker, A.: Transcendental Number Theory. Cambridge University Press, Cambridge (1990)
3. Bernard, S., Bertot, Y., Rideau, L., Strub, P.Y.: Formal proofs of transcendence for e and pi as an application of multivariate and symmetric polynomials. In: Proceedings of the 5th ACM SIGPLAN Conference on Certified Programs and Proofs, pp. 76–87. ACM (2016)
4. Bingham, J.: Formalizing a proof that e is transcendental. J. Formaliz. Reason. 4(1), 71–84 (2011)
5. Boldo, S., Lelay, C., Melquiond, G.: Coquelicot: a user-friendly library of real analysis for Coq. Math. Comput. Sci. 9(1), 41–62 (2015)
6. Cohen, C.: Finmap library. http://github.com/math-comp/finmap

7. Cohen, C.: Formalized algebraic numbers: construction and first-order theory. Ph.D. thesis, Citeseer (2013)
8. Coq development team: the Coq proof assistant (2008). http://coq.inria.fr
9. Gonthier, G.: Formal proof – the four-color theorem. Not. AMS **55**(11), 1382–1393 (2008)
10. Gonthier, G., et al.: A machine-checked proof of the odd order theorem. In: Blazy, S., Paulin-Mohring, C., Pichardie, D. (eds.) ITP 2013. LNCS, vol. 7998, pp. 163–179. Springer, Heidelberg (2013). doi:10.1007/978-3-642-39634-2_14
11. Hermite, C.: Sur la fonction exponentielle. In: Comptes-Rendus de l'Académie des Sciences, vol. 77, pp. 18–24, 74–79, 226–233, 285–293. Paris (1873)
12. Lang, S.: Algebra. Graduate Texts in Mathematics, vol. 211, 3rd edn. Springer, New York (2002)
13. Lindemann, F.: Über die zahl π. Math. Ann. **20**(2), 213–225 (1882)
14. Liouville, J.: Sur des classes très-étendues de quantités dont la valeur n'est ni algébrique, ni même réductible à des irrationnelles algébriques. J. de mathématiques pures et appliquées **16**, 133–142 (1851)
15. Muñoz, C., Narkawicz, A.: Formalization of a representation of Bernstein polynomials and applications to global optimization. J. Autom. Reason. **51**(2), 151–196 (2013)
16. Weierstrass, K.: Zu Lindemann's Abhandlung: "Über die Ludolph'sche Zahl.". Akademie der Wissenschaften (1885)

CompCertS: A Memory-Aware Verified C Compiler Using Pointer as Integer Semantics

Frédéric Besson[1]([⊠]) , Sandrine Blazy[2]([⊠]) , and Pierre Wilke[3]([⊠])

[1] Inria, Rennes, France
frederic.besson@inria.fr
[2] Université Rennes 1 - CNRS - IRISA, Rennes, France
sandrine.blazy@irisa.fr
[3] Yale University, New Haven, USA
pierre.wilke@yale.edu

Abstract. The COMPCERT C compiler provides the formal guarantee that the observable behaviour of the compiled code improves on the observable behaviour of the source code. In this paper, we present a formally verified C compiler, COMPCERTS, which is essentially the COMPCERT compiler, albeit with a stronger formal guarantee: it gives a semantics to more programs and ensures that the memory consumption is preserved by the compiler. COMPCERTS is based on an enhanced memory model where, unlike COMPCERT but like GCC, the binary representation of pointers can be manipulated much like integers and where, unlike COMPCERT, allocation may fail if no memory is available.

The whole proof of COMPCERTS is a significant proof-effort and we highlight the crux of the novel proofs of 12 passes of the back-end and a challenging proof of an essential optimising pass of the front-end.

Keywords: Verified compilation · Low-level code · Optimisations · Pointer as integer

1 Introduction

Over the past decade, the COMPCERT compiler has established a milestone in compiler verification. COMPCERT is a formally verified C compiler written with the COQ proof assistant, which initially targeted safety-critical embedded software. The compiler comes with a machine-checked proof that it does not introduce bugs during compilation [2]. This semantic preservation proof relies on the formal semantics of the source and target languages of the compiler, and requires that the source program has a defined semantics. Therefore, COMPCERT only provides formal guarantees for programs that do not exhibit undefined behaviours – a property that is in general undecidable.

COMPCERT's memory model is a central component of the compiler. In this paper, we show how to adapt COMPCERT for a more expressive memory model which lifts two main limitations. First, memory allocation in COMPCERT always

© Springer International Publishing AG 2017
M. Ayala-Rincón and C.A. Muñoz (Eds.): ITP 2017, LNCS 10499, pp. 81–97, 2017.
DOI: 10.1007/978-3-319-66107-0_6

succeeds, therefore modelling infinite memory. As a consequence, the compiler does not guarantee anything on the memory consumption of the compiled program. In particular, the compiled program may exhibit a stack overflow. Second, COMPCERT's memory model limits pointer arithmetic: every implementation-defined operations on pointers results in an undefined behaviour of the memory model. This may seem restrictive but this is compliant with the C standard.

In previous work [3], we proposed a more concrete memory model inspired by COMPCERT where memory is finite and pointers can be used as integers. On that basis, we have adapted the proof of 3 passes of COMPCERT's front-end [4].

In this work, we present a fully verified COMPCERT compiler where 12 remaining passes have been ported to our new memory model. This compiler is called COMPCERTS (for COMPCERT with Symbolic values). COMPCERTS gives much stronger guarantees about the behaviour of arbitrary pointer arithmetic, thus avoiding the miscompilation of programs performing bit-level manipulation of pointers.

COMPCERTS also provides strong guarantees about the relative memory usage of the source and target programs. This is challenging because it is unclear how to even define the memory usage at the C level. We show how to tackle this challenge using oracles, aiming at ensuring that compiled programs use no more memory than source programs. In particular, this ensures that the absence of memory overflow is preserved by compilation.

All the results presented in this paper have been mechanically verified using the Coq proof assistant. The development is available online [1]. Additionally, we include links to the online documentation for several definitions and theorems in this paper under the form of Coq logos 🦊. Our contribution is COMPCERTS, which is safer than COMPCERT in the following sense: (1) COMPCERTS offers guarantees for a wider class of programs; (2) COMPCERTS also offers guarantees about the memory usage of the compiled program. More precisely, we make the following technical contributions:

- We present the proof of the compiler back-end (i.e. 12 compiler passes) including constant propagation, common sub-expression elimination and dead-code elimination. In particular, we detail how the existing alias analyses of COMPCERT [15] benefit from our more defined semantics.
- We show how to instrument the C semantics with oracles specifying the memory usage of functions, so that the compiler only reduces the memory usage of the program. We thus ensure that the absence of memory overflow is preserved by compilation.

The rest of the paper is organised as follows. First, Sect. 2 gives background information on COMPCERT and the symbolic memory model of our previous work [4]. Section 3 highlights the proof challenges related to treating pointers as integers. In particular, we explain the impact on optimisations and on the proof of one important pass of the front-end of COMPCERT. Section 4 shows how we ensure that the compiler reduces the memory usage of programs and proves that the absence of memory overflows is preserved. Section 5 mentions related work and finally, Sect. 6 concludes.

2   Background on CompCert

This section describes the architecture of the COMPCERT compiler [12]. It also summarises the main features and properties of our memory model [3,4].

2.1   Architecture of the CompCert Compiler

COMPCERT compiles C programs into assembly code, through 8 intermediate languages. The same memory model is shared by all the languages of the compiler. Each language is given a formal semantics in the form of a state transition system. Every transformation from one language to another is proved to be semantics preserving using simulation relations, stating that every step in the source language can be *simulated* by a number of steps in the target language, such that some matching relation between program states is preserved by those steps. The composition of all the simulation lemmas for the individual compiler passes forms the semantic preservation theorem given below. For the sake of simplicity, we consider that the semantics observe behaviours that are either defined behaviours, with a trace of I/O events, or undefined behaviours.

Theorem 1. *Suppose that tp is the result of the successful compilation of the program p. If bh' is a behaviour of tp then there exists a behaviour bh such that bh is a behaviour of p and bh' improves on the behaviour bh.*

$$bh' \in ASem(tp) \Rightarrow \exists bh.bh \in CSem(p) \land bh \subseteq bh'$$

In the theorem, *CSem* gives the semantics of C programs and *ASem* gives the semantics of assembly programs. Moreover, a behaviour bh' improves on a behaviour bh (written $bh \subseteq bh'$) if either bh and bh' are the same, or undefined behaviours in bh are replaced by defined behaviours in bh'.

2.2   The Memory Model of CompCert

The memory model of COMPCERT is the cornerstone of the semantics of all the intermediate languages. It consists of a collection of separated *blocks*, where blocks are arrays of a given size. A value $v \in val$ (see Fig. 1) can be either a 32-bit integer $int(i)$, a pointer or the token undef. A pointer is a pair $ptr(b, o)$ consisting of a block identifier b and an offset o. COMPCERT also features 64-bit integers, single and double precision floating-point numbers, which we ignore in this paper for the sake of simplicity. To allow fine-grained access to the memory, COMPCERT does not store values directly in the memory. Rather, values are encoded as sequences of byte-sized *memory values* called memval that describe the content of a memory block. They are either concrete 8-bit integers Byte (b), a special Undef byte that represents uninitialised memory, or a byte-sized fragment of a pointer value Pointer (b, o, n) (read: n-th byte of pointer $ptr(b, o)$). Therefore, a pointer $ptr(b, o)$ is encoded in memory as a sequence of 4 memvals, from Pointer$(b, o, 0)$ to Pointer$(b, o, 3)$. The memory model exports four operations: load reads values from the memory at a given address (a block and an offset), store writes values into the memory at a given address, alloc allocates a new block and free frees a given block.

$$val \ni v := \mathtt{int}(i) \mid \mathtt{ptr}(b, o) \mid \mathtt{undef}$$
$$memval \ni mv := \mathtt{Byte}(b) \mid \mathtt{Pointer}(b, o, n) \mid \mathtt{Undef}$$

Fig. 1. Run-time and memory values

2.3 A Symbolic Memory Model for CompCert

In previous work [3,4], we extended COMPCERT's memory model and gave semantics to pointer operations by replacing the value domain *val* by a more expressive domain *sval* of symbolic values. This low-level memory model enables reasoning about the bit-level encoding of pointers within COMPCERT. In this section, we first give a motivating example; then we recall the principles of symbolic values and their normalisation.

Motivation for Pointers as Integers. Figure 2 shows an example of C code that benefits from our low-level memory model. This is an implementation of red-black trees which belongs to the Linux kernel. A node in a red-black tree (type rb_node) contains an integer rb_parent_color and two pointers to its children nodes. The integer rb_parent_color encodes both the color of the node and a pointer to the parent node. The rationale for this encoding is as follows: (1) pointers to rb_nodes are at least 4-byte aligned, therefore the two trailing bits are zeros; and (2) the color of a node can be encoded with a single bit. Retrieving each piece of information from this encoding is implemented by the two macros rb_color and rb_parent shown in Fig. 2. To get the parent pointer, the macro clears the two trailing bits using a bitwise & with ~ 3 (i.e. $0b1 \ldots 100$). In COMPCERT, these operations are undefined because of the bitwise operations on pointers. In COMPCERTS, these operations are defined and therefore this kernel code can be safely compiled without fear of any miscompilation.

```
struct rb_node {
    uintptr_t rb_parent_color;
    struct rb_node *rb_right;
    struct rb_node *rb_left;  };
#define rb_color(rb) (((rb)-> rb_parent_color) & 1)
#define rb_parent(r) \
    ((struct rb_node *) ((r)-> rb_parent_color & ~3))
```

Fig. 2. Red-black tree implementation in Linux

Symbolic Values. A symbolic value $sv \in sval$ (see Fig. 3) is either a value v or an expression built from unary and binary C operators over symbolic values. Memory values memval are also generalised into symbolic memory values smemval, which have a single constructor $\mathtt{Symbolic}(sv, n)$, denoting the n-th

$$sval \ni sv := val \mid \mathbf{unop}(u, sv) \mid \mathbf{binop}(b, sv_1, sv_2)$$
$$\texttt{smemval} \ni smv := \texttt{Symbolic}(sv, n)$$

Fig. 3. Symbolic run-time and memory values

byte of a symbolic value sv. This constructor is inspired from the $\texttt{Pointer}\ (\cdot, \cdot, \cdot)$ constructor of COMPCERT (see Fig. 1) and subsumes the three existing cases.

Building symbolic values instead of the token **undef** for undefined operations delays the challenge of giving more semantics to C expressions. However, symbolic values cannot be kept symbolic indefinitely. To perform memory accesses at an address represented by the symbolic value $addr$, the address $addr$ must be *normalised* into a genuine pointer $\texttt{ptr}(b, o)$. Similarly, the condition cond of a conditional statement must be normalised into an integer $\texttt{int}(i)$ to decide which branch to follow. The normalisation is specified as a function $\texttt{normalise}$ which takes as input a memory state m and a symbolic value sv, and outputs a value v. Its specification relies on the notions of concrete memories valid for a memory state m, and of evaluation of expressions that we recall below.

Concrete Memories and Evaluation. A concrete memory is a mapping from blocks to concrete addresses, represented as 32-bit integers. Each memory block b has a size $size$ and an alignment constraint al; a pointer $\texttt{ptr}(b, o)$ is *valid* if the offset o is within the bounds $[0, size[$, written $\texttt{valid}(m, b, o)$. We can retrieve the alignment of a block b with the accessor $\texttt{align}(m, b)$.

Definition 1. ✎ *A concrete memory cm is valid for a memory state m $(cm \vdash m)$ if the following conditions hold:*

1. *Valid addresses lie within the address space, i.e.*
   $\forall b\ o,\ \texttt{valid}(m, b, o) \Rightarrow cm(b) + o \in\ ]0; 2^{32} - 1[.$
2. *Valid pointers from distinct blocks do not overlap, i.e.*
   $\forall b\ b'\ o\ o',\ b \neq b' \wedge\ \texttt{valid}(m, b, o) \wedge \texttt{valid}(m, b', o') \Rightarrow cm(b) + o \neq cm(b') + o'.$
3. *Addresses are properly aligned, i.e.* $\forall b,\ 2^{\texttt{align}(m,b)} \mid cm(b).$

The evaluation of a symbolic value sv in a concrete memory cm (written $[\![sv]\!]_{cm}$) consists in replacing pointers with their integer value (according to cm) and then evaluating the resulting expression with standard integer operations.

Example 1. Consider for example a concrete memory cm_1 that maps a block b to the address 32. The evaluation of the symbolic value $sv = \texttt{ptr}(b, 5)\ \&\ \texttt{int}(1)$ results in $\texttt{int}(1)$ because $[\![sv]\!]_{cm} = (cm(b) + 5)\ \&\ 1 = (32 + 5)\ \&\ 1 = 37\ \&\ 1 = 1.$

Specification of the Normalisation. The normalisation of sv in m returns a value v if for every $cm \vdash m$, sv and v evaluate identically in cm.

$$(\forall cm \vdash m \Rightarrow [\![sv]\!]_{cm} = [\![v]\!]_{cm}) \Rightarrow \texttt{normalise}(m, sv) = v$$

If no such value v can be found, the normalisation returns **undef**.

Example 2. Consider a program which stores information in the 2 least significant bits of a 4-byte aligned pointer (cf. Fig. 2). The symbolic value after setting the last 2 bits of a pointer $\mathtt{ptr}(b,0)$ is $sv = \mathtt{ptr}(b,0) \mid 3$. To recover the original pointer, the last two bits can be cleared by the following bitwise manipulation: $sv' = sv \, \& \sim 3$. We have that sv' normalises into pointer $\mathtt{ptr}(b,0)$ because for any valid concrete memory cm:

$$[\![sv']\!]_{cm} = [\![(\mathtt{ptr}(b,0) \mid 3) \, \& \sim 3]\!]_{cm} = (cm(b) \mid 3) \, \& \sim 3 = cm(b)$$

The last rewriting step is justified by the alignment constraints of block b. Since $[\![\mathtt{ptr}(b,0)]\!]_{cm} = cm(b)$ for any cm, then sv' normalises into $\mathtt{ptr}(b,0)$.

2.4   Memory Injections

Memory injections are COMPCERT's central notion to formalise the effect of merging blocks together; they are used to specify the passes that transform the memory layout. The stereotypical example is the construction of stack frames, which happens during the transformation from C♯minor to Cminor. At the C♯minor level, each local variable is allocated in its own block. In Cminor, a single block contains all the local variables, stored at different offsets. This mapping from local variable blocks in C♯minor to offsets in the stack block in Cminor is captured by a memory injection. A memory injection is characterised by an injection function $f : block \rightarrow \lfloor block \times \mathbb{Z} \rfloor$ that optionally associates with each block a new block and an offset within that block. For example, in Fig. 4, the blocks b_1, b_2 and b_3 are injected by f into the single block b', at different offsets.

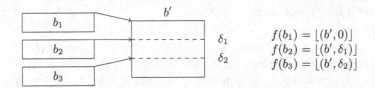

Fig. 4. Injecting several blocks into one

In addition to reflecting the structural relation between memory states, injections also relate the contents of the memory states. Values that are stored at corresponding locations are required to be *in injection*. Two values v_1 and v_2 are in injection if (1) v_1 is $\mathtt{undef}$, or (2) v_1 and v_2 are the same non-pointer value, or (3) v_1 is $\mathtt{ptr}(b,o)$, v_2 is $\mathtt{ptr}(b',o+\delta)$ and $f(b) = \lfloor (b',\delta) \rfloor$[1]. For example, in Fig. 4, the pointer $\mathtt{ptr}(b_2,o)$ is in injection with the pointer $\mathtt{ptr}(b',o+\delta_1)$.

Two symbolic values are in injection (see [4]) if they have the same structure (the same operators are applied) and the values at the leaves of each symbolic value are in injection. We proved a central result that relates injections and normalisations, recalled in Theorem 2.

---

[1] $\lfloor \cdot \rfloor$ denotes the option type. We write $\lfloor v \rfloor$ for Some(v) and $\emptyset$ for None.

Theorem 2. ❡ *For any total injection f, for any memory states m_1 and m_2 in injection by f, for any symbolic values sv_1 and sv_2 in injection by f, the normalisations of sv_1 in m_1 and of sv_2 in m_2 are in injection by f.*

This theorem has the precondition that f must be a *total* injection, i.e. all non-empty blocks must be injected (i.e. $f(b) \neq \emptyset$). In this paper, one of our contributions is a generalisation of Theorem 2, which covers the case of more general injections. As we shall see in Sect. 3.1, it is required to prove the SimplLocals pass of COMPCERT.

3   Proof Challenges for Pointers as Integers

This section presents the proof challenges that we tackle for porting COMPCERT to a semantics with symbolic values, where pointer operations behave as integer operations, e.g. bitwise operators are defined on pointers. The first challenge concerns the SimplLocals pass of COMPCERT, which modifies the structure of the memory. The second challenge is related to optimisations, and in particular the notion of pointer provenance. The existing pointer analysis in COMPCERT needs to be refined, so that it is correct in our symbolic setting.

3.1   Proving the Correctness of SimplLocals

The SimplLocals compiler pass is one of the earliest in COMPCERT. Its source language is Clight, a stripped-down dialect of C where expressions are side-effect-free. The purpose of this pass is to pull out of memory the local variables that do not need to reside in memory: those whose address is never taken. Those variables are transformed into *temporaries*, i.e. pseudo-registers, upon which most subsequent optimisations operate.

Arguments for the Correctness of SimplLocals. In COMPCERT, the correctness of this compiler pass relies on memory injections. The blocks corresponding to variables that are not transformed into temporaries are injected into themselves (i.e. $f(b) = \lfloor b, 0 \rfloor$), while the blocks corresponding to variables that are transformed into temporaries are not injected (i.e. $f(b) = \emptyset$).

The core difficulty of porting the proof of SimplLocals to the symbolic setting resides in proving that normalisations are preserved by injections. In previous work, we have established Theorem 2 which proves this preservation for total injections. Here, the injection is partial (i.e. some blocks are not injected) and therefore Theorem 2 does not apply. The following example illustrates the challenge of dealing with partial injections.

Example 3. For the sake of simplicity, consider a memory size of 32 bytes and a memory state m_1 with two blocks b and b' which are both 4-byte aligned: b of size 8 and b' of size 16. We show in Fig. 5a the only two possible concrete memories, where b is the darker block and b' is the lighter one. Note that no block can be assigned the address 0 nor the address 28, as per Definition 1.

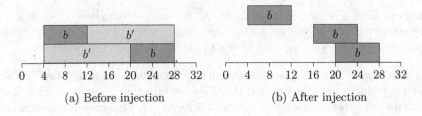

(a) Before injection   (b) After injection

Fig. 5. Concrete memories and partial injections

Consider the symbolic value $sv = \mathtt{ptr}(b, 0)! = 16$. It normalises into 0 in m_1, because b is never allocated at address 16 in any concrete memory valid for m_1. Indeed, this address is always occupied by block b'. Now consider a memory state m_2 where the block b' has been pulled out of memory. Figure 5b shows that in m_2 it is, of course, still possible to allocate block b at addresses 4 and 20. However, there is a new possible configuration where block b can be allocated at address 16. The normalisation of sv is now undefined because sv evaluates to different values (1 or 0) depending on the concrete memory used.

The essence of the problem illustrated by the above example is that blocks may have more allowed positions after the injection than before, meaning that the set of valid concrete memories is larger after the injection. Therefore, the normalisation may be less defined after a partial injection and Theorem 2 cannot be generalised for arbitrary partial injections.

Well-Behaved Injections. We identify a restricted class of *well-behaved* injections functions f, for which we show that blocks that are injected by f (those for which $f(b) \neq \emptyset$) do not gain new valid concrete addresses after the injection. The criterion for well-behavedness of injection functions f is defined in Definition 2.

Definition 2 (Well-behaved injection). *An injection function f is said to be* well-behaved *if only the blocks that are at most 8-byte wide and at most 8-byte aligned may be forgotten by f. Formally,*

$$\mathtt{well\_behaved}\,(f, m) \triangleq \forall\, b,\ f(b) = \emptyset \Rightarrow \mathtt{size}(m, b) \leq 8 \wedge \mathtt{align}(m, b) \leq 8.$$

The injection used for the correctness proof of SimplLocals satisfies this constraint because only scalar variables may be removed from the memory, i.e. the largest are long-typed variables that are 8-byte wide and 8-byte aligned. Using such well-behaved injections, we can prove Lemma 1, from which a generalised version of Theorem 2 can be derived, as we explain at the end of this section.

Lemma 1. *Let f be a well-behaved injection function. Let m_1 and m_2 be memory states in injection by f. For every concrete memory cm_2 valid for m_2, there is a corresponding concrete memory cm_1 valid for m_1, such that every non-forgotten block has the same address in cm_1 and cm_2. Formally,*

$\forall f,$ well_behaved $f \Rightarrow$

$\quad \forall\ m_1\ m_2,$ mem_inject $f\ m_1\ m_2 \Rightarrow \forall\ cm_2 \vdash m_2, \exists\ cm_1 \vdash m_1 \wedge cm_1 \equiv_f cm_2$

$\quad where\ cm_1 \equiv_f cm_2 \triangleq \forall\ b\ b', f(b) = \lfloor (b', 0) \rfloor \Rightarrow cm_1(b) = cm_2(b')$

The problem that Lemma 1 solves can be thought of as follows: for every concrete memory cm_2 valid for m_2 ($cm_2 \vdash m_2$), it is possible to insert back all the blocks that have been forgotten by f, without moving the others. In other words, all block positions that are allowed in m_2 were already allowed in m_1, therefore we avoid the problems illustrated by Example 3. The proof of Lemma 1 goes by counting 8-byte wide and 8-byte aligned regions of memory that we call *boxes*. We call nbox(cm) the number of used boxes for a given concrete memory cm. Our allocation algorithm [4] entails that for every memory state m, there exists a concrete memory cm that we call the canonical concrete memory of m and write canon_cm(m), that is built by allocating all the blocks of m at *maximally-aligned*, i.e. 8-byte aligned, addresses. Thanks to alignment constraints, we have that for any concrete memory cm valid for m, cm uses no more boxes than canon_cm(m), i.e. nbox(cm) $\leq$ nbox(canon_cm(m)).

Consider now two memory states m_1 and m_2 in injection by some well-behaved injection function f, such that m_2 is the result of *forgetting* F blocks from m_1. We have that nbox(canon_cm(m_2)) = nbox(canon_cm(m_1)) $- F$. Starting from a concrete memory $cm_2 \vdash m_2$, we derive that nbox(cm_2) $+ F \leq$ nbox(canon_cm(m_1)). In other words, it is possible to find F free boxes in cm_2. Because the blocks we forgot each fit in a box, all we have to do at this point is use each of these F boxes to contain the F forgotten variables.

Theorem 3 is the generalised version of Theorem 2 for well-behaved injections.

Theorem 3. 🔖 *For any well-behaved injection f, for any memory states m_1 and m_2 in injection by f, for any symbolic values sv_1 and sv_2 in injection by f, the normalisations of sv_1 in m_1 and of sv_2 in m_2 are in injection by f.*

Proof. The proof is performed in two steps.

- First, we exhibit some value v such that the normalisation of sv_1 injects into v. This shows that if the normalisation of sv_1 is a pointer, then this pointer is injected by f. This is a consequence of the fact that sv_1 is injected into another symbolic value.
- Then, we show that this v is necessarily the normalisation of sv_2 in m_2. This boils down to showing that: $\forall\ cm_2 \vdash m_2,\ [\![v]\!]_{cm_2} = [\![sv_2]\!]_{cm_2}$. Using Lemma 1 and the specification of the normalisation, we conclude this proof.

This theorem is a central piece of the proof of the SimplLocals pass, which is now fully proved in CompCertS.

3.2   Optimisations

CompCert features several standard optimisations. Among them, constant propagation, strength reduction and common sub-expression elimination exploit the result of a dataflow analysis computing the combination of an interval analysis and an alias analysis. In this section, we explain why the existing dataflow transfer functions are not sound for COMPCERTS and how to fix them. This demonstrates that the semantics of COMPCERTS is a provably strong safeguard preventing the miscompilations of low-level pointer arithmetic.

The Abstract Value Domain of CompCert is made of the sum of a pointer domain and a numeric domain. One purpose of the pointer domain is to distinguish pointers to the current stack frame from other pointers. A representative but simplified abstract pointer-domain ($aptr$) is given below. Its semantics is given by its concretisation function γ_{sb} where sb stands for the memory block of the current stack frame. The empty set of pointers is denoted by $\bot$. $Stk\ ofs$ represents the stack pointer $\texttt{ptr}(sb, ofs)$. The set of all pointers to the current stack frame (block sb at any offset) is captured by $Stack$. All pointers to blocks different from the stack block sb are abstracted by $\neg Stack$. Finally, $\top$ is the set of all pointers.

$$aptr ::= \bot \mid Stk\ ofs \mid Stack \mid \neg Stack \mid \top$$

The numeric domain $anum$ is standard: it tracks intervals of integers and floating-point constants. The domain of abstract values $aval = aptr \times anum$ is the sum domain such that $\gamma_{sb}(ap, an) = \gamma_{sb}(ap) \cup \gamma(an)$. The sum domain is relevant because a value can be either a pointer or an integer but not both.

In COMPCERT, the transfer functions are written with *prudence* in order to avoid miscompilations and *"[Track] leakage of pointers through arithmetic operations"*.[2] This is done by computing carefully crafted transfer functions which are purposely non-optimal in order to prevent aggressive optimisations (which are sound by rely on undefined behaviours of the COMPCERT semantics). For instance, the most precise transfer function for a bitwise & is such that

$$(\neg Stack, \top)\ \&\ (\bot, \top) = (\bot, \top).$$

For the pointer part, it returns $\bot$ because a bitwise & between pointers returns **undef** (it cannot be a pointer). For the integer part, it returns $\top$ because a bitwise & between arbitrary integers is still an arbitrary integer. This formulation is semantically sound but is not *prudent* because several bits of the pointer may leak through the bitwise &.

Example 4. To illustrate the severe consequence of not tracking the leakage of pointers, consider the red-black tree code of Fig. 6. The code is annotated by an aggressive dataflow analysis and a *prudent* dataflow analysis, both being semantically sound. When both analyses differ (e.g. Lines 5 and 7), we write the aggressive result first. At function entry, the current stack frame has just been created and is therefore free of aliases. As a result, the parameter r and the local variable rpc can be abstracted by $(\neg Stack, \top)$. Line 5, the aggressive analysis is using the previous transfer function for the bitwise & and obtain $(\bot, \top)$ for the

---

[2] See https://github.com/AbsInt/CompCert/blob/a968152051941a0fc50a86c3fc15e90
e22ed7c47/backend/ValueDomain.v#L707.

abstraction of p. This makes the reasoning that p can only be an integer. As the dereference of an integer has no semantics, the aggressive analysis infers that the rest of the code is not reachable. Line 7, this is encoded by the abstraction $(\bot, \bot)$ for the variable rchild. Based on this information, a live-variable analysis and an aggressive dead-code removal could replace the whole function body by a no-op which is obviously a miscompilation.

```
1  rb_node* get_parents_right_child(rb_node* r){ //   r: (¬Stack, ⊤)
2  uintptr_t rpc = r->rb_parent_color;//get the parent/color field
3  // rpc: (¬Stack, ⊤)
4  rb_node* p = (rb_node*) (rpc & ~3);//get the parent of r
5  // p: (⊥, ⊤), (¬Stack, ⊤)
6  rb_node* rchild = p->rb_right;       // access its right child
7  // rchild: (⊥, ⊥), (¬Stack, ⊤)
8  return rchild; }
```

Fig. 6. Dataflow analysis for red-black trees

A Formally Prudent Dataflow Analysis. With our semantics, the aggressive dataflow analysis of Example 4 is not sound and therefore such miscompilations cannot occur. The reason is that our semantics computes symbolic values for arithmetic operations (e.g. the bitwise &) that need to be captured by the concretisation function. Interestingly, we eventually noticed that, to get a concretisation that is both sound and robust to syntactic variations, what was needed was a formal account of pointer tracking. It is formalised by a notion of pointer *dependence* of a symbolic value sv with respect to a set S of memory blocks. We say that sv depends at most on the set of blocks S if sv evaluates identically in concrete memories that are identical for all the blocks in S; they may differ arbitrarily for other blocks. Formally, $dep(sv, S) = \forall\ cm \equiv_S cm', [\![sv]\!]_{cm} = [\![sv]\!]_{cm'}$, where $cm \equiv_S cm' = \forall\ b \in S, cm(b) = cm'(b)$. The concretisation function γ_{sb}, where sb is the current stack block, is defined in Fig. 7 ?. Intuitively, Cst represents any symbolic value which always evaluates to the same value whatever the concrete memory (*i.e.*, it does not depends on pointers); $Stack$ represents any symbolic value which depends at most on the current stack block sb and $\neg Stack$ represents any symbolic value which may depend on any block except the current stack block sb.

Our abstract domain is still a pair of values $(ap, an) \in aptr \times anum$ but it represents a (reduced) product of domains. For symbolic values, there is no syntactic distinction between pointer and integer values. Hence, the concretisation is given by an intersection of concretisations (instead of a union)

$$\gamma_{sb}(ap, an) = \gamma_{sb}(ap) \cap \gamma(an),$$

where the concretisation of the numeric abstract domain is defined in terms of the evaluation of symbolic expressions: $\gamma(an) = \{sv \mid \forall cm, [\![sv]\!]_{cm} \in \gamma(an)\}$.

$$\begin{aligned}
\gamma_{sb}(\bot) &= \{\} \quad \gamma_{sb}(\top) = sval \\
\gamma_{sb}(Cst) &= \{sv \mid dep(sv, \emptyset)\} \\
\gamma_{sb}(Stk\ o) &= \{sv \mid \forall cm, [\![sv]\!]_{cm} = cm(sb) + o\} \\
\gamma_{sb}(Stack) &= \{sv \mid dep(sv, \{sb\})\} \\
\gamma_{sb}(\neg Stack) &= \{sv \mid dep(sv, block \setminus \{sb\})\}
\end{aligned}$$

Fig. 7. COMPCERTS concretisation for alias analysis

With this formulation, the most precise transfer function for a bitwise & is given by $(\neg Stack, \top)$ & $(\bot, \top) = (\neg Stack, \top)$.

For the pointer part, it returns $\neg Stack$ because the resulting expression may still depends on a $\neg Stack$ pointer. For the integer part, it returns $\top$ because (like before) a bitwise & between arbitrary integers is still an arbitrary integer. As a result, the aggressive transfer function of COMPCERTS implements the informally *prudent* transfer functions of COMPCERT. It follows that, for our semantics, miscompilation due to pointer leaking (e.g. Example 4) is impossible.

While adapting the proof, we found and fixed several minor but subtle *bugs* in COMPCERT related to pointer tracking, where the existing transfer functions were unsound for our low-level memory model. Though unlikely, each of them could potentially be responsible for a miscompilation. Note that COMPCERTS generates the right code not by chance but really because our semantics forbids program transformations that are otherwise valid for COMPCERT. In general, we believe that our semantics provides the right safeguard for avoiding any miscompilation of programs performing arbitrary arithmetic operations on pointers.

4   Preservation of Memory Consumption

The C standard does not impose a model of memory consumption. In particular, there is no requirement that a conforming implementation should make a disciplined use of memory. A striking consequence is that the possibility of stack overflow is not mentioned. From a formal point of view, COMPCERT models an unbounded memory and therefore, as the C standard, does not impose any limit on stack consumption of the binary code. As a result, the existing COMPCERT theorem is oblivious of memory consumption of the assembly code. Though COMPCERT makes a wise usage of memory this is not explicit in the correctness statement and can only be assessed by a close inspection of the code. COMPCERTS provides a stronger formal guarantee. It ensures that if the source code does not exhaust the memory, then neither does the assembly code. Said otherwise, the compilation ensures that the assembly code consumes no more memory than the source code does.

4.1   Evolution of Stack Memory Usage Throughout Compilation

Figure 8 shows the evolution of the size of stack frames across compiler passes. The figure distinguishes the three passes which modify the memory usage. First,

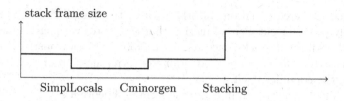

Fig. 8. Evolution of the size of stack frames

the SimplLocals pass introduces pseudo-registers for certain variables, which are pulled out of memory. This pass reduces the memory usage of functions and therefore satisfies our requirement that compilation reduces memory usage. The Cminorgen pass allocates a unique stack frame containing all the remaining variables of a function. This pass makes the memory usage grow because some padding is inserted to ensure proper alignment. However, because our allocation strategy considers maximally aligned blocks, this pass still preserves the memory usage. The remaining problematic pass is the Stacking pass which builds activation records from stack frames. This pass makes explicit some low-level data (e.g. return address or spilled locals) and is responsible for an increase of the memory usage. In the following, we explain how to solve this discordance and ensure nonetheless a decreasing usage of memory across the compiler passes.

4.2 The Stacking Compiler Pass

Stacking transforms Linear programs into Mach code. The Linear stack frame consists of a single block which contains local variables. The Mach stack frame embeds the Linear stack frame together with additional data, namely the return address of the function, spilled pseudo-registers that could not be allocated in machine registers, callee-save registers, and outgoing arguments to function calls.

Provisioning Memory. In order to fit the Stacking pass into the *decreasing memory usage* framework, our solution is to provision memory from the beginning of the compilation chain. Hence, we instrument the semantics of all intermediate languages, from C to Linear, with an oracle ns which specifies, for each function f, the additional space that is needed. The semantics therefore include special operations that reserve some space at function entry and release it at function exit. To justify that the Mach stack frame fits into our finite memory space, we can now leverage the fact that at the Linear level, there was enough space for the Linear stack frame plus $ns(f)$ additional bytes. Provided that the oracle ns is correct, this entails that the Mach stack frame fits in memory.

It may be possible to derive an over-approximation of the needed stack space for each function from a static analysis. However, the estimate would probably be very rough as, for instance, it seems unlikely that the impact of register allocation could be modelled accurately. Instead, as the exact amount of additional memory space is known during the Stacking pass, we construct the oracle ns as a byproduct of the compilation. In other words, the compiler returns not only an assembly program but also a function that associates with each function the quantity of additional stack space required. Note that the construction is not circular since the oracle is only needed for the correctness proof of the compiler and not by the compiler itself.

COMPCERTS' final theorem takes the form of Theorem 4.

Theorem 4. ✤ *Suppose that* (tp, ns) *is the result of the successful compilation of the program* p. *If* tp *has the behaviour* bh', *then there exists a behaviour* bh *such that* bh *is a behaviour of* p *with oracle* ns *and* bh' *improves on the behaviour* bh.

$$bh' \in ASem(tp) \Rightarrow \exists bh.bh \in CSem(p, ns) \wedge bh \subseteq bh'.$$

The only difference with COMPCERT is that the C semantics is instrumented by the oracle ns computed by the compiler. Though not completely explicit, Theorem 4 ensures that the absence of memory overflows is preserved by compilation. The fundamental reason is that the failure to allocate memory results in an observable going wrong behaviour. On the contrary, if the source code does not have a going wrong behaviour, neither does the assembly. It follows that if the C source succeeds at allocating memory, so does the assembly. Hence, COMPCERTS ensures that the absence of memory overflows is preserved by compilation.

Recycling Memory. Because our semantics are now parameterised by a bound on the memory usage of functions, this bound should be as low as possible so that as many programs as possible can be given a defined semantics.

In order to give a smaller bound, we notice that the SimplLocals pass forgets some blocks and therefore throws away some memory space. We can reuse this freed space and therefore have a weaker requirement on the source semantics.

Example 5. Consider a function with long-integer local variables x and y. During SimplLocals, x is transformed into a temporary while y is kept and allocated on the stack. During Stacking, say 20 additional bytes are needed to build the Mach activation record from the Linear stack frame. Then, we must reserve those 20 bytes from the beginning, i.e. from the C semantics. However, we can recycle the space from the local variable x, therefore saving 8 bytes and we only require 12 bytes at the C level, therefore making it easier to have a C semantics.

5   Related Work

Formal Semantics for C. The first formal realistic semantics of C is due to Norrish [14]. More recent works [7,9,10] aim at providing a formal account of the subtleties of the C standard. Hathhorn *et al.* [7] present an executable C semantics within the K framework which precisely characterise the undefined behaviours of C. Krebbers [9,10] gives a formal account of sequence points and non-aliasing. These notions are probably the most intricate of the ISO C standard. Memarian *et al.* [13] realise a survey among C experts, in which they aim at capturing the *de facto* semantics of C. They consider problems such as uninitialised values and pointer arithmetic.

Our work builds upon the COMPCERT C compiler [12]. The semantics and the memory model used in the compiler are close to ISO C. Our previous works [3,4] show how to extend the support for pointer arithmetic and adapt most of the front-end of COMPCERT to this extended semantics.

COMPCERT *and Memory Consumption.* Carbonneaux *et al.* [6] propose a logic for reasoning, at source level, on the resource consumption of target programs compiled by COMPCERT. They instrument the event traces to include resource consumption events that are preserved by compilation, and use the compiler itself to determine the actual size of stack frames. We borrow from them the idea of using a compiler-generated oracle. Their approach to finite memory is more lightweight than ours. However, our ambition to reason about symbolic values in COMPCERT requires more intrusive changes.

COMPCERTTSO [16] is a version of COMPCERT implementing a TSO relaxed memory model. It also models a finite memory where pointers are pairs of integers. Their soundness theorem is oblivious of out-of-memory errors. They remark that they could exploit memory bounds computed by the compiler, but do not implement it. In terms of expressiveness, their semantics and ours seem to be incomparable. For instance, CompCertTSO gives a defined semantics to the comparison of arbitrary pointers, we do not. Yet, the example of Sect. 2.3 is not handled by the formal semantics of CompCertTSO.

Pointers as Integers. Kang *et al.* [8] propose a hybrid memory model where an abstract pointer is mapped to a concrete address at pointer-integer cast time. Their semantics may get stuck at cast-time if there is not enough memory available. For our semantics, a cast is a no-op and our semantics may get stuck at allocation time. They study aggressive program optimisations but do not preserve memory consumption. In COMPCERTS, we consider simpler optimisations but implemented in a working compiler for a real language. Moreover, we ensure that the memory consumption is preserved by compilation.

6 Conclusion

We present COMPCERTS, an extension of the COMPCERT compiler that is based on a more defined semantics and provides additional guarantees about the compiled code. Programs performing low-level bitwise operations on pointers are now covered by the semantics preservation theorem, and can thus be compiled safely. COMPCERTS also guarantees that the compiled program does not require more memory than the source program. This is done by instrumenting the semantics with an oracle providing, for each function, the size of the stack frame.

COMPCERTS compiles down to assembly; compared to COMPCERT, we adapted all the 4 passes of the front-end and 12 out of 14 passes of the back-end.

This whole work amounts to more than 210 k lines of Coq code, which is 60 k more than the original COMPCERT 2.4 we started with. COMPCERTS does not feature the two following optimization passes. First, the inlining optimisation makes functions use potentially more stack space after the transformation than before. This disagrees with our decreasing memory size policy, but we should be able to provision memory in a similar way as we did for the Stacking pass, as described in Sect. 4.2. Second, the tail call recognition transforms regular function calls into tail calls when appropriate. Its proof cannot be adapted in a straightforward way because of the additional stack space we introduced for the Stacking pass: the release of those blocks does not happen at the same place before and after the transformation. We need to investigate further the proof of this optimisation and come up with a more complex invariant on memory states.

As future work, we shall investigate how security-related program transformations would benefit from the increased expressiveness of COMPCERTS. Kroll et al. [11] implement software isolation within COMPCERT. However, the transformation they define depends on a pointer masking operation which has no COMPCERT semantics and is therefore axiomatised. In COMPCERTS, pointer masking is defined and the isolated program could benefit from all the existing optimisations. Recently, Blazy and Trieu [5] pioneered the integration of an obfuscation pass within COMPCERT. Our semantics paves the way for aggressive obfuscations, which cannot be proved sound for pointers with COMPCERT.

Lastly, currently every function stores its stack frame in a distinct block, even in assembly. An ultimate compiler pass that merges blocks into a concrete stack would be possible with our finite memory and would bring even more confidence in COMPCERTS.

Acknowledgments. This work has been partially funded by the French ANR project AnaStaSec ANR-14-CE28-0014, NSF grant 1521523 and DARPA grant FA8750-12-2-0293.

References

1. Companion website. http://www.cs.yale.edu/homes/wilke-pierre/itp17/
2. Bedin Franca, R., Blazy, S., Favre-Felix, D., Leroy, X., Pantel, M., Souyris, J.: Formally verified optimizing compilation in ACG-based flight control software. In: ERTS 2012: Embedded Real Time Software and Systems (2012)
3. Besson, F., Blazy, S., Wilke, P.: A precise and abstract memory model for C using symbolic values. In: Garrigue, J. (ed.) APLAS 2014. LNCS, vol. 8858, pp. 449–468. Springer, Cham (2014). doi:10.1007/978-3-319-12736-1_24
4. Besson, F., Blazy, S., Wilke, P.: A concrete memory model for CompCert. In: Urban, C., Zhang, X. (eds.) ITP 2015. LNCS, vol. 9236, pp. 67–83. Springer, Cham (2015). doi:10.1007/978-3-319-22102-1_5
5. Blazy, S., Trieu, A.: Formal verification of control-flow graph flattening. In: CPP. ACM (2016)
6. Carbonneaux, Q., Hoffmann, J., Ramananandro, T., Shao, Z.: End-to-end verification of stack-space bounds for C programs. In: PLDI. ACM (2014)

7. Hathhorn, C., Ellison, C., Rosu, G.: Defining the undefinedness of C. In: PLDI. ACM (2015)
8. Kang, J., Hur, C., Mansky, W., Garbuzov, D., Zdancewic, S., Vafeiadis, V.: A formal C memory model supporting integer-pointer casts. In: PLDI (2015)
9. Krebbers, R.: Aliasing restrictions of C11 formalized in Coq. In: Gonthier, G., Norrish, M. (eds.) CPP 2013. LNCS, vol. 8307, pp. 50–65. Springer, Cham (2013). doi:10.1007/978-3-319-03545-1_4
10. Krebbers, R.: An operational and axiomatic semantics for non-determinism and sequence points in C. In: POPL. ACM (2014)
11. Kroll, J.A., Stewart, G., Appel, A.W.: Portable software fault isolation. In: CSF. IEEE (2014)
12. Leroy, X.: Formal verification of a realistic compiler. C. ACM 52(7), 107–115 (2009)
13. Memarian, K., Matthiesen, J., Lingard, J., Nienhuis, K., Chisnall, D., Watson, R.N., Sewell, P.: Into the depths of C: elaborating the de facto standards. In: PLDI. ACM (2016)
14. Norrish, M.: C formalised in HOL. Ph.D. thesis, University of Cambridge (1998)
15. Robert, V., Leroy, X.: A formally-verified alias analysis. In: Hawblitzel, C., Miller, D. (eds.) CPP 2012. LNCS, vol. 7679, pp. 11–26. Springer, Heidelberg (2012). doi:10.1007/978-3-642-35308-6_5
16. Ševčík, J., Vafeiadis, V., Zappa Nardelli, F., Jagannathan, S., Sewell, P.: CompCertTSO: a verified compiler for relaxed-memory concurrency. J. ACM 60(3), 22:1–22:50 (2013)

Formal Verification of a Floating-Point Expansion Renormalization Algorithm

Sylvie Boldo[1]([⊠]), Mioara Joldes[2], Jean-Michel Muller[3], and Valentina Popescu[4]

[1] Inria, LRI, CNRS, Université Paris-Sud, Université Paris-Saclay, Bâtiment 650,
Université Paris-Sud, 91405 Orsay Cedex, France
sylvie.boldo@inria.fr
[2] LAAS-CNRS, 7 Avenue du Colonel Roche, 31077 Toulouse, France
joldes@laas.fr
[3] LIP Laboratory, CNRS, 46 Allée d'Italie, 69364 Lyon Cedex 07, France
jean-michel.muller@ens-lyon.fr
[4] LIP Laboratory, ENS Lyon, 46 Allée d'Italie, 69364 Lyon Cedex 07, France
valentina.popescu@ens-lyon.fr

Abstract. Many numerical problems require a higher computing precision than the one offered by standard floating-point formats. A common way of extending the precision is to use floating-point expansions. As the problems may be critical and as the algorithms used have very complex proofs (many sub-cases), a formal guarantee of correctness is a wish that can now be fulfilled, using interactive theorem proving. In this article we give a formal proof in Coq for one of the algorithms used as a basic brick when computing with floating-point expansions, the renormalization, which is usually applied after each operation. It is a critical step needed to ensure that the resulted expansion has the same property as the input one, and is more "compressed". The formal proof uncovered several gaps in the pen-and-paper proof and gives the algorithm a very high level of guarantee.

Keywords: Floating-point arithmetic · Floating-point expansions · Multiple-precision arithmetic · Formal proof · Coq

1 Introduction

Many numerical problems require higher precisions than the standard *single*-(*binary32*) or *double*-precision (*binary64* [8]). Examples can be found in dynamical systems [1,12], planetary orbit dynamics [14], computational geometry, etc. Several examples are given by Bailey and Borwein [2]. These calculations rely on arbitrary-precision libraries. A crucial design point of these libraries is the way numbers are represented. A first solution is the *multiple-digit* representation: each number is represented by one exponent and a sequence of possibly high-radix digits. The digits follow a regular pattern, which greatly facilitates the

The authors would like to thank Région Rhône-Alpes and ANR FastRelax Project for the grants that support this activity.

© Springer International Publishing AG 2017
M. Ayala-Rincón and C.A. Muñoz (Eds.): ITP 2017, LNCS 10499, pp. 98–113, 2017.
DOI: 10.1007/978-3-319-66107-0_7

analysis of the associated arithmetic algorithms. GNU MPFR [6] is a C library that uses such a representation.

A second solution is the *multiple-term* representation in which a number is expressed as the unevaluated sum of several floating-point (FP) numbers. This sum is usually called a *FP expansion* (FPE). Since each term of this expansion has its own exponent, the term's positions can be quite irregular. The QD library [7] uses this approach and supports double-double and quad-double formats, i.e. numbers are represented as the sum of 2 or 4 double-precision FP numbers. Campary [10,11] supports FPEs with an arbitrary number of terms, and also targets GPU implementation. Other libraries that manipulate FPEs are presented in [17,20].

The intrinsic irregularity of FP expansions makes the design and proof of algorithms very difficult. Many algorithms are published without a proof, or with a proof so complex that is difficult to fully trust it. Obtaining formal proof of the critical parts of the algorithms that manipulate FPEs would bring much more confidence in them.

To make sure that a FPE carries enough information we need to ensure that it is *non-overlapping* (see Sect. 2). Even if the input FPEs satisfy this requirement, this property is often "broken" during the calculations, so after each operation we need to perform a *renormalization*. It is a basic brick for manipulating FPEs. While several renormalization algorithms have been proposed, until recently, Priest's [17] was the only one provided with a complete correctness proof. However it uses many conditional branches, which makes it slow in practice. To overcome this problem, some of us developed in [10] a new algorithm (Algorithm 3 below), that takes advantage of the machine pipeline and is provided with a correctness proof. However, the proof is complex, hence errors may have been left unnoticed. We decided to build a formal proof of that algorithm, using the Coq proof assistant and the Flocq library [5]. We also rely on a new iterator on lists that behaves better for functions that do not have an identity element [3]. There have already been some formal proofs on expansions [4]. Basic operations were formally proved in the PFF library (a library which preceded Flocq). The induction proofs were tedious, partly due to the formalization of FP numbers that was less convenient than that of Flocq.

The paper is organized as follows. Section 2 gives the pen-and-paper and the formal definitions of FPEs. Section 3 presents the renormalization algorithm and the wanted properties. Section 4 explains the formal verification of the various levels of the algorithm. Finally, Sect. 5 concludes this work.

2 Floating-Point Expansions

2.1 Pen-and-Paper Definitions

We begin with several literature definitions for FP expansions and their properties.

Definition 2.1. *A normal binary precision-p floating-point (FP) number has the form* $x = M_x \cdot 2^{e_x - p + 1}$, *with* $2^{p-1} \leq |M_x| \leq 2^p - 1$. *The integer* e_x *is the exponent of* x, *and* $M_x \cdot 2^{-p+1}$ *is the* significant *of* x. *We denote* $\text{ulp}(x) = 2^{e_x - p + 1}$ *(unit in the last place)* [15, Chap. 2], *and* $\text{uls}(x) = \text{ulp}(x) \cdot 2^{z_x}$, *where* z_x *is the number of trailing zeros at the end of* M_x *(unit in the last significant place).*

Definition 2.2. *A FP expansion (FPE) u with n terms is the unevaluated sum of n FP numbers* $u_0, u_1, \ldots, u_{n-1}$, *in which all nonzero terms are ordered by magnitude (i.e., if v is the sequence obtained by removing all zeros in the sequence u, and if sequence v contains m terms,* $|v_i| \geq |v_{i+1}|$, *for all* $0 \leq i < m - 1$).

Arithmetics on FP expansions have been introduced by Priest [17], and later on by Shewchuk [20].

To make sure that an FPE carries enough information, it is required that the u_i's do not "overlap". The notion of *non-overlapping* varies depending on the authors. We give different definitions: $\mathcal{P}$-nonoverlapping and $\mathcal{S}$-nonoverlapping are common in the literature. The third definition allows for a relatively relaxed handling of the FPEs and keeps the redundancy to a minimum. An FPE may contain interleaving zeros: the definitions that follow apply only to the non-zero terms of the expansion (i.e., the array v in Definition 2.2).

Definition 2.3. *Assuming x and y are normal numbers with representations* $M_x \cdot 2^{e_x - p + 1}$ *and* $M_y \cdot 2^{e_y - p + 1}$ *(with* $2^{p-1} \leq |M_x|, |M_y| \leq 2^p - 1$), *they are* $\mathcal{P}$*-nonoverlapping (non-overlapping according to Priest's definition* [18]*) if we have* $|e_y - e_x| \geq p$.

Definition 2.4. *An expansion is* $\mathcal{P}$*-nonoverlapping if all its components are mutually* $\mathcal{P}$*-nonoverlapping.*

Shewchuk [20] weakens this into:

Definition 2.5. *An FPE* $u_0, u_1, \ldots, u_{n-1}$ *is* $\mathcal{S}$*-nonoverlapping (non-overlapping according to Shewchuk's definition* [20]*) if for all* $0 < i < n$, *we have the inequality* $e_{u_{i-1}} - e_{u_i} \geq p - z_{u_{i-1}}$, *i.e.,* $|u_i| < \text{uls}(u_{i-1})$.

In general, a $\mathcal{P}$-nonoverlapping expansion carries more information than an $\mathcal{S}$-nonoverlapping one with the same number of components. Intuitively, the stronger the sense of the non-overlapping definition, the more difficult it is to guarantee it in the output. In practice, the $\mathcal{P}$-nonoverlapping property proved to be quite difficult to obtain and the $\mathcal{S}$-nonoverlapping is not strong enough, this is why we chose to compromise by using a different sense of non-overlapping, referred to as ulp-*nonoverlapping*, that we define below.

Definition 2.6. *A FPE* $u_0, \ldots, u_{n-1}$ *is ulp-nonoverlapping if for all* $0 < i < n$, $|u_i| \leq \text{ulp}(u_{i-1})$.

In other words, the components are either $\mathcal{P}$-nonoverlapping or they overlap by one bit, in which case the second component is a power of two.

Remark 2.7. *Note that for* $\mathcal{P}$*-nonoverlapping expansions, we have* $|u_i| \leq \frac{2^p - 1}{2^p} \text{ulp}(u_{i-1})$.

When using standard FP formats, the exponent range forces a constraint on the number of terms in a non-overlapping expansion. The largest expansion can be obtained when the largest term is close to overflow and the smallest is close to underflow. When using Definition 2.4 or 2.6, the maximum expansion size is 39 for *double*-precision, and 12 for *single*-precision.

The algorithms performing arithmetic operations on FP expansions use the so-called *error-free transforms* (such as the algorithms 2Sum, Fast2Sum, Dekker's product and 2MultFMA presented for instance in [15]), that make it possible to compute both the result and the error of a FP addition or multiplication. This implies that in general, each such algorithm, applied to two FP numbers, still returns two FP numbers.

In this article we make use of two algorithms that compute the exact sum of two FP numbers a and b and return the result under the form $s + e$, where s is the result rounded to nearest and e is the rounding error.

Algorithm 1. 2Sum (a, b).

$s \leftarrow \mathrm{RN}(a + b)$ // RN stands for performing the operation in rounding to nearest mode.
$t \leftarrow \mathrm{RN}(s - b)$
$e \leftarrow \mathrm{RN}(\mathrm{RN}(a - t) + \mathrm{RN}(b - \mathrm{RN}(s - t)))$
return (s, e) such that $s = \mathrm{RN}(a + b)$ and $s + e = a + b$

Algorithm 2. Fast2Sum (a, b).

Input: exponent of a larger than or equal to exponent of b
$s \leftarrow \mathrm{RN}(a + b)$
$z \leftarrow \mathrm{RN}(s - a)$
$e \leftarrow \mathrm{RN}(b - z)$
return (s, e) such that $s = \mathrm{RN}(a + b)$ and $s + e = a + b$

The *2Sum* (Algorithm 1) algorithm requires 6 *flops*, which it was proven to be optimal in [13], if we have no information on the ordering of a and b. However, if the ordering of the two inputs is known, a better alternative is *Fast2Sum* (Algorithm 2), that uses only 3 native FP operations. The latter one requires the exponent of a to be larger than or equal to that of b in order to return the correct result. This condition might be difficult to check, but of course, if $|a| \geq |b|$, it will be satisfied.

2.2   Coq Definitions

We formalize FPEs as lists of real numbers with a property. We could have used arrays, but lists have an easy-to-use induction that was extensively used. This is not the data structure used for computing, but only for proving. As seen previously, FPEs are a bunch of FP numbers with some property (such as for instance $|u_i| \leq |u_{i-1}|$ or $|u_i| \leq \mathrm{ulp}(u_{i-1})$). In a formal setting, we are trying to generalize this definition in order to cover many definitions, and have theorems powerful enough to handle both $\mathcal{S}$-nonoverlapping and $\mathcal{P}$-nonoverlapping FPEs for instance. The property between u_i and u_{i-1} will therefore be generic by default: it is a $P : \mathbb{R} \to \mathbb{R} \to$ Prop, meaning a property linking two real numbers. This P will be $\leq$ for ordered lists. For ulp-nonoverlapping, it is:

$$\text{fun}\, x\, y \Rightarrow |x| \leq \text{ulp}(y).$$

For $\mathcal{S}$-nonoverlapping, it is slightly more complex (see also Sect. 4.2):

$$\text{fun}\, x\, y \Rightarrow \exists e \in \mathbb{Z}, \exists n \in \mathbb{Z}, x = n \times 2^e \wedge |y| < 2^e.$$

We could have assumed that a FPE is a list where each value has this property with its successor (when it exists). Unfortunately, this fails when zeros appear in the FPE, therefore we have to account for the fact that we may have intermediate zeros inside our FPEs. This case was not considered in [10], so nothing was proved when a zero is involved. We get rid of this flaw and handle possible intermediate zeros in the proofs; in particular, we want to prove the FPE has the wanted property P, even if we remove the zeros. The Coq listing for defining a FPE with the P property is as follows:

```
Inductive Exp_P (P:R->R->Prop) : list R -> Prop :=
  | Exp_Nil: Exp_P P List.nil
  | Exp_One: forall x : R, Exp_P  P (x :: nil)
  | Exp_Z1: forall l: list R, Exp_P P l -> Exp_P  P (0 :: l)
  | Exp_Z2: forall x: R, forall l: list R, Exp_P P (x::l)
        -> Exp_P P (x :: ( 0 :: l))
  | Exp_NZ: forall x y: R, forall l: list R,
      x <> 0 -> y <> 0
    -> Exp_P P (y:: l)
    -> P x y -> Exp_P P (x :: (y :: l)).
```

Any empty or 1-element list is a FPE, and zeros in first or second position are useless. To prove that the list $x :: y :: \ell$ is a FPE, it suffices to prove that x and y are non-zero reals having the P property, and that $y :: \ell$ is a FPE. As common with formal developments, many lemmas need to be proved in order to use this definition (e.g., proving that a part of a FPE is also a FPE, or reversing a FPE). The most useful lemma is the one that proves that a given list is a FPE:

```
Lemma nth_Exp_P: forall (P:R->R->Prop) l,
        (forall i j, (0 <= i < length l)%nat
        -> (0 <= j < length l)%nat -> (i < j)%nat
        -> nth i l 0 <> 0 -> nth j l 0 <> 0
        -> P  (nth i l 0) (nth j l 0))
      -> Exp_P P l.
```

where nth i l 0 is the i-th element of the list ℓ or 0 if ℓ is too short. This lemma means that if, whatever i and j such that $i < j$, and the i-th and j-th elements of the list are non-zero, and they have the P property, then the list is a FP expansion with the property P. The proof is straightforward, and it provides an easy way to handle intermediate zeros inside the formal verification of the algorithm. Let us now describe the renormalization algorithm published in [10], before describing its formal proof in Sect. 4.

3   Renormalization Algorithm for FPEs

In [10] an algorithm with $m+1$ levels that would render the result as an m-term $\mathcal{P}$-nonoverlapping FP expansion was presented. After testing it, we concluded that using only ulp-nonoverlapping expansions greatly diminishes the cost of the algorithm while keeping the overlapping bits to a minimum, i.e., only one bit. This is why we only focus on the first two levels of the initial algorithm, that allow us to achieve the desired property.

Algorithm 3. Renormalization algorithm

Input: FP expansion $x = x_0 + \ldots + x_{n-1}$ consisting of FP numbers that overlap by at most d digits, with $d \leq p - 2$; m length of output FP expansion.

Output: FP expansion $f = f_0 + \ldots + f_{m-1}$ with $f_{i+1} \leq \mathrm{ulp}(f_i)$, for all $0 \leq i < m - 1$.

1: $e[0 : n - 1] \leftarrow VecSum(x[0 : n - 1])$
2: $f[0 : m - 1] \leftarrow VecSumErrBranch(e[0 : n - 1], m)$
3: **return** FP expansion $f = f_0 + \ldots + f_{m-2} + f_{m-1}$.

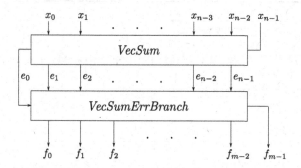

Fig. 1. Renormalization of n term-FPEs. The *VecSum* box performs Algorithm 4 of Fig. 2, and the *VecSumErrBranch* box, Algorithm 5 of Fig. 3.

The reduced renormalization algorithm (Algorithm 3, illustrated in Fig. 1) is made up with two different layers of chained *2Sum*. It receives as input an array with terms that overlap by at most d digits, with $d \leq p - 2$. Let us first define this concept.

Definition 3.1. *Consider an array of FP numbers:* $x_0, x_1, \ldots, x_{n-1}$. *According to Priest's [17] definition, they overlap by at most d digits $(0 \leq d < p)$ if and only if $\forall i, 0 \leq i \leq n - 2, \exists k_i, \delta_i$ such that:*

$$2^{k_i} \leq |x_i| < 2^{k_i+1}, \tag{1}$$

$$2^{k_i-\delta_i} \leq |x_{i+1}| \leq 2^{k_i-\delta_i+1}, \tag{2}$$

$$\delta_i \geq p - d, \tag{3}$$

$$\delta_i + \delta_{i+1} \geq p - z_{i-1}, \tag{4}$$

where z_{i-1} is the number of trailing zeros at the end of x_{i-1} and for $i = 0$, $z_{-1} := 0$.

It was proven in [10] that the algorithm returns an ulp-nonoverlapping expansion:

Proposition 3.2. *Consider an array* $x_0, x_1, \ldots, x_{n-1}$ *of FP numbers that overlap by at most* $d \leq p - 2$ *digits and let* m *be an input parameter, with* $1 \leq m \leq n - 1$. *Provided that no underflow/overflow occurs during the calculations, Algorithm 3 returns a "truncation" to* m *terms of an* ulp-*nonoverlapping FP expansion* $f = f_0 + \ldots + f_{n-1}$ *such that* $x_0 + \ldots + x_{n-1} = f$.

For the sake of simplicity, the 2 levels are represented as variations of the Algorithm *VecSum* of Ogita et al. [16, 20], and treated separately. The proof was done using intermediate properties for each layer. Also, at each step it was proved that all the *2Sum* blocks can be replaced by *Fast2Sum* ones.

First Level (Line 1, Algorithm 3). The error-free transforms can be extended to work on several inputs by chaining, resulting in "distillation" algorithms [19]. The most frequently used one is *VecSum* (Fig. 2 and Algorithm 4). It is a chain of *2Sum* that performs an error-free transformation on n FP numbers.

Algorithm 4. VecSum $(x_0, \ldots, x_{n-1})$ [20, 16].

Input: $x_0, \ldots, x_{n-1}$ FP numbers.
Output: $e_0 + \ldots + e_{n-1} = x_0 + \ldots + x_{n-1}$.

$s_{n-1} \leftarrow x_{n-1}$
for $i \leftarrow n - 2$ **to** 0 **do**
    $(s_i, e_{i+1}) \leftarrow \mathit{2Sum}(x_i, s_{i+1})$
end for
$e_0 \leftarrow s_0$
return $e_0, \ldots, e_{n-1}$

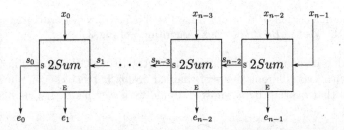

Fig. 2. *VecSum* with n terms. Each *2Sum* [15] box outputs the sum to the left and the error downwards.

The first level consists in applying Algorithm 4 on the input array, from where we obtain the array $e = (e_0, e_1, \ldots, e_{n-1})$.

In [10] the following theorem was proved:

Proposition 3.3. *After applying the* VecSum *algorithm on an input array* $x = (x_0, x_1, \ldots, x_{n-1})$ *of FP numbers that overlap by at most* $d \leq p - 2$ *digits, the output array* $e = (e_0, e_1, \ldots, e_{n-1})$ *is* $\mathcal{S}$-*nonoverlapping and may contain interleaving zeros.*

Also, it was shown that for all arrays that satisfy the input requirements *2Sum* (6 FP operations) can be replaced by *Fast2Sum* (3 FP operations).

Second Level (Line 2, Algorithm 3). This level is applied on the array e obtained previously. This is also a chain of *2Sum*, but now we start from the most significant component. Also, instead of propagating the sums we propagate the errors. If however, the error after a *2Sum* block is zero, then we propagate the sum (this is shown in Fig. 3). In what follows we will refer to this algorithm by *VecSumErrBranch* (see Algorithm 5).

Algorithm 5. VecSumErrBranch $(e_0, \ldots, e_{n-1})$

Input: S-*nonoverlapping* FP expansion $e = e_0 + \ldots + e_{n-1}$; m length of the output expansion.
Output: FP expansion $f = f_0 + \ldots + f_{m-1}$ with $f_{j+1} \le \mathrm{ulp}(f_j)$, $0 \le j < m - 1$.
 1: $j \leftarrow 0$
 2: $\varepsilon_0 = e_0$
 3: **for** $i \leftarrow 0$ to $n - 2$ **do**
 4:     $(f_j, \varepsilon_{i+1}) \leftarrow 2Sum(\varepsilon_i, e_{i+1})$
 5:     **if** $\varepsilon_{i+1} \neq 0$ **then**
 6:         **if** $j \ge m - 1$ **then**
 7:             **return** FP expansion $f = f_0 + \ldots + f_{m-1}$.
 8:         **end if**
 9:         $j \leftarrow j + 1$
10:     **else**
11:         $\varepsilon_{i+1} \leftarrow f_j$
12:     **end if**
13: **end for**
14: **if** $\varepsilon_{n-1} \neq 0$ **and** $j < m$ **then**
15:     $f_j \leftarrow \varepsilon_{n-1}$
16: **end if**
17: **return** FP expansion $f = f_0 + \ldots + f_{m-1}$.

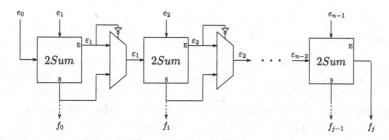

Fig. 3. *VecSumErrBranch* with n terms. Each *2Sum* box outputs the sum downwards and the error to the right. If the error is zero, the sum is propagated to the right.

For this algorithm we can also use *Fast2Sum* instead of *2Sum*. The following property holds:

Proposition 3.4. *Let e be an input array $(e_0, \ldots, e_{n-1})$ of S-nonoverlapping terms and $1 \leq m \leq n$ the required number of output terms. After applying VecSumErrBranch, the output array $f = (f_0, \ldots, f_{m-1})$, with $0 \leq m \leq n-1$ satisfies $|f_{i+1}| \leq \mathrm{ulp}(f_i)$ for all $0 \leq i < m-1$.*

Figure 4 gives an intuitive drawing showing the different constraints between the FP numbers before and after the two levels of Algorithm 3. The notation is the same as in Fig. 1.

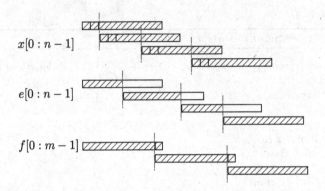

Fig. 4. The effect of Algorithm 3: x is the input FPE, e is the sequence obtained after the 1st level and f is the sequence obtained after the 2nd level.

Remark 3.5. *In the worst case, Algorithm 3 performs $n-1$ Fast2Sum calls in the first level and $n-2$ Fast2Sum calls plus $n-1$ comparisons in the second one. This accounts for a total of $7n-10$ FP operations.*

4   Formal Proof

Let us now dive into what exactly is formally proved and how. We assume a FP format with radix 2 and precision p, that includes subnormals.

4.1   Prerequisites

A big difference between pen-and-paper proofs and formal proofs is the use of previous theorems. The proof in [10] relies on a result by Jeannerod and Rump [9] about an error bound on the sum of n FP numbers. In a formal proof, we need the previous result to be also formally proved (in the same proof assistant). As the Coq formal library of facts does not grow as fast as the pen-and-paper library of facts, we usually end up proving more results than expected, in particular, the result of Jeannerod and Rump does not belong to the Flocq library.

Theorem 4.1 (error_sum_n, from [9]). *Let ℓ be a list of n FP numbers. Define $e = \sum_{i=0}^{n} \ell_i$ and $a = \sum_{i=0}^{n} |\ell_i|$. Let $f = \oplus_{i=0}^{n} \ell_i$ be the computed sum. Then $|f - e| \leq (n-1)2^{-p}a$.*

The formal proof follows exactly the pen-and-paper proof from [9] without any problem. Note that the pen-and-paper result is more generic than the Coq one: it does not enforce the parenthesis, while we precisely choose where the parentheses are. As underflowing additions are correct, this holds even when subnormals are involved.

4.2   Formal Proof of the First Level

As explained in Sect. 3, Algorithm 3 has two steps. Let us deal with its first step, the *VecSum* algorithm (Algorithm 4).

Formalization of the *VecSum* Algorithm. A generic VecSum_g algorithm is defined. Indeed, *VecSum* is an iteration of a *2Sum* operator, where the sum is given to the next step, but we may imagine a *VecSum* with another operator, or with a *2Sum* operator, where the error is given to the next step (this is the case of the third step of the renormalization algorithm described in [10]). Many basic lemmas apply on all these examples and it is better to factorize them.

The Coq definition is a fixed-point operator applying on a list. The variant (for the termination) is n, that will be the length of the list l.

```
Fixpoint VecSum_g_aux (n:nat) (l:list R)
      {struct n} : list R := match n,l with
 | 0, nil => nil
 | 1, x :: nil => x :: nil
 | S n, x :: y :: l =>
        (f x y) :: (VecSum_g_aux n (x+y-(f x y) :: l))
 | _, _ => nil
  end.
```

```
Fixpoint VecSum_g (l:list R) : list R
      := VecSum_g_aux (length l) l.
```

We prove various lemmas, including the fact that the output list has the same size as the input list and that the sum of the values of the input list is the same as the sum of the values of the output list.

Another point is that each value of the output list is exactly known from the input list (in the second level, we have tests that make this statement more complex). Taking the notations from Fig. 2, the value e_0 (the last one computed) is the FP sum of all the x_is. Using our formal definition, this means that the last element of the list is the iterated of the function fun x y => x+y-(f x y). As for the i-th element, it is proved to be $f(s_{i+1}, x_{i+1})$, where x_k is the k-th element of the list and s_k is the iterated of the function fun x y => x+y-(f x y) on the first k elements of the list. Note that lists are "numbered" from 0 to their size minus 1.

VecSum Algorithm Property. Now let us deal with what is to be proved for this algorithm. In this case f will be the addition error function. We first look into the requirements on the input and output expansions. Let d be a positive integer such that $d \leq p - 2$, and the input an expansion with numbers that overlap by at most d digits. Formally, it is an expansion with a certain predicate *IVS_P* (for Input of *VecSum* Property). This property linking two reals x and y can be stated as follows: $|y| < 2^d \operatorname{ulp}(x)$. As $d \leq p - 2$, it would intuitively mean that the expansion is clearly decreasing (as seen in Fig. 4), but this is not the case when dealing with subnormal numbers. For instance, consider the smaller positive (subnormal) FP number η, then this property holds for η and himself, for η and 2η, and for 2η and η.

This is the reason why we choose:

$$IVS\_P(x, y) = 2^{p-d} \operatorname{ulp}(x) \leq \operatorname{ulp}(y)$$

It is equivalent with the previous definition for normal numbers, but implies $|x| \leq |y|$, even when subnormals appear. Another interesting property is the fact that $IVS\_P(x, y)$ implies that y is normal.

The output of the expansion is supposed to be $\mathcal{S}$-nonoverlapping (see Definition 2.5). We choose the simple definition for OVS_P (for Output of *VecSum* Property):

$$OVS\_P(x, y) := \exists e, n : \mathbb{Z}, \; y = n \cdot 2^e \wedge |x| < 2^e.$$

Said otherwise, there exists an exponent e such that y can be expressed with exponent e and $|x| < 2^e$. This e is an overestimation of the $\operatorname{uls}(x)$ of Definition 2.4.

Proof of the *VecSum* Algorithm Property. Here is some excerpt of the proof in [10]:

$$|x_{j+1}| + |x_{j+2}| + \cdots \leq$$
$$\leq [2^d + 2^{2d-p} + 2^{3d-2p} + 2^{4d-3p} + \dots] \operatorname{ulp}(x_j)$$
$$\leq 2^d \cdot 2^p / (2^p - 1) \cdot \operatorname{ulp}(x_j).$$

This is partly wrong! In fact the geometric series should be bounded by:

$$2^d + 2^{2d-p} + 2^{3d-2p} + 2^{4d-3p} + \cdots \leq 2^d / (1 - 2^{d-p}).$$

The proof can be fixed as the two inequalities were coarse enough, so there is no problem at this point.

A small gap in [10] is that it does not handle underflow. We did that in the formal proof without many problems. Given that the input list is an expansion with the *IVS_P* property, there can be at most one subnormal FP number (at the end). This implies a special treatment for one-element lists (which is not difficult) or lists with only one non-zero element (also not difficult) and taking care of the last element in all other cases. Furthermore, [10] proves that the output is non-overlapping assuming two successive outputs are non-zeros. Unfortunately, this does not cover the most complex case: a non-zero output followed by one (or more) zero output, due to (partial) cancellation, and then another non-zero output. When considering the outputs, if e_i and e_{i+1} are non-zero, the corresponding s_i is bigger than s_{i+1} so that the exponent of s_i is bigger than that of s_{i+1} (property used in the proof in [10]). Yet, if e_i and e_{i+2} are non-zero, but $e_{i+1} = 0$, then the exponents of s_i and s_{i+2} may be in any order, and the proof does not hold anymore. We therefore needed several additional lemmas.

Lemma 4.2. *We assume we have an expansion* $\ell = (x_i)$ *with the IVS_P property with the length smaller than* $2 + 2^{p-1}$. *Then*

- *Except if the* $(x_k)_{k=i \text{ to } (n-1)}$ *are all zeros, then* $s_i \neq 0$.
- *If* $x_i \times s_{i+1} \geq 0$, *then* $\mathrm{ulp}(s_{i+1}) \leq \mathrm{ulp}(s_i)$.
- *If* $x_i \times s_{i+1} < 0$ *and* $|x_i| \leq 2|s_{i+1}|$, *then* $e_{i+1} = 0$.
- *If* $x_i \times s_{i+1} < 0$ *and* $2|s_{i+1}| < |x_i|$, *then* $\mathrm{ulp}(s_{i+1}) \leq \mathrm{ulp}(s_i)$.
- *If* $x_i \times s_{i+1} < 0$ *and* $|x_i| \leq 2|s_{i+1}|$, *then* $\mathrm{ulp}(s_i) \leq \mathrm{ulp}(s_{i+1})$.
- *If* $x_i \times s_{i+1} < 0$ *and* $|x_i| \leq 2|s_{i+1}|$ *and* $x_{i-1} \neq 0$, *then* $\mathrm{ulp}(s_{i+2}) \leq \mathrm{ulp}(s_i)$.

The lemma corresponds to various possibilities, including partial cancellation. The main result is then the following one:

Theorem 4.3 (incr_exp_sj). *Assuming an expansion* $\ell = (x_i)$ *with the IVS_P property, such that all* $x_k \neq 0$, *and such that its length is smaller than* $2 + 2^{p-1}$. *For all* $0 \leq i \leq j \leq \mathrm{length}(\ell)$, *we have* $\mathrm{ulp}(s_i) \leq \max(\mathrm{ulp}(s_j), \mathrm{ulp}(s_{j+1}))$.

In other words, if the exponent of s_j does not have the wanted property, then the exponent of s_{j+1} does. From this result, we deduce the main theorem of the *VecSum* algorithm:

Theorem 4.4 (VecSum_correct). *Assuming an expansion* ℓ *with the IVS_P property and such that its length is smaller than* $2 + 2^{p-1}$, *then VecSum*(ℓ) *has the OVS_P property.*

To prove the *OVS_P* property, we just have to exhibit an exponent and prove its properties. Using the previous theorem, we exhibit the maximum between the exponents of s_j and s_{j+1}, and the rest of the proof is straightforward. There are special cases, e.g., small lists, or handling e_0, which is not the same as the others e_i, or input lists with zeros. Yet, in the formal proof this is handled (with care) more easily than the exponent problem explained above. Note that we need a maximum size of the list: here $2 + 2^{p-1}$, meaning more than 4.5×10^{15} in *binary64* and more than 8 million in *binary32*. As done in [10], we also prove that a *Fast2Sum* can be used instead of a *2Sum* at every level of the algorithm.

4.3 Formal Proof of the Second Level

Contrary to the formal proof of the first level, where we uncovered several difficulties, the formal proof of the second level was much simpler. Two additional features are the handling of subnormal numbers, which was trivial, and some discussions about the last two terms of the output list.

Formalization of the *VecSumErrBranch* Algorithm. As was done for *Vec-Sum*, we define in Coq the formal version of *VecSumErrBranch*. It is more complicated and was therefore not put here. Refer to Fig. 3 for a more understandable version. We proved several lemmas, including the fact that the sum of the values of the input list is the same as the sum of the values of the output list. We also proved that the output list has a smaller or equal size than the input list.

VecSumErrBranch Algorithm Property. A difficulty is the need to reverse the list. VecSum was taking care of the values from the smallest to the largest, while *VecSumErrBranch* manipulates the values from the largest to the smallest. The input property of the expansion is the OVS_P property on the reverse list, meaning an expansion with the property

$$\text{fun}\, x\, y \Rightarrow OVS\_P(y, x).$$

As for the output property called $OVSB\_P$ (for Output of *VecSumErrBranch* Property), we only have the ulp-nonoverlapping property, with the definition

$$OVSB\_P(x, y) := |y| \leq \text{ulp}(x).$$

Proof of the *VecSumErrBranch* Algorithm Property. The input expansion has the S-nonoverlapping property, hence we can apply this lemma that will be helpful for our induction:

Lemma 4.5. *Let us assume that $(a :: b :: \ell)$ is an expansion with the $OVS\_P(y, x)$ property. Then*

- *$(a + b :: \ell)$ is an expansion with the $OVS\_P(y, x)$ property.*
- *$(\circ(a + b) :: \ell)$ is an expansion with the $OVS\_P(y, x)$ property.*
- *$(a + b - \circ(a + b) :: \ell)$ is an expansion with the $OVS\_P(y, x)$ property.*

Now we need to exhibit the first value output by VecSumErrBranch. It is the sum of several of the e_is, in unknown number:

Lemma 4.6. *Let $\ell = (e_i)$ be an expansion with the $OVS\_P(y, x)$ property. We assume that it is non-empty and has no zeros. Define $x = e_0 \oplus \sum_{i=1}^{n} e_i$. There exists an integer n and a list ℓ' such that:*

$$either\ x \neq 0\ or\ (x = 0\ \wedge\ \ell'\ is\ empty)$$
$$and\ VecSumErrBranch(\ell) = x :: \ell'.$$

Note that the sum of $(e_i)_{1 \leq i \leq n}$ is exact. This is because e_i is kept to be propagated in the test when the error is zero, meaning the addition was exact.

All there is left to prove is that the $OVSB\_P$ (ulp-nonoverlapping) property is obtained with such terms:

Lemma 4.7. *Let us assume that* $(b :: a :: \ell)$ *is an expansion with the OVS_P* (y, x) *property and that* $b + a - b \oplus a \neq 0$. *Let* $\ell = (e_i)$. *Then*

$$OVSB\_P \left(b \oplus a, \ \circ \left((b + a - b \oplus a) + \sum_{i=0}^{n} e_i \right) \right).$$

Then, we can prove the main theorem of the *VecSumErrBranch* algorithm:

Theorem 4.8. *Assume an expansion* ℓ *with the OVS_P* (y, x) *property. Then,* $VecSumErrBranch(\ell)$ *has the OVSB_P property.*

As before, we take care of inputs with zeros using a special treatment. We also formally prove that the *Fast2Sum* operator can be used at every level.

A last point is that the output list is (nearly) non-zero. Due to the test, most output values cannot be zeros. All the $(f_i)_{i=0 \text{ to } j-2}$ are non-zeros, but the last two terms f_{j-1} and f_j may be zeros. Note that f_j is zero if and only if the list is composed only of zeros.

5 Conclusion

The renormalization algorithm is a call to the previously-defined functions, and the proofs are successive calls to the previous proofs. Here is a final Coq excerpt for the renormalization algorithm formal verification, available at http://fastrelax.gforge.inria.fr/files/Renormalization.tgz.

```
1  Context { prec_gt_0_ : Prec_gt_0 p }.
2  Variable d:Z.
3  Hypothesis d_betw: (0 < d <= p-2)%Z.
4
5  Variable l: list R.
6  Hypothesis Fl: Forall format l.
7  Hypothesis Hl: InputVecSum d l.
8
9  Let res := VecSumErrBranch (rev (VecSum l)).
10
11 Lemma Renorm_1: Forall format res.
12 Lemma Renorm_2:
13        fold_right Rplus 0 l = fold_right Rplus 0 res.
14 Lemma Renorm_3: INR (length l) <= 2 + bpow (p-1) ->
15        OutputVecSumErrBranch res.
```

The hypotheses are as follows: $p > 0$ (Line 1), $0 < d \leq p - 2$ (Lines 2–3). The input is a list of reals called ℓ (Line 5). This list is a list of FP numbers (Line 6) and has the *IVS_P* property (Line 7).

The result of the renormalization is then **res** which is the application of *VecSumErrBranch* on the reverse of the application of VecSum to ℓ (Line 9).

We then have the three proved lemmas. The first one says that the result is a list of FP numbers (Line 11), the second one says that the sum of the values resulted is equal to the sum of the values of the input ℓ (Line 12–13), and the third one is the expected property: it is an expansion with the $OVSB\_P$ property (ulp-*nonoverlapping*) (Line 14–16). This was fully proved in the Coq proof assistant within a few thousand lines:

Spec	Proof	File
74	497	AboutFP.v
200	1049	AboutLists.v
19	24	Renormalization.v
168	1162	VecSum1.v
93	975	VecSumErrBranch.v
554	3707	Total

AboutFP includes the result of the error of FP addition from [9]. AboutLists includes results about parts of lists (such as n first elements, n last elements).

A missing part in the formal verification is the handling of overflow, as Flocq does not have an upper bound on the exponent range. It is not a real flaw, as an overflow would produce an infinity or a NaN at the end of the algorithm as only additions are involved.

This work was unexpectedly complicated: the formal verification was tedious due to many gaps, both expected (e.g., underflow) or unexpected (error in [10], handling of intermediate zeros that greatly modifies the proofs). More precisely about the intermediate zeros, let us assume that the result of the renormalization is $(y_0, y_1, y_2, 0, y_3, y_4, 0, y_5, 0, 0, 0)$ with the $y_i \neq 0$. Then the pen-and-paper theorem of [10] ensures that $OVS\_P(y_0, y_1)$, $OVS\_P(y_1, y_2)$, and $OVS\_P(y_3, y_4)$. Our theorem formally proved in Coq ensures that $OVS\_P(y_0, y_1)$, $OVS\_P(y_1, y_2)$, $OVS\_P(y_2, y_3)$, $OVS\_P(y_3, y_4)$, and $OVS\_P(y_4, y_5)$. Interleaving zeros may seem a triviality as zeros could easily be removed, except that this is costly as it involved testing each value. For this basic block, it is much more efficient to handle zeros inside the proof than to remove them in the algorithm.

We had some hope to be able to automate that kind of proofs, as many such algorithms exist in the literature, but this experiment has shown that, even from a detailed and reasonable pen-and-paper proof, much remains to be done to get a formal one. This paper shows that complex algorithms in FP arithmetic really need formal proofs: it is otherwise impossible to be certain that special cases have not been forgotten or overlooked in the pen-and-paper proof, as was the case here.

References

1. Abad, A., Barrio, R., Dena, A.: Computing periodic orbits with arbitrary precision. Phys. Rev. E **84**, 016701 (2011)
2. Bailey, D.H., Borwein, J.M.: High-precision arithmetic in mathematical physics. Mathematics **3**(2), 337 (2015)

3. Boldo, S.: Iterators: where folds fail. In: Workshop on High-Consequence Control Verification, Toronto, Canada, July 2016
4. Boldo, S., Daumas, M.: A mechanically validated technique for extending the available precision. In: 35th Asilomar Conference on Signals, Systems, and Computers, Pacific Grove, California, pp. 1299–1303 (2001)
5. Boldo, S., Melquiond, G.: Flocq: a unified library for proving floating-point algorithms in Coq. In: Proceedings of the 20th IEEE Symposium on Computer Arithmetic, Tübingen, Germany, pp. 243–252, July 2011
6. Fousse, L., Hanrot, G., Lefèvre, V., Pélissier, P., Zimmermann, P.: MPFR: a multiple-precision binary floating-point library with correct rounding. ACM Trans. Math. Softw. **33**(2) (2007). http://www.mpfr.org/
7. Hida, Y., Li, X.S., Bailey, D.H.: Algorithms for quad-double precision floating-point arithmetic. In: Burgess, N., Ciminiera, L. (eds.) Proceedings of the 15th IEEE Symposium on Computer Arithmetic (ARITH 2016), Vail, CO, pp. 155–162, June 2001
8. IEEE Computer Society: IEEE Standard for Floating-Point Arithmetic. IEEE Standard 754-2008, August 2008
9. Jeannerod, C.P., Rump, S.M.: Improved error bounds for inner products in floating-point arithmetic. SIAM J. Matrix Anal. Appl. **34**(2), 338–344 (2013)
10. Joldes, M., Marty, O., Muller, J.M., Popescu, V.: Arithmetic algorithms for extended precision using floating-point expansions. IEEE Trans. Comput. **PP**(99), 1 (2015)
11. Joldes, M., Muller, J.-M., Popescu, V., Tucker, W.: CAMPARY: cuda multiple precision arithmetic library and applications. In: Greuel, G.-M., Koch, T., Paule, P., Sommese, A. (eds.) ICMS 2016. LNCS, vol. 9725, pp. 232–240. Springer, Cham (2016). doi:10.1007/978-3-319-42432-3_29
12. Joldes, M., Popescu, V., Tucker, W.: Searching for sinks for the Hénon map using a multipleprecision GPU arithmetic library. SIGARCH Comput. Archit. News **42**(4), 63–68 (2014)
13. Kornerup, P., Lefèvre, V., Louvet, N., Muller, J.M.: On the computation of correctly-rounded sums. In: Proceedings of the 19th IEEE Symposium on Computer Arithmetic (ARITH 2019), Portland, OR, June 2009
14. Laskar, J., Gastineau, M.: Existence of collisional trajectories of mercury, mars and venus with the earth. Nature **459**(7248), 817–819 (2009)
15. Muller, J.M., Brisebarre, N., de Dinechin, F., Jeannerod, C.P., Lefèvre, V., Melquiond, G., Revol, N., Stehlé, D., Torres, S.: Handbook of Floating-Point Arithmetic. Birkhäuser, Boston (2010)
16. Ogita, T., Rump, S.M., Oishi, S.: Accurate sum and dot product. SIAM J. Sci. Comput. **26**(6), 1955–1988 (2005)
17. Priest, D.M.: Algorithms for arbitrary precision floating point arithmetic. In: Kornerup, P., Matula, D.W. (eds.) Proceedings of the 10th IEEE Symposium on Computer Arithmetic, pp. 132–144. IEEE Computer Society Press, Los Alamitos (1991)
18. Priest, D.M.: On properties of floating-point arithmetics: numerical stability and the cost of accurate computations. Ph.D. thesis, University of California at Berkeley (1992)
19. Rump, S.M., Ogita, T., Oishi, S.: Accurate floating-point summation part I: faithful rounding. SIAM J. Sci. Comput. **31**(1), 189–224 (2008)
20. Shewchuk, J.R.: Adaptive precision floating-point arithmetic and fast robust geometric predicates. Discret. Comput. Geom. **18**, 305–363 (1997)

How to Simulate It in Isabelle: Towards Formal Proof for Secure Multi-Party Computation

David Butler[1]([✉]), David Aspinall[1], and Adrià Gascón[2]

[1] The Alan Turing Institute, University of Edinburgh, Edinburgh, UK
dbutler@turing.ac.uk
[2] The Alan Turing Institute, University of Warwick, Coventry, UK

Abstract. In cryptography, secure Multi-Party Computation (MPC) protocols allow participants to compute a function jointly while keeping their inputs private. Recent breakthroughs are bringing MPC into practice, solving fundamental challenges for secure distributed computation. Just as with classic protocols for encryption and key exchange, precise guarantees are needed for MPC designs and implementations; any flaw will give attackers a chance to break privacy or correctness. In this paper we present the first (as far as we know) formalisation of some MPC security proofs. These proofs provide probabilistic guarantees in the computational model of security, but have a different character to machine proofs and proof tools implemented so far—MPC proofs use a *simulation* approach, in which security is established by showing indistinguishability between execution traces in the actual protocol execution and an ideal world where security is guaranteed by definition. We show that existing machinery for reasoning about probabilistic programs can be adapted to this setting, paving the way to precisely check a new class of cryptography arguments. We implement our proofs using the CryptHOL framework inside Isabelle/HOL.

Keywords: Oblivious transfer · Cryptography · Simulation-based proof · Formal verification

1 Introduction

Correctness guarantees are essential for cryptographic protocols and it is an area where formalisation continues to have impact. Older work was restricted to the *symbolic (Dolev-Yao) model* [11], where cryptographic primitives are modelled as abstract operations and assumed to be unbreakable. The symbolic model provides a baseline for correctness but modern cryptography is based on the more realistic *computational model* [1]. Adversaries are now allowed to break primitives, but are assumed to have limited computational power—typically, polynomial time in a security parameter n, such as a key size. Proofs in the

This work was supported by The Alan Turing Institute under the EPSRC grant EP/N510129/1.

M. Ayala-Rincón and C.A. Muñoz (Eds.): ITP 2017, LNCS 10499, pp. 114–130, 2017.
DOI: 10.1007/978-3-319-66107-0_8

computational model provide probabilistic guarantees: an adversary can break a security property only with negligible probability, i.e. probability bounded by a negligible function $\mu(n)$. There are two main proof styles, the *game-based* approach [21] and the *simulation-based* approach sometimes called the real/ideal world paradigm [14].

The simulation-based approach is a general proof technique especially useful for arguing about security of Multi-Party Computation (MPC) protocols. MPC is an area of cryptography concerned with enabling multiple parties to jointly evaluate a public function on their private inputs, without disclosing unnecessary information (that is, without leaking any information about their respective inputs that cannot be deduced from their sizes or the result of the computation). Several generic techniques can be used for that goal including Yao's garbled circuits [15,22], the GMW protocol [12], and other protocols based on secret-sharing [8,16]. These differ in whether they are designed for an arbitrary or fixed number of parties, how the computed function is represented (e.g., Boolean vs. arithmetic circuits), which functions can be represented (e.g., bounded-degree polynomials vs. arbitrary polynomials), as well trade-offs regarding communication, computation requirements, and security guarantees.

In the last decade, groundbreaking developments have brought MPC closer to practice. Efficient implementations of the protocols listed above are available [9,13,17,23], and we are now seeing the beginning of general solutions to fundamental security challenges of distributed computation. Security in these settings is proved by establishing a simulation between the *real world*, where the protocol plays out, and an *ideal world*, which is taken as the definition of security. This formalises the intuition that a protocol is secure if it can be simulated in an ideal environment in which there is no data leakage by definition.

A central protocol in MPC is Oblivious Transfer (OT), which allows a *sender* to provide several values and a *receiver* to choose some of them to receive, without learning the others, and without the sender learning which has been chosen. In this paper we build up to a security proof of the Naor-Pinkas OT [19], a practically important 1-out-of-2 oblivious transfer protocol (the receiver chooses one out of two messages). This can be used as a foundation for more general MPC, as secure evaluation of arbitrary circuits can be based on OT [12].

Contribution. As far as we know, this is the first formalisation of MPC proofs in a theorem prover. Our contributions are as follows.

- Starting from the notion of computational indistinguishablity, we formalise the simulation technique following the general form given by Lindell [14].
- Lindell's method spells out a process but leaves details of reasoning to informal arguments in the cryptographer's mind; to make this fully rigorous, we use probabilistic programs to encode *views* of the real and ideal worlds which can be successively refined to establish equivalence. This is a general method which can be followed for other protocols and in other systems; it corresponds to *hybrid arguments* often used in cryptography.
- As examples of the method, we show information-theoretic security for a two-party secure multiplication protocol that uses a trusted initialiser, and a

proof of security in the semi-honest model of the Naor-Pinkas OT protocol. The latter involves a reduction to the DDH assumption (a computational hardness assumption).

- Finally, we demonstrate how a formalisation of security of a 1-out-of-2 OT can be extended to formalising the security of an AND gate.

We build on Andreas Lochbihler's recent *CryptHOL* framework [18], which provides tools for encoding probabilistic programs using a shallow embedding inside Isabelle/HOL. Lochbihler has used his framework for game-based cryptographic proofs, along similar lines to proofs constructed in other theorem provers [2, 20] and dedicated tools such as EasyCrypt [3].

Outline. In Sect. 2 we give an overview of the key parts of CryptHOL that we use and extend. Section 3 shows how we define computational indistinguishability in Isabelle and Sect. 4 shows how it is used to define simulation-based security. In Sect. 4.1 we demonstrate how we use a probabilistic programming framework to do proofs in the simulation-based setting. Section 5 gives the proof of security of a secure multiplication protocol as a warm up and Sect. 6 shows the proof of security of the Naor-Pinkas OT protocol. In Sect. 7 we show how an OT protocol can be used to securely compute an AND gate, paving the way towards generalised protocols. Our formalisation is available online at https://github. com/alan-turing-institute/isabelle-mpc.

2   CryptHOL and Extensions

CryptHOL is a probabilistic programming framework based around *subprobability mass functions* (spmfs). An spmf encodes a discrete (sub) probability distribution. More precisely, an spmf is a real valued function on a finite domain that is non negative and sums to at most one. Such functions have type α *spmf* for a domain which is a set of elements of type α. We use the notation from [18] and let $p!x$ denote the subprobability mass assigned by the spmf p to the event x. The weight of an spmf is given by $||p|| = \sum_y p!y$ where the sum is taken over all elementary events of the corresponding type; this is the total mass of probability assigned by the spmf p. If $||p|| = 1$ we say p is *lossless*. Another important function used in our proofs is *scale*. The expression *scale* r p scales, by r, the subprobability mass of p. That is, we have *scale* r $p!x = r.(p!x)$ for $0 \leq r \leq \frac{1}{||p||}$.

Probabilistic programs can be encoded as sequences of functions that compute over values drawn from spmfs. The type α *spmf* is used to instantiate the polymorphic monad operations $return_{spmf} :: \alpha \Rightarrow \alpha$ *spmf* and $bind_{spmf} :: \alpha$ *spmf* $\Rightarrow (\alpha \Rightarrow \beta$ *spmf*$) \Rightarrow \beta$ *spmf*.

This gives a shallow embedding for probabilistic programs which we use to define simulations and views, exploiting the monadic do notation. As usual, $do \{x \leftarrow p; f\}$ stands for $bind_{spmf}\ p\ (\lambda x.\ do\ f)$.

We note that $bind_{spmf}$ is commutative and constant elements cancel. In particular if p is a lossless spmf, then

$$bind_{spmf}\ p\ (\lambda_{\text{-}}.\ q) = q. \tag{1}$$

Equation 1 can be shown using the lemma *bind_spmf_const*,

$$bind_{spmf} \; p \; (\lambda x. \; q) = scale_{spmf} \; (weight_{spmf} \; p) \; q \tag{2}$$

and the fact *sample_uniform* is lossless and thus has weight equal to one. In Eq. 2, $weight_{spmf} \; p$ is $\|p\|$ described above.

The monad operations give rise to the functorial structure, $map_{spmf} :: (\alpha \Rightarrow \beta) \Rightarrow \alpha \; spmf \Rightarrow \beta \; spmf$.

$$map_{spmf} \; f \; p = bind_{spmf} \; p \; (\lambda x. \; return_{spmf}(f \; x)) \tag{3}$$

CryptHOL provides an operation, *sample_uniform* :: *nat* $\Rightarrow$ *nat spmf* where *sample_uniform* $n = spmf\_of\_set \; \{.. < n\}$, the lossless spmf which distributes probability uniformly to a set of n elements. Of particular importance in cryptography is the uniform distribution *coin_spmf* $= spmf\_of\_set \; \{\mathsf{True}, \mathsf{False}\}$. Sampling from this corresponds to a coin flip.

We also utilise the function *assert_spmf* :: *bool* $\Rightarrow$ *unit spmf* which takes a predicate and only allows the computation to continue if the predicate holds. If it does not hold the current computation is aborted. It also allows the proof engine to pick up on the assertion made.

One way we extend the work of CryptHOL is by adding one time pad lemmas needed in our proofs of security. We prove a general statement given in Lemma 1 and instantiate it prove the one time pads we require.

Lemma 1. *Let f be injective and surjective on $\{.. < q\}$. Then we have*

$$map_{spmf} \; f \; (sample\_uniform \; q) = sample\_uniform \; q.$$

Proof. By definition, *sample_uniform* $q = spmf\_of\_set \; \{.. < q\}$. Then $map_{spmf} \; f \; (spmf\_of\_set \; \{.. < q\}) = spmf\_of\_set(f \, ` \{.. < q\})$ follows by simplification and the injective assumption (the infix $`$ is the image operator). Simplification uses the lemma *map_spmf_of_set_inj_on*:

$$inj\_on \; f \; A \implies map_{spmf} \; (spmf\_of\_set \; A) = spmf\_of\_set \; (f \, ` A).$$

We then have $map_{spmf} \; f \; (spmf\_of\_set \; \{.. < q\}) = spmf\_of\_set(\{.. < q\})$ by using the surjectivity assumption. The lemma then follows from the definition of *sample_uniform*. □

We note a weaker assumption, namely $f \, ` \{.. < q\} \subseteq \{.. < q\}$ can be used instead of the surjectivity assumption. To complete the proof with this assumption we use the *endo_inj_surj* rule which states

$$finite \; A \implies f \, ` A \subseteq A \implies inj\_on \; f \; A \implies f \, ` A = A.$$

For the maps we use we prove injectivity and show surjectivity using this.

Lemma 2 (Transformations on uniform distributions).

1. $map_{spmf} \; (\lambda b. \; (y - b) \; mod \; q) \; (sample\_uniform \; q) = sample\_uniform \; q.$

2. $map_{spmf}\ (\lambda b.\ (y + b)\ mod\ q)\ (sample\_uniform\ q) = sample\_uniform\ q$.
3. $map_{spmf}\ (\lambda b.\ (y + x.b)\ mod\ q)\ (sample\_uniform\ q) = sample\_uniform\ q$.

Proof. These follow with the help of Lemma 1. Case 3 holds only under the additional assumption that x and q are coprime. This will always be the case in the applications we consider as $x \in \mathbb{Z}_q$ and q is a prime. $\square$

3   Computational Indistinguishability in Isabelle

We introduce the notion of computational indistinguishability as the definitions of security we give in Sect. 4 rely on it. We use the definition from [14].

Definition 1. *A probability ensemble $X = \{X(a, n)\}$ is a sequence of random variables indexed by $a \in \{0, 1\}^*$ and $n \in \mathbb{N}$. Two ensembles X and Y are said to be computationally indistinguishable, written $X \overset{c}{\equiv} Y$, if for every non-uniform polynomial-time algorithm D there exists a negligible function[1] ϵ such that for every a and every $n \in \mathbb{N}$,*

$$|Pr[D(X(a, n)) = 1] - Pr[D(Y(a, n)) = 1]| \leq \epsilon(n)$$

The original definition restricts $a \in \{0, 1\}^*$, but we generalise this to an arbitrary first-order type, α. We model a probability ensemble as having some input of this type, and a natural number security size parameter. The space of events considered depends on the *view*; also of arbitrary first-order type, ν.

$$type\_synonym\ (\alpha, \nu)\ ensemble = \alpha \Rightarrow nat \Rightarrow \nu\ spmf$$

We do not formalise a notion of polynomial-time programs in Isabelle as we do not need it to capture the following proofs. In principle this could be done with a deep embedding of a programming language, its semantic denotation function and a complexity measure. Instead, we will assume a family of constants giving us the set of all polynomial-time distinguishers for every type ν, indexed by a size parameter.

A polynomial-time distinguisher "characterises" an arbitrary spmf.

$$consts\ polydist :: nat \Rightarrow (\nu\ spmf \Rightarrow bool\ spmf)\ set$$

Now we can formalise Definition 1 directly as:
$comp\_indist :: (\alpha, \nu)\ ensemble \Rightarrow (\alpha, \nu)\ ensemble \Rightarrow bool$
$where\ comp\_indist\ X\ Y \equiv$
$\quad \forall (D :: \nu\ spmf \Rightarrow bool\ spmf).$
$\qquad \exists\ (\epsilon :: nat \Rightarrow real).\ negligible\ \epsilon\ \wedge$
$\qquad\quad (\forall\ (a :: \alpha)\ (n :: nat).$
$\qquad\qquad (D \in polydist\ n) \longrightarrow$
$\qquad\qquad\quad |spmf\ (D\ (X\ a\ n))\ True - spmf\ (D\ (Y\ a\ n))\ True| \leq \epsilon\ n))$

---

[1] A negligible function is a function $\epsilon :: \mathbb{N} \to \mathbb{R}$ such that for all $c \in \mathbb{N}$ there exists $N_c \in \mathbb{N}$ such that for all $x > N_c$ we have $|\epsilon(x)| < \frac{1}{x^c}$.

4   Semi-honest Security and Simulation-Based Proofs

In this section we first define security in the semi-honest adversary model using the simulation-based approach. We then show how we use a probabilistic programming framework to formally prove security.

A protocol is an algorithm that describes the interaction between parties and can be modelled as a set of probabilistic programs. A two party protocol π computes a map from pairs of inputs to pairs of outputs. This map is called the protocol's *functionality* as it represents the specification of what the protocol should achieve. It can be formalised as a pair of (potentially probabilistic) functions

$$f_1 : input_1 \times input_2 \longrightarrow output_1$$

$$f_2 : input_1 \times input_2 \longrightarrow output_2$$

which represent each party's output independently. The composed pairing is the functionality, f, of type

$$f : input_1 \times input_2 \longrightarrow output_1 \times output_2$$

where $f = (f_1, f_2)$. That is, given inputs (x, y) the functionality outputs $(f_1(x, y), f_2(x, y))$. This indicates that party one gets $f_1(x, y)$ and party two gets $f_2(x, y)$ as output. In general the types of inputs and outputs can be arbitrary. For our instantiation we use concrete types depending on the functionality concerned.

For the initial example secure multiplication protocol we consider in Sect. 5 we have the probabilistic functionality $f(x, y) = (s_1, s_2)$ where $s_1 + s_2 = x.y$. Each party obtains an additive share of the multiplication. The protocol is run using a publicly known field $\mathbb{Z}_q$ where q is a prime number dependent on the security parameter. To ensure neither of the outputs alone reveal the value of $x.y$, we uniformly sample one of the outputs in the functionality

$$f(x, y) = (s_1, x.y - s_1), s_1 \xleftarrow{\$} \mathbb{Z}_q \tag{4}$$

The notation $s_1 \xleftarrow{\$} \mathbb{Z}_q$ means we sample s_1 uniformly from $\mathbb{Z}_q$. The Isabelle definition of the functionality is given below. It makes use of the do notation:

$$f \; x \; y = do \; \{$$
$$s_1 \leftarrow sample\_uniform \; q;$$
$$return_{spmf}(s_1, x.y - s_1)\}$$

This functionality is easy to compute if one does not consider security; the parties can share their inputs and compute it. But with the security requirement that neither party learns anything about the others' input the problem becomes harder. We will give a protocol that securely computes this functionality later. We first introduce the notions used to define security. Security is based on *views* which capture the information known by each party. We follow the definitions given by Lindell in [14] to define security in the semi-honest model.

Definition 2. *Let π be a two party protocol with inputs (x, y) and with security parameter n.*

- *The real view of the ith party (here $i \in \{1, 2\}$) is denoted by*

$$view_i^\pi(x, y, n) = (w, r^i, m_1^i, ..., m_t^i)$$

where $w \in \{x, y\}$ and is dependent on which view we are considering, r^i accumulates random values generated by the party during the execution of the protocol, and the m_j^i are the messages received by the party.
- *Denote the output of the ith party, $output_i^\pi(x, y, n)$, and the joint output as*

$$output^\pi(x, y, n) = (output_1^\pi(x, y, n), output_2^\pi(x, y, n)).$$

Definition 3. *A protocol π is said to securely compute f in the presence of a semi-honest adversary if there exist probabilistic polynomial time algorithms (simulators) S_1, S_2 such that*

$$\{S_1(1^n, x, f_1(x, y)), f(x, y)\} \stackrel{c}{\equiv} \{view_1^\pi(x, y, n), output^\pi(x, y, n)\}$$

$$\{S_2(1^n, y, f_2(x, y)), f(x, y)\} \stackrel{c}{\equiv} \{view_2^\pi(x, y, n), output^\pi(x, y, n)\}.$$

A semi-honest adversary is one that follows the protocol description. The simulator is given a unary encoding of the security parameter.

This definition formalises the idea that a protocol is secure if whatever can be computed by a party can also be computed from only the input and output of the party meaning that nothing extra is learned from the protocol.

For the secure multiplication protocol and the receiver's security in the Naor-Pinkas OT we prove security in an information theoretic sense. This means even computationally unbounded adversaries cannot gain extra information from the protocol. This is shown by proving the two sets of distributions above are equal. Information theoretic security is a stronger notion of security than computational indistinguishability and Isabelle proves the former implies the latter with ease.

A functionality is deterministic if given inputs always produce the same output. For a deterministic protocol it is shown in [14] that the above definition can be relaxed. We require correctness and

$$\{S_1(1^n, x, f_1(x, y))\} \stackrel{c}{\equiv} \{view_1^\pi(x, y, n)\} \tag{5}$$

$$\{S_2(1^n, y, f_2(x, y))\} \stackrel{c}{\equiv} \{view_2^\pi(x, y, n)\} \tag{6}$$

For a protocol to be correct we require that for all x, y and n there exists a negligible function μ such that

$$Pr[output^\pi(x, y, n) \neq f(x, y)] \leq \mu(n).$$

The Naor-Pinkas OT protocol, and the OT we use in the AND gate protocol given later, are both deterministic. The secure multiplication protocol however is not. For the deterministic cases we will focus on the more interesting property, showing the views are equal. As such when we refer to a deterministic protocol as being secure we explicitly show Eqs. 5 and 6 and assume correctness. For the non-deterministic secure multiplication protocol we must show exactly the property given in Definition 3.

4.1 Probabilistic Programming Used for Simulation-Based Proofs

CryptHOL provides a strong foundation from which to manipulate and show equivalence between probabilistic programs. So far it has only been used to prove security in the game-based setting. The game-based definitions of security use a game played between an adversary and a benign challenger. The players are modelled as probabilistic programs and communicate with each other. The definition of security is tied to some event which is defined as the output of the security game. In general, proofs describe a reduction of a sequence of games (probabilistic programs) that end in a game where it can be shown the adversary has the same advantage of winning over the challenger as it would have against a problem assumed to be hard. The games in the sequence are then shown to be equivalent. This is shown on the left hand side of Fig. 1. We use a probabilistic programming framework to construct simulation-based proofs. Our method of proof models the simulator and the real view of the protocol as probabilistic programs. In the right hand side of Fig. 1 we start with the real view of the protocol, R^1, and the simulator, S^1. We define a series of intermediate probabilistic programs (R^i, S^i) which we show to be computationally indistinguishable (or equal in the case of information theoretic security)—this is referred to as the *hybrid argument* in cryptography. This sequence ends in R^n and S^n which we show to be computationally indistinguishable (or equal). We have shown the diagram for the simulation-based approach in Fig. 1 is transitive.

Game-based		Simulation-based
G_1	R^1	S^1
$\wr\wr$	$\overset{c}{\equiv}$	$\overset{c}{\equiv}$
$\vdots$	$\vdots$	$\vdots$
$\wr\wr$	$\overset{c}{\equiv}$	$\overset{c}{\equiv}$
G_n	R^n $\overset{c}{\equiv}$ S^n	

Fig. 1. A comparison between the game-based and simulation-based approaches. The game-based approach uses reductions (denoted $\preceq$) whereas in the simulation approach we show computational indistinguishability between probabilistic programs.

Lemma 3. *Let X, Y and Z be probability ensembles then we have*

$$[X \overset{c}{\equiv} Y; \ Y \overset{c}{\equiv} Z] \implies X \overset{c}{\equiv} Z.$$

For the non-deterministic secure multiplication protocol we will construct the protocol and functionality outputs in the real and simulated views, instead of constructing them separately and combining them to form the ensembles.

5   Secure Multiplication Protocol

We now present a protocol that computes the functionality in Eq. 4. The protocol requires some pre-generation of elements to be distributed to the parties. This is known in MPC as the preprocessing model [5], where the parties run an offline phase to generate correlated random triples—sometimes called Beaver triples—that are used to perform fast secure multiplications in an online phase. For this task we assume a trusted initialiser that aids in the computation. We denote the assignment of variables by $a \leftarrow b$ and all operations are taken modulo q. The claim of security is:

$$a, b, r \xleftarrow{\$} \mathbb{Z}_q$$
$$(c_1, d_1) \leftarrow (a, r), (c_2, d_2) \leftarrow (b, a.b - r)$$

P_1

$(c_1, d_1), x \in \mathbb{Z}_q$

$e_2 \leftarrow x + c_1$ $\xrightarrow{\hspace{2cm} e_2 \hspace{2cm}}$

e_1 $\xleftarrow{\hspace{2cm} e_1 \hspace{2cm}}$

$s_1 \leftarrow x.e_1 - d_1$

P_2

$(c_2, d_2), y \in \mathbb{Z}_q$

e_2

$e_1 \leftarrow y - c_2$

$s_2 \leftarrow e_2.c_2 - d_2$

Fig. 2. A protocol for secure multiplication

Theorem 1. *The protocol in Fig. 2 securely computes the functionality given in Eq. 4 in the semi-honest adversary model.*

Intuitively, security results from the messages being sent in the protocol always being masked by some randomness. In the message party one sends, e_2, the input (x) is masked by the uniform sample, c_1. Likewise in the message party two sends, e_1, the input (y) is masked by the uniform sample, c_2.

5.1   Formal Proof of Security

The simulator and the real view of party one are defined in Isabelle as in Fig. 3. Recall that the protocol output is output by the real view and the functionality is output by the simulated view for this non-deterministic case. Thus we sample b and r twice (second time as b', r') in the real view. The outputs o_1 and o_2 refer to the output $(output^\pi(x, y, n)$ in Definition 3) of the protocol. Note that the simulator S_1 takes x and y as inputs. The simulated view however is constructed using only party one's input, x, according to the definition in Sect. 4. The second input, y, is used to construct the functionality output at the same time.

To show information theoretic security we prove that the two probabilistic programs given in Fig. 3 are equal. This involves a series of small equality steps between intermediate probabilistic programs as shown in Fig. 1. In particular, in the series of intermediate programs we manipulate the real and simulated views

$S_1\ x\ y = do\ \{$
  $a,\ b,\ c \leftarrow sample\_uniform\ q;$
  $s_1 \leftarrow sample\_uniform\ q;$
  $let\ z = (x.b - c)\ mod\ q;$
  $let\ s_2 = (x.y - s_1)\ mod\ q;$
  $return_{spmf}(x, a, z, b, s_1, s_2)\}$

$R_1\ x\ y = do\ \{$
  $a,\ b,\ r \leftarrow sample\_uniform\ q;$
  $b',\ r' \leftarrow sample\_uniform\ q;$
  $let\ z = (y - b)\ mod\ q;$
  $let\ o_1 = (x(y - b') - r')\ mod\ q;$
  $let\ o_2 = (x.y - (x(y - b') - r'))\ mod\ q;$
  $return_{spmf}(x, a, r, z, o_1, o_2)\}$

Fig. 3. Probabilistic programs to output the real and simulated views for party one.

into a form where we can apply Lemma 2(1). To do this we mainly use existing lemmas from CryptHOL, two of which are given in Eqs. 2 and 3.

This gives us the first half of formal security which can be seen in Lemma 4

Lemma 4. *For all inputs x and y we have, $S_1\ x\ y = R_1\ x\ y$. This implies the definition of security we gave in Sect. 4, $S_1\ x\ y \overset{c}{\equiv} R_1\ x\ y$.*

The proof of security for party two is analogous and together, Lemmas 4 and 5 establish Theorem 1.

Lemma 5. *For all inputs x and y we have, $S_2\ x\ y = R_2\ x\ y$. This implies the definition of security we gave in Sect. 4, $S_2\ x\ y \overset{c}{\equiv} R_2\ x\ y$.*

6   Naor-Pinkas Protocol

In the Naor-Pinkas OT protocol [19] we work with a cyclic group $\mathbb{G}$ of order q where q is a prime, for which the DDH assumption holds. The Decisional Diffie Hellman (DDH) assumption [10] is a computational hardness assumption on cyclic groups. Informally, the assumption states that given g^a and g^b, where a and b are uniform samples from $\mathbb{Z}_q$, the group element $g^{a.b}$ looks like a random element from $\mathbb{G}$. A triple of the form $(g^a, g^b, g^{a.b})$ is called a DDH triple. In the protocol, given in Fig. 4, the Sender (party one) begins with input messages $(m_0, m_1) \in \mathbb{G}^2$ and the Receiver (party two) begins with $v \in \{0, 1\}$, the choice bit. At the end of the protocol the receiver will know m_v but will learn nothing about m_{1-v} and the sender will not learn v.

We prove information theoretic security in the semi-honest model for the receiver. Security for the sender is proven with a reduction to the DDH assumption. In particular, the receiver is only able to decrypt m_v as the corresponding ciphertext is a valid ElGamal ciphertext, while m_{1-v} is garbage.

In the protocol description, given in Fig. 4, DDH-SR refers to a DDH *random self reduction* operation which takes DDH triples to DDH triples and non DDH triples to non DDH triples. The reduction is defined as follows. Given an input tuple (g, g^x, g^y, g^z), one picks a, b uniformly from $\mathbb{Z}_q$ and outputs $(g, g^{(x+b)a}, g^y, g^{(z+b.y)a})$. The role of the DDH random self reduction is to destroy any partial information in the message the Receiver sends to the Sender.

P_1 (Sender)

$(m_0, m_1) \in \mathbb{G}^2$

P_2 (Receiver)

$v \in \{0, 1\}$

$a, b \xleftarrow{\$} \mathbb{Z}_q$

$c_v = a.b, \; c_{1-v} \xleftarrow{\$} \mathbb{Z}_q$

$x \leftarrow g^a, \; y \leftarrow g^b$

$z_0 \leftarrow g^{c_0}, \; z_1 \leftarrow g^{c_1}$

$A = (g, x, y, z_0)$

$\xleftarrow{\qquad (x, y, z_0, z_1) \qquad}$

$B = (g, x, y, z_1)$

verifies $z_0 \neq z_1$

$(g, x_1, y_1, z_0') \xleftarrow{\text{DDH-SR}} A$

$(g, x_2, y_2, z_1') \xleftarrow{\text{DDH-SR}} B$

$CT_0 = (y_0, m_0.z_0')$

$CT_1 = (y_1, m_1.z_1')$

$\xrightarrow{\hspace{4cm}}$

CT_0, CT_1

decrypts CT_v

Fig. 4. The Naor-Pinkas OT protocol

Theorem 2. *The protocol defined in Fig. 4 securely computes a 1-out-of-2 OT in the semi-honest adversary model.*

6.1 The Formal Proof

We have a deterministic protocol and so do not include the overall functionality as part of the views. We must first consider the DDH-SR. In particular the two cases, when the input tuple is a DDH triple and when it is not. In both cases we simplify the operation that is performed. The simplified definitions are given in Fig. 5 and the formal statements in Lemmas 6 and 7:

$DDH\_SR\_triple \; x \; y \; z = do \; \{$
  $x_1 \leftarrow sample\_uniform \; q;$
  $return_{spmf}(g, g^{x_1}, g^y, g^{y.x_1 \bmod q})\}$

$DDH\_SR\_non\_triple \; x \; y \; z = do \; \{$
  $x_1, x_2 \leftarrow sample\_uniform \; q;$
  $return_{spmf}(g, g^{x_1}, g^y, g^{x_2})\}$

Fig. 5. The two simplified probabilistic programs for the DDH triples and non-triples.

Lemma 6. *For all x, y, z such that $z = y.x \bmod q$ we have*

$$DDH\_SR \; x \; y \; z = DDH\_SR\_triple \; x \; y \; z.$$

Lemma 7. *For all x, y, z such that $z \neq y.x \bmod q$ we have*

$$DDH\_SR \; x \; y \; z = DDH\_SR\_non\_triple \; x \; y \; z.$$

The Simulators and Views. First we consider party two. In constructing the real and simulated views we use the assert function to ensure the condition given in the protocol in Fig. 4, $z_0 \neq z_1$, holds. This ensures that only one of A and B is a DDH triple; the other is not and hence the corresponding ciphertext CT_0 or CT_1 cannot be decrypted. The simulator may take as inputs $v \in \{0, 1\}$ and CT_v (although does not require it). We use $\otimes$ to denote multiplication in the group (as in Isabelle). The real view and simulator are shown below.

$S_2 \, v = do \, \{$
  $a, b \leftarrow sample\_uniform \, q;$
  $let \, c_v = a.b;$
  $c'_v \leftarrow sample\_uniform \, q;$
  $\_ \leftarrow assert\_spmf \, (c'_v \neq b.a \bmod q);$
  $x_0 \leftarrow sample\_uniform \, q;$
  $x_1 \leftarrow sample\_uniform \, q;$
  $return_{spmf} \, (v, a, b, c'_v, g^b, g^{x_1}, g^b, g^{x_2}) \}$

$R_2 \, m_0 \, m_1 \, v = do \, \{$
  $a, b \leftarrow sample\_uniform \, q;$
  $let \, c_v = a.b;$
  $c'_v \leftarrow sample\_uniform \, q;$
  $\_ \leftarrow assert\_spmf \, (c'_v \neq b.a \bmod q);$
  $(g, x_0, y_0, z'_0) \leftarrow DDH\_SR \, a \, b \, c_v;$
  $(g, x_1, y_1, z'_1) \leftarrow DDH\_SR \, a \, b \, c'_v;$
  $let \, e_0 = z'_0 \otimes m_0;$
  $let \, e_1 = z'_1 \otimes m_1;$
  $return_{spmf} \, (v, a, b, c'_v, y_0, e_0, y_1, e_1) \}$

For party one, the simulator, S_1, takes in the two messages (m_0, m_1) (again, it does not use them) and the Sender's output - which amounts to nothing. We break this proof down into cases on v, the receivers input, however it is important to stress that the simulator must stay the same in both cases. Below we give the simulator and the real view for the non trivial case, namely when $v = 1$.

$S_1 \, m_0 \, m_1 = do \, \{$
  $a, b, c_1 \leftarrow sample\_uniform \, q;$
  $return_{spmf} \, (g^a, g^b, g^{a.b}, g^{c_1}) \}$

$R_1\_v\_eq\_1 \, m_0 \, m_1 = do \, \{$
  $a, b, c_0 \leftarrow sample\_uniform \, q;$
  $return_{spmf} \, (g^a, g^b, g^{c_0} \otimes g^{a.b}) \}$

Proof of Security for the Receiver. From the construction of the real view one can see the triple (a, b, c_v) is a DDH triple and (a, b, c'_v) is not. Thus we are able to rewrite the real view using Lemmas 6 and 7.

The only components of the outputs of R_2 and S_2 which differ, up to unfolding of definitions are the encryptions. In the real view they are of the form $g^z \otimes m_i$ where z is uniformly sampled and in the simulator they are of the form g^z. We utilise a lemma from CryptHOL which states that if $c \in carrier \, \mathbb{G}$ then:

$$map_{spmf} \, (\lambda x. \, g^x \otimes c) \, (sample\_uniform \, q)$$
$$= map_{spmf} \, (\lambda x. \, g^x) \, (sample\_uniform \, q)$$

This allows us to show our security result stated in Lemma 8.

Lemma 8. *For all inputs m_0, m_1 and v we have, $S_2 \, v = R_2; m_0 \, m_1 \, v$. This implies the definition of security we gave in Sect. 4, $S_2 \, v \stackrel{c}{=} R_2; m_0 \, m_1 \, v$.*

Proof of Security for the Sender. For $v = 0$, the proof is trivial as the simulator and real views are constructed in exactly the same way.

Lemma 9. *The case of $v = 0$ for party one implies for all inputs m_0 and m_1,*

$$R_1\_v\_eq\_0 \; m_0 \; m_1 = S_1 \; m_0 \; m_1.$$

The proof for $v = 1$ is equivalent to showing the distributions $(g^a, g^b, g^{a.b}, g^c)$ and $(g^a, g^b, g^c, g^{a.b})$ are computationally indistinguishable, when a, b, c are uniformly sampled. Here we provide a high level view of the pencil and paper. Our formalisation can be found at https://github.com/alan-turing-institute/isabelle-mpc.

To show security we provide a reduction to the DDH assumption, which implies the two distributions are computationally indistinguishable. In particular we show that if there exists a D that can distinguish the above two 4-tuples then one can construct an adversary that breaks the DDH assumption. In order to prove this formally we provide a way of formalising the DDH advantage.

Definition 4. *The DDH advantage for a distinguisher D is defined as*

$$adv\_ddh(D) = Pr[D((g^a, g^b, g^{a.b}), (g^a, g^b, g^c)) = 1]$$
$$- Pr[D((g^a, g^b, g^c), (g^a, g^b, g^{a.b})) = 1]$$

where $a, b, c \xleftarrow{\$} \mathbb{Z}_q$.

We assume that no efficient distinguisher has an advantage greater than a negligible function of the security parameter. We define the advantage of a 4-tuple distinguisher, D, as follows.

Definition 5. *The 4-tuple distinguisher's advantage is given by*

$$adv\_dist(D) = Pr[D((g^a, g^b, g^{a.b}, g^c), (g^a, g^b, g^c, g^{a.b})) = 1]$$
$$- Pr[D((g^a, g^b, g^c, g^{a.b}), (g^a, g^b, g^{a.b}, g^c)) = 1]$$

where $a, b, c, d \xleftarrow{\$} \mathbb{Z}_q$.

The adversary we use to break the DDH assumption, which uses D is constructed below.

DDH Adversary, inputs: $(g^a, g^b, g^{a.b})$ and (g^a, g^b, g^c).

- The adversary constructs $a_1 = (g^a, g^b, g^c, g^{a.b})$ and $a_2 = (g^a, g^b, g^{a.b}, g^c)$ and gives them to D.
- The adversary outputs whatever D outputs.

We show the DDH advantage of the adversary (using D) is the same as the 4-tuple advantage of D. Thus we have reduced the security of party one to a known hard problem. In particular we show

Lemma 10. *For any 4-tuple distinguisher D we have,*

$$adv\_ddh(A(D)) = adv\_dist(D).$$

This along with showing information theoretic security (Lemma 8) for the receiver means we have shown the protocol to be secure in the semi-honest model.

7   Towards Evaluating Arbitrary Functionalities

Several MPC techniques allow for the secure joint evaluation of *any* functionality represented as a Boolean circuit or an arithmetic circuit. At a high level, these protocols proceed by evaluating the circuit gate by gate while always keeping a secret share of the partial evaluation. In particular the GMW protocol relies on OT to securely evaluate AND gates.

In this section we use a basic OT protocol (Fig. 6) to construct a protocol to compute the output of an AND gate. The OT protocol we use employs a trusted initialiser, like the secure multiplication protocol of Sect. 5. The trusted initialiser pre-distributes correlated randomness to the parties so they can carry out the protocol. In particular r_0 and r_1 are uniformly sampled and given to party one, and d is uniformly sampled and given to party two along with r_d. The AND gate protocol then uses OT, this is done in a similar way as in the GMW protocol. The AND gate protocol we use here is taken from [6] and is described in Fig. 7. This demonstrates that OT can be used in powerful ways to construct protocols to compute fundamental functions securely.

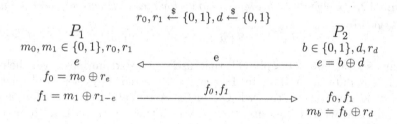

Fig. 6. Single bit OT

Initially we show information theoretic security for the OT construction given in Fig. 6. That is we construct simulators S_1^{OT} and S_2^{OT} such that for the appropriately defined views R_1^{OT} and R_2^{OT} the result in Lemma 11 holds. To do this we define an appropriate XOR function ($\oplus$) on Booleans and prove a one time pad lemma on the XOR function.

Lemma 11. $R_1^{OT} \ m_0 \ m_1 \ b = S_1^{OT} \ m_0 \ m_1$ and $R_2^{OT} \ m_0 \ m_1 \ b = S_2^{OT} \ b$.

We now define a protocol (Fig. 7) to compute an AND gate. The protocol uses OT as a black box to transfer m_b. Each party outputs an additive share of the desired AND gate output. This protocol is proved secure using the simulation-based approach. We use Lemma 11 to prove security of this protocol in the semi-honest model. The real view and the simulator for party A are given in Fig. 8. The simulator for party B, S_B, is constructed in an analogous way. Using these simulators we are able to show the AND gate protocol in Fig. 7 is information theoretically secure.

$$
\begin{array}{cc}
\text{A} & \text{B} \\
a \in \{0,1\} & b \in \{0,1\} \\
u \xleftarrow{\$} \{0,1\} & \\
(m_0, m_1) = (\neg u, a \oplus \neg u) \quad \xrightarrow{\quad OT((m_0, m_1), b) \quad} & m_b \\
\text{output } u & \text{output } m_b
\end{array}
$$

Fig. 7. A protocol to compute an AND gate

$R_A \; a \; b = do \;\{$
  $u \leftarrow coin\_spmf;$
  $let \; m_0 = \neg u;$
  $let \; m_1 = a \oplus (\neg u);$
  $r \leftarrow R_1^{OT} \; m_0 \; m_1 \; b;$
  $return_{spmf}(u, r)\}$

$S_A \; a = do \;\{$
  $u \leftarrow coin\_spmf;$
  $let \; m_0 = \neg u;$
  $let \; m_1 = a \oplus (\neg u);$
  $r \leftarrow S_1^{OT} \; m_0 \; m_1;$
  $return_{spmf}(u, r)\}$

Fig. 8. Simulator and real view of party A

Lemma 12. *Information theoretic security for the AND gate protocol is shown by the equalities*

$$
R_A \; a \; b = S_A \; a \;\; and \;\; R_B \; a \; b = S_B \; b.
$$

We have shown how a simple OT that uses a trusted initialiser can help to securely compute an AND gate. In general a trusted initialiser would not be necessary as one can use the N-P OT in the AND gate protocol. There is one technical issue with doing this. In the N-P OT we work with a group with multiplication but the AND gate protocol requires addition. In practice this is overcome by implementing the N-P OT using a ring (which has both operations), for which the DDH assumption holds. The proof would follow as in the proof given above, but an extension of the theory of rings in Isabelle is required for this - something we plan to develop in future work.

8   Conclusion

We have shown a general approach for capturing simulation-based cryptographic proofs in the computational model, building on Lochbihler's CryptHOL framework, and giving a proof of the Naor-Pinkas OT protocol. We also have shown how out technique can be used to formally prove security of a simple two party protocol for an AND gate based on OT.

Future Work. The work presented here is only a starting point for the development of theory and examples of simulation-based proofs. Oblivious Transfer is a fundamental cryptographic primitive which can be used to construct generic protocols for MPC. For example, Yao's garbled circuits use OT as a sub-protocol to exchange garbled inputs, while the GMW protocol relies on OT for computing

AND gates. Section 7 took a first step towards a formal proof of the GMW protocol. Section 7 took a first step towards a formal proof of the GMW protocol. We plan to extend this work towards formalising general MPC protocols.

Related Work. Many formal techniques and tools have been devised which use the symbolic model. Work on formalising proofs in the computational model has begun more recently and is more challenging, requiring mathematical reasoning about probabilities and polynomial functions, besides logic. The CertiCrypt [2] tool built in Coq helped to capture the reasoning principles that were implemented directly in the dedicated interactive EasyCrypt tool [3]. Again in Coq, the Fundamental Cryptographic Framework [20] provides a definitional language for probabilistic programs, a theory that is used to reason about programs, and a library of tactics for game-based proofs. Interactive tools seem invaluable for complex protocols or exploring new techniques, but automatic tools are more practical when things become routine. CryptoVerif [7] is a tool with a high level of automation but its scope only stretches to secrecy and authentication in protocols. AutoG&P [4] is another automated tool dedicated to security proofs for pairing-based cryptographic primitives. So far, all of these tools have been used to perform game-based cryptographic proofs and not simulation-based proofs.

Acknowledgements. We are deeply grateful to Andreas Lochbihler for providing and continuing to develop CryptHOL and for his kind help given with using it. Also we are thankful to the reviewers for their comments regarding the presentation of our work.

References

1. Abadi, M., Rogaway, P.: Reconciling two views of cryptography (the computational soundness of formal encryption). J. Cryptol. **20**(3), 395 (2007)
2. Barthe, G., Grégoire, B., Béguelin, S.Z.: Formal certification of code-based cryptographic proofs. In: POPL, pp. 90–101. ACM (2009)
3. Barthe, G., Grégoire, B., Heraud, S., Béguelin, S.Z.: Computer-aided security proofs for the working cryptographer. In: Rogaway, P. (ed.) CRYPTO 2011. LNCS, vol. 6841, pp. 71–90. Springer, Heidelberg (2011). doi:10.1007/978-3-642-22792-9_5
4. Barthe, G., Grégoire, B., Schmidt, B.: Automated proofs of pairing-based cryptography. In: ACM Conference on Computer and Communications Security, pp. 1156–1168. ACM (2015)
5. Beaver, D.: Efficient multiparty protocols using circuit randomization. In: Feigenbaum, J. (ed.) CRYPTO 1991. LNCS, vol. 576, pp. 420–432. Springer, Heidelberg (1992). doi:10.1007/3-540-46766-1_34
6. Bennett, C.H., Brassard, G., Crépeau, C., Skubiszewska, M.-H.: Practical quantum oblivious transfer. In: Feigenbaum, J. (ed.) CRYPTO 1991. LNCS, vol. 576, pp. 351–366. Springer, Heidelberg (1992). doi:10.1007/3-540-46766-1_29
7. Blanchet, B.: A computationally sound mechanized prover for security protocols. IEEE Trans. Dependable Secur. Comput. **5**(4), 193–207 (2008)
8. Bogdanov, D., Laur, S., Willemson, J.: Sharemind: a framework for fast privacy-preserving computations. In: Jajodia, S., Lopez, J. (eds.) ESORICS 2008. LNCS, vol. 5283, pp. 192–206. Springer, Heidelberg (2008). doi:10.1007/978-3-540-88313-5_13

9. Demmler, D., Schneider, T., Zohner, M.: ABY - a framework for efficient mixed-protocol secure two-party computation. In: NDSS. The Internet Society (2015)
10. Diffie, W., Hellman, M.E.: New directions in cryptography. IEEE Trans. Inf. Theory **22**(6), 644–654 (1976)
11. Dolev, D., Yao, A.: On the security of public key protocols. IEEE Trans. Inf. Theory **29**(2), 198–207 (1983)
12. Goldreich, O., Micali, S., Wigderson, A.: How to play any mental game or a completeness theorem for protocols with honest majority. In: STOC, pp. 218–229. ACM (1987)
13. Keller, M., Orsini, E., Scholl, P.: MASCOT: faster malicious arithmetic secure computation with oblivious transfer. In: ACM Conference on Computer and Communications Security, pp. 830–842. ACM (2016)
14. Lindell, Y.: How to simulate it - a tutorial on the simulation proof technique. IACR Cryptology ePrint Archive 2016:46 (2016)
15. Lindell, Y., Pinkas, B.: A proof of security of Yao's protocol for two-party computation. J. Cryptol. **22**(2), 161–188 (2009)
16. Lindell, Y., Pinkas, B., Smart, N.P., Yanai, A.: Efficient constant round multi-party computation combining BMR and SPDZ. In: Gennaro, R., Robshaw, M. (eds.) CRYPTO 2015. LNCS, vol. 9216, pp. 319–338. Springer, Heidelberg (2015). doi:10.1007/978-3-662-48000-7_16
17. Liu, C., Wang, X.S., Nayak, K., Huang, Y., Shi, E.: ObliVM: a programming framework for secure computation. In: IEEE Symposium on Security and Privacy, pp. 359–376. IEEE Computer Society (2015)
18. Lochbihler, A.: Probabilistic functions and cryptographic oracles in higher order logic. In: Thiemann, P. (ed.) ESOP 2016. LNCS, vol. 9632, pp. 503–531. Springer, Heidelberg (2016). doi:10.1007/978-3-662-49498-1_20
19. Naor, M., Pinkas, B.: Efficient oblivious transfer protocols. In: SODA, pp. 448–457. ACM/SIAM (2001)
20. Petcher, A., Morrisett, G.: The foundational cryptography framework. In: Focardi, R., Myers, A. (eds.) POST 2015. LNCS, vol. 9036, pp. 53–72. Springer, Heidelberg (2015). doi:10.1007/978-3-662-46666-7_4
21. Shoup, V.: Sequences of games: a tool for taming complexity in security proofs. IACR Cryptology ePrint Archive 2004:332 (2004)
22. Yao, A.: How to generate and exchange secrets (extended abstract). In: FOCS, pp. 162–167. IEEE Computer Society (1986)
23. Zahur, S., Evans, D.: Obliv-C: a language for extensible data-oblivious computation. IACR Cryptology ePrint Archive 2015:1153 (2015)

FoCaLiZe and Dedukti to the Rescue
for Proof Interoperability

Raphaël Cauderlier[1](✉) and Catherine Dubois[2]

[1] University Paris Diderot, Irif, Paris, France
`raphael.cauderlier@irif.fr`
[2] ENSIIE, Samovar, Évry, France
`catherine.dubois@ensiie.fr`

Abstract. Numerous contributions have been made for some years to allow users to exchange formal proofs between different provers. The main propositions consist in ad hoc pointwise translations, e.g. between HOL Light and Isabelle in the Flyspeck project or uses of more or less complete certificates. We propose in this paper a methodology to combine proofs coming from different theorem provers. This methodology relies on the Dedukti logical framework as a common formalism in which proofs can be translated and combined. To relate the independently developed mathematical libraries used in proof assistants, we rely on the structuring features offered by FoCaLiZe, in particular parameterized modules and inheritance to build a formal library of transfer theorems called Math-Transfer. We finally illustrate this methodology on the Sieve of Eratosthenes, which we prove correct using HOL and Coq in combination.

1 Introduction

According to the IEEE Standards Glossary, interoperability can be considered as *the ability of computer systems or software to exchange and make use of information*. Prover interoperability as a way for exchanging formal proofs between different theorem provers is a research topic that received many contributions along years. The most successful approach is probably the integration of automatic theorems provers (ATP) in interactive proof assistants (ITP) like Coq [1] or Isabelle [6]. In that case more or less detailed witnesses are provided and proofs can be imported or re-built. Furthermore many ad hoc pairwise translations have been proposed e.g. between HOL Light and Isabelle in the Flyspeck project [18], between HOL Light and Coq [12,19,25] or between HOL and Nuprl [15]. To avoid the quadratic blowup in the number of translators to develop, proof formats are emerging either for proofs in a specific logic such as the OpenTheory format [17] for ITPs in the HOL family or relying on logical frameworks [14,20,24] such as λ-prolog and Twelf to represent proofs in several logics. We propose to combine proofs coming from different theorem provers relying on the Dedukti logical

This work has been supported in part by the VECOLIB project of the French national research organization ANR (grant ANR-14-CE28-0018).

M. Ayala-Rincón and C.A. Muñoz (Eds.): ITP 2017, LNCS 10499, pp. 131–147, 2017.
DOI: 10.1007/978-3-319-66107-0_9

framework [23], a typed λ-calculus with dependent types and rewriting, as a common formalism in which proofs can be translated and combined.

In [5], Assaf and Cauderlier describe a manual attempt of interoperability between HOL and Coq where they prove in Coq the correctness of the insertion sort algorithm on polymorphic lists and instantiate it with HOL natural numbers. This experiment relies on a translation to Dedukti for both the sorting function and the definition of HOL natural numbers (using respectively Coqine and Holide) and the result is checked by Dedukti. The interaction between both parts only happens at the level of booleans. However, for such a simple fact the proof is very long and verbose (around 700 Dedukti lines).

The goal of this paper is to make prover interoperability reach a new scale. We can notice that the art specific to the case study of Assaf and Cauderlier required a lot of work that could be automated and *has to* be automated to scale up. For this task, we use Zenon Modulo [8], an automated theorem prover outputting Dedukti proofs.

In this work, we go beyond simple boolean interaction. When a type and operations over this type, such as natural numbers and arithmetic operations, are independently defined in two ITPs, we can translate them but we end up with distinct isomorphic structures A and B in Dedukti. A theorem φ_A proved for A does not give us for free the corresponding theorem φ_B about B in which we are interested. Two solutions to this problem have been proposed:

- modify one of the translators to make it use the type and operations of the other structure thus identifying structures A and B,
- keep structures A and B distinct and use tactics to automatically prove *transfer* theorems of the form $\varphi_A \rightarrow \varphi_B$.

The first solution is favored in several ad hoc interoperability proposals [17,19]. The main limitation of this solution is the complexification of the translators which lacks scalability: for interoperability between n proof systems independently defining a mathematical structure, $n - 1$ translators need to be modified to become customizable and point to the definition of the nth proof system. The second solution has first been proposed in the context of formalization in Isabelle/HOL of quotient structures [16] and recently ported to Coq [26]. Its main limitation is that the definitions of the morphisms and the proofs that operations are preserved by morphisms are left to the human user. We propose a compromise between these two solutions: we prove transfer theorems in FoCaLiZe [21], an ITP featuring a customizable Dedukti translator and use them to relate independent developments coming from uncustomizable translators.

The first contribution of this paper is a FoCaLiZe library of mathematical structures, morphisms, and transfer theorems called MathTransfer. The second contribution is a proposed methodology for scalable interoperability relying on Dedukti, Zenon Modulo, FoCaLiZe, and MathTransfer. The third contribution is the correctness proof of the Sieve of Erathostenes considered as the combination of a lemma coming from HOL and another coming from Coq. This proof illustrates our methodology.

The paper is structured as follows. In Sects. 2, 3, and 4 we present very briefly resp. the Dedukti logical framework, the FoCaLiZe system, and the MathTransfer library. These tools form the basis of our approach to interoperability presented in Sect. 5. Sections 6 and 7 are devoted to a case study illustrating it on a correctness proof of the Sieve of Eratosthenes. Finally, we conclude and discuss in Sect. 8 the generality and reusability of our development.

The MathTransfer library and our interoperability case study are distributed together at the following URL: https://gitlab.math.univ-paris-diderot.fr/cauderlier/math_transfer.

2 Dedukti, a Universal Proof Language

Dedukti [23] is a variant of the dependently-typed λ-calculus Twelf, a logical framework based on the Curry-Howard correspondence. Logics are encoded in Dedukti by providing a signature and then proof checking in the encoded logic is reduced to type checking in the encoding signature. For example, the conjunction in natural deduction can be encoded by the following signature:

```
Prop: Type.
proof: Prop -> Type.
and: Prop -> Prop -> Prop.
and_intro: A: Prop -> B: Prop -> proof A -> proof B ->
           proof (and A B).
and_elim1: A: Prop -> B: Prop -> proof (and A B) -> proof A.
and_elim2: A: Prop -> B: Prop -> proof (and A B) -> proof B.
```

The type Prop of logical propositions is first declared, then to each proposition A we associate the dependent type of its proofs proof A. The conjunction and is then declared and so are finally the usual elimination and introduction rules.

The dependent product $\Pi x : A. B$ is written x: A -> B in Dedukti. It is used to encode universal quantification. Dependent products and arrow types are introduced by λ-abstractions and eliminated by applications. The λ-abstraction $\lambda x : A. b$ is written x: A => b in Dedukti. For example, a proof of commutativity of conjunction in Dedukti is the term

```
A: Prop => B: Prop => H: proof (and A B) =>
and_intro B A (and_elim2 A B H) (and_elim1 A B H)
```

of type A: Prop -> B: Prop -> proof (and A B) -> proof (and B A)

Dedukti also features rewriting which is used to express computational higher-order logics such as the Calculus of Inductive Constructions implemented in the Coq proof assistant [2].

Translators from various ITPs to Dedukti have been developed [4]. In particular, Holide [3], Coqine [2], and Focalide [10] are translators from respectively the OpenTheory format for ITPs in the HOL family, the Coq proof assistant and the FoCaLiZe framework. Some ATPs also produce Dedukti files, e.g. iProver Modulo [7] and Zenon Modulo [8,11] which is used in this work.

Dedukti is a mere proof checker for a wide variety of logics, it is not intended for direct human use and it intentionally lacks features commonly found in similar systems such as modularity, type inference and implicit arguments. While these features are not needed in a proof checker, they are crucial for scalability of interoperability developments. We propose to compensate this lack by using FoCaLiZe as an interoperability framework for linking mathematical libraries.

3   FoCaLiZe, Zenon Modulo, and Focalide to the Rescue

FoCaLiZe (http://focalize.inria.fr) has been designed as a formal environment for developing certified programs and libraries. It provides a set of tools to formally specify and implement functions and prove logical statements. FoCaLiZe comes with three backends, a backend to OCaml for execution and two backends for formal verification. The historic one produces Coq code and requires the use of the ATP Zenon which can output proofs as Coq terms. A more recent backend, called Focalide, produces Dedukti code [10] and requires to use Zenon Modulo [8], an extension of Zenon which produces Dedukti proofs [11]. In this work, we only use the Focalide backend.

We present here very briefly the main ingredients of FoCaLiZe. For more details please consult [21].

In FoCaLiZe, specifications are written in a typed version of first-order logics; implementations are written with the help of a pure functional programming language very close to ML with algebraic datatypes, first class citizens functions, polymorphic types, recursion and pattern-matching. FoCaLiZe proposes a high-level proof language and discharges the logical details to Zenon or Zenon Modulo (according to the used backend). A proof in this language consists of intermediate lemmas and hints to the prover. When a proof is out of scope of the prover, a manual proof expressed in the backend language, Coq or Dedukti, is required.

A FoCaLiZe unit, named a species, is made of signatures, properties, definitions of functions and types and also proofs of user-defined properties. A species mearly defines a set of values and functions manipulating them where the meaning of the functions are given by properties. Inside a species, the type Self denotes the type of these values, it is usually abstract early in the development and made concrete later. When a species is complete, that is every function is definied and every property is proven, it can be turned into a collection which is close to an abstract data type. FoCaLiZe features modularity, more precisely multiple inheritance. Thus a species can be defined by inheriting from some others, allowing the reuse of all the signatures, definitions and proofs coming from them. When the definition of a function is inherited, it is possible to give it a new definition overriding the inherited one. This feature is not used here. A FoCaLiZe development appears as a hierarchy of species linked by inheritance, such as the one described in Fig. 1. Moreover species can be parameterized by collections. In that case, inside a species, the user is allowed to use functions and properties as black boxes as in a functor. In the following, we say a species is *instanciated* when it is applied to a particular collection.

Similarly to the possibility to prove directly a theorem in one of the target logical languages, FoCaLiZe allows the definition of global symbols by custom external expressions of the target languages (OCaml, Coq, and Dedukti). It is, with modularity, a key feature for our interoperability application. For example, addition of integers is defined in FoCaLiZe standard library as follows. It is declared with its type in the FoCaLiZe side, each branch in the definition maps + to a function written in the corresponding target language:

```
let ( + ) = internal int -> int -> int
    external
    | caml -> {* Ml_builtins.bi__int_plus *}
    | coq -> {* coq_builtins.bi__int_plus *}
    | dedukti -> {* dk_int.plus *};;
```

In this article, we use FoCaLiZe as an interoperability framework to provide the features missing in Dedukti for this task: modularity offered by FoCaLiZe inheritance, and proof automation provided by Zenon Modulo.

4   MathTransfer, a Library of Transfer Theorems

If A and B are two isomorphic mathematical structures, then for any formula φ_A expressed in the language of A, the formula $\varphi_A \to \varphi_B$ is a theorem where φ_B is the formula corresponding to φ_A in the language of B. Theorems of the form $\varphi_A \to \varphi_B$ are called transfer theorems. The use of transfer theorems is a way to formalize rigorously the mathematical habit of *reasoning modulo isomorphism*.

MathTransfer is a FoCaLiZe library of transfer theorems about natural numbers. More precisely, the MathTransfer library contains:

- definitions of the mathematical structures obtained by adding common arithmetic operations on natural numbers,
- definitions of (iso)morphisms between abstract representations of natural numbers,
- proofs that all operations are preserved by the morphisms, and
- 84 transfer theorems.

Each structure is defined as a FoCaLiZe species. Because the definitions of some operations depend on other operations, these species are organized in a hierarchy presented in Fig. 1 (where frames represent species and an arrow goes from a species S_1 to a species S_2 if S_1 directly inherits from S_2).

The species in this hierarchy contain only the axiomatisations of the operations, not their other properties. For example, the species corresponding to the multiplication ($\times$ frame in Fig. 1) contains:

- a new binary operation $\times$ representing multiplication,
- two first-order axioms: $\forall n.\ 0 \times n = 0$ and $\forall m\ n.\ \text{succ}(m) \times n = n + (m \times n)$.

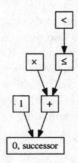

Fig. 1. The FoCaLiZe species hierarchy of MathTransfer structures

This species is written as follows in FoCaLiZe:

```
species NatTimes =
  inherit NatPlus;
  signature times : Self -> Self -> Self;
  property zero_times : all n : Self, times(zero, n) = zero;
  property succ_times : all m n : Self,
    times(succ(m), n) = plus(n, times(m, n));
end;;
```

On top of this small hierarchy, we build two orthogonal extensions: (a) a list of 84 statements about the arithmetic operations and (b) a hierarchy of morphisms between the structures.

The 84 chosen statements are a FoCaLiZe copy of the theorems about the operations of Fig. 1 that are proved in OpenTheory base library. Among them, 7 statements are properties of multiplication:

```
species NatTimesThm =
  inherit NatTimes;
  property times_zero : all m : Self,
    times(m, zero) = zero;
  property times_succ : all m n : Self,
    times(m, succ(n)) = plus(times(m, n), m);
  property times_assoc : all m n p : Self,
    times(times(m, n), p) = times(m, times(n, p));
  property times_commutes : all m n : Self,
    times(m, n) = times(n, m);
  property times_regular_left : all m n p : Self,
    times(m, n) = times (m, p) <-> (n = p \/ m = zero);
  property times_regular_right : all m n p : Self,
    times(m, p) = times (n, p) <-> (m = n \/ p = zero);
  property times_is_zero : all m n : Self,
    times(m, n) = zero <-> (m = zero \/ n = zero);
end;;
```

Morphisms on the other hand form a parameterized hierarchy of species. A morphism from a representation A of natural numbers is defined by a function `morph` of type `A -> Self` preserving zero and successors. From Peano axioms, assumed both in `A` and in the current species, we prove that `morph` is a bijection preserving all the operations. For example, here is the parameterized species proving that multiplication is preserved by the morphism (proof is omitted):

```
species NatTimesMorph (A is NatTimes) =
  inherit NatTimes , NatPlusMorph(A);
  theorem morph_times : all a1 a2 : A,
          morph(A!times(a1, a2)) = times(morph(a1), morph(a2))
  proof = ...;
end;;
```

These proofs of preservation of operations are not fully automatized because they require reasoning by induction which is not handled by Zenon Modulo but Zenon Modulo is extensively used for the subproofs.

By inheriting from both the morphism hierarchy and the list of statements, we can state and automatically prove the transfer theorems. Below is a fragment of the species containing the 7 transfer theorems relative to the previous 7 theorems about multiplication:

```
species NatTimesTransfer (A is NatTimesThm) =
  inherit NatTimesMorph(A), NatTimesThm;
  proof of times_zero =
  by property A!times_zero , morph_zero , morph_times ,
              morph_injective , morph_surjective;
  proof of times_succ =
  by property  A!times_succ , morph_succ , morph_times ,
              morph_injective , morph_surjective;
  proof of times_assoc =
  by property A!times_assoc , morph_times ,
              morph_injective , morph_surjective;
...
end;;
```

Each transfer proof relies on three ingredients:

- the corresponding theorem in the parameter `A`,
- bijectivity of `morph` (hypotheses `morph_injective` and `morph_surjective`),
- preservation of some operations by the morphism (hypotheses `morph_zero`, `morph_succ`, `morph_times`).

These transfer proofs are not automatically found by Zenon Modulo but are generated by a specialized transfer tactic written in Dedukti and similar to the transfer tactics for Isabelle and Coq [16,26].

5 Methodology for Dedukti-Based Interoperability

In this section, we propose an interoperability methodology based on Dedukti and MathTransfer. More precisely we detail below the different steps which must be followed when we want to use a lemma from a tool/formalism A in a formal proof of a theorem in another formalism B. The statements of the lemma in A and B do not need to be syntactically identical but thanks to the ATP Zenon Modulo some degree of rephrasing of the lemma is tolerated.

Some prerequisites about A and B are required before applying the process. First translators from A and B to Dedukti must exist. Then we rely on the fact that formalisms A and B have already been merged in Dedukti, it means that the logical linking of both logics has been done (sources of inconsistencies have been identified and fixed).

The steps are the following ones (between parentheses appears the formalism or the tool to be used to realize the step):

1. identify the lemma L to exchange between A and B and prove it (A);
2. prove in B the target theorem with the exported A lemma L considered as an hypothesis (B);
3. translate both the A lemma L and the B development T in Dedukti (use the corresponding translators);
4. if needed, extend the FoCaLiZe hierarchies of the MathTransfer library with the operations appearing in the statement of the lemma L (FoCaLiZe);
5. instantiate the FoCaLiZe hierarchies with external definitions and proofs from A and B; if the statements do not exactly match, use Zenon Modulo (FoCaLiZe with the help of Zenon Modulo);
6. automatically transfer the lemma L (FoCaLiZe);
7. translate the whole FoCaLiZe development in Dedukti (use Focalide);
8. write the proof of the final target theorem (a trivial Dedukti proof).

In Sects. 6 and 7, we apply this methodology to the correctness proof of the Sieve of Eratosthenes which is a small but typical case study where A is HOL and B is Coq.

6   Presentation of the Example: An Incomplete Coq Proof of the Sieve of Eratosthenes

In [5], Assaf and Cauderlier managed to link a Coq development with an HOL development directly in Dedukti because the example was chosen to minimize the interaction between Coq and HOL types. We now consider a more complicated example: a proved version of the Sieve of Eratosthenes. In this new proof of concept of interoperability in Dedukti, HOL and Coq have to agree on the type of natural numbers despite having slightly different definitions for it:

– in Coq, the type of natural numbers is defined as an inductive type;
– in HOL, inductive types are not primitive and natural numbers are encoded.

The Sieve of Eratosthenes is a well-known algorithm for listing all the prime numbers smaller than a given bound. In this section, we propose a certified implementation of this algorithm in the Coq proof assistant. We decompose this task in three: we have to program the sieve in Coq, to specify its correctness, and to prove it. In Sect. 6.1, we program the sieve in Coq and in Sect. 6.2 we specify it and sketch a proof of the correctness of the algorithm. We highlight the mathematical theorems on which this proof relies. In order to experiment with interoperability, we will not prove these mathematical results in Coq but import them from the OpenTheory libraries[1].

6.1   Programming the Sieve of Eratosthenes in Coq

Divisibility plays two purposes in our development: we need a divisibility test inside the definition of the algorithm and we also need divisibility to define primality and specify the algorithm. In order to get a simple definition of primality, we introduce *strict* divisibility: we say that a is a strict divisor of b if a divides b and $1 < a < b$. Using Euclidean division, we define strict divisibility as a boolean function (sd in Coq, definition omitted here). A natural number $p > 1$ is then called a *prime* number if and only if it has no strict divisor.

We now have all the prerequisites for defining the sieve's core function. We use the usual *fuel* trick for avoiding a termination proof. In the following definition, filter p l computes the list of elements of l that satisfy the boolean function p and negb is boolean negation.

```
Fixpoint Sieve (l : list)(fuel : nat) {struct fuel} : list :=
  match fuel with
  | 0 => Nil
  | S fuel => match l with
    | Nil => Nil
    | Cons a l =>
      Cons a (Sieve (filter (fun b => negb (sd a b)) l) fuel)
    end
  end.
```

When fuel is bigger than the length of l, Sieve l fuel gives the expected result so the length of l is a convenient default value for fuel. Finally, the prime numbers smaller than $2 + n$ can be computed by the following function where interval 2 n computes the interval $[2, 2 + n]$.

```
Definition eratosthenes n := Sieve (interval 2 n) n.
```

6.2   Specification and Correctness Proof

The specification of the Sieve of Eratosthenes is quite simple: a number p is a member of the list returned by eratosthenes n if and only if p is a prime number smaller than 2 + n.

We first define the prime predicate to be satisfied when its argument is a prime natural number:

---

[1] The purpose is to illustrate the methodology previously presented. Of course, this example is simple enough to be completely realized within Coq or done by reusing e.g. the translation from Hol Light to Coq proposed by Keller and Werner [19].

```
Inductive Istrue : bool -> Prop := ITT : Istrue true.

Definition prime p :=
  2 <= p /\ forall d, Istrue (negb (sd d p)).
```

We state the specification of the Sieve of Eratosthenes as the following three lemmata (where In is the list membership predicate).

A natural number returned by the function erathostenes is a prime number and is lower than the·bound:

```
Lemma sound_1 p n : In p (eratosthenes n) -> p <= 2 + n.
Lemma sound_2 p n : In p (eratosthenes n) -> prime p.
```

Any prime number lower than the bound will be returned by the function erathostheses:

```
Lemma complete p n :
  prime p -> p <= 2 + n -> In p (eratosthenes n).
```

For completeness, it is enough to prove that the Sieve function preserves prime numbers (assuming it received enough fuel).

The first soundness lemma also relies on an invariant of the Sieve function, namely that the members of Sieve l fuel are all members of l. The proof is then concluded by a simple soundness property of intervals.

The second soundness lemma is where arithmetic is required. Let p be a member of eratosthenes n, we can easily prove that $2 \leq p$ by an argument similar to the proof of the first soundness lemma. To prove that p has no strict divisor, we use the following standard arithmetic result:

Lemma 1. *Let n be a natural number greater than 2, n has a prime divisor.*

For the sake of our proof of concept, we shall not prove this result in Coq. Fortunately, the prime divisor lemma is proved in OpenTheory natural-prime library so item number 1 on our interoperability checklist presented in Sect. 5 is skipped.

We prove the correctness of the Sieve of Eratosthenes in Coq when Lemma 1 is considered as a parameter thus completing item number 2 on our checklist. This development can be split into three parts of approximately the same size:

- straightforward arithmetic results such as commutativity of addition and multiplication, these results are proved in both Coq standard library and OpenTheory but they are so straightforward that they are easier to reprove than to import and Coqine is not yet able to translate the part of the standard library in which they are proved,
- correctness of auxiliary functions which could be reused in other developments (modaux, strict divisibility and functions manipulating lists), and
- correctness of the functions Sieve and eratosthenes which are specific to this problem.

As in [5], the results that we want to import from HOL are hypotheses of the final theorem that has to be provided in Dedukti.

7 Mixing the Proofs

In this section, we follow the steps outlined in Sect. 5 to import in our Coq development the prime divisor lemma from HOL. The prerequisites for the methodology to apply are met thanks to the work of Assaf and Cauderlier [5] that we summarize in Sect. 7.1. The various steps of the methodology are then followed in Sects. 7.2 to 7.5. These steps are also pictured in Fig. 2.

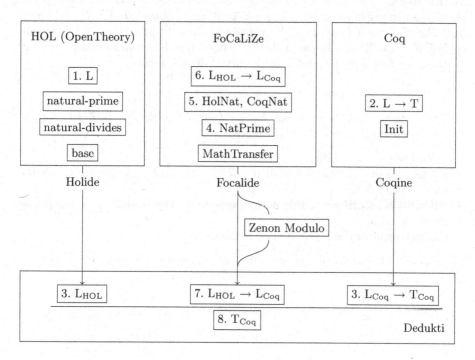

Fig. 2. The methodology in action for HOL/Coq interoperability

7.1 Linking HOL and Coq in Dedukti

In [5], Assaf and Cauderlier propose a first proof of concept of interoperability in Dedukti between HOL and Coq. The goal of this experiment was to study the logical linking of HOL and Coq logics.

Two sources of inconsistencies were identified. First, Coq and HOL do not agree on the question of type inhabitation: Coq allows empty types whereas we can prove in HOL that all types are inhabited. Second, the notions of booleans and logical propositions are identified in HOL and distinguished in Coq.

Type inhabitation is solved in [19] and [5] by identifying HOL types not with Coq types but with Coq inhabited types (in the Coq type $\Sigma A : \mathrm{Type}_0 . A$).

The difference between HOL booleans and Coq propositions is solved by identifying the type of HOL booleans with the type of Coq booleans, which are reflected as proposition by the symbol `hol_to_coq.Is_true` of type `hol.bool -> coq.prop`. This symbol is used to express provability in HOL as a special case of provability in Coq.

7.2   Extension of the MathTransfer Hierarchies up to the Prime Divisor Lemma

MathTransfer, as we have seen, contains transfer theorems corresponding to the most common arithmetic operations and relations such as found in OpenTheory base library. OpenTheory does also contain arithmetic definitions and theorems outside its base library. In particular, it defines divisibility and primality and it contains the following statement of the prime divisor lemma:

$$\forall n, n \neq 1 \rightarrow \exists p, (\text{prime}(p) \land p \mid n)$$

Following item number 4 on our checklist, we extend the FoCaLiZe hierarchies that we presented in Sect. 4 by four blocks:

- a definition of divisibility,
- a definition of strict (non-trivial) divisibility, this notion is used in the definition of primality,
- a definition of primality, this notion appears in the statement of the prime divisor lemma,
- the statement of the prime divisor lemma.

The extended hierarchy of operation definitions is shown in Fig. 3.

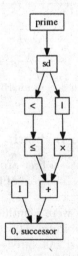

Fig. 3. The FoCaLiZe hierarchy of MathTransfer structures extended up to primality

Divisibility is required because this notion appears in the statement of the prime divisor theorem. It is defined as a binary relation | defined by $m \mid n \leftrightarrow$

$\exists p, m \times p = n$. Strict divisibility is used to define primality. There is also a binary relation sd defined by m sd $n \leftrightarrow (1 < m < n \wedge m \mid n)$. Primality is defined as the absence of strict divisor for numbers greater than 1. The corresponding predicate prime is defined by $\mathrm{prime}(p) \leftrightarrow (1 < p \wedge \forall d, \neg(d \text{ sd } p))$.

It is not required to state and transfer all the HOL lemmas dealing with divisibility and primality, it is enough to do so for the few ones that we are interested in such as the prime divisor lemma. The notion of isomorphism between representations of natural numbers is extended to take the new operations into account and the prime divisor lemma is automatically transferred.

7.3 Instantiation of Coq Natural Numbers

We can instantiate the hierarchy of species on the Coq side using FoCaLiZe external Dedukti definitions mapping directly the symbols to their Coqine translation in Dedukti. All the proofs required to instantiate the axioms characterizing the operations are trivial Dedukti proofs of reflexivity. For example, the species NatTimes is instantiated as follows:

```
species CoqTimes =
  inherit NatTimes , CoqPlus;
  let times(m : coq_nat, n : coq_nat) = internal coq_nat
    external
    | dedukti -> {* Coq__Init__Peano.mult m n *};
  proof of zero_times =
  dedukti proof definition of zero, times
  {* (n : cc.eT abst_T => logic.eq_refl abst_T abst_zero). *};
  proof of succ_times =
  dedukti proof definition of succ, plus, times
  {* (m : cc.eT abst_T => n : cc.eT abst_T =>
      logic.eq_refl abst_T (abst_times (abst_succ m) n)). *};
end;;
```

7.4 Instantiation of HOL Natural Numbers

Thanks to FoCaLiZe external definitions again, we can import in FoCaLiZe the HOL definitions of natural numbers and arithmetic operations. All the required proofs are found in the OpenTheory libraries. For example, the species NatTimes is instantiated as follows:

```
species HolTimes =
  inherit NatTimes , HolPlus;
  let times (p : hol_nat, q : hol_nat) = internal hol_nat
  external
  | dedukti -> {* HolNaturals.Number_2ENatural_2E_2A p q *};

  proof of zero_times =
  dedukti proof definition of zero, times
    {* HolNaturals.thm_117. *};
```

```
theorem hol_succ_times : all m n : Self,
  times(succ(m), n) = plus(times(m, n), n)
proof = dedukti proof definition of succ, plus, times
  {* HolNaturals.thm_157. *};
proof of succ_times =
<1>1 assume m n : Self,
     prove times(succ(m), n) = plus(n, times(m, n))
     <2>1 prove times(succ(m), n) = plus(times(m, n), n)
          by property hol_succ_times
     <2>2 prove plus(times(m, n), n) = plus(n, times(m, n))
          by property plus_commutes
     <2>f conclude
<1>f conclude;
end;;
```

The theorems number 117 and 157 in the Holide output of OpenTheory base library respectively state $\forall n.\ 0 \times n = 0$ and $\forall m\ n.\ succ(m) \times n = (m \times n) + n$. The first one is exactly the statement of zero_times but the statement of succ_times is $\forall m\ n.\ succ(m) \times n = n + (m \times n)$. The gap is filled by Zenon Modulo thanks to a previous import of the commutativity of addition (property plus_commutes).

The hierarchy is fully implemented and can be turned in a collection, that is a species where every signature received a definition and every property has been proved.

```
species HolPrimeDiv =            collection HolPrimeDivColl =
  inherit NatPrimeDiv, HolPrime;   implement HolPrimeDiv;
  ...                              end;;
end;;
```

7.5   Instantiation of the Morphism

If f is a function of type $\alpha \to \alpha$ and n is a natural number, we note f^n the nth iteration of the function f ($f^0 = Id$, $f^n = f \circ f \circ \ldots f$, n times).

Both the Coq Init library[2] and the OpenTheory base library define this polymorphic iteration of a function f. We use them to define the isomorphism between HOL natural numbers and Coq ones. The morphism from HOL natural numbers to Coq ones is defined by an HOL iteration of the Coq successor function $morph(n) := coq\_succ^n(coq\_zero)$ (coq_zero and coq_succ are mapped to the Dedukti translation of the Coq definitions) and its inverse is defined by a Coq iteration of the HOL successor function $inv\_morph(n) := hol\_succ^n(hol\_zero)$.

By instantiating all the morphisms and transfer hierarchies (items 5 and 6 of our methodology), we finally obtain in FoCaLiZe the prime divisor theorem on the Coq formulation of arithmetic structures. Once translated in Dedukti by Focalide, this theorem matches the assumption of the correctness proof of the

---

[2] The Coq Init library is the part of Coq standard library defining logical connectives and basic datatypes such as natural numbers and lists.

Sieve of Eratosthenes translated from Coq so we obtain a Dedukti proof of the correctness of the Sieve of Eratosthenes (item number 8 of our methodology).

Quantitatively, the size of the various parts of this development are given in Fig. 4. The HOL part of the development consists in a fragment of the OpenTheory library that was developed independently and contains thousands of theorems irrelevant to our case study. The Coq development however is of reasonable size and was specifically developed for this case study. In the case of FoCaLiZe, more than half of the generated code is produced by Zenon Modulo; this shows how useful proof automation has been in this development. Finally, the small Dedukti development is taken from the merging of Coq and HOL logics in [5].

	HOL	Coq	FoCaLiZe	Zenon Modulo	Dedukti
Source Code	3.2M	31K	129K		9K
Generated Dedukti Code	90M	828K	1.3M	1.7M	

Fig. 4. Size of the various parts of the development

8 Conclusion

We achieved our goal of certifying a Coq implementation of the Sieve of Eratosthenes using arithmetic results from OpenTheory. FoCaLiZe inheritance and parametrization allowed us to devise MathTransfer, a library of mathematical structures and transfer theorems. Zenon Modulo was of great help during this formalization since a lot of small steps of equational reasoning were needed and proving them in Dedukti would have been painful. We tried to do as much work as possible in a system independent way. The MathTransfer library is independent of HOL and Coq. Thanks to the symmetry in the roles of Coq and HOL, we can not only import lemmas from HOL to Coq but also in the other direction. Moreover, the definitions of the operations do not need to be identical in both systems. It is usual in FoCaLiZe to limit the dependencies to the definitions of methods thanks to late binding [22]. For small differences Zenon Modulo can fill the gap, for bigger ones such as divisibility (which is derived from Euclidean division on the Coq side) the equivalence of the definitions can be proved in either system.

This working example of interoperability needs to be reproduced with bigger proofs but also with proofs coming from some other systems if their underlying logics can be encoded within Dedukti. We believe that the methodology illustrated in this paper is scalable. However more automation is required in particular for the extension of MathTransfer. We expect the work of Gauthier and Kaliszyk [13] on automatic discovering of isomorphic structures from different formal libraries to adapt for this task. A limitation of our approach to interoperability in Dedukti is the trust we can have in the final proof because it is expressed in an uncommon logic whose consistency is not yet proved. Users of ITPs might expect from an interoperability development to obtain a proof in their trusted system. In order to translate back the proof in the combined logic to one of the original systems, we need to remove from the proof the use of unnecessary axioms of the other system. Preliminary work in this topic has been proposed in [9] where Cauderlier uses Dedukti rewriting to automatically remove classical axioms in Zenon proofs.

References

1. Armand, M., Faure, G., Grégoire, B., Keller, C., Théry, L., Werner, B.: A modular integration of SAT/SMT solvers to Coq through proof witnesses. In: Jouannaud, J.-P., Shao, Z. (eds.) CPP 2011. LNCS, vol. 7086, pp. 135–150. Springer, Heidelberg (2011). doi:10.1007/978-3-642-25379-9_12

2. Assaf, A.: A framework for defining computational higher-order logics. Ph.D. thesis, École Polytechnique (2015)

3. Assaf, A., Burel, G.: Translating HOL to Dedukti. In: Kaliszyk, C., Paskevich, A. (eds.) Proceedings Fourth Workshop on Proof eXchange for Theorem Proving, EPTCS, Berlin, Germany, 2–3 August 2015, vol. 186, pp. 74–88 (2015)

4. Assaf, A., Burel, G., Cauderlier, R., Delahaye, D., Dowek, G., Dubois, C., Gilbert, F., Halmagrand, P., Hermant, O., Saillard, R.: Expressing theories in the $\lambda\Pi$-calculus modulo theory and in the Dedukti system (2016). http://www.lsv.ens-cachan.fr/dowek/Publi/expressing.pdf

5. Assaf, A., Cauderlier, R.: Mixing HOL and Coq in Dedukti. In: Kaliszyk, C., Paskevich, A. (eds.) 4th Workshop on Proof eXchange for Theorem Proving, EPTCS, Berlin, Germany, 2–3 August 2015, vol. 186, pp. 89–96 (2015)

6. Blanchette, J.C., Bulwahn, L., Nipkow, T.: Automatic proof and disproof in Isabelle/HOL. In: Tinelli, C., Sofronie-Stokkermans, V. (eds.) FroCoS 2011. LNCS (LNAI), vol. 6989, pp. 12–27. Springer, Heidelberg (2011). doi:10.1007/978-3-642-24364-6_2

7. Burel, G.: Experimenting with deduction modulo. In: Bjørner, N., Sofronie-Stokkermans, V. (eds.) CADE 2011. LNCS (LNAI), vol. 6803, pp. 162–176. Springer, Heidelberg (2011). doi:10.1007/978-3-642-22438-6_14

8. Bury, G., Delahaye, D., Doligez, D., Halmagrand, P., Hermant, O.: Automated deduction in the B set theory using typed proof search and deduction modulo. In: LPAR 20 : 20th International Conference on Logic for Programming, Artificial Intelligence and Reasoning, Suva, Fiji, November 2015

9. Cauderlier, R.: A rewrite system for proof constructivization. In: Proceedings of the 2016 International Workshop on Logical Frameworks and Meta-languages: Theory and Practice, pp. 2:1–2:7. ACM (2016)

10. Cauderlier, R., Dubois, C.: ML pattern-matching, recursion, and rewriting: from FoCaLiZe to Dedukti. In: Sampaio, A., Wang, F. (eds.) ICTAC 2016. LNCS, vol. 9965, pp. 459–468. Springer, Cham (2016). doi:10.1007/978-3-319-46750-4_26

11. Cauderlier, R., Halmagrand, P.: Checking Zenon modulo proofs in Dedukti. In: Kaliszyk, C., Paskevich, A. (eds.) Proceedings 4th Workshop on Proof eXchange for Theorem Proving, EPTCS, Berlin, Germany, 2–3 August 2015, vol. 186, pp. 57–73 (2015)

12. Denney, E.: A prototype proof translator from HOL to Coq. In: Aagaard, M., Harrison, J. (eds.) TPHOLs 2000. LNCS, vol. 1869, pp. 108–125. Springer, Heidelberg (2000). doi:10.1007/3-540-44659-1_8

13. Gauthier, T., Kaliszyk, C.: Matching concepts across HOL libraries. In: Watt, S.M., Davenport, J.H., Sexton, A.P., Sojka, P., Urban, J. (eds.) CICM 2014. LNCS (LNAI), vol. 8543, pp. 267–281. Springer, Cham (2014). doi:10.1007/978-3-319-08434-3_20

14. Horozal, F., Rabe, F.: Representing model theory in a type-theoretical logical framework. Theor. Comput. Sci. **412**, 4919–4945 (2011)

15. Howe, D.J.: Importing mathematics from HOL into Nuprl. In: Goos, G., Hartmanis, J., Leeuwen, J., Wright, J., Grundy, J., Harrison, J. (eds.) TPHOLs 1996. LNCS, vol. 1125, pp. 267–281. Springer, Heidelberg (1996). doi:10.1007/BFb0105410

16. Huffman, B., Kunčar, O.: Lifting and transfer: a modular design for quotients in Isabelle/HOL. In: Gonthier, G., Norrish, M. (eds.) CPP 2013. LNCS, vol. 8307, pp. 131–146. Springer, Cham (2013). doi:10.1007/978-3-319-03545-1_9

17. Hurd, J.: The opentheory standard theory library. In: Bobaru, M., Havelund, K., Holzmann, G.J., Joshi, R. (eds.) NFM 2011. LNCS, vol. 6617, pp. 177–191. Springer, Heidelberg (2011). doi:10.1007/978-3-642-20398-5_14

18. Kaliszyk, C., Krauss, A.: Scalable LCF-style proof translation. In: Blazy, S., Paulin-Mohring, C., Pichardie, D. (eds.) Interactive Theorem Proving. number 7998 in LNCS, pp. 51–66. Springer, Heidelberg (2013)

19. Keller, C., Werner, B.: Importing HOL light into Coq. In: Kaufmann, M., Paulson, L.C. (eds.) ITP 2010. LNCS, vol. 6172, pp. 307–322. Springer, Heidelberg (2010). doi:10.1007/978-3-642-14052-5_22

20. Miller, D., Certificates, F.P.: Making proof universal and permanent. In: Momigliano, A., Pientka, B., Pollack, R. (eds.) Proceedings of the Eighth ACM SIGPLAN International Workshop on Logical Frameworks & Meta-languages: Theory & Practice, LFMTP 2013, Boston, Massachusetts, USA, 23 September 2013, pp. 1–2. ACM (2013)

21. Pessaux, F.: FoCaLiZe: inside an F-IDE. In: Dubois, C., Giannakopoulou, D., Méry, D. (eds.) Proceedings 1st Workshop on Formal Integrated Development Environment, F-IDE 2014, EPTCS, Grenoble, France, 6 April 6 2014, vol. 149, pp. 64–78 (2014)

22. Prevosto, V., Jaume, M.: Making proofs in a hierarchy of mathematical structures. In: Proceedings of Calculemus, September 2003

23. Saillard, R.: Type checking in the Lambda-Pi-Calculus modulo: theory and practice. Ph.D. thesis, MINES Paritech (2015)

24. Schürmann, C., Stehr, M.-O.: An executable formalization of the HOL/Nuprl connection in the metalogical framework twelf. In: Hermann, M., Voronkov, A. (eds.) LPAR 2006. LNCS, vol. 4246, pp. 150–166. Springer, Heidelberg (2006). doi:10.1007/11916277_11

25. Wiedijk, F.: Encoding the HOL light logic in Coq (2007, unpublished notes)

26. Zimmermann, T., Herbelin, H.: Automatic and transparent transfer of theorems along isomorphisms in the coq proof assistant. CoRR, abs/1505.05028 (2015)

A Formal Proof in Coq of LaSalle's Invariance Principle

Cyril Cohen(✉) and Damien Rouhling(✉)

Université Côte d'Azur, Inria, Sophia Antipolis, France
{cyril.cohen,damien.rouhling}@inria.fr

Abstract. Stability analysis of dynamical systems plays an important role in the study of control techniques. LaSalle's invariance principle is a result about the asymptotic stability of the solutions to a nonlinear system of differential equations and several extensions of this principle have been designed to fit different particular kinds of system. In this paper we present a formalization, in the Coq proof assistant, of a slightly improved version of the original principle. This is a step towards a formal verification of dynamical systems.

Keywords: Formal proofs · Coq · Dynamical systems · Stability

1 Introduction

Computer softwares are increasingly used to control moving objects: robots, planes, self-driving cars... This raises security issues, especially for human beings that are in the surroundings of such objects, or even inside them. Control theory brings answers by providing techniques to control the behaviour of dynamical systems. Control theoreticians focus on the mathematical foundation of their techniques. But another important aspect is to check that the implementations of such techniques respect their theoretical semantics.

The Coq proof assistant [23] provides a framework for both implementing functional programs and checking their correctness. It has also proven to be a convenient tool for the formalization of mathematics, for instance through the formalizations of the Four-Color Theorem [9] and of the Odd Order Theorem [10] based on the MATHEMATICAL COMPONENTS library[1] and the SSREFLECT extension of Coq's tactic language [11].

In this paper we present a formalization in Coq[2] of a mathematical result about the asymptotic stability of dynamical systems defined by a nonlinear system of differential equations: LaSalle's invariance principle [15]. Stability is an important notion for the control of nonlinear systems [14] and LaSalle's invariance principle or extensions of it have been successfully used to prove stability of different kinds of system [1,8,18,19].

---

[1] https://math-comp.github.io/math-comp/.
[2] https://github.com/drouhling/LaSalle.

© Springer International Publishing AG 2017
M. Ayala-Rincón and C.A. Muñoz (Eds.): ITP 2017, LNCS 10499, pp. 148–163, 2017.
DOI: 10.1007/978-3-319-66107-0_10

For this formalization, we used the SSREFLECT tactic language and the COQUELICOT library [3], which extends CoQ's standard library for real analysis [20]. We first present our improvements on the statement of LaSalle's invariance principle (Sect. 2), obtained by relaxing constraints on the original statement by LaSalle. Then we discuss details of the formalization (Sect. 3). Finally, we give the formal statement of the result we proved (Sect. 4) before pointing out the parts of the proof where classical reasoning was necessary (Sect. 5).

2 A Stronger Result

The original statement of LaSalle's invariance principle [15] contains hypotheses that can be relaxed and draws a conclusion which is weaker than what has been really proved. In this section, we first state LaSalle's invariance principle in its original form and then we explain how to strengthen it.

2.1 LaSalle's Invariance Principle

LaSalle's invariance principle [15] is a result about the asymptotic stability of the solutions to a system of differential equations in $\mathbb{R}^n$. The notion of asymptotic stability is expressed as "remain[ing] near the equilibrium state and in addition tend[ing] to return to the equilibrium". In fact, LaSalle proves that under some conditions the solutions approach a given (bounded) region of space when time goes to infinity (see Definition 1) and he uses this result on examples where the properties of this region imply that it is the equilibrium.

Definition 1. *A function of time $y(t)$ approaches a set A as t approaches infinity, denoted by $y(t) \to A$ as $t \to +\infty$, if*

$$\forall \varepsilon > 0, \exists T > 0, \forall t > T, \exists p \in A, \|y(t) - p\| < \varepsilon.$$

This definition is an easy generalization of the notion of convergence to a point to convergence to a set.

In its original form, LaSalle's invariance principle concerns only autonomous systems, i.e. where the behaviour of the system only depends on its position. Thus, we consider the following vector differential equation:

$$\dot{y} = F \circ y \tag{1}$$

where y is a function of time and F is a vector field in $\mathbb{R}^n$.

It is often not possible to remain near the equilibrium nor to converge to it regardless of the perturbation from it. It is thus important to determine the equilibrium's basin of attraction, or at least a region around it which is invariant with respect to (1).

Definition 2. *A set A is said to be invariant with respect to a differential equation $\dot{y} = F \circ y$ if every solution to this equation starting in A remains in A.*

In the remainder of this paper, since (1) is the only differential equation we consider, "invariant" will stand for "invariant w.r.t. (1)".

LaSalle's argument is that Lyapunov's second method [17] is a good means of studying asymptotic stability. This method requires the existence of a scalar function V, what we call today a Lyapunov function, which satisfies some properties. These properties are sign conditions on $\tilde{V}$, which is defined as follows:

Definition 3. *Let V be a scalar function with continuous first partial derivatives. Define:*
$$\tilde{V}(p) = \langle (\operatorname{grad} V)(p), F(p) \rangle$$
where $\langle .,. \rangle$ is the scalar product of $\mathbb{R}^n$.

We are now ready to state LaSalle's invariance principle, illustrated by Fig. 1.

Theorem 1 (LaSalle's invariance principle). *Assume F has continuous first partial derivatives and $F(0) = 0$. Let K be an invariant compact set. Suppose there is a scalar function V which has continuous first partial derivatives in K and is such that $\tilde{V}(p) \leqslant 0$ in K. Let E be the set of all points $p \in K$ such that $\tilde{V}(p) = 0$. Let M be the largest invariant set in E. Then for every solution y starting in K, $y(t) \to M$ as $t \to +\infty$.*

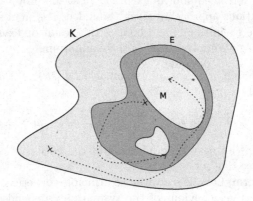

Fig. 1. Illustration of LaSalle's invariance principle

2.2 Relaxing Hypotheses

Some of LaSalle's hypotheses are unnecessary, although they are justified by the context. In his paper [15], he allows himself any assumption which makes it easier to focus on the method.

First, LaSalle assumes there is an equilibrium at the origin because his method is designed to show convergence to this equilibrium. The fact that it is the origin is just a convenience allowed by an easy translation. What is more, the existence of an equilibrium plays no role in the validity of Theorem 1. Thus, we removed the hypothesis $F(0) = 0$.

Still regarding the vector field F, the assumption "F has continuous first partial derivatives" is also a convenience. What is truly needed, as LaSalle puts it, is "any other conditions that guarantee the existence and uniqueness of solutions and the continuity of the solutions relative to the initial conditions". We can even go further and assume these properties only on the subset K of the ambient space. Indeed, for some systems the vector field is valid only in a restricted area, for instance when using a control function which has singularities (see e.g. [18]).

Then, the ambient space does not need to be $\mathbb{R}^n$, nor does it need to be a finite-dimensional vector space. A normed module over $\mathbb{R}$ was sufficient to prove this result. Since we work in an abstract normed module, we cannot express $\tilde{V}$ using the gradient of V. However, in $\mathbb{R}^n$ we know that for any points p and q, the scalar product between q and the gradient of V at point p is the value of the differential of V at point p applied to q. Thus, $\tilde{V}(p)$ can be expressed as the differential of V at point p applied to $F(p)$, which generalizes the definition of $\tilde{V}$ to normed modules.

$$\tilde{V}(p) = \langle (\mathrm{grad}\, V)(p), F(p) \rangle = (dV_p \circ F)(p)$$

Finally, concerning the Lyapunov function V, the assumption of continuous first partial derivatives is again a convenience. It is sufficient for V to be differentiable in K. Indeed, when y is a solution to (1), a step in the proof of Theorem 1 is to show that $V \circ y$ is non increasing using the assumption $\tilde{V}(p) \leqslant 0$ in K. Remarking that

$$(\tilde{V} \circ y)(t) = (dV_{y(t)} \circ F \circ y)(t) = (dV_{y(t)} \circ \dot{y})(t)$$

so that $\tilde{V} \circ y$ is the derivative of $V \circ y$, only the existence of this derivative is required to conclude this step.

2.3   Strengthening the Conclusion

While studying LaSalle's proof [15], we noticed it proves more than the result stated by Theorem 1. Indeed, the largest invariant subset M of the set $\left\{ p \in K \mid \tilde{V}(p) = 0 \right\}$ we called E is not interesting in itself: it is the fact that M is an invariant subset of E which gives M the nice property of being reduced to the equilibrium in LaSalle's applications.

The maximality of M plays a minor role in LaSalle's proof: given a solution y starting in K, this function happens to approach an invariant subset of E, **which depends on** y, as time goes to infinity, thus y approaches any of its supersets and M in particular. This set depending on y is in fact the positive limiting set of y, defined as follows:

Definition 4. *Let y be a function of time. The positive limiting set of y, denoted by $\Gamma^+(y)$, is the set of all points p such that*

$$\forall \varepsilon > 0, \forall T > 0, \exists t > T, \|y(t) - p\| < \varepsilon.$$

In other terms, $\Gamma^+(y)$ is the set of limit points of y at infinity. The fact that a function with values in a compact set approaches its limit points as time goes to infinity is intuitive and easy to prove. The fact that this set is invariant is a consequence of the continuity of solutions relative to initial conditions. The core of LaSalle's proof is thus to show that for all solution y starting in K, we have

$$\Gamma^+(y) \subseteq \left\{ p \in K \mid \tilde{V}(p) = 0 \right\}.$$

Let us give an intuition of proof of this point. The first step is to remark that it is in fact sufficient to prove that V is constant on $\Gamma^+(y)$ thanks to the interpretation of $\tilde{V}$ in terms of derivative. Then, the second step is to reduce this statement to the fact that $V \circ y$ converges at infinity. Finally, this last statement is just a consequence of the fact that $V \circ y$ is a bounded non increasing function.

Now, to remove the dependency in y, it is sufficient to take the union of all $\Gamma^+(y)$ for y solution starting in K, which is still an invariant subset of E and is thus smaller than the largest of them.

Ultimately, we proved the following result, illustrated by Fig. 2:

Theorem 2. *Assume F is such that we have the existence and uniqueness of solutions to (1) and the continuity of solutions relative to initial conditions on an invariant compact set K. Suppose there is a scalar function V, differentiable in K, such that $\tilde{V}(p) \leqslant 0$ in K. Let E be the set of all points $p \in K$ such that $\tilde{V}(p) = 0$ and L be the union of all $\Gamma^+(y)$ for y solution starting in K. Then, L is an invariant subset of E and for all solution y starting in K, $y(t) \to L$ as $t \to +\infty$.*

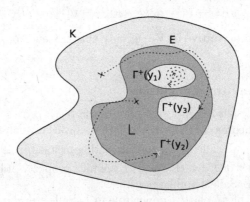

Fig. 2. Illustration of the refined version of LaSalle's invariance principle

3   Formalization

We present in this section the notations we used to make our formalization more readable and intuitive. Then we discuss details on how we worked with differential equations and how we expressed topological notions such as convergence and compactness.

3.1   Real Analysis and Notations

Our formalization is based on the COQUELICOT library [3], which is itself compatible with CoQ's standard library on classically axiomatized real numbers [20]. The COQUELICOT library exploits the notion of filter to develop a theory of convergence. It was inspired by the work of Hölzl et al. on the analysis library of ISABELLE/HOL [13].

Let us first recall some mathematical background and give some intuition. In topology, a *filter* is a set of sets, which is nonempty, upward closed, and closed under intersection. In this work, we use extensively two filters on real numbers: the set $\{N \mid \exists \varepsilon > 0, B_\varepsilon(p) \subseteq N\}$ of neighbourhoods of a point p and the set of neighbourhoods of $+\infty$ i.e. the set of sets that contain $[M, +\infty)$ for some M. The former is denoted by (locally p) in COQUELICOT and the latter by (Rbar_locally p_infty). With these definitions, f converges to q at point p iff the image by f of the filter of neighbourhoods of p is a subset of the filter of neighbourhoods of q. Even though this definition unfolds to the elementary characterization of convergence $\forall \varepsilon > 0, \exists \delta > 0, \forall r, |r - p| < \delta \Rightarrow |f(r) - q| < \varepsilon$, keeping the abstraction in terms of filters as much as possible yields more concise proofs and is also well supported by the library.

In this work, we experimented with notations to overload the ones in COQUELICOT, so that they read a bit closer to textbook mathematics. First, since sets are represented as predicates over a type, we pose set T := T -> Prop and we define MATHEMATICAL COMPONENTS-like notations to denote set theoretic operations. Indeed set0, setT and [set p] are respectively the empty set, the total set and the singleton $\{p\}$. Also, (A '&' B), (A '|' B), (~' A) and (A '\' B) are respectively the intersection, union, complement and difference. We write (A '<=' B) for $A \subseteq B$ and (A !=set0) for $\exists p \in A$, note that !=set0 is a token here. We also introduce set comprehension notations [set p | A p] (which is a typed alias for (fun p => A)) and the big operators \bigcup_(i in A) F i and \bigcap_(i in A) F i respectively denoting union and intersection of families indexed by A.

Secondly, COQUELICOT introduces layers over filters to abbreviate convergence. For example the predicate (is_lim f t p) is specialized to t and p in Rbar (i.e. $\mathbb{R} \cup \{\pm\infty\}$) and is defined in terms of filterlim which expands to the definition of this section. Since we introduce other notions of convergence in Sect. 3.3, adding more alternative definitions for approximately the same notion would only clutter the formalization, so we decided to remove this extra-layer. Instead, we provide a unique mechanism to infer which notion of convergence is required, by inspecting the form of the arguments and their types.

We now write f @ p --> q whatever the types of p and q: our mechanism selects the appropriate filters for p and q. This is in fact the composition of two independent notations: f @ p selects a filter for p and applies f to it, while q' --> q selects filters for q' and q and compares them.

We provide the notation +oo for the element p_infty of Rbar, so that a limit p of f at $+\infty$ now reads f @ +oo --> p. Moreover, although we do not use it in this part of the development, we also cast functions from nat, i.e. sequences, to the only sensible filter on nat (named eventually in COQUELICOT), so that one can write u --> p where u : nat -> U is a sequence.

CoQ's coercion mechanism is not powerful enough to handle casts from an arbitrary term to an appropriate filter. Hence, the mechanism to automatically infer a filter from an arbitrary term and its type is implemented using canonical structures. More precisely, we provide three structures: the first one to recognize terms that could be cast to filters, the second one to recognize types whose elements could be cast to filters, and the third one to recognize arrow types which could be cast to filters. If the first canonical structure fails to cast a given term to a filter, it gives its type to the second canonical structure and if it is an arrow it tries to match the source of the arrow using the third canonical structure.

3.2   On Differential Equations

To deal with systems of differential equations, we also use the COQUELICOT library [3], which already contains convenient definitions for functional analysis. This library also contains a hierarchy of topological structures, among which is the structure of normed module we used for the ambient space.

The property of being a solution to the differential system (1) is then easily expressed as follows: y is a solution if at each time t, the derivative of y at point t is $(F \circ y)(t)$. In CoQ:

```
Definition is_sol (y : R -> U) :=
  forall t, is_derive y t (F (y t)).
```

Here, U is the ambient space and R the set of real numbers. We could have considered only non negative times, since our goal is to describe physical systems. However, for reasons we will give later, we stick to this more constrained version in this paper.

Now, what we need is a way to express the existence and uniqueness of solutions to (1) and the continuity of solutions relative to initial conditions. In our work, we assumed the existence of a function sol : U -> R -> U which represents all the solutions to (1). Its first argument corresponds to the initial condition and the second one to time. Thus, when time is equal to 0, the result of this function must be equal to the initial condition.

```
Hypothesis sol0 : forall p, sol p 0 = p.
```

We combined the conditions of existence (for all p in K, sol p is a solution) and uniqueness of solutions into the following hypothesis: a function y starting in K is a solution to (1) if and only if it is equal to the function sol (y 0), that is the solution which has same initial condition.

```
Hypothesis solP : forall y, K (y 0) -> is_sol y <-> y = sol (y 0).
```

Note that here we wrote an equality between functions. This assumption together with the axiom of functional extensionality made our proofs more natural and shorter. Indeed, by using solP we can replace any solution by an application of the function sol, which removes all the hypotheses of the form is_sol y from our theorems. Moreover, proof scripts in the SSREFLECT tactic language [11] heavily rely on the rewriting of equalities.

This hypothesis would not be satisfiable if, as mentioned before, we constrained the derivative of solutions only for non negative times. Indeed, we could not control the values of these solutions for negative times, hence there would be (infinitely) many solutions y different from sol (y 0). Adapting naively this hypothesis by considering equality only on non negative times would cancel all the benefits of this formulation.

One solution could be to change the type of the functions in order to work with functions whose domain is the set of non negative real numbers. This would require to construct the type for this set and to develop its theory. A compromise which would be easier to implement and more ligthweight in our context is to keep functions on $\mathbb{R}$, but only require the solution to satisfy the differential equation for non negative times, and fix its value for negative times. For example we could ask the function to be constant equal to $y(0)$ for negative values, or make the solution symmetric with regard to its initial value (i.e. $y(-t) = 2y(0) - y(t)$), which would keep the solution derivable everywhere.

Finally, the continuity of solutions relative to initial conditions on K is expressed as the continuity on K of the function fun p : U => sol p t for all t.

3.3 Convergence to a Set

To generalize the notion of convergence to a point to convergence to a set, what we called "approaching a set when time goes to infinity" in Sect. 2, we need to generalize the notion of neighbourhood filter to a set. Recall the definition of neighbourhood filter for a point (Sect. 3.1): a neighbourhood of a point is a set that contains a ball with positive radius centered on this point. In COQUELICOT [3], this definition is not restricted to real numbers but applies to any uniform space.

```
Definition locally (p : U) :=
  [set A | exists eps : posreal, ball p eps '<=' A].
```

For a set, there is no ball anymore. However, we can extend the set with a band of fixed width ε, which is in fact the union of all balls of radius ε centered on points of the set.

```
Definition ball_set (A : set U) (eps : posreal) :=
  \bigcup_(p in A) ball p eps.
```

The neighbourhood filter for a set then has a very analogous definition to the one of neighbourhood filter for a point.

```
Definition locally_set (A : set U) :=
  [set B | exists eps : posreal, ball_set A eps '<=' B].
Instance locally_set_filter (A : set U) :
  Filter (locally_set A).
```

We can prove that it is a generalization of the notion of neighbourhood filter for a point by proving that we define the same filter on the singleton [set p] as with COQUELICOT's definition on p.

```
Lemma locally_set1P p A : locally p A <-> locally_set [set p] A.
```

In our notation mechanism from Sect. 3.1, we declare locally_set to be the canonical filter to use for sets over a uniform space, so that we can write y @ +oo --> A where A is a set. Thus, the notion of convergence to a set is expressed thanks to COQUELICOT's notion of limit using this particular filter. And again, we can prove that it generalizes COQUELICOT's notion of limit when applied on singletons.

```
Lemma cvg_to_set1P y p : y @ +oo --> [set p] <-> y @ +oo --> p.
```

3.4 Compactness

To express compactness, we decided to experiment with a definition of compact sets using filters. In fact, this definition involves the notion of clustering, which is closely related to convergence and limit points. Indeed, a filter clusters to a point if each of its elements intersects each neighbourhood of the point. We say that a filter clusters if there is a point to which it clusters.

```
Definition cluster (F : set (set U)) p :=
  forall A B, F A -> locally p B -> A :&: B !=set0
```

To see the link with limit points of a function y, consider the filter (y @ +oo), which is the set of sets which ultimately contain all images of y (formerly filtermap y (Rbar_locally p_infty) in COQUELICOT). The set of points to which (y @ +oo) clusters is exactly the positive limiting set of y i.e. the set of limit points of y (recall Definition 4).

```
Definition pos_limit_set (y : R -> U) :=
  \bigcap_(eps : posreal) \bigcap_(T : posreal)
    [set p | Rlt T '&' (y @^-1' ball p eps) !=set0].
Lemma plim_set_cluster (y : R -> U) :
  pos_limit_set y = cluster (y @ +oo).
```

Note that we wrote an equality between sets i.e. between functions with propositions as value. We used the axiom of propositional extensionality (on top of functional extensionality) to be able to prove this. Again, this makes our code closer to textbook mathematics.

We were already using this equality to prove some properties of the positive limiting set of a function y. Consequently, we decided to state each property of this set as a property of cluster (y @ +oo). Thanks to this strategy, we managed to shorten some of our proofs.

A set A is compact if every proper filter on A clusters in A.

```
Definition compact A :=
   forall (F : set (set U)), F A -> ProperFilter F ->
   (A :&: cluster F) !=set0.
```

Note how the hypothesis "on A" has been translated into "A is an element of F". This is possible thanks to the properties of filters: every filter on A is a filter base in U whose completion is a filter containing A, and every filter on U containing A defines a filter on A when restricted to sets contained in A. Thanks to this, we do not have to consider the subspace topology, which would add complications. Indeed, the type classes `Filter` and `ProperFilter` of COQUELICOT are defined on sets of sets in a uniform space i.e. on functions of type `(U -> Prop) -> Prop` for U a uniform space. The `UniformSpace` structure of COQUELICOT requires an element of type `Type`, while in our context A is of type `U -> Prop`. Canonically transfering structures to subsets would then require wrapping functions into types, while our solution is simpler.

This notion of compact set is quite convenient to use to work with convergence and limit points: the only hard part is finding the right filter on your compact set and then the cluster point this hypothesis gives you is usually the point your are looking for. However for other proofs this notion is quite complicated to use. Proving that a set is compact requires finding a cluster point for any abstract proper filter on this set, or going through a proof by contradiction. Moreover, to prove that any compact set is bounded, we had to go through the definition of compactness based on open covers. We proved the equivalence between both definitions following the proof in Wilansky's textbook on topology [24] (see Sect. 5 for more details on the proof).

4   The Formal Statement of LaSalle's Invariance Principle

As explained in Sect. 2, we formalized a slightly stronger version of LaSalle's invariance principle [15]. In particular, we proved the convergence of solutions to (1) to a more constrained set: the union of the positive limiting sets of the solutions starting in K.

```
Definition limS (A : set U) :=
   \bigcup_(q in A) cluster (sol q @ +oo).
```

Recall that we require K to be compact and invariant (see Definition 2). Both these hypotheses are used to prove the convergence of solutions to `limS K`.

```
Definition is_invariant A :=
   forall p, A p -> forall t, 0 <= t -> A (sol p t).
Lemma cvg_to_limS (A : set U) : compact A -> is_invariant A ->
   forall p, A p -> sol p @ +oo --> limS A.
```

This is in fact an "easy" part of LaSalle's invariance principle. It is indeed sufficient for a function to ultimately have values in a compact set in order for it to converge to the set of its limit points, hence to any superset of its positive limiting set.

```
Lemma cvg_to_pos_limit_set y (A : set U) :
  (y @ +oo) A -> compact A -> y @ +oo --> cluster (y @ +oo).
Lemma cvg_to_superset A B y : A '<=' B ->
  y @ +oo --> A -> y @ +oo --> B.
```

The invariance of K is a strong way to force the solutions to ultimately have values in K. However, since in our proof of LaSalle's invariance principle we need to use the uniqueness of solutions for initial conditions which are values of solutions starting in K, the invariance of K is required anyway.

There are two other aspects to our version of LaSalle's invariance principle: limS K is invariant and it is a subset of the set of points p for which $\tilde{V}(p) = 0$.

The first point does not need any hypothesis: the positive limiting set of any solution starting in K is invariant, hence any union of such sets is invariant too.

```
Lemma invariant_pos_limit_set p :
  K p -> is_invariant (cluster (sol p @ +oo)).
Lemma invariant_limS A : A '<=' K -> is_invariant (limS A).
```

As explained in Sect. 2.3, the core of LaSalle's proof is thus the second point. To state this part, we need to use the differential of the Lyapunov function V. Indeed, as mentioned in Sect. 2.2, since we work in an abstract normed module, we cannot express $\tilde{V}$ using the gradient of V. We express differentials using COQUELICOT [3]: filterdiff f (locally p) g expresses the fact that g is the differential of f at point p. Thus, we assume a function V' : U -> U -> R which is total, together with the hypothesis that for any p in K, V' p is the differential of V at point p (i.e. forall p : U, K p -> filterdiff V (locally p) (V' p)). All hypotheses on $\tilde{V}$ can then be expressed by replacing it with the function fun p => (V' p \o F) p.

Assuming a total function V' is a way to mimic COQUELICOT's proof style on derivatives. Indeed, to represent the derivative of a real function f : R -> R in COQUELICOT, one has access to a total function Derive f. All theorems then concern Derive f, with the hypothesis that f admits a derivative at some point x, written ex_derive f x, when it is needed. This is a very convenient way to deal with derivatives. However, because COQUELICOT lacks such a total function for differentials, we had to introduce the differential as a parameter.

Finally, the last part of our version of LaSalle's invariance principle, i.e. the set limS K is a subset of the set of points p for which $\tilde{V}(p) = 0$, is stated as follows:

```
Lemma stable_limS (V : U -> R) (V' : U -> U -> R) :
  (forall p : U, K p -> filterdiff V (locally p) (V' p)) ->
  (forall p : U, K p -> (V' p \o F) p <= 0) ->
  limS K '<=' [set p | (V' p \o F) p = 0].
```

Note that the proof in COQ follows exactly the same steps as in the paper proof we sketched in Sect. 2.3.

5   On Classical Reasoning

Several proofs in our work required classical reasoning, although we tried to remain as constructive as possible. Indeed, while proofs and statements in constructive analysis are very different from the ones in classical analysis, we believe some of our constructive proofs could still be used in a purely constructive context. In particular, in our development we use a classical axiomatization of real numbers, but we hope some results are actually independent from the representation of real numbers. For example, most results in topology, like our constructive theorems on sets, filters, closures, and compactness are phrased in a way which does not make real numbers appear.

Hence, we redefined some notions, namely closed sets, closures, compactness and Hausdorff separation, to fit our purposes, and the proofs that these are equivalent to preexisting definitions were often classical. We also list here two other main theorems for which we could only give a classical proof.

First, the notion of closure was very practical to use whenever closed sets appeared. In CoQUELICOT [3], a set A is closed if it contains all points for which the complement of A is not a neighbourhood. A point is in the closure of a set if all its neighbourhoods intersect the set. A set is closed if and only if its closure is included in it (the other inclusion always holds).

```
Definition closed (A : set U) :=
  forall p, ~ (locally p (~' A)) -> A p.
Definition closure (A : set U) p :=
  forall B, locally p B -> A '&' B !=set0.
Lemma closedP (A : set U) : closed A <-> closure A '<=' A.
```

The right implication of `closedP` was proved constructively while the other direction required classical reasoning. The notion of closure proved to be very practical, especially in our settings since it is related to clustering: a filter clusters to a point if and only if this point is in the closure of each element of the filter.

```
Lemma clusterE F : cluster F = \bigcap_(A in F) (closure A).
```

Then, the proof of equivalence between the filter-based and open covers-based definitions of compactness is classical. In fact, we prove this equivalence by going through a third definition of compactness: a set A is compact if every family of closed sets of A with the finite intersection property has a nonempty intersection. This definition is very close to the filter-based one, and we proved constructively that they are equivalent. Indeed, the set of all finite intersections in such a family is a filter base defining a proper filter which clusters. Conversely, the family of closures of the elements of a proper filter which clusters has the finite intersection property. It is the equivalence between this third definition and the open covers-based definition which is classical. More precisely, both directions in this equivalence are proved by contraposition and classical steps are required to push negations under existential quantifiers and to remove double negations.

Similarly, we worked with a different definition of a Hausdorff space. We proved that normed modules are necessarily Hausdorff spaces. This allowed us to prove that the compact set K is also closed, which was needed to show that the positive limiting set of any solution starting in K is included in K. We could have used the usual notion of Hausdorff space (whenever you have two different points, you can find two respective neighbourhoods of these points which do not intersect), but in fact, its contrapositive was more practical in our settings because it admits a nice statement using clustering: if two points p and q are such that all their neighbourhoods intersects, i.e. the neighbourhood filter of p clusters to q (and vice-versa), then they are equal.

```
Definition hausdorff (U : UniformSpace) :=
  forall p q : U, cluster (locally p) q -> p = q.
Lemma hausdorffP (U : UniformSpace) :
  hausdorff U <-> forall p q : U, p <> q -> exists A B,
    locally p A /\ locally q B /\ forall r, ~ (A '&' B) r.
```

Again, the proof of equivalence between both definitions required classical reasoning to push and remove negations.

Another classical proof we did is the proof of convergence of a function with values ultimately in a compact set to its positive limiting set.

```
Lemma cvg_to_pos_limit_set y (A : set U) :
  (y @ +oo) A -> compact A -> y @ +oo --> cluster (y @ +oo).
```

We proved this theorem in two ways. The first proof is by contradiction, as in LaSalle's paper [15]. The second one goes through a generalization of this result: any proper filter on a compact set contains any neighbourhood of its set of cluster points.

```
Lemma filter_cluster (F : set (set U)) (A : set U) :
  ProperFilter F -> F A -> compact A ->
  forall eps : posreal, F (ball_set (cluster F) eps).
```

We proved this lemma using yet another definition of compactness: the contrapositive of the definition based on families of closed sets. Going back and forth between emptiness and nonemptiness properties once again introduced classical reasoning steps. Similarly, as mentioned in Sect. 3.4, we had to use a classical equivalence between two definitions of compactness to prove that compact sets are bounded.

Finally, we had to prove classically that a monotonic bounded real function admits a finite limit at infinity. For instance in the case of a non decreasing function, one has to prove that the lowest upper bound of its values is the aforementioned limit. What is classical is the proof that, if l is the least upper bound of the set A, then for all $\varepsilon > 0$ there exists $p \in A$ such that $l - \varepsilon \leqslant p \leqslant l$. This last example illustrates the problem, already noticed by A. Mahboubi and G. Melquiond, that CoQ's axiomatization of real numbers is not expressive enough to give an arbitrary approximation of a least upper bound.

6   Related Work

Several formalizations in topology already exist: in Coq [4], in PVS [16], in ISABELLE/HOL [13] or in MIZAR [7,22] for instance. All of them express compactness using open covers. We adapted Cano's formalization [4] for our proof of equivalence with the filter-based definition. We could not use it directly since it relies on the eqType structure of the MATHEMATICAL COMPONENTS library and COQUELICOT's structures [3] are not based on this structure.

Note that in the work of Hölzl et al. [13] there is a definition of compactness in terms of filters which is slightly different from ours: a set A is compact if for each proper filter on A there is a point $p \in A$ such that a neighbourhood of p is contained in the filter. This is a bit less convenient to use than clustering since you cannot choose the neighbourhood. To our knowledge, our work is the first attempt to exploit the filter-based definition of compactness to get simple proofs on convergence.

We must also mention COQUELICOT's definition of compactness, which is based on gauge functions, and COQ's topology library by Schepler[3]. Both are unfortunately unusable in our context: COQUELICOT's definition is specialized to $\mathbb{R}^n$ while we are working on an abstract normed module, and Schepler's library does not interface with COQUELICOT, since it redefines filters for instance. Schepler's library contains a proof of equivalence between the filter-based and open covers-based definitions of compactness, which is very close to ours. However, these definitions concern topological spaces whereas, as mentionned in Sect. 3.4, we focus on subsets of such spaces without referring to the subspace topology.

Concerning formalizations on stability and Lyapunov functions, Chan et al. [5] used a Lyapunov function to prove in COQ the stability of a particular system. They have however no proof of a general stability theorem. Mitra and Chandy [21] formalized in PVS stability theorems using Lyapunov-like functions in the particular case of automata. Herencia-Zapana et al. [12] took another approach to stability proofs: stability proofs using Lyapunov functions, under the form of Hoare triples annotations on C code, are used to generate proof obligations for PVS.

We are definitely not the first to generalize LaSalle's invariance principle. We decided to prove a version of the principle which is close to the original statement but several generalizations were designed to make it available in more complex settings. Chellaboina et al. [6] weakened further the regularity hypothesis on the Lyapunov function at the cost of sign conditions and a boundedness hypothesis on the Lyapunov function along the trajectories. Barkana [1] restricted the hypotheses on the Lyapunov function to hypotheses along the trajectories in order to generalize LaSalle's invariance principle to nonautonomous systems. Mancilla-Aguilar and García [19] generalized LaSalle's invariance principle to switched autonomous systems by adding further conditions related to the switching, but removed the conditions of existence and uniqueness of solutions and of continuity of the solutions relative to initial conditions by working on a set of

---

[3] http://www.lix.polytechnique.fr/coq/pylons/contribs/view/Topology/v8.4.

admissible trajectories. Fischer et al. [8] also weakened the hypotheses on the solutions of a nonautonomous system by using a generalized notion of solution.

7   Conclusion and Future Work

In this paper we presented our formalization of LaSalle's invariance principle, a theorem about the asymptotic stability of solutions to a nonlinear system of differential equations. We proved a version of this theorem which is very close to its original statement but we removed unnecessary hypotheses and chose a more precise conclusion.

Our use of set theoretic notations in this formalization made our proofs more readable, closer to the intuition of filters as sets of sets. Functional extensionality also gave us a convenient way to write proofs on the solutions to a differential equation, allowing us to use a single function to represent all of them. We used propositional extensionality to be able to prove equalities between sets, but it is not as critical as functional extensionality: all the proofs were written without this axiom before we decided to add it.

Our experiment with filter-based compactness is partially conclusive: filters are really adapted to proofs on convergence but we had to use other definitions of compactness for other purposes.

All in all, our formalization of LaSalle's invariance principle takes around 1250 lines. Around 250 lines were devoted to the proofs of the properties on the positive limiting set and of LaSalle's invariance principle. The remaining 1000 lines contain the definitions of notations, the generalization of convergence notions to sets and proofs about closed sets, compact sets and monotonic functions.

This formalization is a step towards a formal verification of dynamical systems. LaSalle's invariance principle and its extensions play an important role in the study of control techniques. We plan to use this work to formally verify a control law for the swing-up of an inverted pendulum [18].

Acknowledgements. We thank the anonymous reviewers for their useful feedback.

References

1. Barkana, I.: Defending the beauty of the Invariance Principle. Int. J. Control **87**(1), 186–206 (2014). http://dx.doi.org/10.1080/00207179.2013.826385
2. Blazy, S., Paulin-Mohring, C., Pichardie, D. (eds.): ITP 2013. LNCS, vol. 7998. Springer, Heidelberg (2013). doi:10.1007/978-3-642-39634-2
3. Boldo, S., Lelay, C., Melquiond, G.: Coquelicot: A User-Friendly Library of Real Analysis for Coq. Math. Comput. Sci. **9**(1), 41–62 (2015). http://dx.doi.org/10.1007/s11786-014-0181-1
4. Cano, G.: Interaction entre algèbre linéaire et analyse en formalisation des mathématiques. (Interaction between linear algebra and analysis in formal mathematics). Ph.D. thesis, University of Nice Sophia Antipolis, France (2014). https://tel.archives-ouvertes.fr/tel-00986283
5. Chan, M., Ricketts, D., Lerner, S., Malecha, G.: Formal Verification of Stability Properties of Cyber-Physical Systems, January 2016

6. Chellaboina, V., Leonessa, A., Haddad, W.M.: Generalized Lyapunov and invariant set theorems for nonlinear dynamical systems. Syst. Control Lett. **38**(4–5), 289–295 (1999). http://www.sciencedirect.com/science/article/pii/S0167691199000766
7. Darmochwał, A.: Compact Spaces. Formaliz. Math. **1**(2), 383–386 (1990). http://fm.mizar.org/1990-1/pdf1-2/compts_1.pdf
8. Fischer, N.R., Kamalapurkar, R., Dixon, W.E.: LaSalle-Yoshizawa Corollaries for Nonsmooth Systems. IEEE Trans. Automat. Control **58**(9), 2333–2338 (2013). http://dx.doi.org/10.1109/TAC.2013.2246900
9. Gonthier, G.: Formal Proof - The Four-Color Theorem. Notices AMS **55**(11), 1382–1393 (2008)
10. Gonthier, G., et al.: A Machine-Checked Proof of the Odd Order Theorem. In: Blazy et al. [2], pp. 163–179 (2013). doi:10.1007/978-3-642-39634-2_14
11. Gonthier, G., Mahboubi, A., Tassi, E.: A Small Scale Reflection Extension for the Coq system. Research Report RR-6455, Inria Saclay Ile de France (2015). https://hal.inria.fr/inria-00258384
12. Herencia-Zapana, H., Jobredeaux, R., Owre, S., Garoche, P.-L., Feron, E., Perez, G., Ascariz, P.: PVS linear algebra libraries for verification of control software algorithms in C/ACSL. In: Goodloe, A.E., Person, S. (eds.) NFM 2012. LNCS, vol. 7226, pp. 147–161. Springer, Heidelberg (2012). doi:10.1007/978-3-642-28891-3_15
13. Hölzl, J., Immler, F., Huffman, B.: Type Classes and Filters for Mathematical Analysis in Isabelle/HOL. In: Blazy et al. [2], pp. 279–294 (2013). doi:10.1007/978-3-642-39634-2_21
14. Khalil, H.: Nonlinear Systems. Pearson Education, Prentice Hall (2002). https://books.google.fr/books?id=t_d1QgAACAAJ
15. LaSalle, J.: Some Extensions of Liapunov's Second Method. IRE Trans. Circ. Theory **7**(4), 520–527 (1960)
16. Lester, D.R.: Topology in PVS: Continuous Mathematics with Applications. In: Proceedings of the Second Workshop on Automated Formal Methods, AFM 2007, pp. 11–20. ACM, New York (2007). http://doi.acm.org/10.1145/1345169.1345171
17. Liapounoff, A.: Problème général de la stabilité du mouvement. In: Annales de la Faculté des sciences de Toulouse: Mathématiques, vol. 9, pp. 203–474 (1907). http://eudml.org/doc/72801
18. Lozano, R., Fantoni, I., Block, D.: Stabilization of the inverted pendulum around its homoclinic orbit. Syst. Control Lett. **40**(3), 197–204 (2000)
19. Mancilla-Aguilar, J.L., García, R.A.: An extension of LaSalle's invariance principle for switched systems. Syst. Control Lett. **55**(5), 376–384 (2006). http://dx.doi.org/10.1016/j.sysconle.2005.07.009
20. Mayero, M.: Formalisation et automatisation de preuves en analyses réelle et numérique. Ph.D. thesis, Université Paris VI (décembre 2001)
21. Mitra, S., Chandy, K.M.: A Formalized Theory for Verifying Stability and Convergence of Automata in PVS. In: Mohamed, O.A., Muñoz, C., Tahar, S. (eds.) TPHOLs 2008. LNCS, vol. 5170, pp. 230–245. Springer, Heidelberg (2008). doi:10.1007/978-3-540-71067-7_20
22. Padlewska, B., Darmochwał, A.: Topological Spaces and Continuous Functions. Formaliz. Math. **1**(1), 223–230 (1990). http://fm.mizar.org/1990-1/pdf1-1/pre_topc.pdf
23. The Coq Development Team: The Coq proof assistant reference manual, version 8.6. (2016). http://coq.inria.fr
24. Wilansky, A.: Topology for Analysis. Dover Books on Mathematics. Dover Publications, New York (2008). http://cds.cern.ch/record/2222525

How to Get More Out of Your Oracles

Luís Cruz-Filipe[✉], Kim S. Larsen, and Peter Schneider-Kamp

Department Mathematics and Computer Science, University Southern Denmark,
Campusvej 55, 5230, Odense M, Denmark
{lcf,kslarsen,petersk}@imada.sdu.dk

Abstract. Formal verification of large computer-generated proofs often relies on certified checkers based on oracles. We propose a methodology for such proofs, advocating a separation of concerns between formalizing the underlying theory and optimizing the algorithm implemented in the checker, based on the observation that such optimizations can benefit significantly from adequately adapting the oracle.

1 Introduction

During the last decade, we have seen the advent of larger and larger computer-generated proofs, often based on exhaustive case analysis. To allow for independent verification, the programs performing such proofs also generate a trace, detailing their reasoning steps. These *proof witnesses* have been growing significantly in size, from a few MB [15] to a few GB [5,13], culminating with the impressive 200 TB proof of the Boolean Pythagorean Triples conjecture [12].

Formal verification of such proofs amounts to checking whether the proof witnesses can be used to reconstruct a proof. Directly importing the witnesses into a theorem prover [4] does not scale to the size of recent proofs due to memory and computational requirements. Instead, the witnesses can be obtained from an external untrusted source, the *oracle*, and checked for correctness before use [14]. The formal proof is thus split between two components: the untrusted oracle and the proven correct proof checker. The latter needs to be correct regardless of its input data, typically by ignoring or rejecting incorrect data.

The benefit of using the oracle is efficiency: since its results are not trusted, it can be optimized for performing difficult computations efficiently. (The point is, of course, that these results should be correct, but there is no burden of proof of this fact.) The certified checker, on the other hand, typically constitutes the computational bottleneck of the overall system. Thus, in order to minimize execution time, it is natural to try to shift as much computation as possible from the checker to the oracle.

Traditionally, this path has not been explored to its full potential. Often oracles are queried when needed [9,11,14], computing witnesses at runtime. In other cases [2,15,16], the oracle is pre-computed and responsible for controlling the flow of the algorithm; in this case, the checker's queries amount to asking "What should I do next?". Our simple observation is: *the overall system of untrusted oracle and trusted checker can be optimized by utilizing the oracle*

© Springer International Publishing AG 2017
M. Ayala-Rincón and C.A. Muñoz (Eds.): ITP 2017, LNCS 10499, pp. 164–170, 2017.
DOI: 10.1007/978-3-319-66107-0_11

maximally. We have identified a successful strategy for approaching this, which we feel deserves to be communicated.

Based on this observation, we propose a systematic methodology for using oracles in large-scale proofs (Sect. 2), modulizing the cooperation between the untrusted oracle and the certified checker. We identify the characteristics of problems whose proofs could and should profit from this methodology, and illustrate it (Sect. 3) using two cases: an optimality proof on sorting networks [6], which inspired this methodology, and a formalized checker for unsatisfiability proofs [7], obtained by directly applying this methodology. The latter ultimately was able to verify the 200 TB proof from [12], as described in [8].

2   Methodology

We first identify the characteristics a problem should have in order to benefit from our methodology. We motivate and illustrate these requirements by small toy examples. Then we present the methodology as a simple step-by-step procedure, with a focus on its separation of concerns.

Problem requirements. A common element to most oracles is that they relate to problems where finding proof witnesses is computationally much more expensive than checking their correctness. Because of this commonality, there is a significant pay-off in being able to write highly optimized untrusted code for finding witnesses.

Example 1. As an example, consider a proof that involves properties of the image of a function. Assume that in the process, one needs to find a pre-image of this object. If the function is easier to compute than its inverse, then an *ad hoc* way of finding pre-images can greatly improve performance.

A concrete extreme example is the inversion of a hash function, for which there is no realistic way of computing pre-images of given concrete hashes. An oracle might use large distributed Internet databases to try to find them, though. Such an oracle would by its nature be both incomplete (it fails when a pre-image exists, but is not in the database) and untrustworthy (the database could be erroneous), and therefore impossible to implement and prove correct inside a theorem prover such as Coq. However, its result is very simple to check.   □

Requirement 1 (Existential Subproblems). *The problem contains multiple occurrences of existential statements as subproblems, for which witness checking is computationally easier than witness generation.*

By "computationally easier", we simply mean that more efficient algorithmic solutions are known for one of the problems; we are not claiming that the problems provably belong to different complexity classes. If the condition in the requirement is met, this is an indication that the use of an oracle may be beneficial.

In general, a pre-computed oracle cannot be omniscient, since it can only provide finitely many different answers. Even if the problem domain is finite, it is still typically prohibitive to precompute all possible answers. Therefore, our methodology requires the set of problems that the oracle may be called upon to be sufficiently restricted (for the answers to fit into memory, for example).

Requirement 2 (Known Subproblems). *There is a manageable set of subproblems that contains all subproblems encountered during the proof.*

Our last requirement is that changes to the answers provided by the oracle should have an impact on the control flow and (consequently) on the efficiency of the remainder of the proof. We illustrate this point by an example.

Example 2. Imagine a proof that requires factorizing certain composite numbers into sorted lists of their prime factors as a recurring step. Suppose also that we have an efficient oracle that, given a composite number, delivers one of its prime factors. The oracle will have to be called multiple times in order to obtain the list of all factors, and this list has to be sorted (either at construction time or after obtaining all factors).

If we compute all prime factors, sort them, and have the oracle provide them in sorted order, we can replace the sorting step in the proof by a simple verification that the list provided is sorted, making the proof both simpler and more efficient. Note that this potentially changes the set of subproblems the oracle will be called upon, since it may change the control flow of the checker; a fact that needs to be taken into consideration in the implementation.   □

Requirement 3 (Data-Dependent Proof). *The structure of the proof is dependent on the answers provided by the oracle.*

In Example 2, this requirement would not be satisfied if the subproblems consisted of just showing that certain numbers were composite. The case studies in Sect. 3 illustrate all three requirements in realistic settings.

Step-by-step guide to verifying large proofs. We now describe our methodology for verifying large proofs that fit the requirements discussed above. This consists of four phases.

- *Formalize* the theory underlying the results
- *Implement* a naive checker (using an oracle) and prove it correct
- *Optimize* the checker in lock-step with adapting the oracle
- *Reprove* the correctness of the checker

In the *Formalize* phase, the focus is on the mathematical theory needed to prove the soundness of the algorithm used in the checker. The key aspect here is to separate concerns by *not* worrying about how these results will be used in the actual implementation. In other words, we advocate formalizing the theory as close as possible to the mathematical definitions and other formal elements of the algorithm.

In the *Implement* phase, the goal is to implement a checker as simple as possible. The algorithm of the checker should do as little work as possible, using the information in the proof witnesses as much as possible. This straightforward implementation must then be proven correct.

The *Optimize* phase is the most complex and most interesting one. In this phase, we analyze the complexity of the checker to determine possible local improvements. These can be of two kinds. The first kind is to use standard computer science techniques to optimize performance – for example, by using binary search trees instead of lists, or by enriching the proof witnesses to lower the complexity of checking their correctness. The second is to use the fact that all answers needed from the oracle are available beforehand to implement a more efficient algorithm, as illustrated by Example 2. In both cases, this may also require changes to the implementation of the oracle.

The *Reprove* phase consists of reproving the correctness of the optimized checker. This phase may be interleaved with the previous one, as optimizations tend to be localized and, thus, only require localized changes to the soundness proof. This is the case for optimizations of the implementation, in particular, where soundness is a property of the algorithm, and thus not significantly effected by the low-level choice of data structures. By applying one optimization at a time and immediately proving its soundness, it is easier to connect the necessary changes in the formalization to the original change in the algorithm.

The key observation in this methodology is the realization that the formalizations involved in different stages are of very different natures, and benefit from being treated separately. In the *Formalize* phase, the emphasis is on the underlying theory, and it will present the challenges present in any formalization – choosing the correct definitions, adapting proofs that are not directly formalizable, etc. In the *Implement* phase, the results that need to be formalized deal directly with the correctness of the algorithm being implemented, and will use the results proved in the earlier stage. Typically, the complexity of these proofs will arise from having to consider different cases or finding the right invariants throughout program execution, but not from mathematical issues pertaining to the underlying theory.

This is particularly relevant for the *Reprove* phase, where the modularity of the approach will have an impact in two ways. First, the formalization of the underlying theory for its own sake (rather than as a library tailored towards proving the correctness of the original algorithm) will make it more likely that all the needed results are readily available, and that they have been stated in forms making them more generically applicable. Second, changes to the algorithm will typically require different inductive proofs, but their correctness will likely use the same underlying arguments, which will already be available from previous phases. For example: if an algorithm iterating over lists is replaced by one iterating over trees, the structure of the soundness proof changes, but the key inductive argument (regarding how an additional element is processed) is unchanged. Therefore, the iterative steps in alternating *Optimize* and *Reprove* phases will likely be much simpler and faster than the original *Implement* phase.

As a consequence, the final library will also be more likely to be reusable in future proofs.

The requirements identified earlier are essential for this methodology to be meaningful. In general, existential subproblems indicate that using an untrusted oracle can be a good strategy, since verifying the proof witnesses it provides is easier than implementing a certified algorithm for computing them in the theorem prover. The known subproblems requirement guarantees that the oracle can pre-compute all proof witnesses that will be needed in the proof, so that they can be suitably processed before the checker is executed. Finally, data dependency ensures that changing the implementation of the oracle is meaningful, as it can improve the overall performance of the checker.

3   Case Studies

We illustrate our methodology by referring to two previously published formalizations. While we used Coq [1] as the theorem prover in both, our methodology should be portable to other formal generic proof environments.

Optimal sorting networks. In [5], we describe a computer-generated proof of the optimality of 25 comparisons for data-independently sorting 9 inputs. This proof is based on an exhaustive case analysis, made feasible by a powerful, but computationally expensive (NP-complete) subsumption relation [6]. A proof witness consists of two comparator networks and a permutation under which the relation holds. While known algorithms for solving the *existential subproblem* (by finding a permutation) have worst-case exponential runtime, checking the relation given a permutation is much easier.

The subsumption relation is used to eliminate comparator networks that are subsumed by others. The structure of the proof is thus highly *data-dependent*, with the order in which proof witnesses are provided by the oracle influencing the set of subproblems encountered during the remainder of the proof. This is a challenge for the *known subproblems* requirement, which is solved by oracle pre-processing based on transitivity of subsumption.

In the *Formalize* and *Implement* phases, we made a direct implementation of the algorithm in the original proof from [5], obtaining a checker with an expected runtime of 20 years to process the 27 GB of proof witnesses. In the *Optimize* and *Reprove* phases, we optimized this algorithm by changing the order in which the oracle provides proof witnesses, which allowed us to use asymptotically better algorithms and data structures. These optimizations reduced the execution time to just under 1 week [6]. Separating the formalization of the theory and the correctness proof of the checker meant that the cost of *Reprove* was marginal – at most 1 day per major change – compared to *Formalize*, which took approx. 3 months.

Unsatisfiability proofs. More and more applications are relying on SAT solvers for exhaustive case analysis, including computer-generated proofs [12,13]. While formally verifying satisfiability results is trivial given a witness, verifying unsatisfiability results returned by untrusted solvers requires showing that the original formula entails the empty clause. To this end, many SAT solvers provide the entailed clauses they learn during execution as proof witnesses. Finding meaningful such clauses is clearly a non-trivial *existential subproblem*.

To check an unsatisfiability proof, the clauses provided by the oracle are added to the original set of clauses after their entailment has been checked by reverse unit propagation. This addition of clauses determines which further clauses can be added, i.e., the structure of the proof is *data-dependent*. Since the proof checker simply follows the information provided by the oracle, the *known subproblems* requirement is trivially satisfied.

By applying our methodology directly, we were able to improve the state-of-the-art of formally verifying unsatisfiability [10,16] by several orders of magnitude. Here, the *Formalize* phase consisted simply of building a deep encoding of propositional logic in the theorem prover Coq and defining notions of entailment and satisfiability, and the *Implement* phase yielded a simple checker based on reverse unit propagation. In the *Optimize* phase, we achieved a performance improvement of several orders of magnitude by observing that the core algorithm for checking reverse unit propagation can be simplified significantly by enriching the proof witnesses with information about the clauses used [7]. This improvement in the performance of the checker was obtained at the cost of a noticeable (yet manageable) increase in computation time on the oracle side, due to the need to enrich the proof witnesses, but this shift ultimately allowed us to verify the 200 TB proof from [12], as described in [7,8].

4   Concluding Remarks

We have introduced a methodology based on distilling the key idea behind our two case studies: the overall system of proof checker and oracle can profit from shifting the computational burden from the trusted, inefficient proof checker to the untrusted, efficient oracle implementation. In other words, we let the proof checker be implemented as efficiently as possible, doing as little work as possible, while pre-processing the oracle information such that the right amount information was provided in the right order. Since all the data provided by the oracle is verified by the proof checker, this does not affect the reliability of the results. By revisiting the case studies in this unifying presentation, we hope to inspire others to obtain similar performance gains when formally verifying other large-scale proofs.

Acknowledgments. We would like to thank Pierre Letouzey for his suggestions and help with making our extracted code more efficient.

The authors were supported by the Danish Council for Independent Research, Natural Sciences, grant DFF-1323-00247, and by the Open Data Experimentarium at the University of Southern Denmark. Computational resources were generously provided by the Danish Center for Scientific Computing.

References

1. Bertot, Y., Castéran, P.: Interactive Theorem Proving and Program Development. Texts in Theoretical Computer Science. Springer, Heidelberg (2004)
2. Blanqui, F., Koprowski, A.: CoLoR: a Coq library on well-founded rewrite relations and its application to the automated verification of termination certificates. Math. Struct. Comp. Sci. **21**, 827–859 (2011)
3. Blazy, S., Paulin-Mohring, C., Pichardie, D. (eds.): ITP 2013. LNCS, vol. 7998. Springer, Heidelberg (2013)
4. Claret, G., González-Huesca, L., Régis-Gianas, Y., Ziliani, B.: Lightweight proof by reflection using a posteriori simulation of effectful computation. In: Blazy et al. [3], pp. 67–83
5. Codish, M., Cruz-Filipe, L., Frank, M., Schneider-Kamp, P.: Sorting nine inputs requires twenty-five comparisons. J. Comput. Syst. Sci. **82**(3), 551–563 (2016)
6. Cruz-Filipe, L., Larsen, K.S., Schneider-Kamp, P.: Formally proving size optimality of sorting networks. J. Autom. Reason. Accepted for publication. doi:10.1007/s10817-017-9405-9
7. Cruz-Filipe, L., Marques-Silva, J., Schneider-Kamp, P.: Efficient certified resolution proof checking. In: Legay, A., Margaria, T. (eds.) TACAS 2017. LNCS, vol. 10205, pp. 118–135. Springer, Heidelberg (2017). doi:10.1007/978-3-662-54577-5_7
8. Cruz-Filipe, L., Schneider-Kamp, P.: Formally proving the boolean triples conjecture. In: Eiter, T., Sands, D. (eds.) LPAR-21. EPiC Series in Computing, vol. 46, pp. 509–522. EasyChair Publications (2017)
9. Cruz-Filipe, L., Wiedijk, F.: Hierarchical reflection. In: Slind, K., Bunker, A., Gopalakrishnan, G. (eds.) TPHOLs 2004. LNCS, vol. 3223, pp. 66–81. Springer, Heidelberg (2004). doi:10.1007/978-3-540-30142-4_5
10. Darbari, A., Fischer, B., Marques-Silva, J.: Industrial-strength certified SAT solving through verified SAT proof checking. In: Cavalcanti, A., Deharbe, D., Gaudel, M.-C., Woodcock, J. (eds.) ICTAC 2010. LNCS, vol. 6255, pp. 260–274. Springer, Heidelberg (2010). doi:10.1007/978-3-642-14808-8_18
11. Fouilhé, A., Monniaux, D., Périn, M.: Efficient generation of correctness certificates for the abstract domain of polyhedra. In: Logozzo, F., Fähndrich, M. (eds.) SAS 2013. LNCS, vol. 7935, pp. 345–365. Springer, Heidelberg (2013). doi:10.1007/978-3-642-38856-9_19
12. Heule, M.J.H., Kullmann, O., Marek, V.W.: Solving and verifying the boolean pythagorean triples problem via cube-and-conquer. In: Creignou, N., Le Berre, D. (eds.) SAT 2016. LNCS, vol. 9710, pp. 228–245. Springer, Cham (2016). doi:10.1007/978-3-319-40970-2_15
13. Konev, B., Lisitsa, A.: A SAT attack on the Erdős discrepancy conjecture. In: Sinz, C., Egly, U. (eds.) SAT 2014. LNCS, vol. 8561, pp. 219–226. Springer, Cham (2014). doi:10.1007/978-3-319-09284-3_17
14. Leroy, X.: Formal verification of a realistic compiler. Commun. ACM **52**(7), 107–115 (2009)
15. Sternagel, C., Thiemann, R.: The certification problem format. In: Benzmüller, C., Paleo, B. (eds.) UITP, EPTCS, vol. 167, pp. 61–72 (2014)
16. Wetzler, N.D., Heule, M.J.H., Hunt Jr., W.A.: Mechanical verification of SAT refutations with extended resolution. In: Blazy et al. [3], pp. 229–244

Certifying Standard and Stratified Datalog Inference Engines in SSReflect

Véronique Benzaken[1], Évelyne Contejean[2], and Stefania Dumbrava[3(✉)]

[1] Université Paris Sud, LRI, 91 405 Orsay, France
veronique.benzaken@u-psud.fr
[2] CNRS, LRI, Université Paris Sud, 91 405 Orsay, France
evelyne.contejean@lri.fr
[3] LIRIS, Université Lyon 1, Lyon, France
stefania-gabriela.dumbrava@univ-lyon1.fr

Abstract. We propose a SSReflect library for logic programming in the Datalog setting. As part of this work, we give a first mechanization of standard Datalog and of its extension with stratified negation. The library contains a formalization of the model theoretical and fixpoint semantics of the languages, implemented through bottom-up and, respectively, through stratified evaluation procedures. We provide corresponding soundness, termination, completeness and model minimality proofs. To this end, we rely on the Coq proof assistant and SSReflect. In this context, we also construct a preliminary framework for dealing with stratified programs. We consider this to be a necessary first step towards the certification of security-aware data-centric applications.

1 Introduction

Datalog [7] is a deductive language capturing the function-free fragment of Horn predicate logic. Syntactically a subset of Prolog [22], Datalog has the advantage of guaranteed termination (in PTIME [33]). Expressivity-wise, it extends relational algebra/calculus with recursion; also, it allows for computing transitive closures and, generally, for compactly formulating graph traversals and analytics. Comparatively, more popular query languages, such as SQL, XPath, or SPARQL, are either not capable of expressing these directly/inherently or do so with limitations. Given the present ubiquity of interconnected data, seamlessly supporting such recursive queries has regained relevance. For example, these are key to Web infrastructure, being used by web-crawlers and PageRank algorithms. Efficiently querying and reasoning about graph topologies is, in fact, fundamental in various areas, including, but not limited to: the Semantic Web; social, communication and biological networks; and geographical databases.

Due to its purely declarative nature and simplicity (few primitives), Datalog lends itself particularly well to domain-specific extensions. As surveyed in the literature [2,28], multiple additions to its core language have been introduced, e.g.,

This work was supported by the Datacert project (ANR-15-CE39-0009) of the French ANR.

M. Ayala-Rincón and C.A. Muñoz (Eds.): ITP 2017, LNCS 10499, pp. 171–188, 2017.
DOI: 10.1007/978-3-319-66107-0_12

with negation, existentials, aggregates, functions, updates, etc. Indeed, Datalog can be seen as the lingua franca for a plethora of custom query languages, e.g., Starlog [25], Overlog [24], Netlog [17], DATALOG± [6], SociaLite [30], LogiQL [4], etc. In a recent resurge of interest [1], marked by the *Datalog 2.0 Workshop*, such tailored versions of Datalog found new applications in data integration, security [10], program analysis [34], cloud computing, parallel and distributed programming [18], etc. An overview is given in [19].

These applications have not only sparked interest in the academic setting, but also in industry. Indeed, commercial Datalog engines have started to gain popularity, with LogicBlox [23], Google's Yedalog [8], Datomic [9], Exeura [12], Seemle [29], and Lixto [16], as prominent examples. Moreover, their scope has extended to include safety-critical, large-scale use cases. A case in point is the LogicBlox platform, which underpins high-value web retail and insurance applications. Its Datalog-based engine unifies the modern enterprise software stack (encompassing bookkeeping, analytics, planning, and forecasting) and runs with impressive efficiency [5]. Also, more recently, Datalog has been proposed as tool for automating the verification of railway infrastructure high-security regulations against its CAD design [26].

We argue that, given the role Datalog is starting to play in data-centric and security-sensitive applications, obtaining the strong guarantees Coq certification provide is an important endeavor. We envisage a methodology aimed at ultimately certifying a realistic Datalog engine by *refinement*. This would encompass: (1) a high-level formalization suitable for proof-development and, thus, employing more inefficient algorithms, (2) a mechanization of the real-world engine implementation, and (3) (refinement) proofs of their extensional equivalence.

This paper describes the first necessary step towards realizing this vision. As such, we propose a deep specification of a stratified Datalog inference engine in the SSReflect extension [15] of the Coq proof-assistant [27]. With respect to the scope of our formalization, the chosen fragment is the one used by LogicBlox and it is the most expressive one that retains termination[1]. Our chosen evaluation heuristic is bottom-up, as ensuring that top-down/more optimized heuristics do not get stuck in infinite loops is harder to establish. Also, this allows us to modularly extend and reuse our standard Datalog inference engine in the stratified setting. We do envisage supporting, for example, magic-sets rewriting.

The choice of using SSReflect is due to the fact that the model-theoretic semantics of Datalog is deeply rooted in *finite model theory*. To quote [21]: "*For many years, finite model theory was viewed as the backbone of database theory, and database theory in turn supplied finite model theory with key motivations and problems. By now, finite model theory has built a large arsenal of tools that can easily be used by database theoreticians without going to the basics*". The Mathematical Components library[2], built on top of SSReflect, is especially well-suited for our purposes, as it was the basis of extensive formalizations of finite model

---

[1] Arithmetic predicates and skolem functions destroy this guarantee.
[2] http://math-comp.github.io/math-comp/.

theory, in the context of proving the Feit-Thompson theorem [14], central to finite group classification. Moreover, as detailed next, our proof-engineering efforts were greatly eased by our reuse of the `fintype`, `finset` and `bigop` libraries.

Contributions. Our key modeling choice is properly setting up the base types to make the most of the finite machinery of SSReflect. Heavily relying on type finiteness ensures desirable properties, such as decidability. As every Datalog program has a finite model [2], i.e., its Herbrand Base (Sect. 2.1), this does not restrict generality. The paper's contributions are, as documented online in http://datacert.lri.fr/datalogcert/datalogcert.html:

1. a certified "positive" inference engine for standard Datalog: We give a scalable formalization of the syntax, semantics and bottom-up inference of Datalog. The latter consists of mechanizing a matching algorithm for terms, atoms and clause bodies and proving corresponding soundness and completeness results. We formally characterize the engine, by establishing *soundness, termination, completeness* and *model minimality*, based on *monotonicity, boundedness* and *stability* proofs.
2. a certified "negative" inference engine for stratified Datalog: We extend the syntax and semantics of Datalog with negation and mechanize its stratified evaluation. We model program stratification and "slicing", embed negated literals as flagged positive atoms and extend the notion of an interpretation to that of a "complemented interpretation". The crux of stratified evaluation is the reuse of the "positive" engine, for each program "slice". When formally characterizing the "negative engine", this required us to precisely identify additional properties, i.e., *incrementality* and *modularity*, and to correspondingly extend the previous library. We establish *soundness, termination, completeness* and *model minimality*.

Lastly, we extract our *certified* standard Datalog engine in OCaml as a proof-of-concept. This is a first step towards building a realistic (high-performance), correct-by-construction Datalog engine. To realize this, we plan to replace the generic SSReflect algorithms used in our implementation by more efficient ones and to prove that the results we established are preserved.

Organization. The paper is organized as follows. In Sect. 2, we give a concise theoretical summary of standard and stratified Datalog. In Sects. 3 and 4, we present the corresponding SSReflect inference engine mechanizations. Section 5 describes related work. We conclude in Sect. 6.

2 Preliminaries

We review the theory of standard and stratified Datalog in Sects. 2.1 and 2.2.

2.1 Standard Datalog

Syntax. `Given the sets $\mathcal{V}$, $\mathcal{C}$ and $\mathcal{P}$ of variables, constants and predicate symbols, a program is a *finite* collection of clauses, as captured by the grammar:[3]

Programs	P	$::=$	$C_1, \ldots, C_k$
Clauses	C	$::=$	$A_0 \leftarrow A_1, \ldots, A_m$
Atoms	A	$::=$	$p(\vec{t})$, where $p \in \mathcal{P}$ is denoted $sym(A)$ and has arity $ar(p) = |\vec{t}|$
Terms	t	$::=$	$x \in \mathcal{V} \mid c \in \mathcal{C}$

A clause is a sentence separating hypotheses *body* atoms from the conclusion *head* atom. Clauses allow inferring new facts (true ground atoms) from existing ones. The restriction below ensures *finitely* many facts are inferred.

Definition 1 (Safety). *A standard Datalog program is safe iff all of its clauses are safe, i.e., all of their head variables appear in their body. Consequently, safe program facts are ground*[4].

Semantics. Let $\mathbf{B}(P)$ be the Herbrand Base of P, i.e., the ground atom set built from its predicates and constants. By the Herbrand semantics, an *interpretation* I is a subset of $\mathbf{B}(P)$. For a valuation ν, mapping clause variables to program constants[5], and a clause C equal to $p_0(\vec{t_0}) \leftarrow p_1(\vec{t_1}), \ldots, p_m(\vec{t_m})$, the *clause grounding*[6] νC is $p_0(\nu\vec{t_0}) \leftarrow p_1(\nu\vec{t_1}), \ldots, p_m(\nu\vec{t_m})$. Note that variables in clauses are implicitly universally quantified. C is satisfied by I iff, *for all* valuations ν, if $\{p_1(\nu\vec{t_1}), \ldots, p_m(\nu\vec{t_m})\} \subseteq I$ then $p_0(\nu\vec{t_0}) \in I$. I is a *model* of P iff all clauses in P are satisfied by I. The *intended semantics* of P is $\mathbf{M_P}$, its *minimal model* w.r.t set inclusion. This model-theoretic semantics indicates *when* an interpretation is a model, but not *how* to construct such a model. Its *computational* counterpart centers on the *least fixpoint* of the following operator.

Definition 2 (The T_P Consequence Operator). *Let P be a program and I an interpretation. The T_P operator is the set of program consequences F: $T_P(I) = I \cup \{F \in \mathbf{B}(P) \mid F = hd(\nu C), \text{for } C \in P, \nu : \mathcal{V} \rightarrow \mathcal{C}, bd(\nu C) \subseteq I\}$.*

Definition 3 (Fixpoint Evaluation). *The iterations of the T_P operator are: $T_P \uparrow 0 = \emptyset$, $T_P \uparrow (n+1) = T_P(T_P \uparrow n)$. Since T_P is monotonous and bound by $\mathbf{B}(P)$, the Knaster-Tarski theorem [31] ensures $\exists \omega, T_P \uparrow \omega = \bigcup_{n \geq 0} T_P \uparrow n$, where $T_P \uparrow \omega = lfp(T_P)$. The* least fixpoint evaluation *of P is thus defined as $lfp(T_P)$.*

Note that, by Van Emden and Kowalski [32], $lfp(T_P) = \mathbf{M_P}$.

---

[3] Term sequences $t_1, \ldots, t_n$ are abbreviated as $\vec{t}$ and $|\vec{t}| = n$ denotes their length.
[4] We call language constructs that are variable-free, *ground* and, otherwise, *open*.
[5] The set of program constants is also called its *active domain*, denoted $adom(P)$.
[6] Also called *clause instantiation*.

Example 1. Let

$$P = \begin{cases} e(1,3).\ e(2,1).\ e(4,2).\ e(2,4). \\ t(X,Y) \leftarrow e(X,Y). \\ t(X,Y) \leftarrow e(X,Z), t(Z,Y). \end{cases}$$

$T_P \uparrow 0 = \emptyset$; $T_P \uparrow 1 = \{e(1,3), e(2,1), e(4,2), e(2,4)\}$; $T_P \uparrow 2 = T_P \uparrow 1 \cup \{t(1,3), t(2,1), t(4,2), t(2,4)\}$; $T_P \uparrow 3 = T_P \uparrow 2 \cup \{t(2,3), t(4,1), t(4,4), t(2,2)\}$. The minimal model of P is $\mathbf{M_P} = lfp(T_P) = T_P \uparrow 4 = T_P \uparrow 3 \cup \{t(4,3)\}$.

2.2   Stratified Datalog

Syntax. Adding stratified negation amounts to extending the syntax of standard Datalog, by introducing *literals* and adjusting the definition for clauses.

$$\begin{array}{llll} Clauses & C & ::= & A \leftarrow L_1, \ldots, L_m \\ Literals & L & ::= & A \mid \neg A \end{array}$$

Definition 4 (Predicate Definitions). *Let P be a program. The definition define(p) of program predicate $p \in \mathcal{P}$ is $\{C \in P \mid sym(hd(C)) = p\}$.*

Definition 5 (Program Stratification and Slicing). *Let P be a program with clauses C of the form $H \leftarrow A_1, \ldots, A_k, \neg A_{k+1}, \ldots, \neg A_l$, where $bd^+(C) = \{A_1, \ldots, A_k\}$ and $bd^-(C) = \{A_{k+1}, \ldots, A_l\}$. Consider a mapping $\sigma : \mathcal{P} \to \mathbb{N}$, such that: (1) $\sigma(sym(A_j)) \leq \sigma(sym(H))$, for $j \in [1, k]$, and (2) $\sigma(sym(A_j)) < \sigma(sym(H))$, for $j \in [k+1, l]$. σ induces a partitioning[7] $P = \bigsqcup_{j \in [1,n]} P_{\sigma_j}$ with*

$\sigma_j = \{p \in \mathcal{P} \mid \sigma(p) = j\}$ *and* $P_{\sigma_j} = \bigcup_{p \in \sigma_j} define(p)$. *We have that, for $C \in P_{\sigma_j}$:*

(1) if $p \in \{sym(A) \mid A \in bd^-(C)\}$, then $define(p) \subseteq \bigcup_{1 \leq k < j} P_{\sigma_k}$, and (2) if $p \in \{sym(A) \mid A \in bd^+(C)\}$, then $define(p) \subseteq \bigcup_{1 \leq k \leq j} P_{\sigma_k}$.

We call P *stratified*; σ, *a stratification*; σ_j, *a stratum*; the set $\{P_{\sigma_1}, \ldots, P_{\sigma_n}\}$, *a program slicing*[8] *and* P_{σ_j}, *a program slice, henceforth denoted* P_j.

Stratification ensures program slices P_j are *semipositive* programs [2] that can be evaluated *independently*. Indeed, checking if their negated atoms belong to some interpretation I is equivalent to checking that their positive counterparts belong to the *complement* of I w.r.t the Herbrand Base $\mathbf{B}(P_j)$.

Semantics. The model of a stratified Datalog program is given by the step-wise, bottom-up computation of the least fixpoint model for each of its slices.

Definition 6 (Stratified Evaluation). *For $P = P_1 \sqcup \ldots \sqcup P_n$, the model[9], $M_n = T_{P_n} \uparrow \omega(M_{n-1})$[10], with $M_j = T_{P_j} \uparrow \omega(M_{j-1})$, $j \in [2, n]$, $M_1 = T_{P_1} \uparrow \omega(\emptyset)$.*

---

[7] $\sqcup$ denotes the pairwise disjoint set union.

[8] A program can have multiple stratifications.

[9] As proven by Apt et al. [3], M_n is *independent from the choice of stratification.*

[10] By abuse of notation, we use the same ω for the different numbers of T_P iterations needed to reach a fixpoint, when evaluating each program slice.

Example 2. Let

$$P = \begin{cases} q(a).\ s(b).\ t(a).\ r(X) \leftarrow t(X). \\ p(X) \leftarrow \neg q(X), r(X). \\ p(X) \leftarrow \neg t(X), q(X). \\ q(X) \leftarrow s(X), \neg t(X). \end{cases}$$

for which a *stratification* $\sigma(s) = 1$, $\sigma(t) = 1$, $\sigma(r) = 1$, $\sigma(q) = 2$, $\sigma(p) = 3$, with the *strata* $\sigma_1 = \{s, t, r\}$, $\sigma_2 = \{q\}$, $\sigma_3 = \{p\}$, induces the partitioning $P = P_1 \sqcup P_2 \sqcup P_3$, with the *slices*

$$P_1 = \begin{cases} s(b).\ t(a). \\ r(X) \leftarrow t(X). \end{cases} \quad P_2 = \begin{cases} q(a). \\ q(X) \leftarrow s(X), \neg t(X). \end{cases} \quad P_3 = \begin{cases} p(X) \leftarrow \neg q(X), r(X). \\ p(X) \leftarrow \neg t(X), q(X). \end{cases}$$

$M_1 = T_{P_1} \uparrow \omega(\emptyset) = \{r(a), s(b), t(a)\}$; $M_2 = T_{P_2} \uparrow \omega(M_1) = M_1 \cup \{q(a), q(b)\}$; $\mathbf{M_P} = M_3 = T_{P_0} \uparrow \omega(M_2) = M_2 \cup \{p(b)\} = \{r(a), s(b), t(a), q(a), q(b), p(b)\}$.

3   A Mechanized Standard Datalog Engine

In Sect. 3.1, we present our formalization of the syntax and semantics of standard Datalog. Next, in Sect. 3.2, we detail the bottom-up evaluation heuristic of its inference engine. We formally characterize the engine in Sect. 3.3.

3.1   Formalizing Standard Datalog

Syntax. We assume the `symtype` and `constype` finite types for predicate *symbols* and *constants*, as well as an `arity` finitely-supported function (`ffun`) that associates a corresponding positive value to each symbol.

Variables (`symtype constype : finType`) (`arity : {ffun symtype → nat}`).

Terms are encoded by an inductive joining (1) variables, of ordinal type `'I_n`, bound by a computable maximal value n, and (2) constants.

```
Inductive constant := C of constype.
Inductive term : Type := Var of 'I_n | Val of constant.
```

To avoid redundant case analyses, we henceforth distinguish between *ground* and *open* (non-ground) atoms and clauses. Intuitively, this dichotomy is desirable as the former are primitives of the *semantics*, while the latter, of the *syntax*. As such, *ground atoms* are modeled with `gatom` records, joining the `rgatom` base type and the boolean well-formedness condition `wf_rgatom`. The first packs a symbol and its arguments, i.e., a list (`seq` in SSReflect notation) of *constants*; the second ensures symbol arity and argument size match. Note that, as we set up the `gatom` subtyping predicate to be inherently proof-irrelevant, checking ground atom equality can be conveniently reduced to checking the equality of their underlying base types. *Atoms* are encoded similarly, except that their base type packs, as an argument, a *term* list instead.

```
Inductive  rgatom := RawGAtom of symtype & seq constant.
Definition wf_rgatom rga := size (arg rga) == arity (sym rga).
Structure gatom := GAtom {rga :> rgatom; _ : wf_rgatom rga}.
```

(Ground) clauses pack a distinguished (ground) atom, (head_gcl), respectively, head_cl, and a (ground) atom list, (body_gcl), respectively, body_cl. *Programs* are clause lists. The safety condition formalization mirrors Definition 1.

Semantics. An *interpretation* i is a finite set of ground atoms. Note that, since its type, interp, is finite, the latter has a lattice structure, whose top element, setT, is the set of all possible ground atoms. The *satisfiability* of a ground clause gcl w.r.t i is encoded by gcl_true. As in Sect. 2.1, we define i to be a *model* of a program p, if, for all grounding substitutions ν, it satisfies all corresponding clause instantiations, gr_c ν cl. We discuss the encoding of grounding substitutions next.

```
Notation interp := {set gatom}.
Definition gcl_true gcl i := (* i satisfies gcl *)
  all (mem i) (body_gcl gcl) ==⇒ (head_gcl gcl ∈ i).
Definition prog_true p i := (* i is a model of p *)
  ∀ν : gr, all (fun cl ⇒ gcl_true (gr_cl ν cl) i) p.
```

3.2 Mechanizing the Bottom-Up Evaluation Engine

The inference engine iterates the logical consequence operator from Definition 2. To build a model of an input program, it maintains a current "candidate model" interpretation, which it iteratively tries to "repair". The repair process first identifies clauses that violate satisfiability, i.e., whose ground instance bodies are in the current interpretation, but whose heads are not. The current interpretation is then "fixed", adding to it the missing facts, i.e., the head groundings. This is done by a *matching algorithm*, incrementally constructing *substitutions* that homogeneously instantiate all clause body atoms to "candidate model" facts. As safety ensures all head variables appear in the body, these substitutions are indeed grounding. Hence, applying them to the head produces *new facts*. Once the current interpretation is "updated" with all facts inferable in one forward chain step, the procedure is repeated, until a fixpoint is reached. We prove this to be a *minimal model* of the input program. As outlined, the mechanization of the engine centers around the encoding of *substitutions* and of *matching* functions.

Groundings and Substitutions. Following a similar reasoning to that in Sect. 3.1, we define a separate type for grounding substitutions (groundings). Both groundings and substitutions are modeled as finitely-supported functions from variables to constants[11], except for the latter being partial[12].

```
Definition gr  := {ffun 'I_n → constant}.
Definition sub := {ffun 'I_n → option constant}.
```

---

[11] Since Datalog does not have function symbols and interpretations are ground, we can restrict substitution codomains to the set of program constants, w.l.o.g.

[12] Groundings can be coerced to substitutions and substitutions can be lifted to groundings, by padding with a default element def.

We account for the engine's gradual extension of substitutions, by introducing a *partial ordering*[13] over these. To this end, using finitely-supported functions was particularly convenient, as they can be used both as functions and as lists of bindings. We say a substitution σ_2 *extends* a substitution σ_1, if all variables bound by σ_1 appear in σ_2, bound to the same values. We model this predicate as sub_st[14] and the *extension* of a substitution σ, as the add finitely-supported function.

```
Definition sub_st σ₁ σ₂ := (* henceforth denoted as σ₁ ⊆ σ₂ *)
  [∀v : 'I_n, if σ₁ v is Some c then (v, c) ∈ σ₂ else true].
Definition add σ v c :=
  [ffun u ⇒ if u == v then Some c else σ u].
```

Term Matching. Matching a term t to a constant d under a substitution σ, will either: (1) return the *input substitution*, if t or $\sigma\, t$ equal d, (2) return the *extension* of σ with the corresponding binding, if t is a variable not previously bound in σ, or (3) *fail*, if t or $\sigma\, t$ differ from d.

```
Definition match_term d t σ : option sub :=
  match t with
  | Val e ⇒ if d == e then Some σ else None
  | Var v ⇒ if σ v is Some e
              then (if d == e then Some σ else None)
              else Some (add σ v d)
  end.
```

Atom Matching. We define the match_atom and match_atom_all functions that return substitutions and, respectively, substitution sets, instantiating an atom to a ground atom and, respectively, to an interpretation. To compute the substitution matching a raw-atom ra to a ground one rga, we first check their symbols and argument sizes agree. If such, we extend the initial substitution σ, by iterating term matching over the item-wise pairing of their terms zip arg2 arg1. As term matching can fail, we wrap the function with an option binder extracting the corresponding variable assignments, if any. Hence, match_raw_atom is a monadic option fold that either fails or returns substitutions extending σ. Atom matching equals raw atom matching, by coercion to raw_atom.

```
Definition match_raw_atom rga ra σ : option sub :=
  match ra, rga with RawAtom s1 arg1, RawGAtom s2 arg2 ⇒
    if (s1 == s2) && (size arg1 == size arg2)
    then foldl (fun acc p ⇒ obind (match_term p.1 p.2) acc)
           (Some σ) (zip arg2 arg1)
    else None
  end.
Definition match_atom σ a ga := match_raw_atom σ a ga.
```

---

[13] We establish corresponding reflexivity, antisymmetry and transitivity properties.
[14] We use the boolean quantifier, as the ordinal type of variables is finite.

Next, we compute the substitutions that can match an atom a to a fact in an interpretation i. This is formalized as the set of substitutions σ that belong to the set gathering all substitutions matching a to ground atoms ga in i.

```
Definition match_atom_all i a σ :=
  [set σ' | Some σ' ∈ [set match_atom ga a σ | ga ∈ i]].
```

While the match_term and match_atom functions are written as Gallina algorithms, we were able to cast the match_atom_all algorithm mathematically as: $\{\sigma' \mid \sigma' \in \{\text{match\_atom ga a }\sigma \mid \text{ga} \in \text{i}\}\}$. The function is key for expressing forward chain and fixpoint evaluation. Hence, its declarative, high-level implementation propagates to all the underlying functions of the bottom-up engine. We could thus "reduce" soundness and completeness proofs to set theory ones. In this setting, it was particularly convenient we could rely on **finset** properties.

Body Matching. The match_body function extends an initial substitution set ssb with bindings matching all atoms in the atom list tl, to an interpretation i. These are built using match_atom_all and *uniformly* extending substitutions matching each atom to i. We model this based on our definition of foldS, a monadic fold for the set monad. This iteratively composes the applications of a seeded function to all the elements of a list, flattening intermediate outputs.

```
Definition match_body i tl ssb := foldS (match_atom_all i) ssb tl.
```

The T_P Consequence Operator. We model the logical consequences of a clause cl w.r.t an interpretation i as the set of *new* facts inferable from cl by matching its body to i. Such a fact, gr_atom_def def σ (head_cl cl), is the head instantiation with the *grounding* matching substitution σ[15].

```
Definition emptysub : sub        := [ffun _ ⇒ None].
Definition cons_clause def cl i :=
  [set gr_atom_def def σ (head_cl cl) |
       σ ∈ match_body i (body_cl cl) [set emptysub]].
```

One-Step Forward Chain. One inference engine iteration computes the set of *all* consequences inferable from a program p and an interpretation i. This amounts to taking the union of i with all the program clause consequences. The encoding mirrors the mathematical expression i $\cup$ $\bigcup_{cl \in p}$ cons_clause def i cl[16].

```
Definition fwd_chain def p i :=
  i ∪ \bigcup_(cl ← p) cons_clause def cl i.
```

---

[15] gr_atom_def lifts substitutions to groundings, by padding with the **def** constant.
[16] Thanks to using the **bigcup** operator from the SSReflect **bigop** library.

3.3   Formal Characterization of the Bottom-Up Evaluation Engine

We first state the main intermediate theorems, leading up to the key Theorem 7. The first two results are established based on analogous ones for terms and atoms. We assume an interpretation i and a seed substitution set ssb.

Theorem 1 (Matching Soundness). *Let* tl *be an atom list. If a substitution* σ *is in the output of* match_body, *extending* ssb *with bindings matching* tl *to* i, *then there exists a ground atom list* gtl *such that: (1)* gtl *is the instantiation of* tl *with* σ *and (2) all* gtl *atoms belong to* i.
Proof by induction on tl.

Theorem 2 (Matching Completeness). *Let* cl *be a clause and* ν *a grounding* compatible *with any substitution* σ *in* ssb. *If* ν *makes the body of* gcl *true in* i, *then* match_body *outputs a compatible substitution smaller or equal to* ν.
Proof by induction on tl.

Theorem 3 (T_P Stability). *Let* cl *be a clause and* i *an interpretation satisfying it. The facts inferred by* cons_clause *are in* i.
Proof by Theorem 1.

Theorem 4 (T_P Soundness). *Let* cl *be a safe clause and* i *an interpretation. If the facts inferred by* cons_clause *are in* i, *then* i *is a model of* cl.
Proof by Theorems 2 and 3.

Theorem 5 (Forward Chain Stability and Soundness). *Let* p *be a safe program. Then, an interpretation* i *is a model of* p *iff it is a* fwd_chain *fixpoint.*[17]
Proof by Theorems 3 and 4.

Theorem 6 (Forward Chain Fixpoint Properties). *The* fwd_chain *function is* monotonous, increasing *and* bound *by* $\mathbf{B}(P)$.
Proof by compositionality of set-theoretical properties.

Theorem 7 (Bottom-up Evaluation Soundness and Completeness). *Let* p *be a safe program. By iterating forward chain as many times as there are elements in* $\mathbf{B}(P)$, *the engine terminates and outputs a* minimal *model for* p.
Proof by Theorems 5 and 6, using a corollary of the Knaster-Tarski result, as established in Coq by [11].

4   A Mechanized Stratified Datalog Engine

We summarize the formalization of the syntax and semantics of stratified Datalog in Sect. 4.1. In Sect. 4.2 we present the mechanization of the stratified Datalog engine. We outline its formal characterization in Sect. 4.3.

---

[17] We state this as the fwd_chainP reflection lemma.

4.1 Formalizing Stratified Datalog

Syntax. We extend the syntax of positive Datalog with *literals*, reusing the definitions of ground/non-ground atoms. As before, we distinguish ground/non-ground literals and clauses. The former are encoded enriching ground/non-ground atoms with a boolean flag, marking whether they are negated.

```
Inductive glit := GLit of bool * gatom.
Inductive lit  := Lit  of bool * atom.
```

(Ground) clauses pack (ground) atoms and (ground) literal lists. The encodings of programs and their safety condition are the same as in Sect. 3.1.

Semantics. The only additions to Sect. 3.1 concern ground literals and clauses. The `glit_true` definition captures the fact that an interpretation i satisfies a ground literal `gl`, by casing on the latter's flag. If it is true, i.e. the literal is positive, we check if the underlying ground atom is in i; otherwise, validity holds if the underlying ground atom is *not* in i. The definition for the satisfiability of a ground clause w.r.t i is analogous to that given in Sect. 3.1.

```
Definition gatom_glit  gl := let: GLit (_, ga) := gl in ga.
Definition flag_glit   gl := let: GLit (b, _)  := gl in b.
Definition glit_true i gl := if flag_glit gl then gatom_glit gl ∈ i
                                             else gatom_glit gl ∉ i.
```

4.2 Mechanizing the Stratified Evaluation Engine

Stratification. We model a stratification as a list of symbol sets, implicitly assuming the first element to be its lowest stratum. As captured by `is_strata_rec`, the characteristic properties mirror those in Definition 5. Namely, these are (1) *disjointness*: no two strata share symbols, (2) *negative-dependency*: stratum symbols can only refer to negated symbols in strictly lower strata, and (3) *positive-dependency*: stratum symbols only depend on symbols from lower or equal strata. We can give an effective, albeit inefficient algorithm for computing a stratification satisfying the above, by exploring the finite set of all possible program stratifications. Hence, we use the finite search infrastructure of SSReflect, i.e., the [pick e : T | P e] construct that, among all inhabitants of a finite type T, when possible, picks an element e, satisfying a predicate P.

Positive Embedding. To enable the reuse of the forward chain operator in Sect. 3.2, we will embed the Coq representation of Datalog programs with stratified negation into that of standard Datalog programs, used by the positive engine. This is realized via functions that *encode/decode* constructs *to/from* their "positive" counterparts; we denote these as $\ulcorner \cdot \urcorner / \llcorner \cdot \lrcorner$. To the end, we augment symbol types with a boolean flag, marking if the original atom is negated. For example, $\ulcorner s(a) \urcorner = (s, \top)(a)$, $\ulcorner \neg s(a) \urcorner = (s, \bot)(a)$, $\llcorner (\top, s)(a) \lrcorner = s(a)$ and $\llcorner (\bot, s)(a) \lrcorner = \neg s(a)$. We show literal encoding/decoding are inverse w.r.t each other and, hence, injective, by proving the corresponding cancellation lemmas. For clauses, encoding is inverse to decoding and, hence, injective, only when the flag of its encoded head atom is positive. This is expressed by a *partial* cancellation lemma; for the converse direction the cancellation lemma holds. Based on these injectivity properties, we prove Theorem 9.

Stratified Evaluation. Let p be a program and str, a strata. The evalp stratified evaluation of p traverses str, accumulating the processed strata, $str_<$. It then computes the minimal model, cf. Theorem 7, for each program slice, $p_{str_<}$. The main modeling choice - for convenience and modular reuse - is to construct the *complemented interpretation* for $p_{str_<}$. This accounts for the all "negative" facts that hold, by absence from the current model. These will be collected, decoded, in a second interpretation. The corresponding cinterp type is thus defined as an interp pairing. To bookkeep the accumulated strata $str_<$, we wrap cinterp and the symbol set type of $str_<$ in a *cumulative interpretation* type, sinterp.

```
Notation   cinterp := (interp * interp)%type.
Definition sinterp := (cinterp * {set symtype})%type.
```

At an intermediate step, having already processed $str_<$, we encode the p_curr program slice *up to* the current stratum ss. We feed it, together with the previous *complemented interpretation* ci, to the positive engine pengine_step. Since this operates on *positive interpretations*[18], we have to relate the two. As such, we define the c2p_bij bijection between them, i.e., mutually inverse functions c2p and p2c, and apply it to obtain the needed types. The positive engine iterates the forward chain operator, as many times as there are elements in the program bound bp[19]. It adds the facts inferable from the current stratum and outputs a *positive* interpretation. It does not add the *implicitly* true *negated* ground atoms.

```
Definition bp : pinterp := setT.
Definition pengine_step def (pp : pprogram) (ci : cinterp) : cinterp :=
  p2c (iter #|bp| (P.fwd_chain pdef pp) (c2p ci)).
```

Hence, the ciC complementation function augments m_next.2 with the complement of m_next.1 w.r.t setT[20]; the complement is filtered to ensure only atoms with symbols in ss are retained (see the encoding of ic_ssym).

```
Variables (def : constant) (p : program) (psf : prog_safe p).
Fixpoint evalp (str : strata) ((ci, str<) : sinterp) :=
  match str with [::] ⇒ (ci, str<) | ss :: str> ⇒
    let p_curr := slice_prog p (str< ∪ ss) in
    let m_next := pengine_step def (encodep p_curr) ci in
    let m_cmpl := ciC ss m_next in evalp str> (m_cmpl, str< ∪ ss)
  end.
```

---

[18] "Positive" interpretations are sets of ground atoms with a **true** flag.

[19] This corresponds to the set of all "positive" ground atoms.

[20] This is the top element of interp cf. Sect. 3.1.

The resulting m_cmpl is thus *well-complemented*. As encoded by ci_wc, the property states that, for *any* ci of cinterp type and *any* symbol set ss, the ci components partition the slicing of setT with ss, i.e., the set of all ground atoms with symbols in ss. The next strata, i.e., str$_>$, are processed by the recursive call.

Example 3. Revisiting Example 2, the slice encodings, marked by $\ulcorner\cdot\urcorner$, are:

$$\ulcorner\mathbf{P_1}\urcorner = \left\{ \begin{array}{l} (\top,s)(b). \ (\top,t)(a). \\ (\top,r)(X) \leftarrow (\top,t)(X). \end{array} \right. \qquad \ulcorner\mathbf{P_2}\urcorner = \left\{ \begin{array}{l} (\top,q)(a). \\ (\top,q)(X) \leftarrow (\top,s)(X),(\bot,t)(X). \end{array} \right.$$

$$\ulcorner\mathbf{P_3}\urcorner = \left\{ \begin{array}{l} (\top,p)(X) \leftarrow (\bot,q)(X),(\top,r)(X). \\ (\top,p)(X) \leftarrow (\bot,t)(X),(\top,q)(X). \end{array} \right.$$

The positive engine computes the *minimal model* of $\ulcorner P_1 \urcorner$: $M_1 = T_{\ulcorner P_1 \urcorner} \uparrow \omega(\emptyset) = \{(\top,r)(a),(\top,s)(b),(\top,t)(a)\}$; complementing it w.r.t the Herbrand Base $\mathbf{B}(\ulcorner\mathbf{P_1}\urcorner)$ yields: $M_1 = \{(\bot,r)(b),(\bot,s)(a),(\bot,t)(b)\}$). Next, when passing the resulting *positive interpretation* $M_1 \cup M_1$ to the positive engine: $M_2 = T_{P_2} \uparrow \omega(M_1 \cup M_1) = M_1 \cup \{(\top,q)(a),(\top,q)(b)\}$. Its complement w.r.t $\mathbf{B}(\ulcorner\mathbf{P_2}\urcorner)$ is $M_2 = \emptyset$. Finally, $M_3 = T_{P_3} \uparrow \omega(M_2 \cup M_2) = M_2 \cup \{(\top,p)(b)\}$, whose complement w.r.t $\mathbf{B}(\ulcorner\mathbf{P_3}\urcorner)$ is $M_3 = \{(\bot,p)(a)\}$. The *stratified model* $\mathbf{M}(P)$ of P is the decoding of M_3, i.e., $\{r(a),s(b),t(a),q(a),q(b),p(b)\}$.

4.3 Formal Characterization of the Stratified Evaluation Engine

We first state the main intermediate results, leading up to the key Theorem 16. We assume p to be a program; pp, pp1 and pp2, "positive" programs; ci, an initial complemented interpretation and pdef, the default "positive" constant.

Theorem 8 (Complementation Preserves Satisfiability). *If symbols of a stratum* ss *do not appear negated in the body of* p *clauses, then the satisfiability of* pp *w.r.t* (c2p ci) *is preserved when complementing* ci *w.r.t* ss.

Theorem 9 (Encoding/Decoding Preserves Satisfiability). *In the following, assume* ci *is well-complemented. If* ci.1 *is a model of* p *and all* p *symbols are in* ss*, then* (c2p ci) *is a model of* $\ulcorner p \urcorner$*. If* (c2p ci) *is a model of* pp *and all* pp *body symbols are in* ss*, then* ci.1 *is a model of* $\llcorner pp \lrcorner$*.*

Intuitively, this is captured by the relations in the informal diagram below:[21]

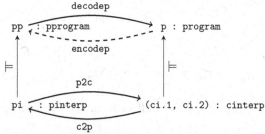

---

[21] The dashed encodep arrow marks the *partiality* of the cancellation lemma.

Theorem 10 (Preservation Properties). *If* pp *is safe, its* pengine_step *evaluation w.r.t* ci *is* sound, bound *by its Herbrand Base,* increasing *and* stable.

Theorem 11 (Symbol Stratifiability). *The atoms outputted by* pengine_step *are either in* ci *or have symbols appearing in the* head *of* pp *clauses.*

Theorem 12 (Positivity). *The "negative" component of* ci, *i.e.,* ci.2, *is not modified by* pengine_step, *i.e.,* (pengine_step pdef pp ci).2 = ci.2.

Theorem 13 (Injectivity). *If* pp1 *and* pp2 *are extensionally* equal, *their corresponding* pengine_step *evaluations w.r.t* ci *are equal.*

Theorem 14 (Modularity). *If* pp1 *does not contain head symbols in* pp2 *and* (c2p ci) *is a model of* pp1, *then evaluating the concatenation of* pp1 *and* pp2 *w.r.t* ci *equals the union of their respective evaluations w.r.t* ci.

Theorem 15 (Incrementality). *Let* p *be a stratifiable program;* $(ci, str_\leq)$, *a cumulative interpretation of* $p_{str_\leq}$, *and* ss, *a stratum. Assume that:* (1) $\ulcorner p_{str_\leq} \urcorner$ *symbols are not head symbols in* $\ulcorner p_{ss} \urcorner$, (2) $p_{str_\leq}$ *symbols are in* $str_\leq$, (3) ci *is* well-complemented *w.r.t* $str_\leq$, *and* (4) ci.1 *is a model of* $p_{str_\leq}$. *The* pengine_step *evaluation of* $\ulcorner p_{str_\leq \cup ss} \urcorner$ *increments* ci.1 *with facts having symbols in* ss.

Stratified Evaluation Invariant. Let p be a stratifiable program and (ci, $str_\leq$), a cumulative interpretation of $p_{str_\leq}$. The *invariant* of *stratified evaluation* si_invariant states: (1) ci.1 is a model of $p_{str_\leq}$, (2) $p_{str_\leq}$ symbols are in $str_\leq$, (3) ci is *well-complemented* with respect to $str_\leq$, and (4) ci symbols are in $str_\leq$.

Theorem 16 (Stratified Evaluation Soundness and Completeness). *Let* p *be a program,* str, *a strata - consisting of lower and upper strata,* $str_\leq$ *and* $str_>$[22] *- and* ci, *a complemented interpretation. If the input* cumulative *interpretation* (ci, $str_\leq$) *satisfies the above invariant conditions, then the output interpretation of the one-step evaluation of* $p_{str_>}$ *also satisfies them.*
Proof by induction on $str_>$.

As a corollary of Theorem 16, the encoded evaluation engine computes a *model* for a stratifiable program p. A more subtle discussion concerns its *minimality*:

Example 4 Let $P = \begin{cases} p \leftarrow q. \\ r \leftarrow \neg q. \\ s \leftarrow \neg q. \\ t \leftarrow \neg q. \end{cases} = P_1 \sqcup P_2$, $P_1 = \{p \leftarrow q.\}$, $P_2 =$

$\begin{cases} r \leftarrow \neg q. \\ s \leftarrow \neg q. \\ t \leftarrow \neg q. \end{cases}$. As $M_1 = T_{P_1} \uparrow \omega(\emptyset) = \emptyset$, $M_2 = T_{P_2} \uparrow \omega(M_1) = \{r, s, t\}$,

the *computed model* $M_P = \{r, s, t\}$ differs from the *cardinality-minimal model* $M_P^{min} = \{p, q\}$.

---

[22] i.e, $str_\leq$ stratifies $p_{str_\leq}$ and $str_>$ stratifies $p_{str_>}$.

This is because the minimality of a computed stratified model depends on *fixing* its input. Specifically, a model is minimal w.r.t others, if they *agree on the submodel relative to the accumulated stratification*. Since we need to consider both the previous and current candidate model, we cannot state the corresponding is_min_str_rec minimality condition within the strata invariant. We thus define it independently and prove it by induction on $str_>$.

5   Related Work

The work of [20] provides a Coq formalization of the correctness and equivalence of forward and backward, top-down and bottom-up semantics, based on a higher-order abstract syntax for Prolog. Related to our work, as it provides formal soundness proofs regarding the fixpoint semantics, it nonetheless differs in perspective and methodology. Also, while we do not support function symbols and other evaluation heuristics, we do support negation and manage to establish correctness and completeness for the underlying *algorithms* of bottom-up inference. The work in [10] gives a Coq mechanization of standard Datalog in the context of expressing distributed security policies[23]. The development contains the encoding of the language, of bottom-up evaluation and decidability proofs. In our corresponding formalizations, we did not need to explicitly prove the latter, as we set up our types as finite. While we did not take into account modelling security policies, the scope of our established results is wider.

6   Conclusion, Lessons and Perspectives

The exercise of formalizing database aspects has been an edifying experience. It helped clarify both the fundamentals underlying theoretical results and the proof-engineering implications of making these machine readable and user reusable.

On the database side, it quickly became apparent that, while foundational theorems appeared intuitively clear, if not obvious, understanding their *rigorous* justification required deeper reasoning. Resorting to standard references (even comprehensive ones, such as [2]), led at times to the realization that low-level details were either glanced over or left to the reader. For instance, to the best of our knowledge, no *scrupulous* proofs exist for the results we established. Indeed, as these are theoretically uncontroversial, their proofs are largely taken for granted and, understandably so, as they ultimately target database practitioners. Hence, these are mostly *assumed* in textbook presentations or when discussing further language extensions. It was only by mechanizing these proofs "from the ground up", in a proof assistant, that the relevance of various properties (e.g., safety and finiteness), the motivation behind certain definitions (e.g., predicate intensionality/extensionality, strata restrictions, logical consequence,

---

[23] http://www.cs.nott.ac.uk/types06/slides/NathanWhitehead.pdf.

stratified evaluation), or the precise meaning of ad-hoc notions/notations (e.g., "substitution compatibility", $\mathbf{B}(P)$, model restrictions) became apparent.

As it is well known, database theory is based on solid mathematical foundations, from model theory to algebra. This suggests that, when compared to off-the-shelf program verification, verification in the database context requires that proof systems have good support for mathematics. It was an interesting to discover, in practice, the extent to which database theory proofs could be recast into mathematical ones. To exemplify, by expressing forward chain as an elegant set construct, we transferred proofs about Datalog inference engines into set-theory ones, which are more natural to manipulate. Conversely, when formalizing the stratified semantics of Datalog with negation, we were compelled to resort to some ad-hoc solutions to handle the lack of native library support for lattice theory. For example, proving that the type inhabited by models is a lattice would automatically imply that stratification and complementation retain partial ordering. Thus, we could do away with current lemmas concerning symbol membership and well-complementation. Indeed, textbooks largely omit explanations as to why and how it is necessary to reason about such structures when proving properties of stratified evaluation. To this end, we were led to introduce specialized notions, such as interpretation complementation. Also, we had to *explicitly* establish that, at each evaluation step, the Herbrand Base of the program's restriction w.r.t the set of already processed strata symbols was a *well-complemented* lattice.

On the theorem proving side, a crucial lesson is the importance of relying on infrastructure that is well-tailored to the nature of the development. This emerged as *essential* while working on the formalization of standard Datalog. The triggering realization was that, as we could, w.l.o.g, restrict ourselves to the active domain, models could be reduced to the finite setting and atoms could be framed as finite types. Hence, the Mathematical Components library, prominently used in carrying out finite model theory proofs, stood out as best suited for our purposes. As we could heavily rely on the convenient properties of finite types and on already established set theory properties, proofs were rendered much easier and more compact. Another key aspect is the impact that type encoding choices can have on the size and complexity of proofs. For example, while having too many primitives is undesirable in programming language design, it turned out to be beneficial to opt for greater base granularity. Separating the type of ground/non-ground constructs helped both conceptually, in understanding the relevance of standard range restrictions, and practically, in facilitating proof advancement. Another example concerns the mechanization of substitutions. Having the option to representing them as finitely supported functions, together with all the useful properties this type has, was instrumental to finding a suitable phrasing for the soundness and completeness of the matching algorithm. Indeed, as the algorithm incrementally constructs groundings, it seemed natural to define an ordering on substitutions leading up to these. Being able to have a type encoding allowing to regard substitutions both as functions and as lists was essential for this purpose. A final example regards the formalization of models. As previously mentioned, setting up the type of ground atoms as finite payed off in that we could use many results and properties from the fintype library, when reasoning about models - which was often the case. In particular, we took advantage of the inherent lattice structure of such types and did not need to *explicitly* construct $\mathbf{B}(P)$. Finally, relying on characteristic properties (the SSReflect P-lemmas), many of which are conveniently stated as reflection lemmas, led to leaner proofs by compositionality. In cases in which induction would have been the default approach, these provided a shorter alternative (also, see [13], which gives a comprehensive formalization of linear algebra *without induction*).

We are working on modularly enriching the development with further language features, e.g., existentials, function symbols and update constructs. Envisaged applications of such extensions target security policy inference, data integration algorithms, and the certified property-based testing of realistic engines.

References

1. Barceló, P., Pichler, R. (eds.): Datalog in Academia and Industry. LNCS, vol. 7494. Springer, Heidelberg (2012)
2. Abiteboul, S., Hull, R., Vianu, V.: Foundations of Databases. Addison-Wesley, Boston (1995)
3. Apt, K.R., Blair, H.A., Walker, A.: Foundations of Deductive Databases and Logic Programming. Morgan Kaufmann Publishers Inc., San Francisco (1988)
4. Aref, M., ten Cate, B., Green, T.J., Kimelfeld, B., Olteanu, D., Pasalic, E., Veldhuizen, T.L., Washburn, G.: Design and implementation of the LogicBlox system. In: SIGMOD ACM Proceedings of ICMD, pp. 1371–1382 (2015)
5. Bagan, G., Bonifati, A., Ciucanu, R., Fletcher, G.H.L., Lemay, A., Advokaat, N.: gMark: schema-driven generation of graphs and queries. IEEE TKDE **29**, 856–869 (2017)
6. Calì, A., Gottlob, G., Lukasiewicz, T.: Datalog±: a unified approach to ontologies and integrity constraints. In: Fagin, R. (ed.) ACM Proceedings of ICDT, vol. 361, pp. 14–30 (2009)
7. Ceri, S., Gottlob, G., Tanca, L.: Logic Programming and Databases. Springer, Heidelberg (1990)
8. Chin, B., von Dincklage, D., Ercegovac, V., Hawkins, P., Miller, M.S., Och, F.J., Olston, C., Pereira, F.: Yedalog: exploring knowledge at scale. In: Ball, T., Bodk, R., Krishnamurthi, S., Lerner, B.S., Morrisett, G. (eds.) LIPIcs Proceedings of SNAPL, vol. 32, pp. 63–78 (2015)
9. Datomic. http://www.datomic.com/
10. DeTreville, J.: Binder, a logic-based security language. In: IEEE Proceedings of the Symposium on Security and Privacy, Washington, DC, USA, pp. 105–115. IEEE Computer Society (2002)
11. Doczkal, C., Smolka, G.: Completeness and decidability results for CTL in Coq. In: Klein, G., Gamboa, R. (eds.) ITP 2014. LNCS, vol. 8558, pp. 226–241. Springer, Cham (2014). doi:10.1007/978-3-319-08970-6_15
12. Exeura. http://www.exeura.com/
13. Gonthier, G.: Point-free, set-free concrete linear algebra. In: van Eekelen, M., Geuvers, H., Schmaltz, J., Wiedijk, F. (eds.) ITP 2011. LNCS, vol. 6898, pp. 103–118. Springer, Heidelberg (2011). doi:10.1007/978-3-642-22863-6_10

14. Gonthier, G., et al.: A machine-checked proof of the odd order theorem. In: Blazy, S., Paulin-Mohring, C., Pichardie, D. (eds.) ITP 2013. LNCS, vol. 7998, pp. 163–179. Springer, Heidelberg (2013). doi:10.1007/978-3-642-39634-2_14

15. Gonthier, G., Mahboubi, A., Tassi, E.: A small scale reflection extension for the Coq system (2016). https://hal.inria.fr/inria-00258384

16. Gottlob, G., Koch, C., Baumgartner, R., Herzog, M., Flesca, S.: The lixto data extraction project: back and forth between theory and practice. In: ACM SIGMOD-SIGACT-SIGART Proceedings of PODS, pp. 1–12. ACM, New York (2004)

17. Grumbach, S., Wang, F.: Netlog, a rule-based language for distributed programming. In: Carro, M., Peña, R. (eds.) PADL 2010. LNCS, vol. 5937, pp. 88–103. Springer, Heidelberg (2010). doi:10.1007/978-3-642-11503-5_9

18. Hellerstein, J.M.: The declarative imperative: experiences and conjectures in distributed logic. ACM SIGMOD Rec. J. **39**(1), 5–19 (2010)

19. Huang, S.S., Green, T.J., Loo, B.T.: Datalog and emerging applications: an interactive tutorial. In: ACM SIGMOD Proceedings of ICMD, pp. 1213 1216 (2011)

20. Kriener, J., King, A., Blazy, S.: Proofs you can believe in: proving equivalences between prolog semantics in Coq. In: ACM Proceedings of PPDP, pp. 37–48 (2013)

21. Libkin, L.: The finite model theory toolbox of a database theoretician. In: ACM SIGMOD-SIGACT-SIGART Proceedings of PODS, pp. 65–76 (2009)

22. Lloyd, J.W.: Foundations of Logic Programming. Springer, Heidelberg (1987)

23. LogicBlox. http://www.logicblox.com/

24. Loo, B.T., Condie, T., Hellerstein, J.M., Maniatis, P., Roscoe, T., Stoica, I.: Implementing declarative overlays. In: ACM Proceedings of SOSP, pp. 75–90 (2005)

25. Lu, L., Cleary, J.G.: An operational semantics of starlog. In: Nadathur, G. (ed.) PPDP 1999. LNCS, vol. 1702, pp. 294–310. Springer, Heidelberg (1999). doi:10.1007/10704567_18

26. Luteberget, B., Johansen, C., Feyling, C., Steffen, M.: Rule-based incremental verification tools applied to railway designs and regulations. In: Fitzgerald, J., Heitmeyer, C., Gnesi, S., Philippou, A. (eds.) FM 2016. LNCS, vol. 9995, pp. 772–778. Springer, Cham (2016). doi:10.1007/978-3-319-48989-6_49

27. The Coq Development Team: The Coq Proof Assistant. Reference Manual (2016). https://coq.inria.fr/refman/. Version 8.6

28. Ramakrishnan, R., Ullman, J.D.: A survey of research on deductive database systems. J. Log. Program. **23**(2), 125–149 (1993)

29. Semmle. https://semmle.com/

30. Seo, J., Park, J., Shin, J., Lam, M.S.: Distributed socialite: a datalog-based language for large-scale graph analysis. Proc. VLDB Endow. **6**, 1906–1917 (2013)

31. Tarski, A.: A lattice-theoretical fixpoint theorem and its applications. Pac. J. Math. **5**(2), 285–309 (1955)

32. Van Emden, M.H., Kowalski, R.A.: The semantics of predicate logic as a programming language. J. ACM **23**(4), 733–742 (1976)

33. Vardi, M.Y.: The complexity of relational query languages. In: ACM Proceedings of STOC, pp. 137–146 (1982)

34. Whaley, J., Avots, D., Carbin, M., Lam, M.S.: Using datalog with binary decision diagrams for program analysis. In: Yi, K. (ed.) APLAS 2005. LNCS, vol. 3780, pp. 97–118. Springer, Heidelberg (2005). doi:10.1007/11575467_8

Weak Call-by-Value Lambda Calculus
as a Model of Computation in Coq

Yannick Forster[✉] and Gert Smolka[✉]

Saarland University, Saarbrücken, Germany
{forster,smolka}@ps.uni-saarland.de

Abstract. We formalise a weak call-by-value λ-calculus we call L in the constructive type theory of Coq and study it as a minimal functional programming language and as a model of computation. We show key results including (1) semantic properties of procedures are undecidable, (2) the class of total procedures is not recognisable, (3) a class is decidable if it is recognisable, corecognisable, and logically decidable, and (4) a class is recognisable if and only if it is enumerable. Most of the results require a step-indexed self-interpreter. All results are verified formally and constructively, which is the challenge of the project. The verification techniques we use for procedures will apply to call-by-value functional programming languages formalised in Coq in general.

1 Introduction

We study a minimal functional programming language L realising a subsystem of the λ-calculus [3] known as weak call-by-value λ-calculus [8]. As in most programming languages, β-reduction in weak call-by-value λ-calculus is only applicable if the redex is not below an abstraction and if the argument is an abstraction. Our goal is to formally and constructively prove the basic results from computability theory [9,11] for L. The project involves the formal verification of self-interpreters and other procedures computing with encodings of procedures. The verification techniques we use will apply to call-by-value functional programming languages formalised in Coq in general. We base our work on the constructive type theory of Coq [15] and provide a development verifying all results.

The results from computability theory we prove for L include (1) semantic properties of procedures are undecidable (Rice's theorem), (2) the class of total procedures is not recognisable, (3) a class is decidable if it is recognisable, corecognisable, and logically decidable (Post's theorem), and (4) a class is recognisable if and only if it is enumerable.

We prove that procedural decidability in L implies functional decidability in Coq. The converse direction cannot be shown in Coq since Coq is consistent with the assumption that every class is functionally decidable and procedurally undecidable classes always exist. The same will be true for any Turing-complete model of computation formalised in Coq.

© Springer International Publishing AG 2017
M. Ayala-Rincón and C.A. Muñoz (Eds.): ITP 2017, LNCS 10499, pp. 189–206, 2017.
DOI: 10.1007/978-3-319-66107-0_13

The result that procedural decidability implies functional decidability seems contradictory at first since procedures come with unguarded recursion while functions are confined to guarded recursion. The apparent paradox disappears once one realises that procedural decidability means that termination of a decision procedure can be shown in Coq's constructive type theory.

Comparing L with the full λ-calculus, we find that L is more realistic as a programming language and simpler when it comes to semantics and program verification. The restrictions L imposes on β-reduction eliminate the need for capture-free substitution and provide for a uniform confluence property [8,13] ensuring that all evaluating reduction sequences of a term have the same length. Uniform confluence simplifies the construction and verification of a self-interpreter by eliminating the need for a reduction strategy like leftmost-outermost. Moreover, uniform confluence for L is easier to prove than confluence for the full λ-calculus.

While L simplifies the full λ-calculus, it inherits powerful techniques developed for the λ-calculus: Procedural recursion can be expressed with self-application, inductive data types can be expressed with Scott encodings [10,12], and program verification can be based on one-step reduction, the accompanying equivalence, and the connecting Church-Rosser property.

One place where the commitment to a constructive theory prominently shows is Post's theorem. The classical formulation of Post's theorem states that a class is decidable if it is recognisable and corecognisable. The classical formulation of Post's theorem is equivalent to Markov's principle and does not hold in a purely constructive setting [7]. We show Post's theorem with the extra assumption that the class is logically decidable. The extra assumption is needed so that we can prove termination of the procedure deciding the class. We refer to the classical formulation of Post's theorem for L as Markov's principle for L and establish two complementary characterisations.

Related Work. There is not much work on computability theory in constructive type theory. We are aware of Asperti and Ricciotti [1,2] who formalise Turing machines in Matita including a verified universal machine and a verified reduction of multi-tape machines to single-tape machines. They do not consider decidable and recognisable classes. Ciaffaglione [6] formalises Turing machines coinductively in Coq and shows the agreement between a big-step and a small-step semantics.

Bauer [4] develops a constructive and anti-classical computability theory abstracting away from concrete models of computation.

There is substantial work on computability theory in classical higher-order logic. Norrish [14] presents a formal development of computability theory in HOL4 where he considers full λ-calculus and partial recursive functions and proves their computational equivalence. Norrish studies decidable and recognisable classes, verifies self-interpreters, and proves basic results including the theorems of Rice and Post.

There are substantial differences between our work and Norrish [14] apart from the fact that Norrish works in a classical setting. Following Barendregt [3],

Norrish works with full λ-calculus and Gödel-Church encodings. We work with L and Scott encodings instead. Church encodings are not possible using weak β-reduction. We remark that Scott encodings are simpler than Gödel-Church encodings (since they don't involve recursion). Norrish proves Rice's theorem for partial recursive functions while we prove the theorem directly for procedures in L.

Xu et al. [16] formalise Turing machines, abacus machines, and partial recursive functions in Isabelle (classical higher-order logic) and show their computational equivalence following Boolos et al. [5]. They prove the existence of a universal function. They do not consider the theorems of Rice and Post.

Dal Lago and Martini [8] consider a weak call-by-value λ-calculus and show that Turing machines and procedures in the calculus can simulate each other with polynomial-time overhead, thus providing evidence that a weak call-by-value λ-calculus may serve as a reasonable complexity model. Their λ-calculus is different from ours in that it employs full substitution and β-reduction is possible if the argument is a variable. Like us, they use Scott encodings of data types. Their work is not formalised.

Main Contributions. Our work is the first formal study of weak call-by-value λ-calculus covering both language semantics and program verification. We are also first in proving results from computability theory for a programming language in constructive type theory.

The development of this paper is carried out in constructive type theory and outlines a machine-checked Coq development. The Coq development is surprisingly compact and consists of less than 2000 lines of code. The theorems in the pdf of the paper are hyperlinked with their formalisations in the Coq development, which can be found at http://www.ps.uni-saarland.de/extras/L-computability.

2   Specification

We start by specifying essential properties of the functional language L we will work with and by describing main results from computability theory we will prove for L.

We assume a discrete type of *terms* and a class of terms called *procedures*. We will use the letters s, t, u, v, w for terms and the letters p, q for classes of terms.

We assume a functional relation $s \triangleright t$ on terms called *evaluation*. We say that a term s *evaluates* and write $\mathcal{E}s$ if there is a term t such that $s \triangleright t$.

We assume a function st from terms to terms called *application*.

We assume two procedures T and F such that $T \neq F$ and $Tst \triangleright s$ and $Fst \triangleright t$ for all procedures s, t. As usual, we omit parentheses in nested applications; for instance, Tst stands for $(Ts)t$.

We assume an injective function $\bar{s}$ from terms to procedures called *term encoding*. The purpose of the encoding function is to encode a term into a procedure providing the term as data to other procedures. This is a subtle point that will become clear later. For now it suffices to know that $\bar{s}$ is a function from terms to procedures.

We now define *decidable*, *recognisable*, and *corecognisable* classes of terms:

- A procedure u *decides* a class p if $\forall s.\ ps \land u\bar{s} \triangleright \mathrm{T} \lor \neg ps \land u\bar{s} \triangleright \mathrm{F}$.
- A procedure u *recognises* a class p if $\forall s.\ ps \leftrightarrow \mathcal{E}(u\bar{s})$.
- A procedure u *corecognises* a class p if $\forall s.\ \neg ps \leftrightarrow \mathcal{E}(u\bar{s})$.

Our assumptions suffice to establish the existence of undecidable and unrecognisable classes.

Fact 1. *Let u decide p. Then $ps \leftrightarrow u\bar{s} \triangleright \mathrm{T}$ and $\neg ps \leftrightarrow u\bar{s} \triangleright \mathrm{F}$.*

Fact 2. $\lambda s.\neg(s\bar{s} \triangleright \mathrm{T})$ *is not decidable, and* $\lambda s.\neg\mathcal{E}(s\bar{s})$ *is not recognisable.*

Proof. Suppose u decides $\lambda s.\neg(s\bar{s} \triangleright \mathrm{T})$. Then $u\bar{s} \triangleright \mathrm{T} \leftrightarrow \neg(s\bar{s} \triangleright \mathrm{T})$ for all s. The equivalence is contradictory for $s := u$. The proof for the unrecognisable class is similar. □

We need different notions of decidability in this paper. We call a class p

- *logically decidable* if there is a proof of $\forall s.\ ps \lor \neg ps$.
- *functionally decidable* if there is a function f such that $\forall s.\ ps \leftrightarrow fs = \mathsf{true}$.
- *procedurally decidable* if there is a procedure deciding p.

If we say decidable without further qualification, we always mean procedurally decidable. Note that functionally decidable classes are logically decidable.

We define two semantic properties of terms. A term s is *total* if the application st evaluates for every term t. *Semantic equivalence* of terms is defined as $s \approx t := \forall uv.\ s\bar{u} \triangleright v \leftrightarrow t\bar{u} \triangleright v$. Note that if $s \approx t$, then s is total iff t is total.

We can now specify major results we will prove in this paper.

- *Rice's theorem.* Every nontrivial class of procedures that doesn't distinguish between semantically equivalent procedures is undecidable.
- *Modesty.* Procedurally decidable classes are functionally decidable.
- *Totality.* The class of total procedures is unrecognisable.
- *Post's Theorem.* A class is decidable if it is recognisable, corecognisable, and logically decidable.

We will also consider enumerable classes and show that they agree with recognisable classes. All results but Rice's theorem require a step-indexed interpreter or step-indexed self-interpreter.

Note the distinction between functions and procedures. While functions are entities of the typed specification language (i.e., Coq's type theory), procedures are entities of the untyped programming language L formalised in the specification language by means of a deep embedding. As we will see, L comes with unbounded recursion and thus admits nonterminating procedures. In contrast, Coq's type theory is designed such that functions always terminate.

3   Definition of L

We will work with the terms of the λ-calculus. We restrict β-reduction such that β-redexes can only be reduced if (1) they are not within an abstraction and (2) their argument term is an abstraction. With this restriction the terms $\lambda x.(\lambda y.y)(\lambda y.y)$ and $(\lambda x.x)x$ are irreducible. We speak of *weak call-by-value β-reduction* and write $s \succ t$ if t can be obtained from s with a single weak call-by-value β-reduction step. We will define the evaluation relation such that $s \rhd t$ holds iff $s \succ^* t$ and t is an abstraction. Procedures will be defined as closed abstractions.

Since we want formal proofs we are forced to formally define the concrete weak call-by-value λ-calculus L we are working with. In fact, there are some design choices. We will work with de Bruijn terms and capturing substitution, two design decisions providing for a straightforward formal development.

We start the formal definition of L with an inductive type of *terms*:

$$s, t ::= n \mid st \mid \lambda s \qquad (n : \mathbf{N})$$

We fix some terms for further use:

$$
\begin{array}{lllll}
\mathrm{I} = \lambda x.x & \mathrm{T} = \lambda xy.x & \mathrm{F} = \lambda xy.y & \omega = \lambda x.xx & \mathrm{D} = \lambda x.\omega\omega \\
:= \lambda 0 & := \lambda(\lambda 1) & := \lambda(\lambda 0) & := \lambda(00) & := \lambda(\omega\omega)
\end{array}
$$

For readability, we will usually write concrete terms with named abstractions, as shown above. The Coq development provides a function translating named abstraction to the implementation using de Bruijn indices. Note that D is reducible in the full λ-calculus but will not be reducible in L.

We define a *substitution function* s_u^k that replaces every free occurrence of a variable k in a term s with a term u. The definition is by recursion on s:

$$
\begin{aligned}
n_u^k &= \text{ if } n = k \text{ then } u \text{ else } n \\
(st)_u^k &= (s_u^k)(t_u^k) \\
(\lambda s)_u^k &= \lambda(s_u^{Sk})
\end{aligned}
$$

A substitution s_u^k may capture free variables in u. Capturing will not affect our development since it doesn't affect confluence and our results mostly concern closed terms.

We now give a formal definition of closed terms. Closed terms are important for our development since procedures are defined as closed abstractions and substitutions do not affect closed terms. Moreover, we need a decider for the class of closed terms.

We define a recursive boolean function *bound k s* satisfying the equations

$$
\begin{aligned}
bound\ k\ n &= \text{ if } n < k \text{ then true else false} \\
bound\ k\ (st) &= \text{ if } bound\ k\ s \text{ then } bound\ k\ t \text{ else false} \\
bound\ k\ (\lambda s) &= bound\ (Sk)\ s
\end{aligned}
$$

Speaking informally, *bound k s* tests whether every free variable in s is smaller than k. We say that s is *bound* by n if *bound n s* = true. We now define *closed* terms as terms bound by 0, and *procedures* as closed abstractions. Note that the terms I, T, F, ω, and D are all procedures. The following fact will be used tacitly in many proofs.

Fact 3. *If s is bound by n and $k \geq n$, then $s_u^k = s$. Moreover, $s_u^k = s$ for closed s.*

We define *evaluation $s \triangleright t$* as an inductive predicate:

$$\frac{}{\lambda s \triangleright \lambda s} \qquad \frac{s \triangleright \lambda u \qquad t \triangleright v \qquad u_v^0 \triangleright w}{st \triangleright w}$$

Recall that we write $\mathcal{E} s$ and say that s *evaluates* if $s \triangleright t$ for some term t.

Fact 4.

1. If $s \triangleright t$, then t is an abstraction.
2. If $s \triangleright t$ and s is closed, then t is closed.
3. If st evaluates, then both s and t evaluate.
4. Fst evaluates if and only if both s and t evaluate.
5. $\omega\omega$ does not evaluate.
6. Ds does not evaluate.

4 Uniformly Confluent Reduction Semantics

To provide for the verification of procedures in L, we complement the big-step semantics obtained with the evaluation predicate with a uniformly confluent reduction semantics.

We define one-step *reduction $s \succ t$* as an inductive predicate:

$$\frac{}{(\lambda s)(\lambda t) \succ s_{\lambda t}^0} \qquad \frac{s \succ s'}{st \succ s't} \qquad \frac{t \succ t'}{st \succ st'}$$

We also define two reduction relations $s \succ^* t$ and $s \succ^n t$ as inductive predicates:

$$\frac{}{s \succ^* s} \qquad \frac{s \succ u \qquad u \succ^* t}{s \succ^* t} \qquad \frac{}{s \succ^0 s} \qquad \frac{s \succ u \qquad u \succ^n t}{s \succ^{Sn} t}$$

Fact 5.

1. $s \succ^* t$ is transitive.
2. If $s \succ^* s'$ and $t \succ^* t'$, then $st \succ^* s't'$.
3. $s \succ^* t$ iff $s \succ^n t$ for some n.
4. If $s \succ^m s'$ and $s' \succ^n t$, then $s \succ^{m+n} t$.
5. If $s \triangleright t$, then $s \succ^* t$ and t is an abstraction.

6. If $s \succ s'$ and $s' \triangleright t$, then $s \triangleright t$.
7. If $s \succ^* s'$ and $s' \triangleright t$, then $s \triangleright t$.
8. If $s \succ^* t$ and t is an abstraction, then $s \triangleright t$.

With the reduction semantics we can specify procedural recursion, which is essential for our goals.

Fact 6 (Recursion Operator). *There is a function ρ from terms to terms such that (1) ρs is a procedure if s is closed and (2) $(\rho u)v \succ^3 u(\rho u)v$ for all procedures u and v.*

Proof. $\rho s := \lambda x.CCsx$ with $C := \lambda xy.y(\lambda z.xxyz)$ does the job. □

We call the function ρ *recursion operator* since it provides for recursive programming in L using well-known techniques from functional programming.

The weak call-by-value λ-calculus in general and L in particular enjoy a strong confluence property [8,13] we call uniform confluence.

Fact 7 (Uniform Confluence).

1. If $s \succ t_1$ and $s \succ t_2$, then either $t_1 = t_2$ or $t_1 \succ u$ and $t_2 \succ u$ for some u.
2. If $s \succ^m t_1$ and $s \succ^n t_2$, then there exist numbers $k \leq n$ and $l \leq m$ and a term u such that $t_1 \succ^k u$ and $t_2 \succ^l u$ and $m + k = n + l$.

Corollary 8. $s \succ t$ is confluent.

We define $s \triangleright^n t := s \succ^n t \wedge$ *abstraction* t and $s \succ^+ t := \exists s'. s \succ s' \wedge s' \succ^* t$.

Corollary 9 (Unique Step Index). If $s \triangleright^m t$ and $s \triangleright^n t$, then $m = n$.

Corollary 10 (Triangle). If $s \triangleright^n t$ and $s \succ^+ s'$, then $s' \triangleright^k t$ for some $k < n$.

We define *reduction equivalence* $s \equiv t$ as the equivalence closure of reduction:

$$\frac{s \succ t}{s \equiv t} \qquad \frac{}{s \equiv s} \qquad \frac{s \equiv t}{t \equiv s} \qquad \frac{s \equiv t \quad t \equiv u}{s \equiv u}$$

Reduction equivalence enjoys the usual Church-Rosser properties and will play a major role in the verification of procedures.

Fact 11 (Church-Rosser Properties).

1. If $s \succ^* t$, then $s \equiv t$.
2. If $s \equiv t$, then $s \succ^* u$ and $t \succ^* u$ for some term u.
3. If $s \equiv s'$ and $t \equiv t'$, then $st \equiv s't'$.
4. $s \equiv t \leftrightarrow s \succ^* t$ if t is a variable or an abstraction.
5. $s \triangleright t$ iff $s \equiv t$ and t is an abstraction.
6. If $s \equiv t$, then $s \triangleright u$ iff $t \triangleright u$.

Proof. Claim 1 follows by induction on $s \succ^* t$. Claim 2 follows by induction on $s \equiv t$ and Corollary 8. Claim 3 follows with Claim 2, Fact 5(2), and Claim 1. The remaining claims follow with Claim 1 and Claim 2. □

Because L employs call-by-value reduction, a conditional if u then s else t needs to be expressed as $u(\lambda s)(\lambda t)I$ in general. We have $T(\lambda s)(\lambda t)I \succ^* s$ and $F(\lambda s)(\lambda t)I \succ^* t$.

5   Scott Encoding of Numbers

Seen as a programming language, L is a language where all values are procedures. We now show how procedures can encode data using a scheme known as Scott encoding [10,12]. We start with numbers, whose Scott encoding looks as follows:

$$\widehat{0} := \lambda ab.a \qquad\qquad \widehat{Sn} := \lambda ab.b\,\widehat{n}$$

Note that $\widehat{n}$ is an injective function from numbers to procedures. We have the equivalences

$$\widehat{0}\,st \equiv s \qquad\qquad \widehat{Sn}\,st \equiv t\,\widehat{n}$$

for all evaluable closed terms s, t and all numbers n. The equivalences tell us that the procedure $\widehat{n}$ can be used as a match construct for the encoded number n.

We define a procedure $\text{Succ} := \lambda xab.bx$ such that $\text{Succ}\,\widehat{n} \equiv \widehat{Sn}$. Note that the procedures $\widehat{0}$ and Succ act as the constructors of the Scott encoding of numbers.

Programming with Scott encodings is convenient in that we can follow familiar patterns from functional programming. We demonstrate the case with a *functional specification*

$$\forall mn. \ \text{Add}\ \widehat{m}\,\widehat{n} \ \equiv \ \widehat{m+n}$$

of a procedure Add for addition. We say that we are looking for a procedure Add *realising* the addition function $m + n$. A well-known *recursive specification* for the addition function consists of the quantified equations $0 + n = n$ and $Sm + n = S(m+n)$. This gives us a recursive specification for the procedure Add (quantification of m and n is omitted):

$$\text{Add}\ \widehat{0}\ \widehat{n} \equiv \widehat{n} \qquad\qquad \text{Add}\ \widehat{Sm}\ \widehat{n} \equiv \text{Succ}\,(\text{Add}\ \widehat{m}\ \widehat{n})$$

With induction on m one can now show that a procedure Add satisfies the functional specification if it satisfies the recursive specification. The recursive specification of Add suggest a recursive definition of Add using L's recursion operator ρ:

$$\text{Add} := \rho(\lambda xyz.yz(\lambda y_0.\text{Succ}(xy_0z)))$$

Using the equivalences for the recursion operator ρ and those for the procedures $\widehat{0}$, $\widehat{Sn}$, and Succ, one easily verifies that Add satisfies the equivalences of the recursive specification. Hence Add satisfies the functional specification we started with.

The functional specification of Add has the virtue that properties of Add like commutativity (i.e., $\text{Add}\ \widehat{m}\ \widehat{n} \equiv \text{Add}\ \widehat{n}\ \widehat{m}$) follow from properties of the addition function $m + n$.

The method we have seen makes it straightforward to obtain a procedure realising a function given a recursive specification of the function. Once we have

Scott encodings for terms and a few other inductive data types, the vast major-ity of procedures needed for our development can be derived routinely from their functional specifications. We are working on tactics that, given a recur-sive function, automatically derive a realising procedure and the corresponding correctness lemma.

6   Scott Encoding of Terms

We define the term encoding function $\bar{s}$ specified in Sect. 2 as follows:

$$\overline{n} := \lambda abc.a\,\widehat{n} \qquad\qquad \overline{st} := \lambda abc.b\,\overline{s}\,\overline{t} \qquad\qquad \overline{\lambda s} := \lambda abc.c\,\overline{s}$$

This definition agrees with the Scott encoding of the inductive data type for terms. We define the constructors for variables, applications, and abstractions such that they satisfy the equivalences

$$\mathrm{V}\,\widehat{n} \equiv \overline{n} \qquad\qquad \mathrm{A}\,\overline{s}\,\overline{t} \equiv \overline{st} \qquad\qquad \mathrm{L}\,\overline{s} \equiv \overline{\lambda s}$$

for all numbers n and all terms s and t.

We will define two procedures N and Q satisfying the equivalences

$$\mathrm{N}\,\widehat{n} \equiv \widehat{\overline{n}} \qquad\qquad\qquad \mathrm{Q}\,\overline{s} \equiv \overline{\overline{s}}$$

for all numbers n and all terms s. The procedure Q is known as *quote* and will be used in the proof of Rice's theorem. The procedure N is an auxiliary procedure needed for the definition of Q. We define the procedures N and Q with the recursion operator realising the following recursive specifications:

$$
\begin{aligned}
\mathrm{N}\,\widehat{0} &\equiv \overline{\widehat{0}} & \mathrm{Q}\,\overline{n} &\equiv \mathrm{L}(\mathrm{L}(\mathrm{L}(\mathrm{A}\,\overline{2}\,(\mathrm{N}\,\widehat{n})))) \\
\mathrm{N}\,\widehat{Sn} &\equiv \mathrm{L}(\mathrm{L}(\mathrm{A}\,\overline{0}\,(\mathrm{N}\,\widehat{n}))) & \mathrm{Q}\,\overline{st} &\equiv \mathrm{L}(\mathrm{L}(\mathrm{L}(\mathrm{A}(\mathrm{A}\,\overline{1}\,(\mathrm{Q}\,\overline{s}))(\mathrm{Q}\,\overline{t})))) \\
& & \mathrm{Q}\,\overline{\lambda s} &\equiv \mathrm{L}(\mathrm{L}(\mathrm{L}(\mathrm{A}\,\overline{0}\,(\mathrm{Q}\,\overline{s}))))
\end{aligned}
$$

Given the definitions of procedures N and Q, one first verifies that they satisfy the equivalences of the recursive specifications. Then one shows by induction on numbers and terms that N and Q satisfy the functional specifications we started with. We summarise the results obtained so far.

Fact 12. *There are procedures V, A, L, and Q such that* $\mathrm{V}\,\widehat{n} \equiv \overline{n}$, $\mathrm{A}\,\overline{s}\,\overline{t} \equiv \overline{st}$, $\mathrm{L}\,\overline{s} \equiv \overline{\lambda s}$, *and* $\mathrm{Q}\,\overline{s} \equiv \overline{\overline{s}}$.

7   Decidable and Recognisable Classes

Now that we have established the term encoding function, we can start prov-ing properties of decidable and recognisable classes. Recall the definitions from Sect. 2. We will prove the following facts: decidable classes are recognisable; the family of decidable classes is closed under intersection, union, and complement; and the family of recognisable classes is closed under intersection. We establish these facts constructively with translation functions.

Fact 13. *Let u decide p and v decide q. Then:*

1. $\lambda x.ux\,\mathrm{I}\,\mathrm{D}\,\mathrm{I}$ *recognises* p.
2. $\lambda x.ux(vx)\mathrm{F}$ *decides* $\lambda s.ps \wedge qs$.
3. $\lambda x.ux\,\mathrm{T}(vx)$ *decides* $\lambda s.ps \vee qs$.
4. $\lambda x.ux\,\mathrm{F}\,\mathrm{T}$ *decides* $\lambda s.\neg ps$.

Fact 14. $\lambda x.\mathrm{F}(ux)(vx)$ *recognises* $\lambda s.ps \wedge qs$ *if u recognise p and v recognise q.*

We now prove Scott's theorem for L following Barendregt's proof [3] of Scott's theorem for the full λ-calculus. Scott's theorem is useful for proving undecidability of classes that do not distinguish between reduction equivalent closed terms.

Fact 15. *Let s be closed. Then there exists a closed term l such that $t = s\bar{t}$.*

Proof. $t := C\overline{C}$ with $C := \lambda x.s(\mathrm{A}x(\mathrm{Q}x))$ does the job. $\qquad\square$

Theorem 16 (Scott).
Every class p satisfying the following conditions is undecidable.

1. *There are closed terms s_1 and s_2 such that ps_1 and $\neg ps_2$.*
2. *If s and t are closed terms such that $s \equiv t$ and ps, then pt.*

Proof. Let p be a class as required and u be a decider for p. Let s_1 and s_2 be closed terms such that ps_1 and $\neg ps_2$. We define $v := \lambda x.ux(\lambda s_2)(\lambda s_1)\,\mathrm{I}$. Fact 15 gives us a closed term t such that $t \equiv v\bar{t} \equiv u\bar{t}(\lambda s_2)(\lambda s_1)\mathrm{I}$. Since u is a decider for p, we have two cases: (1) If $u\bar{t} \equiv \mathrm{T}$ and pt, then $t \equiv s_2$ contradicting $\neg ps_2$; (2) If $u\bar{t} \equiv \mathrm{F}$ and $\neg pt$, then $t \equiv s_1$ contradicting ps_1. $\qquad\square$

Corollary 17. *The class of evaluating terms is undecidable.*

Corollary 18. *For every closed term t the class $\lambda s.s \equiv t$ is undecidable.*

8   Reduction Lemma and Rice's Theorem

The reduction lemma formalises a basic result of computability theory and will be used in our proofs of Rice's theorem and the totality theorem. Speaking informally, the reduction lemma says that a class is unrecognisable if it can represent the class $\lambda s.$ *closed* $s \wedge \neg \mathcal{E}(s\bar{s})$ via a procedurally realisable function.

Fact 19. *The class $\lambda s.$ closed $s \wedge \neg\mathcal{E}(s\bar{s})$ is not recognisable.*

Proof. Suppose u is a recogniser for the class. Then $\mathcal{E}(u\bar{u}) \leftrightarrow$ *closed* $u \wedge \neg\mathcal{E}(u\bar{u})$, which is contradictory. $\qquad\square$

Fact 20. *There is a decider for the class of closed terms.*

Proof. The decider can be obtained with a procedure realising the boolean function *bound k s* defined in Sect. 3. For this we need a procedure realising a boolean test $m < n$. The construction and verification of both procedures is routine using the techniques from Sect. 5.   □

Lemma 21 (Reduction). *A class p is unrecognisable if there exists a function f such that:*

1. $p(fs) \leftrightarrow \neg\mathcal{E}(s\overline{s})$ *for every closed terms s.*
2. *There is a procedure v such that $v\overline{s} \equiv \overline{fs}$ for all s.*

Proof. Let f be a function satisfying (1) and (2) for a procedure v. Suppose u recognises p. Let C be a recogniser for the class of closed terms (available by Fact 20). We define the procedure

$$w := \lambda x.\mathrm{F}(Cx)(u(vx))$$

We have $w\overline{s} \equiv \mathrm{F}(C\overline{s})(u(\overline{fs}))$. Thus $\mathcal{E}(w\overline{s}) \leftrightarrow$ *closed* $s \wedge \mathcal{E}(u(\overline{fs}))$. Since u is a recogniser for p, we have $\mathcal{E}(u(\overline{fs})) \leftrightarrow p(fs)$ for all s. Since $p(fs) \leftrightarrow \neg\mathcal{E}(s\overline{s})$ for closed s by assumption, we have *closed* $s \wedge \mathcal{E}(u(\overline{fs})) \leftrightarrow$ *closed* $s \wedge \neg\mathcal{E}(s\overline{s})$ for all s. Thus w is recogniser for the unrecognisable class of Fact 19. Contradiction.   □

We now come to Rice's theorem. Using the reduction lemma, we will first prove a lemma that is stronger than Rice's theorem in that it establishes unrecognisability rather than undecidability. We did not find this lemma in the literature, but for ease of language we will refer to it as Rice's lemma.

Recall the definition of *semantic equivalence*

$$s \approx t := \forall uv.\ s\overline{u} \triangleright v \leftrightarrow t\overline{u} \triangleright v$$

from Sect. 2. We have $s \equiv t \to s \approx t$ using Fact 11. We say that a class p is *semantic for procedures* if the implication $s \approx t \to ps \to pt$ holds for all procedures s and t.

Lemma 22 (Rice). *Let p be a class that is semantic for procedures such that D is in p and some procedure N is not in p. Then p is unrecognisable.*

Proof. By the reduction lemma. We define fs as a procedure such that for closed s we have $fs \approx D$ if $\neg\mathcal{E}(s\overline{s})$ and $fs \approx N$ if $\mathcal{E}(s\overline{s})$. Here are the definitions of f and the realising procedure v:

$$f := \lambda s.\lambda y.\mathrm{F}(s\overline{s})Ny$$

$$v := \lambda x.\mathrm{L}(\mathrm{A}(\mathrm{A}(\mathrm{A}\,\overline{\mathrm{F}}(\mathrm{A}x(\mathrm{Q}x)))\overline{N})\overline{0})$$

Verifying the proof obligations of the reduction lemma is straightforward.   □

Corollary 23.

1. *The class of non-total terms is unrecognisable.*
2. *The class of non-total closed terms is unrecognisable.*
3. *The class of non-total procedures is unrecognisable.*

Theorem 24 (Rice). *Every nontrivial class of procedures that is semantic for procedures is undecidable.*

Proof. Let p be a nontrivial class that is semantic for procedures. Suppose p is decidable. We proceed by case analysis for pD.

Let pD. Then p is unrecognisable by Rice's Lemma, contradicting the assumption that p is decidable.

Let $\neg pD$. We observe that $\lambda s. \neg ps$ is semantic for procedures and contains D. Thus $\lambda s. \neg ps$ is unrecognisable by Rice's Lemma, contradicting the assumption that p is decidable.   □

Corollary 25. *The class of total procedures is undecidable.*

Rice's theorem looks similar to Scott's theorem but neither can be obtained from the other. Recall that procedures are reduction equivalent only if they are identical.

The key idea in the proof of Rice's lemma is the construction of the procedure v that constructs a procedure that has the right properties. In textbooks this intriguing piece of meta-programming is usually carried out in English using Turing machines in place of procedures. We doubt that there is a satisfying formal proof of Rice's lemma using Turing machines.

9   Step-Indexed Interpreter and Modesty

We will now prove that procedural decidability implies functional decidability. The proof employs a step-indexed interpretation function for the evaluation relation $s \triangleright t$. The interpretation function will also serve as the basis for a step-indexed self-interpreter for L, which is needed for the remaining results of this paper.

We use $\mathbf{T}$ to denote the type of terms, $\mathbf{T}_\emptyset$ to denote the option type for $\mathbf{T}$, and $\lfloor s \rfloor$ and $\emptyset$ to denote the values of $\mathbf{T}_\emptyset$. We define a function $eval : \mathbf{N} \to \mathbf{T} \to \mathbf{T}_\emptyset$ satisfying the following recursive specification.

$$
\begin{aligned}
eval\ n\ k &= \emptyset \\
eval\ n\ (\lambda s) &= \lfloor \lambda s \rfloor \\
eval\ 0\ (st) &= \emptyset \\
eval\ (Sn)\ (st) &= \text{match } eval\ n\ s,\ eval\ n\ t \text{ with} \\
&\qquad |\ \lfloor \lambda s \rfloor,\ \lfloor t \rfloor \Rightarrow eval\ n\ s_t^0 \\
&\qquad |\ \_\_ \Rightarrow \emptyset
\end{aligned}
$$

Fact 26.

1. If eval n s = ⌊t⌋, then eval (Sn) s = ⌊t⌋.
2. If s ≻ s' and eval n s' = ⌊t⌋, then eval (Sn) s = ⌊t⌋.
3. s ▷ t if and only if eval n s = ⌊t⌋ for some n.

Proof. Claim 1 follows by induction on n. Claim 2 follows by induction on n using Claim 1. Claim 3, direction →, follows by induction on $s \succ^* t$ and Claim 2. Claim 3, direction ←, follows by induction on n.   □

Lemma 27. *There is a function of type* ∀s. $\mathcal{E}$ s → Σt. s ▷ t.

Proof. Let s be a term such that $\mathcal{E}s$. Then we have $\exists nt.\ eval\ n\ s = ⌊t⌋$ by Fact 26. Since the predicate $\lambda n.\ \exists t.\ eval\ n\ s = ⌊t⌋$ is functionally decidable, constructive choice for **N** gives us an n such that $\exists t.\ eval\ n\ s = ⌊t⌋$. Hence we have t such that $eval\ n\ s = ⌊t⌋$. Thus $s ▷ t$ with Fact 26.   □

Theorem 28 (Modesty). *Procedurally decidable classes are functionally decidable.*

Proof. Let u be a decider for p. Let s be a term. Lemma 27 gives us a term v such that $u\bar{s} ▷ v$. Now we return true if $v = $ T and false otherwise.   □

We can also show modesty results for procedures other than deciders. For this we need a decoding for the Scott encoding of terms.

Fact 29 (Decoding). *There is a function* $\delta : $ **T** → **T**$_\emptyset$ *such that (1)* $\delta\bar{s} = ⌊s⌋$ *and (2)* $\delta s = ⌊t⌋ → \bar{t} = s$ *for all terms* s *and* t.

Fact 30 (Modesty). *Let* u *be a procedure such that* ∀s∃t. $u\bar{s} ▷ \bar{t}$. *Then there is a function* $f : $ **T** → **T** *such that* ∀s. $u\bar{s} ▷ \overline{fs}$.

Proof. Follows with Lemma 27 and Fact 29.   □

10   Choose

Choose is a procedure that given a decidable test searches for a number satisfying the test. Choose is reminiscent of minimisation for recursive functions [5]. Choose will be the only procedure in our development using truly unguarded recursion. We will use choose to obtain unbounded self-interpreters and to obtain recognisers from enumerators.

A *test* is a procedure u such that for every number n either $u\hat{n} ▷ $ T or $u\hat{n} ▷ $ F. A number n *satisfies* a test u if $u\hat{n} ▷ $ T. A test u is *satisfiable* if it is satisfied by some number.

Theorem 31 (Choose). *There is a procedure* C *such that for every test* u:

1. If u is satisfiable, then $Cu ▷ \hat{n}$ for some n satisfying u.
2. If Cu evaluates, then u is satisfiable.

Proof. We start with an auxiliary procedure H satisfying the recursive specification

$$H\,\widehat{n}\,u \;\equiv\; u\,\widehat{n}\,(\lambda\widehat{n})\,(\lambda(H(\mathrm{Succ}\,\widehat{n})u))\,\mathrm{I}$$

and define $\mathrm{C} := \lambda x.H\,\widehat{0}\,x$. Speaking informally, H realises a loop incrementing n until $u\,\widehat{n}$ succeeds. We say that $H\,\widehat{n}\,u$ is *ok* if $H\,\widehat{n}\,u \rhd \widehat{k}$ for some number k satisfying u and proceed as follows:

1. If n satisfies u, then $H\,\widehat{n}\,u$ is ok.
2. If $H\,\widehat{Sn}\,u$ is ok, then $H\,\widehat{n}\,u$ is ok.
3. If $H\,\widehat{n}\,u$ is ok, then $H\,\widehat{0}\,u$ is ok. Follows by induction on n with (2).
4. Claim 1 follows with (1) and (3).
5. If $H\,\widehat{n}\,u$ evaluates in k steps, then u is satisfiable. Follows by complete induction on k using the triangle property.
6. Claim 2 follows from (5) with $n = 0$.   □

Note that the verification of choose employs in (6) complete induction on the step-index of an evaluation together with the triangle property (Fact 10) to handle the unguarded recursion of the auxiliary procedure H. This is the only time these devices are used in our development.

11 Results Obtained with Self-Interpreters

For the specification of a step-indexed self-interpreter, we define an injective encoding function for term options:

$$\lfloor s \rfloor \;:=\; \lambda ab.a\overline{s}$$
$$\widehat{\emptyset} \;:=\; \lambda ab.b$$

Fact 32. *There is a procedure* E *such that* $\mathrm{E}\,\widehat{n}\,\overline{s} \equiv \widehat{eval\,n\,s}$ *for all n and s.*

Proof. We first construct and verify procedures realising the functions $m=n$ and s_u^k. We then construct and verify the procedure E following the recursive specification of the function *eval* in Sect. 9.   □

Theorem 33 (Step-Indexed Self-Interpreter).

1. *If* $\mathrm{E}\,\widehat{n}\,\overline{s} \rhd \lfloor t \rfloor$, *then* $\mathrm{E}\,\widehat{Sn}\,\overline{s} \rhd \lfloor t \rfloor$.
2. $\forall sn.\;(\mathrm{E}\,\widehat{n}\,\overline{s} \rhd \widehat{\emptyset}) \vee (\exists t.\;\mathrm{E}\,\widehat{n}\,\overline{s} \rhd \lfloor t \rfloor \wedge s \rhd t)$.
3. *If* $s \rhd t$, *then* $\mathrm{E}\,\widehat{n}\,\overline{s} \rhd \lfloor t \rfloor$ *for some n.*

Proof. Follows with Facts 32 and 26.   □

Theorem 34 (Totality). *The class of total procedures is not recognisable.*

Proof By the reduction lemma. We define fs as a procedure that for closed s is total iff $\neg\mathcal{E}(s\overline{s})$. We define fs such that $(fs)\overline{t}$ evaluates if t is an application or an abstraction. If t is a number n, we evaluate $s\overline{s}$ with the step-indexed self-interpreter for n steps. If this succeeds, we diverge using D, otherwise we return I. Here are the definitions of f and the realising procedure v:

$$f := \lambda s.\lambda y.y\,(\lambda z.\mathrm{E}\,z\,(\overline{s\overline{s}})\,\mathrm{D}\,\mathrm{I})\,\mathrm{F}\,\mathrm{I}$$
$$v := \lambda x.\mathrm{L}(\mathrm{A}(\mathrm{A}(\mathrm{A}\,\overline{0}\,(\mathrm{L}(\mathrm{A}(\mathrm{A}(\mathrm{A}(\mathrm{A}\,\overline{\mathrm{E}}\,\overline{0})(\mathrm{Q}\,(\mathrm{A}\,x(\mathrm{Q}\,x))))\,\overline{\mathrm{D}})\,\overline{\mathrm{I}})))\,\overline{\mathrm{F}})\,\overline{\mathrm{I}}) \qquad \square$$

Corollary 35. *The class of total terms is neither recognisable nor corecognisable.*

Proof. Suppose the class of total terms is recognisable. Then the class of total procedures is recognisable since the class of procedures is recognisable (follows with Fact 20). Contradiction with Theorem 34. The other direction is provided by Corollary 23. $\qquad \square$

We now construct an unbounded self-interpreter using the procedure choose and the step-indexed self-interpreter E.

Theorem 36 (Self-Interpreter). *There is a procedure* U *such that:*

1. *If* $s \triangleright t$*, then* $\mathrm{U}\,\overline{s} \triangleright \overline{t}$*.*
2. *If* $\mathrm{U}\,\overline{s}$ *evaluates, then* s *evaluates.*

Proof. $\mathrm{U} := \lambda x.\,\mathrm{E}\,(\mathrm{C}(\lambda y.\mathrm{E}yx(\lambda\mathrm{T})\mathrm{F}))\,x\,\mathrm{I}\,\mathrm{I}$ does the job. The verification uses Theorems 33 and 31. $\qquad \square$

Corollary 37. *The self-interpreter* U *recognises the class of evaluable terms.*

For Post's theorem we need a special self-interpreter considering two terms. We speak of a parallel or operator.

Theorem 38 (Parallel Or). *There is a procedure* O *such that:*

1. *If* s *or* t *evaluates, then* $\mathrm{O}\,\overline{s}\,\overline{t}$ *evaluates.*
2. *If* $\mathrm{O}\,\overline{s}\,\overline{t}$ *evaluates, then either* $\mathcal{E}\,s$ *and* $\mathrm{O}\,\overline{s}\,\overline{t} \triangleright \mathrm{T}$*, or* $\mathcal{E}\,t$ *and* $\mathrm{O}\,\overline{s}\,\overline{t} \triangleright \mathrm{F}$*.*

Proof. $\mathrm{O} := \lambda xy.\,(\lambda z.\mathrm{E}zx(\lambda\mathrm{T})(\mathrm{E}zy(\lambda\mathrm{F})\,\mathrm{I}))\,(\mathrm{C}(\lambda z.\mathrm{E}zx(\lambda\mathrm{T})(\mathrm{E}zy(\lambda\mathrm{T})\,\mathrm{F})))$ does the job. The verification uses Theorems 33 and 31. $\qquad \square$

Corollary 39 (Post). *If* u *recognises* p *and* v *recognises* $\lambda s.\neg ps$*, then the procedure* $\lambda x.\mathrm{O}\,(\mathrm{A}\,\overline{u}\,(\mathrm{Q}x))\,(\mathrm{A}\,\overline{v}\,(\mathrm{Q}x))$ *decides* p *provided* p *is logically decidable.*

With parallel or we can also show that the family of recognisable classes is closed under union.

Corollary 40 (Union). *If* u *recognises* p *and* v *recognises* q*, then the procedure* $\lambda x.\mathrm{O}\,(\mathrm{A}\,\overline{u}\,(\mathrm{Q}x))\,(\mathrm{A}\,\overline{v}\,(\mathrm{Q}x))$ *recognises* $\lambda s.\,ps \vee qs$*.*

12   Enumerable Classes

A class is *enumerable* if there is a procedurally realisable function from numbers to term options that yields exactly the terms of the class. More precisely, a procedure u *enumerates* a class p if:

1. $\forall n.\ (u\widehat{n} \triangleright \widehat{\emptyset}) \vee (\exists s.\ u\widehat{n} \triangleright \lfloor s \rfloor \wedge ps)$.
2. $\forall s.\ ps \rightarrow \exists n.\ u\widehat{n} \triangleright \lfloor s \rfloor$.

Following well-known ideas, we show that a class is recognisable if and only if it is enumerable. We will be content with informal outlines of the proof in the Coq development since we have already seen all necessary formal techniques.

Fact 41. *Given an enumerator for p, one can construct a recogniser for p.*

Proof. Given a term s, the recogniser for p searches for a number n such that the enumerator for p yields s (using the procedure choose).   □

Fact 42. *The class of a all terms is enumerable.*

Proof. One first writes an enumerator function and then translates it into a procedure. The translation to a procedure is routine. Coming up with a compact enumeration function is a nice programming exercise. Our solution is in the Coq development.   □

Fact 43. *Given a recogniser for p, one can construct an enumerator for p.*

Proof. Given n, the enumerator for p obtains the term option for n using the term enumerator. If the option is not of the form $\lfloor ns \rfloor$, the enumerator for p fails. If the option is of the form $\lfloor ns \rfloor$, the recogniser for p is run on $\overline{s}$ for n steps using the step-indexed self-interpreter. If this succeeds, the enumerator for p succeeds with s, otherwise it fails.   □

13   Markov's Principle

Markov's principle is a proposition not provable constructively and weaker than excluded middle [7]. Formulated for L, Markov's principle says that a class is decidable if it is recognisable and corecognisable. We establish two further characterisations of Markov's principle for L using parallel or (Theorem 38) and the enumerability of terms (Fact 42).

Lemma 44. *If p is decidable, then $\lambda\_.\exists s.ps$ is recognisable.*

Proof. Follows with Fact 42 and 31.   □

Theorem 45 (Markov's Principle). *The following statements are equivalent:*

1. *If a class is recognisable and corecognisable, then it is decidable.*
2. *Satisfiability of decidable classes is stable under double negation:*
   $\forall p.\ decidable\ p \to \neg\neg(\exists s.ps) \to \exists s.ps.$
3. *Evaluation of closed terms is stable under double negation:*
   $\forall s.\ closed\ s \to \neg\neg\mathcal{E}s \to \mathcal{E}s.$

Proof. $1 \to 2$. Let p be decidable and $\neg\neg\exists s.ps$. We show $\exists s.ps$. By (1), Lemma 44, and $\neg(\exists s.ps) \leftrightarrow \bot$ we know that the class $\lambda_-.\exists s.ps$ is decidable. Thus we have either $\exists s.ps$ or $\neg\exists s.ps$. The first case is the claim and the second case is contradictory with the assumption.

$2 \to 3$. Let s be a closed term such that $\neg\neg\mathcal{E}s$. We show $\mathcal{E}s$. Consider the decidable class $p := \{\, n \mid eval\ n\ s \neq \emptyset\,\}$. We have $\mathcal{E}s \leftrightarrow \exists t.pt$. By (2) it suffices to show $\neg\neg\exists t.pt$, which follows with the assumption $\neg\neg\mathcal{E}s$.

$3 \to 1$. Let u be a recogniser for p and v be a recogniser for $\lambda s.\neg ps$. We show that $\lambda x.O\,(A\,\overline{u}\,(Qx))\,(A\,\overline{v}\,(Qx))$ is a decider for p. By Theorem 38 it suffices to show that $O\,(\overline{us})\,(\overline{vs})$ evaluates for all terms s. Using (3) we prove this claim by contradiction. Suppose $O\,(\overline{us})\,(\overline{vs})$ does not evaluate. Then, using Theorem 38, neither $u\overline{s}$ nor $v\overline{s}$ evaluates. Thus $\neg ps$ and $\neg\neg ps$. Contradiction. $\square$

We remark that Markov's principle for L follows from a global Markov's principle saying that satisfiability of functionally decidable classes of numbers is stable under double negation. This can be shown with Theorem 45(3) and the equivalence $\mathcal{E}s \leftrightarrow \exists n.\ eval\ n\ s \neq \emptyset$.

References

1. Asperti, A., Ricciotti, W.: Formalizing turing machines. In: Ong, L., Queiroz, R. (eds.) WoLLIC 2012. LNCS, vol. 7456, pp. 1–25. Springer, Heidelberg (2012). doi:10.1007/978-3-642-32621-9_1
2. Asperti, A., Ricciotti, W.: A formalization of multi-tape turing machines. Theoret. Comput. Sci. **603**, 23–42 (2015)
3. Barendregt, H.P., Calculus, T.L.: Its Syntax and Semantics, 2 revised edn. North-Holland, Amsterdam (1984)
4. Bauer, A.: First steps in synthetic computability theory. ENTCS **155**, 5–31 (2006)
5. Boolos, G., Burgess, J.P., Jeffrey, R.C.: Computability and Logic, 5th edn. Cambridge University Press, Cambridge (2007)
6. Ciaffaglione, A.: Towards turing computability via coinduction. Sci. Comput. Program. **126**, 31–51 (2016)
7. Coquand, T., Mannaa, B.: The independence of Markov's principle in type theory. In: FSCD 2016. LIPIcs, vol. 52, pp. 17:1–17:18. Schloss Dagstuhl (2016)
8. Dal Lago, U., Martini, S.: The weak lambda calculus as a reasonable machine. Theor. Comput. Sci. **398**(1–3), 32–50 (2008)
9. Hopcroft, J., Motwani, R., Ullman, J.: Introduction to Automata Theory, Languages, and Computation. Pearson, New York (2013)
10. Jansen, J.M.: Programming in the λ-calculus: from church to scott and back. In: Achten, P., Koopman, P. (eds.) The Beauty of Functional Code. LNCS, vol. 8106, pp. 168–180. Springer, Heidelberg (2013). doi:10.1007/978-3-642-40355-2_12
11. Kozen, D.: Automata and Computability. Springer, New York (1997)

12. Mogensen, T.Æ.: Efficient self-interpretations in lambda calculus. J. Funct. Program. **2**(3), 345–363 (1992)
13. Niehren, J.: Functional computation as concurrent computation. In: POPL 1996, pp. 333–343. ACM (1996)
14. Norrish, M.: Mechanised computability theory. In: Eekelen, M., Geuvers, H., Schmaltz, J., Wiedijk, F. (eds.) ITP 2011. LNCS, vol. 6898, pp. 297–311. Springer, Heidelberg (2011). doi:10.1007/978-3-642-22863-6_22
15. The Coq Proof Assistant. http://coq.inria.fr
16. Xu, J., Zhang, X., Urban, C.: Mechanising turing machines and computability theory in Isabelle/HOL. In: Blazy, S., Paulin-Mohring, C., Pichardie, D. (eds.) ITP 2013. LNCS, vol. 7998, pp. 147–162. Springer, Heidelberg (2013). doi:10.1007/978-3-642-39634-2_13

Bellerophon: Tactical Theorem Proving
for Hybrid Systems

Nathan Fulton[✉], Stefan Mitsch, Rose Bohrer, and André Platzer

Computer Science Department, Carnegie Mellon University, Pittsburgh, PA, USA
{nathanfu,smitsch,aplatzer}@cs.cmu.edu

Abstract. Hybrid systems combine discrete and continuous dynamics, which makes them attractive as models for systems that combine computer control with physical motion. Verification is undecidable for hybrid systems and challenging for many models and properties of practical interest. Thus, human interaction and insight are essential for verification. Interactive theorem provers seek to increase user productivity by allowing them to focus on those insights. We present a tactics language and library for hybrid systems verification, named Bellerophon, that provides a way to convey insights by programming hybrid systems proofs.

We demonstrate that in focusing on the important domain of hybrid systems verification, Bellerophon emerges with unique automation that provides a productive proving experience for hybrid systems from a small foundational prover core in the KeYmaera X prover. Among the automation that emerges are tactics for decomposing hybrid systems, discovering and establishing invariants of nonlinear continuous systems, arithmetic simplifications to maximize the benefit of automated solvers and general-purpose heuristic proof search. Our presentation begins with syntax and semantics for the Bellerophon tactic combinator language, culminating in an example verification effort exploiting Bellerophon's support for invariant and arithmetic reasoning for a non-solvable system.

1 Introduction

Cyber-Physical Systems combine computer control with physical dynamics in ways that are often safety-critical. Reasoning about safety properties of Cyber-Physical Systems requires analyzing the system's discrete and continuous dynamics together in a *hybrid system* [2,13]. For example, establishing safety of an adaptive cruise controller in a car requires reasoning about the computations of the controller together with the resulting physical motion of the car.

Theorem proving is an attractive technique for verifying correctness properties of hybrid systems because it is applicable to a large class of hybrid systems

This material is based upon work supported by the National Science Foundation under NSF CAREER Award CNS-1054246 and NSF CNS-1446712. This research was sponsored by the AFOSR under grant number FA9550-16-1-0288. This research was supported as part of the Future of Life Institute (futureoflife.org) FLI-RFP-AI1 program, grant #2015-143867.

© Springer International Publishing AG 2017
M. Ayala-Rincón and C.A. Muñoz (Eds.): ITP 2017, LNCS 10499, pp. 207–224, 2017.
DOI: 10.1007/978-3-319-66107-0_14

[25]. Verification for hybrid systems is not semidecidable, thus requiring human assistance along two major dimensions. First, general-case hybrid systems proving requires identifying invariants of loops and differential equations, which is undecidable in both theory and practice. Second, the remaining verification tasks consist of first-order logic over the reals with polynomial terms. Decision procedures exist which are complete in theory [7], but are only complete in practice if a human provides additional guidance. Because both these dimensions are essential to hybrid systems proving, innovating along these dimensions benefits a wide array of hybrid systems verification tasks.

We argue that trustworthy and productive hybrid systems theorem proving requires: *(1)* a small foundational core; *(2)* a library of high-level primitives automating common deductions (e.g., computing Lie Derivatives, computing and proving solutions of ODEs, propagating quantities across dynamics in which they do not change, automated application of invariant candidates, and conservation/symmetry arguments); and *(3)* scriptable heuristic search automation.

Even though these ingredients can be found scattered across a multitude of theorem provers, their combination to a tactical theorem proving technique for hybrid systems is non-obvious. Table 1 compares several tools along the dimensions that we identify as crucial to productive hybrid systems verification (SC indicates a soundness-critical dependency on user-defined tactics or on an external implementation of a more scalable arithmetic decision procedure).

Table 1. Comparison to related verification tools and provers

Tool	Small Core	HS Library	HS Auto	Scriptable	External Tools
KeYmaera X	Yes	Yes	Yes	Yes	SC
SpaceEx	No	No	Yes	No	SC
Isabelle,Coq,HOL	Yes	No	No	Yes	No
KeYmaera 3	No	Yes	Yes	SC	SC

General purpose theorem provers, such as Coq [20] and Isabelle [23], have small foundational cores and tactic languages, but their tactic languages and automation are not tailored to the needs of hybrid systems. This paper addresses the problem of getting from a strong mathematical foundation of hybrid systems [27] to a productive hybrid systems theorem proving tool. Reachability analysis tools, e.g. SpaceEx [11], provide automated hybrid systems verification for *linear* hybrid systems, but at the expense of a large trusted codebase and limited ways of helping when automation fails, which is inevitable due to the undecidability of the problem. KeYmaera's [29] user-defined rules are no adequate solution because they enlarge the trusted codebase and are difficult to get right.

KeYmaera X [12] is structured from the very beginning to maintain a small and trustworthy core, upon which this paper builds the Bellerophon tactic language. Using these logical foundations [27], we develop a set of automated deduction procedures. These procedures manifest themselves as a *library of hybrid systems primitives* in which complex hybrid systems can be interactively verified.

Finally, heuristic automation tactics written in Bellerophon automatically apply these primitives to provide automation of hybrid systems reachability analysis.

Contributions. This paper demonstrates how to combine a small foundational core [27], reusable automated deductions, and problem-specific proof-search tactics into a tactical theorem prover for hybrid systems. It presents Bellerophon, a hybrid systems tactics language and library implemented in the theorem prover KeYmaera X [12]. Bellerophon includes a tactics library which provides the decision procedures and heuristics necessary for a productive interactive hybrid systems proving environment. We first demonstrate the interactive verification benefits of Bellerophon through interactive verification of a simple hybrid system, which is optimized to showcase a maximum of features in a minimal example. In the process, we also discuss significant components of the Bellerophon standard library that enable such tactical theorem proving. We then present two examples of proof search procedures implemented in Bellerophon, demonstrating Bellerophon's suitability for implementing reusable proof search heuristics for hybrid systems. Along the way, we demonstrate how the language features of Bellerophon support manual proofs and proof search automation.

2   Background

This section reviews hybrid programs, a programming language for hybrid systems; differential dynamic logic ($d\mathcal{L}$) [24–27] for specifying reachability properties about hybrid programs; and the theorem prover KeYmaera X for $d\mathcal{L}$ [12].

Hybrid (dynamical) systems [2,26] are mathematical models for the interaction between discrete and continuous dynamics, and hybrid programs [24–27] their programming language. The syntax and informal semantics is in Table 2.

Table 2. Hybrid programs

Program statement	Meaning
$\alpha; \beta$	Sequentially composes α and β
$\alpha \cup \beta$	Executes either α or β
α^*	Repeats α zero or more times
$x := \theta$	Evaluates term θ and assigns result to x
$x := *$	Assigns an arbitrary real value to x
$\{x'_1 = \theta_1, ..., x'_n = \theta_n \& F\}$	Continuous evolution within F along this ODE
$?F$	Aborts if formula F is not true

The following[1] hybrid program outlines a simple model of a skydiver who deploys a parachute to land at a safe speed. Here, we illustrate the rough program

---

[1] A continuous evolution along the differential equation system $x'_i = \theta_i$ for an arbitrary duration within the region described by formula F. The $\& F$ is optional so that e.g., $\{x' = \theta\}$ is equivalent to $\{x' = \theta \& true\}$.

structure to become acquainted with the syntax. We will fill in the necessary details (e.g., when to deploy the parachute exactly) for a proof later in Sect. 4.

Example 1 (Skydiver model). $\big(\underbrace{(?Dive \cup r := p)}_{ctrl}; \underbrace{\{x' = v,\ v' = f(v,g,r)\}}_{plant\ (continuous\ dynamics)}\big)^*$

Example 1 describes a skydiver whose *ctrl* chooses nondeterministically to continue diving if the formula *Dive* indicates it is still safe to do so, or to deploy the parachute ($r := p$). The skydiver's altitude x then follows a differential equation, where the velocity v non-linearly depends on v itself, gravity g and drag coefficient r. This process may repeat arbitrarily many times (indicated by the repetition operator *). Because there is no evolution domain constraint on *plant*, each continuous evolution has any arbitrary non-negative duration $e \in \mathbb{R}$.

Differential dynamic logic (d$\mathcal{L}$) [24–27] is a first-order multimodal logic for specifying and proving properties of hybrid programs. Each hybrid program α has modal operators $[\alpha]$ and $\langle \alpha \rangle$, which express reachability properties of program α. The formula $[\alpha]\phi$ expresses that the formula ϕ is true in all states reachable by the hybrid program α. Similarly, $\langle \alpha \rangle \phi$ expresses that the formula ϕ is true after some execution of α. The d$\mathcal{L}$ formulas are generated by the grammar

$$\phi ::= \theta_1 \smile \theta_2 \mid \neg\phi \mid \phi \wedge \psi \mid \phi \vee \psi \mid \phi \rightarrow \psi \mid \phi \leftrightarrow \psi \mid \forall x\, \phi \mid \exists x\, \phi \mid [\alpha]\phi \mid \langle \alpha \rangle \phi$$

where ϕ and ψ are formulas, α ranges over the hybrid programs of Table 2, and $\smile$ is a comparison operator $=, \neq, \geq, >, \leq, <$, and θ is a term of real arithmetic.

Model 1 (Safety specification for the skydiver)

$$\underbrace{x \geq 0 \wedge \ldots}_{initial\ condition} \rightarrow [(\underbrace{(?Dive \cup r := p)}_{ctrl};\ \underbrace{\{x' = v, v' = f(v,g,r)\}}_{plant})^*]\underbrace{(x{=}0 \rightarrow |v|{\leq}|m|)}_{post\ cond.}$$

The formula above expresses that if the skydiver, among other things, starts diving at some non-negative altitude x, then it is always the case that if they touch ground ($x = 0$) they do so softly with a safe descending speed ($|v|{\leq}|m|$, because both v and m are always negative).

KeYmaera X. Bellerophon is part of KeYmaera X, an axiomatic theorem prover for d$\mathcal{L}$ [12]. Its uniform substitution mechanism [27] enables a trusted core of only about 1,700 lines of Scala. This is far smaller than other hybrid systems verification tools and compares favorably even with many other LCF-style provers. While verified real arithmetic solving is possible via witnesses [30], KeYmaera X uses external real arithmetic solvers in practice for their superior performance.

3   The Bellerophon Tactic Language

Bellerophon is a programming language and standard library for automating proof constructions and proof search operations of the KeYmaera X core. As in other LCF-style provers, Bellerophon is not soundness-critical. This frees us to provide courageous reasoning strategies that enable users to perform high-level proofs about hybrid systems while still benefiting from the high degree of trustworthiness that comes from a small soundness-critical core and the cross-verification of d$\mathcal{L}$ in Isabelle and Coq [4]. A basic use of Bellerophon is to recover a convenient sequent calculus for d$\mathcal{L}$ [24] from the simpler Hilbert calculus-based core [27] of KeYmaera X. This demonstrates that Bellerophon is expressive enough to implement the automation capabilities of the predecessor prover KeYmaera [29] from a smaller set of primitives. Beyond that, Bellerophon is used, e.g. for programming both individual proofs and custom proof search procedures.

This section presents the basic constructs of the Bellerophon language. Readers familiar with tactic languages for interactive theorem provers (e.g., [20]) will find many constructs familiar, but should pay particular attention to the discussion of Bellerophon's standard library. For usability, traceability and educational purposes, Bellerophon tactics can be written in a hierarchical structure that maps to the graphical tree structure of the resulting d$\mathcal{L}$ sequent proof [21].

This d$\mathcal{L}$ proof motivates the constructs of our language and standard library.

Proof 1 (Skydiver sequent proof sketch). The proof starts from the initial conjecture (Model 1) at the bottom, phrased as a *sequent*. Each sequent has the shape *assumptions* $\vdash$ *obligations*, which means from the assumptions left of the turnstile $\vdash$, we have to prove any formula on the right. Horizontal lines indicate that the sequent below the horizontal line is proved when the sequent above the horizontal line is proved, justified by the tactic that is annotated left of the horizontal bar (the corresponding operator is highlighted in boldface and red). For example, the first step prop makes all conjuncts left of an implication available as assumptions, so the goal $x \geq 0 \wedge B \rightarrow C$ below the line becomes $x \geq 0, B \vdash C$ above the line. When proof rules (e.g., andR) result in multiple subgoals, each subgoal continues in a separate branch and all need to be proved.

$$
\begin{array}{c}
\cdots \qquad\qquad\qquad\qquad \cdots \\[2pt]
\dfrac{\Gamma \vdash Dive \rightarrow [ode](x{=}0 \rightarrow |v|{\leq}|m|)}{\Gamma \vdash [?Dive][\cdots](x{=}0 \rightarrow |v|{\leq}|m|)} \text{testb} \qquad
\dfrac{\Gamma \vdash [ode(p)](x{=}0 \rightarrow |v|{\leq}|m|)}{\Gamma \vdash [r := p][\cdots](x{=}0 \rightarrow |v|{\leq}|m|)} \text{assignb} \\[6pt]
\dfrac{}{\Gamma \vdash [?Dive][\cdots](x{=}0 \rightarrow |v|{\leq}|m|) \wedge [r := p][\cdots](x{=}0 \rightarrow |v|{\leq}|m|)} \text{andR} \\[6pt]
\dfrac{}{\Gamma \vdash [\{?Dive \cup r := p\}][\{p' = v, v' = f(v,g,r)\}](x{=}0 \rightarrow |v|{\leq}|m|)} \text{choiceb} \\[6pt]
\dfrac{}{x \geq 0, \ldots \vdash [\{?Dive \cup r := p\}; \{x' = v, v' = f(v,g,r)\}](x{=}0 \rightarrow |v|{\leq}|m|)} \text{composeb} \\[6pt]
\dfrac{}{\vdash x \geq 0 \wedge \ldots \rightarrow [\{?Dive \cup r := p\}; \{x' = v, v' = f(v,g,r)\}](x{=}0 \rightarrow |v|{\leq}|m|)} \text{prop}
\end{array}
$$

Each of the steps in the sequent proof above is a built-in tactic:

prop Exhaustively applies propositional proof rules in the sequent calculus.
composeb Splits sequential composition $[\alpha; \beta]P$ into nested modalities $[\alpha][\beta]P$.
choiceb Splits choice $[\alpha \cup \beta]P$ into a conjunction of subsystems $[\alpha]P \wedge [\beta]P$.
andR, implyR, existsL, ... are the right conjunction rule ($\wedge$R), the right implication rule ($\rightarrow$R) and left existential rule ($\exists$L) as usual in sequent calculus. Throughout the paper, we will make use of standard propositional sequent calculus tactics that follow this naming convention.

testb Makes test condition $[?Q]P$ available as assumption $Q \rightarrow P$.
assignb Makes effect of assignment $[x := t]P(x)$ available as update to $P(t)$ or as assumption $x = t$ with proper renaming of other occurrences of x.

Bellerophon programs, called tactics, are functions mapping lists of sequents to (lists of[2]) sequents. Built-in tactics (ranged over by τ) are implemented in Scala. Proof developers can combine existing tactics using the constructs described in Table 3. Built-in programs are implemented as a sequence of operations on a data structure that can only be created or modified by the soundness-critical core of KeYmaera X, thereby ensuring soundness of built-in tactics.

Table 3. Meaning of tactic combinators

Language Primitive	Operational Meaning
$e ::= \tau$	Built-in tactic
$\quad \| \; e(\bar{v})$	Applies a tactic e to a (list of) positions or formulas
$\quad \| \; e_1 \; ; \; e_2$	Sequential Composition: Applies e_2 on the output of e_1
$\quad \| \; e_1 \| e_2$	Either Composition: Applies e_2 if applying e_1 fails
$\quad \| \; e^*$	Saturating Repetition: Repeatedly applies e as long as it is applicable (diverging if it stays applicable indefinitely)
$\quad \| \; ?(e)$	Optional: Applies e if e does not result in an error
$\quad \| \; <(e_1, e_2, \ldots e_n)$	Applies e_1 to the first of n subgoals, e_2 to the second, etc.
$\quad \| \;$ abbrv $P(\bar{x}) = \phi$ in e	Replaces all occurrences of ϕ with $P(\bar{x})$ in the current subgoal, and then applies e. After e, remaining occurrences of $P(\bar{x})$ are uniformly substituted back to ϕ

Built-in Tactics. Bellerophon is both a stand-alone language and a domain-specific language embedded in the Scala programming language. Built-in tactics directly manipulate the KeYmaera X core to transform formulas in a validity-preserving manner. Bellerophon programmers can construct new tactics either by writing new built-in tactics in Scala, or else by combining pre-existing tactics using the combinators described in Table 3. KeYmaera X ships with a large library of tactics for proof construction and proof search. Some built-in tactics – the propositional rules and `choiceb` for example – are straight-forward applications of the axioms in [27]. Others provide a significant amount of automation on top of the axiomatic foundations. For example, `prop` combines propositional sequent calculus rules to an automated proof search procedure that often performs numerous simpler proof steps automatically.

---

[2] Tactics may map a single sequent to a list of sequents; the simplest example of such a tactic andR corresponds to the proof rule $\wedge$R, which maps a single sequent $\Gamma \vdash A \wedge B, \Delta$ to the list of subgoals $\Gamma \vdash A, \Delta$ and $\Gamma \vdash B, \Delta$.

Parameters. Most tactics are parameterized by formulas, locators, or both. Formula parameters are provided whenever the behavior of a tactic is dependent upon a particular formula; for example, the loop and differential induction tactics take an invariant formula as parameter. Locators specify where in a sequent a tactic should be applied. The simplest form of locator is an explicit position. Negative positions refer to formulas to the left of the turnstile ($\vdash$) and positive positions refer to formulas to the right of the turnstile,[3] e.g., $-1:A$, $-2:B$, $-3:C \vdash 1:D$, $2:E$ with annotated formula positions. In addition to explicit positions, Bellerophon provides indirect locators: *(i)* $e(\text{R})$ applies e to the first applicable position[4] in the succedent; *(ii)* $e(\text{Rlast})$ applies e to the last position in the succedent. $e(\text{L})$ and $e(\text{Llast})$ behave accordingly in the antecedent.

Basic Combinators. Tactics are executed sequentially using the ; combinator. In $e;f$, the left tactic e is executed on the current subgoal and then the right tactic f is executed on the result of the left tactic's execution. The | combinator attempts multiple tactics – moving from left to right through a list of alternatives. The $*$ combinator in e^* repeats the tactic e as long as it is applicable. Many proof search procedures are expressible as a repetition of choices.

Branching. Proof search often results in branching. For example, a canonical proof of the induction step of Model 1 decomposes into two cases: a diving case corresponding to the control decision *?Dive* and a deployed parachute case corresponding to the control decision $r := p$. Proof 1 from above in the d$\mathcal{L}$ sequent calculus visually emphasizes the branching structure, which can be helpful for structuring tactics too. The $<$ combinator expresses how a proof decomposes into cases. An explicit tactic directly performing Proof 1 without any search is:

Listing 1.1. A Structured Bellerophon Tactic for a Branching Proof

```
1  prop ; composeb(1) ; choiceb(1) ; andR(1) ; <(
2    testb(1) ; ...,        /* tactic for left  branch of andR */
3    assignb(1) ; ...       /* tactic for right branch of andR */
4  )
```

Equivalently, the proof search tactic `unfold` automates proofs such as Listing 1.1 by applying all propositional and dynamical axioms until encountering a loop program or a differential equation, where cleverness might be needed.

4 Demonstration of Tactical Hybrid Systems Proving

In this section, we demonstrate that the Bellerophon standard library's techniques for invariance properties, conservation properties, and real arithmetic

---

[3] The addressing scheme extends to subformulas and subterms in a straight-forward way. Interested readers may refer to the Bellerophon documentation for details.

[4] Tactic e is applicable at a position *pos* if $e(pos)$ does not result in an error.

simplifications, as implemented in KeYmaera X, make it a convenient mechanism for interactively verifying hybrid systems. The proof developed in this section is at http://web.keymaeraX.org/show/itp17/skydiver.kya.

Model 2 fills in the details of the skydiver model, which guarantees landing at a safe speed if the parachute opens early enough.

Model 2 (Safety specification for the skydiver model).

$$x \geq 0 \wedge g > 0 \wedge 0 < a = r < p \wedge -\sqrt{\frac{g}{p}} < v < 0 \wedge m < -\sqrt{\frac{g}{p}} \wedge T \geq 0 \quad (init)$$

$$\rightarrow [\{ (? \underbrace{\left(r = a \wedge v - g \cdot T > -\sqrt{\frac{g}{p}} \right)}_{Dive} \cup \ r := p); \quad (ctrl)$$

$$t := 0; \ \{x' = v, \ v' = r \cdot v^2 - g \ \& \ x \geq 0 \wedge v < 0 \wedge t \leq T\} \quad (plant)$$

$$\}^*](x = 0 \rightarrow |v| \leq |m|) \quad (post\ cond.)$$

Opening the parachute is a discrete control decision. The diver's physics are modeled as an ODE, accounting for both gravity and drag, which changes when the parachute opens. This example is carefully crafted to demonstrate many of the challenges in hybrid systems reasoning while retaining relatively simple dynamics. Qualitative changes happen to the continuous dynamics after a discrete state transition, the dynamics are non-linear, and the property of interest is not directly inductive.

We model a gravitational force ($g > 0$), a drag coefficient (r) which depends on whether the parachute is closed (air a) vs. open (parachute p), the skydiver's altitude $x \geq 0$ and velocity $v < 0$. The time between control decisions is bounded by the skydiver's reaction time T. We also assume that the diver does not pass through the earth $x \geq 0$ and (to streamline this presentation) that $v < 0$.

The controller contains two options for our skydiver. The left choice lets a closed parachute ($r = a$) stay closed if the speed after one control cycle is definitely safe, computed by over-approximating as if gravity were the only force ($v - g \cdot T > -\sqrt{\frac{g}{p}}$). The right control choice opens the parachute, after which it stays open (as $r \neq a$). For simplicity, we say the parachute opens instantly.

The safety theorem says when the skydiver hits the ground, the velocity is at most a specified safe landing speed $|v| \leq |m|, v < 0$. We assume the parachute is initially closed ($r = a$), the speed initially safe ($v > -\sqrt{\frac{g}{p}}$), and the safe landing speed faster than the limit speed of the parachute ($m < -\sqrt{\frac{g}{p}}$).

Loop Invariants. Verifying a system loop begins with identifying a loop invariant J that is true initially, implies the post-condition and is preserved by the controller. Each formula of the initial condition in Model 2 is invariant except $r = a$; therefore, we will proceed with the following invariant J:

$$\underbrace{(x \geq 0 \wedge v < 0)}_{ev.\,dom.} \wedge \underbrace{\left(g > 0 \wedge 0 < a < p \wedge T \geq 0 \wedge m < -\sqrt{\frac{g}{p}}\right)}_{diff.\,inductive} \wedge \underbrace{v > -\sqrt{\frac{g}{p}}}_{hard} \quad (1)$$

Note that J holds initially and implies formula $|v| \leq |m|$ because $v > -\sqrt{\frac{g}{p}} > m$. These facts prove automatically. Therefore, the core proof needs to prove $J \rightarrow [ctrl; plant]J$. We express the proof thus far with the following tactic:

Listing 1.2. Loop Induction Tactic

```
implyR(1); loop(J, 1); <(QE,QE,nil)
```

The `implyR` tactic corresponds to the right implication rule ($\rightarrow$R) in sequent calculus; the first argument states that we should apply this proof rule at the first position in the succedent. The `loop` tactic uses the d$\mathcal{L}$ axioms about loops to derive three new subgoals: (1) the loop invariant holds initially ($init \rightarrow J$); (2) the loop invariant implies the post condition ($J \rightarrow post\,cond.$); and (3) the loop invariant is preserved throughout a single iteration of the loop ($J \rightarrow [ctrl; plant]J$). The `loop` rule in KeYmaera X is derived in Bellerophon from axioms and automatically retains assumptions about constants that do not change in the system. The `nil` tactic has no effect and is used in <() to keep subgoal (3) unchanged.

The branching combinator <() allows us to isolate each of these three subtasks from one another. Subgoals (1) and (2) are proven using a Real Arithmetic solver (QE, for Quantifier Elimination), since the arithmetic is easy enough here.

Decomposing Control Programs. This model's control program is simple. It checks if it is safe to keep the parachute closed, or sets r to open the parachute (at any time, but at the latest when it is no longer safe to keep it closed). Therefore, we will immediately symbolically execute the control program and consider the two resulting subgoals, both of which are reachability conditions on purely continuous dynamical systems. This splitting could be done manually, as in Listing 1.1. But we decide to split it automatically using the `unfold` tactic.

Listing 1.3. Decomposing Control Programs

```
implyR(1); loop(J, 1); <(QE,QE,unfold)
```

ODE Tactics in the Standard Library. The rest of the proof will make use of several tactics in the Bellerophon standard library:

boxAnd Splits $[\alpha](P \wedge Q)$ into separate postconditions $[\alpha]P$ and $[\alpha]Q$.
dC(R) Proves a new property R of an ODE and then restricts the differential equation to remain within the evolution domain R (differential cut).

dW Proves $[x' = f(x)\&Q]P$ by proving that domain Q implies postcondition P.
dI Proves $[\{x' = f(x)\}]P$ by proving P and its differential P' along $x' = f(x)$.
dG(y'=ay+b,R) Adds new differential equation $y' = ay+b$ to $[x' = f(x)\&Q]P$, and replaces the post condition by equivalent formula R (possibly mentioning the fresh *differential ghost variable* y).

These tactics perform significant automation on top of the d$\mathcal{L}$ axioms. For example, dI performs automatic differentiation via exhaustive left-to-right rewriting of our axiomatization of differentials (e.g., $(s \cdot t)' = s't + st'$) and propagates the local effect of the differential equation. The dI tactic preserves initial value constraints for variables that are not changed by the differential equation. It often performs hundreds of axiom applications automatically. The difference between the sound Differential Induction axiom [27] and the automation provided by the dI tactic is an exemplary demonstration of the difference between a theoretically complete mathematical/logical foundation, and a pragmatically useful tactical library.

We are now ready to consider two purely continuous subgoals of the form $J \rightarrow [plant(r)]J$: one where $r = a$ (the parachute is closed) and one where $r = p$ (the parachute is open), which are both true for different reasons.

Closed Parachute: Chaining Inequalities. We first consider the $r = a$ case, in which the parachute is closed. Symbolically executing the control program results in a remaining subgoal that requires us to prove:

$$J \wedge v - g \cdot T > -\sqrt{g/p} \rightarrow [\{x' = v,\ v' = a \cdot v^2 - g \& x \geq 0 \wedge v < 0 \wedge t \leq T\}]J$$

We use boxAnd to work on the conjuncts of the loop invariant J (1) separately, since each are preserved for different reasons. The proofs for the first two sets of loop invariants in J (labeled *ev. domain* and *diff. induction*) are identical to the $r = p$ case and will be discussed later. Here, we focus on the formula $J \wedge v - g \cdot T > -\sqrt{\frac{g}{p}} \rightarrow [\{x' = v, v' = a \cdot v^2 - g \& x \geq 0 \wedge v < 0 \wedge t \leq T\}]v > -\sqrt{\frac{g}{p}}$, which handles the third conjunct of J (see (1), labeled *hard*).

Compute that $v \geq v_0 - g \cdot t \geq v_0 - g \cdot T > -\sqrt{\frac{g}{p}}$, where v_0 is the value of v before the ODE. In Bellerophon proofs for differential equations, we use $old(v)$ to introduce initial values; you can read $old(v)$ and v_0 inter-changeably here.

Each of the subformulas in the postcondition above is a differentially inductive invariant, or else is valid after the domain constraint is automatically augmented with constants $g > 0 \wedge p > 0$. Therefore, we use a chain of dC justified either by dI or by dW for each inequality in this tactic:

Listing 1.4. A Chain of Inductive Inequalities

```
1  /* Key lemmas                            proofs of lemmas */
2  dC(v>=old(v)-g()*t,1);           <(nil , dI(1));
3  dC(old(v)-g()*t>=old(v)-g()*T,1); <(nil , dW(1);QE);
4  dC(old(v)-g()*T>-c,1);           <(nil , dI(1));
5    dW(1) ; QE
```

The argument is a sequence of differential cuts, each of which has a simple proof, and whose conjunction implies the post-condition. Each of the `nil` tactics in the `<()` passes along a single subgoal to the next tactic, so that at the end we have a long conjunction in our domain constraint containing each of the cuts. This style of proof is pervasive in hybrid systems verification, and easily expressed in Bellerophon. One key feature that makes this proof so concise is the use of `old(v)`, which introduces a variable v_0 that remembers the initial value of v. Tactic `dW;`QE on line 3 proves the cut from the evolution domain constraint.[5]

The inequalities in the evolution domain of the differential equation system are now sufficiently strong to guarantee the postcondition, so we use `dW` to obtain a final arithmetic subgoal: $\Gamma \vdash v \geq v_0 - g \cdot t \geq v_0 - g \cdot T > -\sqrt{\frac{g}{p}} \rightarrow v > -\sqrt{\frac{g}{p}}$, where Γ contains constants propagated by the rule `dW` (unlike the DW axiom).

Although this arithmetic fact is obvious to us, `QE` will take a substantial amount of time to prove this property (at least 15 min on a 32 core machine running version 10 Mathematica and version 4.3.7 of KeYmaera X). This is a fundamental limitation of Real Arithmetic decision procedures, which have extremely high theoretical and practical complexity [9].

The simplest way to help `QE` is to introduce a simpler formula that captures the essential arithmetic argument: e.g., cut in $\forall a, b, c, d\, (a \geq b \geq c > d \rightarrow a > d)$ and then instantiate this formula with our chain of inequalities. We take this approach for demonstration (see the implementation). As an alternative, transforming and abbreviating formulas in Bellerophon achieves a similar effect.

Open Parachute: Differential Ghosts. We now consider case 2, where the parachute is already open $(r = p)$. After executing the discrete program the remaining subgoal is: $J \rightarrow [\{x' = v,\, v' = p \cdot v^2 - g\, \&\, \underbrace{x \geq 0 \wedge v < 0 \wedge t \leq T}\}]J$.

evolution domain constraint

The proof proceeds by decomposing the post-condition J into three separate subgoals, one for each conjunct in (1). In Listing 1.5, the `boxAnd` tactic uses axiom $[\alpha](P \wedge Q) \leftrightarrow [\alpha]P \wedge [\alpha]Q$ from left to right, to rewrite the instance of $[\alpha](P \wedge Q)$ to separate corresponding conjuncts $[\alpha]P \wedge [\alpha]Q$. The first set of formulas in J (labeled *ev. domain*) are *not* differentially inductive, but are trivially invariant because the evolution domain constraint of the system already contains these properties. Differential weakening by `dW` is the appropriate proof technique for these formulas, see line 1 in Listing 1.5. The second set of formulas (labeled *diff. inductive*) are *not* implied by the domain constraint, but are inductive along the ODE because the left and right sides of each inequality have the same time-derivative (0). Differential induction by `dI` is the appropriate proof technique for establishing the invariance of these formulas, see line 2 in Listing 1.5.

---

[5] The attentive reader will notice we use `g()` instead of g. This is to indicate that the model has an arity 0 function symbol `g()`, rather than an assignable variable. This syntactic convention follows KeYmaera X and its predecessors.

Listing 1.5. Differential Weakening and Differential Induction

```
1  boxAnd(1); andR(1); <(dW(1);QE , nil);
2  boxAnd(1); andR(1); <(dI(1) , nil)
```

The third conjunct (labeled *hard*) requires serious effort: we have to show that $v > -\sqrt{\frac{g}{p}}$ is an invariant of the differential equation. This formula is *not* a differentially inductive invariant because it is getting less true over time. To become inductive, we require additional dynamics to describe energy conservation. The Bellerophon library provides a tactic to introduce additional dynamics as *differential ghosts* into a differential equation system. Often, differential ghosts can be constructed systematically. Here, we want to show $v > -c$ where $c = \sqrt{\frac{g}{p}}$, so we need a property with a fresh differential ghost y that entails $v + c > 0$, e.g., $y^2(v + c) = 1$. The formula $y^2(v + c) = 1$ becomes inductively invariant when $y' = -\frac{1}{2}p(v-c)$. In summary, tactic dG in Listing 1.6 introduces $y' = -\frac{1}{2}p(v-c)$ into the system and rewrites the post-condition to $y^2(v + c) = 1$ with the additional assumptions that y does not contain any singularities ($p > 0 \wedge g > 0$).

Listing 1.6. Finishing the parachute open case with a ghost

```
1  dG(y'=-1/2*p*(v-(g()/p)^(1/2)), p>0&g()>0&y^2*(v+c)=1, 1);
2  dI(1.0); QE
```

Tactic dG results in a goal of the form $\exists y[\cdots](p > 0 \wedge g > 0 \wedge y^2(v+c) = 1)$, so in line 2 of Listing 1.6 we apply dI at the first child position 1.0 of succedent 1 *in context* of the existential quantifier to show that the new property $y^2(v + c) = 1$ is differentially invariant with the differential ghost y.

If a system avoids possible singularities, the ODE tactic in the Bellerophon standard library automatically computes the differential ghost dynamics (here $y' = -\frac{1}{2}p(v-c)$) and postcondition (here $y^2(v+c) = 1$) with the resulting proof. Additionally, notice that dG conveniently constructs the axiom instance of DG [27], saving the proof developer from manually constructing such instances.

The proof in Listing 1.6 completes the invariant preservation proof for $r = p$. The full proof artifact for the skydiver demonstrates how Bellerophon addresses each of the major reasoning challenges in a typical hybrid systems verification effort.

5   Automatic Tactics in the Bellerophon Standard Library

This section presents two significant automated tactics in the Bellerophon standard library: a heuristic tactic for invariants of ODEs, and a general-purpose hybrid systems heuristic building upon ODE automation. These tactics use our embedding of Bellerophon as a DSL in Scala, the KeYmaera X host language.[6]

---

[6] Advanced automation generally uses the EDSL. Programs written in the EDSL are executed using the same interpreter as programs written in pure Bellerophon.

The combination of a tactical language and a general-purpose functional language allow us to more cleanly leverage complicated computations, such as integrators and invariant generators, without losing the high-level proof structuring and search strategy facilities provided by Bellerophon. Significant further Bellerophon programs that ship with KeYmaera X include an automated deduction approach to solving differential equations [27], the proof-guided runtime monitor synthesis algorithm ModelPlex [22] and real arithmetic simplification procedures. KeYmaera X provides an IDE [21] for programming Bellerophon tactics and inspecting their effect in a sequent calculus proof.

The purpose of this section is to explain, at a high level, how Bellerophon provides ways of automating hybrid systems proof search. We only present simplified versions of tactics and briefly discuss relevant implementation details.

5.1 Tactical Automation for Differential Equations

Automated reasoning for ODEs is critical to scalable analysis of hybrid systems. Even when human interaction is required, automation for simple reachability problems – such as reachability for solvable or univariate subsystems – streamlines analysis and reduces requisite human effort.

The skydiver example above illustrated the interplay between finding differential invariants and proving with differential induction and differential ghosts. The tactic ODE in the Bellerophon standard library automates this interplay for solvable systems and some unsolvable, nonlinear systems of differential equations, see Listing 1.7. The ODEStep tactic directly proves properties by differential induction, with differential ghosts, and from the evolution domain constraints. The ODEInvSearch tactic cuts additional differential invariants, thereby strengthening the evolution domain constraints for ODEStep to ultimately succeed. Tactic ODE succeeds when ODEStep finds a proof; if ODEStep does not yet succeed, ODEInvSearch provides additional invariant candidates with differential cuts dC or by solving the ODE. This interaction between ODEStep and ODEInvSearch is implemented in Listing 1.7 by mixing recursion and repetition. Repetition is used in ODE so that ODEStep is prioritized over ODEInvSearch each time that a new invariant is added to the system. Recursion is used in ODEInvSearch so that a full proof search is started every time an invariant is successfully added to the domain constraint by dC. The ODEInvSearch tactic calls ODEStep on its second subgoal (the "show" branch of the dC) because differential cuts can be established in the right order without additional cuts.

Listing 1.7. Automated ODE Tactic for Non-Solvable Differential Equations

```
1  ODEStep(pos)       = dI(pos) | dgAuto(pos) | dW(pos) | ...
2  ODEInvSearch(pos)  = dC(nextCandidate); <(ODE(pos), ODEStep(pos
      ))
3                     | solve(pos)
4  ODE(pos)           = ( ODEStep(pos) | ODEInvSearch(pos) )*
```

The ODEStep tactic finds a proof with dI when the post-condition is differentially inductive, meaning that the vector field of the differential equation points into the set described by the differential equation. The dgAuto tactic will also attempt to make properties differentially inductive by constructing differential ghosts for the postcondition, such as the ghosts introduced in the skydiver example. In case the evolution domain of a differential equation system is sufficiently strong, tactic dW succeeds from just the evolution domain constraints. The ODEStep tactic implemented in KeYmaera X contains other proof search techniques (marked ... above) that are guaranteed to terminate but refrain from performing differential cuts.

The invariant search ODEInvSearch constructs candidates for differential invariants heuristically [28], see dC(nextCandidate) in Listing 1.7, or systematically for solvable differential equations with solve. Tactic solve follows an axiomatic ODE solver approach [27] that implements a solver in terms of the differential invariants, cuts, and ghosts reasoning principles to avoid a trusted built-in rule for solving differential equations (such trusted built-in rules are necessary in other hybrid systems tools, e.g., in KeYmaera [29]).

The ODE tactic described above is an idealized version of the ODE tactic implemented in KeYmaera X, which contains additional automated search procedures and specializes proof search based upon the shape of the post-condition.

5.2 Tactical Automation for Hybrid Systems

The solve and ODE tactics provide some automation for continuous systems proofs. The master tactic builds on these to provide a full heuristic for hybrid systems in the canonical form $init \rightarrow [\{ctrl; plant\}^*]safe$. Tactic master combines the three basic reasoning principles that together cover the reasoning tasks arising in hybrid systems models of the above shape: propositional reasoning, symbolic execution of hybrid programs, and reasoning about loops and differential equations.

Listing 1.8. Proof Search Automation for Hybrid Systems

```
1   master = OnAll(prop | step | close | QE | loop | ODE)*
```

In such proofs, branching is prevalent, e.g., due to non-deterministic choices in programs, as well as loop and differential induction. In the proofs so far we specified explicitly how the proof proceeds on each branch using <(). This approach is useful to specifically tailor tactics and provide user insight to certain subgoals. In a general-purpose search tactic, however, we neither know *a priori* how many branches there will be, nor how the specific subgoals on each branch are tackled best. The Bellerophon library lets us specify such general-purpose proof search with tactic alternatives |, repetition *, and continuing proof search on all branches with OnAll. The prop tactic is executed first on each subgoal. Running prop moves *init* into the antecedent in the initial theorem, but also performs propositional reasoning on each new subgoal generated by the proof. This enables propositional simplifications both after symbolic execution and loop/ differential induction, as well as to uncover propositional truths handled by close and thereby avoid potentially expensive arithmetic reasoning in QE. The step tactic picks the canonical dynamical axioms for a formula (by indexing techniques) and applies it in the canonical direction. For example, when applied to $[\alpha \cup \beta]P$, the step tactic will produce a new subgoal $[\alpha]P \land [\beta]P$. The step tactic focuses on the portions of a program that do not need any decisions such as invariants for loops or differential equations. The loop tactic generates loop invariants [28] and performs loop induction for the outer control loops, whereas ODE handles differential equations. The KeYmaera X implementation of master contains several optimizations to the ordering of tactics based upon empirical experience.

The ODE and master tactics demonstrate how Bellerophons's combinators are used to construct proof search procedures out of components available in the Bellerophon standard library.

6   Related Work

The novel contributions of this paper are the design and implementation of a tactics language and library for hybrid systems which have shown themselves to make tactical proving fruitful for realistic hybrid systems verification tasks.

Tactics Programming Languages. Tactics combinators appear in many general-purpose proof assistants, such as NuPRL [8], MetaPRL [15], Isabelle [3], Coq [20], and Lean [1]. However, our goals differ: all of the above aim to work for as many proving domains as possible, while we optimize for hybrid systems proving. In pursuing this aim, we have developed a unique, extensive suite of tactical automation for hybrid systems resting on a small trusted core. We integrate key techniques for continuous systems (ODE solving, invariant generation, and conservation reasoning via differential ghosts) with heuristic simplifications for arithmetic that speed up the use of external real-closed field solvers.

Arithmetic Proving. Proving theorems of first-order real arithmetic should not be confused with formalizing real analysis, though both are valuable. General-purpose proof assistants have been used to formalize much of real analysis [5,14,19,31], and in fact some such formalizations [16,17] have been used to prove the soundness of d$\mathcal{L}$'s proof calculus on which KeYmaera X and Bellerophon rest [4]. However, the style of proof used is different: like other domains in which general-purpose provers excel, formalized analysis benefits from the forms of automation that these provers do well, such as automatically expanding definitions and applying syntactic simplification rules. Because hybrid systems verification is less definition-heavy and because simplification rules alone (e.g. ring axioms) do not make real arithmetic tractable, real arithmetic proofs face problems for which existing automation is insufficient. Since arithmetic proofs do arise in these provers as well, we believe our techniques to be of broader interest. While we provide new automation for important tasks, this does not preclude us from using existing tactical techniques for the subtasks where they are most appropriate, such as propositional reasoning and decomposing composite hybrid systems.

Tactical Proving Styles. A set of patterns and anti-patterns have been proposed for Coq tactic programming in $\mathcal{L}_{\texttt{tac}}$ [6]. The suggestion is to use general-purpose automation as much as possible, conveying any problem-specific details through hints or lemmas. In keeping with this philosophy, the canonical usage of Bellerophon is to provide loop and sequences of differential invariants as hints to the automated `master` tactic. This reduces the proof to arithmetic. At this point the user can steer the proof further, e.g. by using Bellerophon's equational rewriting mechanisms to reduce complex arithmetic to simpler lemmas. This tactical proof-by-hint style can be mixed freely with other styles provided by the KeYmaera X user interface. For example, a user might use the UI to identify and apply an arithmetic simplification, at which point the corresponding tactic is generated automatically. They might then integrate this tactic into a larger proof-search algorithm which then solves similar proof cases automatically.

Analysis Tools for Hybrid Systems. Compared with other hybrid system analysis tools such as PHAver [10], SpaceEx [11], and dReach [18], Bellerophon enjoys the ability to handle a broad class of systems from a small trusted core provided by the host prover KeYmaera X. The addition of Bellerophon to KeYmaera X increases the class of systems for which verification is practical by using proof scripting to solve problems that would be too tedious and time-consuming otherwise.

7   Conclusion and Future Work

Bellerophon and its standard library support both interactive and automated theorem proving for hybrid systems. The library provides users with a clean interface for expressing common insights that are essential in hybrid systems verification tasks. Bellerophon combinators allow users to combine these base tactics in order to implement proofs and proof search procedures. Through Bellerophon, KeYmaera X provides sound tactical theorem proving for hybrid systems.

Bellerophon provides a useful basis upon which further sound hybrid systems verification algorithms can be implemented succinctly. The small core of KeYmaera X is solely responsible for soundness, but provides enough flexibility to reason in many radically different ways about hybrid systems. Bellerophon makes this flexibility easily accessible for programming both high-level hybrid systems verification strategies and concrete case study proofs. Fruitful directions for future work include developing more expressive proof structuring languages and extending the tactic library with more proof techniques that leverage ODE analysis software to produce d$\mathcal{L}$ proofs.

References

1. de Moura, L.M., Kong, S., Avigad, J., Doorn, F., Raumer, J.: The lean theorem prover (system description). In: Felty, A.P., Middeldorp, A. (eds.) CADE 2015. LNCS, vol. 9195, pp. 378–388. Springer, Cham (2015). doi:10.1007/978-3-319-21401-6_26
2. Alur, R., Courcoubetis, C., Henzinger, T.A., Ho, P.-H.: Hybrid automata: an algorithmic approach to the specification and verification of hybrid systems. In: Grossman, R.L., et al. (eds.) [13], pp. 209–229
3. Barras, B., Carmen González Huesca, L., Herbelin, H., Régis-Gianas, Y., Tassi, E., Wenzel, M., Wolff, B.: Pervasive parallelism in highly-trustable interactive theorem proving systems. In: Carette, J., Aspinall, D., Lange, C., Sojka, P., Windsteiger, W. (eds.) CICM 2013. LNCS, vol. 7961, pp. 359–363. Springer, Heidelberg (2013). doi:10.1007/978-3-642-39320-4_29
4. Bohrer, R., Rahli, V., Vukotic, I., Völp, M., Platzer, A.: Formally verified differential dynamic logic. In: Certified Programs and Proofs - 6th ACM SIGPLAN Conference, CPP 2017, pp. 208–221. ACM (2017)
5. Boldo, S., Lelay, C., Melquiond, G.: Coquelicot: a user-friendly library of real analysis for Coq. Math. Comput. Sci. $9(1)$, 41–62 (2015)
6. Chlipala, A.: Certified Programming with Dependent Types - A Pragmatic Introduction to the Coq Proof Assistant. MIT Press, Cambridge (2013)
7. Collins, G.E., Hong, H.: Partial cylindrical algebraic decomposition for quantifier elimination. J. Symb. Comput. $12(3)$, 299–328 (1991)
8. Constable, R.L., Allen, S.F., Bromley, M., et al.: Implementing Mathematics with the Nuprl Proof Development System. Prentice Hall, Upper Saddle River (1986)
9. Davenport, J.H., Heintz, J.: Real quantifier elimination is doubly exponential. J. Symb. Comput. $5(1/2)$, 29–35 (1988)
10. Frehse, G.: PHAVer: algorithmic verification of hybrid systems past HyTech. STTT $10(3)$, 263–279 (2008)
11. Frehse, G., et al.: SpaceEx: scalable verification of hybrid systems. In: Gopalakrishnan, G., Qadeer, S. (eds.) CAV 2011. LNCS, vol. 6806, pp. 379–395. Springer, Heidelberg (2011). doi:10.1007/978-3-642-22110-1_30
12. Fulton, N., Mitsch, S., Quesel, J.-D., Völp, M., Platzer, A.: KeYmaera X: an axiomatic tactical theorem prover for hybrid systems. In: Felty, A.P., Middeldorp, A. (eds.) CADE 2015. LNCS, vol. 9195, pp. 527–538. Springer, Cham (2015). doi:10.1007/978-3-319-21401-6_36
13. Grossman, R.L., Nerode, A., Ravn, A.P., Rischel, H. (eds.): Hybrid Systems. LNCS, vol. 736. Springer, Heidelberg (1993). doi:10.1007/3-540-57318-6
14. Harrison, J.: A HOL theory of euclidean space. In: Hurd, J., Melham, T. (eds.) TPHOLs 2005. LNCS, vol. 3603, pp. 114–129. Springer, Heidelberg (2005). doi:10.1007/11541868_8
15. Hickey, J., et al.: MetaPRL – a modular logical environment. In: Basin, D., Wolff, B. (eds.) TPHOLs 2003. LNCS, vol. 2758, pp. 287–303. Springer, Heidelberg (2003). doi:10.1007/10930755_19
16. Hölzl, J., Immler, F., Huffman, B.: Type classes and filters for mathematical analysis in Isabelle/HOL. In: Blazy, S., Paulin-Mohring, C., Pichardie, D. (eds.) ITP 2013. LNCS, vol. 7998, pp. 279–294. Springer, Heidelberg (2013). doi:10.1007/978-3-642-39634-2_21
17. Immler, F., Traut, C.: The flow of ODEs. In: Blanchette, J.C., Merz, S. (eds.) ITP 2016. LNCS, vol. 9807, pp. 184–199. Springer, Cham (2016). doi:10.1007/978-3-319-43144-4_12

18. Kong, S., Gao, S., Chen, W., Clarke, E.: dReach: δ-reachability analysis for hybrid systems. In: Baier, C., Tinelli, C. (eds.) TACAS 2015. LNCS, vol. 9035, pp. 200–205. Springer, Heidelberg (2015). doi:10.1007/978-3-662-46681-0_15

19. Krebbers, R., Spitters, B.: Type classes for efficient exact real arithmetic in Coq. Log. Methods Comput. Sci. **9**(1) (2011)

20. The Coq Development Team: The Coq proof assistant reference manual. LogiCal Project (2004). http://coq.inria.fr, version 8.0

21. Mitsch, S., Platzer, A.: The KeYmaera X proof IDE: concepts on usability in hybrid systems theorem proving. In: FIDE-3. EPTCS, vol. 240, pp. 67–81 (2016)

22. Mitsch, S., Platzer, A.: ModelPlex: verified runtime validation of verified cyber-physical system models. Form. Methods Syst. Des. **49**(1), 33–74 (2016). Special issue of selected papers from RV'14

23. Nipkow, T., Wenzel, M., Paulson, L.C.: Isabelle/HOL: A Proof Assistant for Higher-Order Logic. Springer, Heidelberg (2002). doi:10.1007/3-540-45949-9

24. Platzer, A.: Differential dynamic logic for hybrid systems. J. Autom. Reason. **41**(2), 143–189 (2008)

25. Platzer, A.: Logical Analysis of Hybrid Systems: Proving Theorems for Complex Dynamics. Springer, Heidelberg (2010). doi:10.1007/978-3-642-14509-4

26. Platzer, A.: Logics of dynamical systems. In: LICS. pp. 13–24. IEEE (2012)

27. Platzer, A.: A complete uniform substitution calculus for differential dynamic logic. J. Autom. Reason. **59**(2), 219–266 (2017)

28. Platzer, A., Clarke, E.M.: Computing differential invariants of hybrid systems as fixedpoints. Form. Methods Syst. Des. **35**(1), 98–120 (2009). Special issue for selected papers from CAV'08

29. Platzer, A., Quesel, J.-D.: KeYmaera: a hybrid theorem prover for hybrid systems (system description). In: Armando, A., Baumgartner, P., Dowek, G. (eds.) IJCAR 2008. LNCS, vol. 5195, pp. 171–178. Springer, Heidelberg (2008). doi:10.1007/978-3-540-71070-7_15

30. Platzer, A., Quesel, J.-D., Rümmer, P.: Real world verification. In: Schmidt, R.A. (ed.) CADE 2009. LNCS, vol. 5663, pp. 485–501. Springer, Heidelberg (2009). doi:10.1007/978-3-642-02959-2_35

31. Solovyev, A., Hales, T.C.: Formal verification of nonlinear inequalities with tay-lor interval approximations. In: Brat, G., Rungta, N., Venet, A. (eds.) NFM 2013. LNCS, vol. 7871, pp. 383–397. Springer, Heidelberg (2013). doi:10.1007/978-3-642-38088-4_26

Formalizing Basic Quaternionic Analysis

Andrea Gabrielli and Marco Maggesi$^{(\boxtimes)}$

University of Florence, Florence, Italy
{andrea.gabrielli,marco.maggesi}@unifi.it
http://www.math.unifi.it/~maggesi/

Abstract. We present a computer formalization of quaternions in the HOL Light theorem prover. We give an introduction to our library for potential users and we discuss some implementation choices.

As an application, we formalize some basic parts of two recently developed mathematical theories, namely, slice regular functions and Pythagorean-Hodograph curves.

1 Introduction

Quaternions are a well-known and elegant mathematical structure which lies at the intersection of algebra, analysis and geometry. They have a wide range of theoretical and practical applications from mathematics and physics to CAD, computer animations, robotics, signal processing and avionics.

Arguably, a computer formalization of quaternions can be useful, or even essential, for further developments in pure mathematics or for a wide class of applications in formal methods.

In this paper, we present a formalization of quaternions in the HOL Light theorem prover. Our aim is to give a quick introduction of our library to potential users and to discuss some implementation choices.

Our code is available from a public Git repository[1] and a significant part of it has been included in the HOL Light library.

The paper is divided into two main parts. The first one (Sect. 3), we describe the core of our library, which is already available in the HOL Light distribution.

Next, in the second part, we outline two applications to recently developed mathematical theories which should serve as further examples and as a testbed for our work. More precisely, we give the definition and some basic theorems about slice regular quaternionic functions (Sect. 4) and Pythagorean-Hodograph curves (Sect. 5).

We thank Graziano Gentili, Carlotta Giannelli and Caterina Stoppato for many enlightening discussions.

This work has been supported by GNSAGA-INdAM and MIUR.

[1] Reachable from the url https://bitbucket.org/maggesi/quaternions/.

© Springer International Publishing AG 2017
M. Ayala-Rincón and C.A. Muñoz (Eds.): ITP 2017, LNCS 10499, pp. 225–240, 2017.
DOI: 10.1007/978-3-319-66107-0_15

2   Related Work

The HOL Light theorem prover furnishes an extensive library about multivariate analysis [7] and complex analysis [8], which has been constantly and steadily extended over the years by Harrison, the main author of the system.

Our objective is to try to further improve this work by adding contributions in (hyper)complex analysis. One previous work along this line was the proof of the Cartan fixed point theorems [1] by Ciolli, Gentili and the second author of this paper.

In a broader context, quaternions are one of the simplest examples of geometric algebra (technically, real Clifford algebra). In this respect, we mention two recent related efforts. Fuchs and Théry [4] devised an elegant inductive data structure for formalizing geometric algebra. More recently, Ma et al. [9], provided a formalization in HOL Light of Conformal Geometric Algebra. In principle, these contributions can be integrated with our work, but at the present stage, we have focused entirely on the specific case of quaternions.

3   The Core Library

Quaternions were "invented" by Hamilton in 1843. From their very inception, they was meant to represent, in an unified form, both scalar and vector quantities. Informally, a quaternion q is expressed as a formal combination

$$q = a + b\mathbf{i} + c\mathbf{j} + d\mathbf{k} \in \mathbb{H} \qquad a, b, c, d \in \mathbb{R},$$

where $\mathbf{i}, \mathbf{j}, \mathbf{k}$ are *imaginary units*. The following identities

$$\mathbf{ij} = \mathbf{k} = -\mathbf{ji}$$
$$\mathbf{jk} = \mathbf{i} = -\mathbf{kj}$$
$$\mathbf{ki} = \mathbf{j} = -\mathbf{ik}$$
$$\mathbf{i}^2 = \mathbf{j}^2 = \mathbf{k}^2 = \mathbf{ijk} = -1$$

together with the distributive law, induce a product that turns the set $\mathbb{H}$ of quaternions into a skew field.

It is useful to consider a number of different possible decompositions for a quaternion q, as briefly sketched in the following schema (here $\mathbb{I} = \mathbb{R}^3$ can be interpreted, depending on the context, as the imaginary part of $\mathbb{H}$ or the 3-dimensional space):

$$q = \underbrace{a}_{\operatorname{Re} q} + \underbrace{b\mathbf{i} + c\mathbf{j} + d\mathbf{k}}_{\operatorname{Im} q} \qquad\qquad \in \mathbb{H} = \mathbb{R} \oplus \mathbb{I}$$

$$= \underbrace{a}_{\text{scalar}} + \underbrace{b\mathbf{i} + c\mathbf{j} + d\mathbf{k}}_{\text{3d-vector}} \qquad\qquad \in \mathbb{R}^4 = \mathbb{R} \oplus \mathbb{R}^3$$

$$= \underbrace{a + b\mathbf{i}}_{z \in \mathbb{C}} + \underbrace{(c + d\mathbf{i})\,\mathbf{j}}_{w \in \mathbb{C}} \qquad\qquad \in \mathbb{H} \simeq \mathbb{C} \oplus \mathbb{C}$$

3.1   The Definition of Quaternion

For the sake of consistency, whenever possible, our development mimics the formalization of complex numbers, due to Harrison, that is present in the HOL Light standard library [8]. Following this path, we define the data type `':quat'` of quaternions as an alias for the type of 4-dimensional vectors `':real^4'`. This approach has the fundamental advantage that we inherit immediately from the general theory of Euclidean spaces the appropriate metric, topology and real-vector space structure.

A set of auxiliary constants for constructing and destructing quaternions is defined to setup a suitable *abstraction barrier*. They are listed in the following table (Table 1).

Table 1. Basic notations for the `':quat'` datatype

Constant name	Type	Description
Hx	:real->quat	Embedding $\mathbb{R} \to \mathbb{H}$
ii, jj, kk	:quat	Imaginary units $\mathbf{i}, \mathbf{j}, \mathbf{k}$
quat	:real#real#real#real->quat	Generic constructor
Hv	:real^3->quat	Embedding $\mathbb{R}^3 \to \mathbb{H}$
Re	:quat->real	Real component
Im1, Im2, Im3	:quat->real	Imaginary components
HIm	:quat->real^3	Imaginary part
cnj	:quat->quat	Conjugation

This is summarized in the following theorem

```
QUAT_EXPAND
|- !q. q = Hx(Re q) + ii*Hx(Im1 q) + jj*Hx(Im2 q) + kk*Hx(Im3 q)
```

With these notations in place, the multiplicative structure can be expressed with an explicit formula

```
let quat_mul = new_definition
  'p * q =
   quat
   (Re p *  Re q - Im1 p * Im1 q - Im2 p * Im2 q - Im3 p * Im3 q,
    Re p * Im1 q + Im1 p *  Re q + Im2 p * Im3 q - Im3 p * Im2 q,
    Re p * Im2 q - Im1 p * Im3 q + Im2 p *  Re q + Im3 p * Im1 q,
    Re p * Im3 q + Im1 p * Im2 q - Im2 p * Im1 q + Im3 p *  Re q)';;
```

and the inverse of a quaternion is defined analogously. Moreover, we also provide auxiliary theorems that express in the same notation the already defined additive and metric structures, e.g.,

```
quat_add
|- p + q =
    quat(Re p + Re q,Im1 p + Im1 q,Im2 p + Im2 q,Im3 p + Im3 q)

quat_norm
|- norm q =
    sqrt(Re q pow 2 + Im1 q pow 2 + Im2 q pow 2 + Im3 q pow 2)
```

Notice that several notations (Re, ii, cnj, ...) overlap in the complex and quaternionic case and, more generally, with the ones of other number systems (+, *, ...). HOL Light features an overloading mechanism that uses type inference to select the right constant associated to a given notation.

3.2   Computing with Quaternions

After settling the basic definitions, we supply a simple automated procedure for proving quaternionic algebraic identities. It consists of just two steps: (1) rewriting the given identity in real components, (2) using an automated procedure for the real field (essentially one involving polynomial normalization, elimination of fractions and Gröbner Basis):

```
let SIMPLE_QUAT_ARITH_TAC =
  REWRITE_TAC[QUAT_EQ; QUAT_COMPONENTS; HX_DEF;
              quat_add; quat_neg; quat_sub; quat_mul;
              quat_inv] THEN
  CONV_TAC REAL_FIELD;;
```

This approach, although very crude, allows us to prove directly nearly 60 basic identities, e.g.,

```
let QUAT_MUL_ASSOC = prove
  ('!x y z:quat. x * (y * z) = (x * y) * z',
   SIMPLE_QUAT_ARITH_TAC);;
```

and it is also occasionally useful to prove ad hoc identities in the middle of more complex proofs. In this way, we quickly bootstrap a small library with the essential algebraic results that are needed for building more complex procedures and theorems.

Next, we provide a conversion RATIONAL_QUAT_CONV for evaluating literal algebraic expressions. This is easily assembled from elementary conversions for each basic algebraic operation (RATIONAL_ADD_CONV, RATIONAL_MUL_CONV, ...) using the HOL mechanism of higher-order conversionals. For instance, the computation

$$\left(1 + 2\mathbf{i} - \frac{1}{2}\mathbf{k}\right)^3 = -\frac{47}{4} - \frac{5}{2}\mathbf{i} + \frac{5}{8}\mathbf{k}$$

is performed with the command

```
# RATIONAL_QUAT_CONV
  `(Hx(&1) + Hx(&2) * ii - Hx(&1 / &2) * kk) pow 3`;;
val it : thm =
  |- (Hx(&1) + Hx(&2) * ii - Hx(&1 / &2) * kk) pow 3 =
      -- Hx(&47 / &4) - Hx(&5 / &2) * ii + Hx(&5 / &8) * kk
```

Finally, we implement a procedure for simplifying quaternionic polynomial expressions. HOL Light provides a general procedure for polynomial normalization, which unfortunately works only for commutative rings. Hence we are forced to code our own solution. In principle, our procedure can be generalized to work with arbitrary (non-commutative) rings. However, at the present stage, it is hardwired to the specific case of quaternions. To give an example, the computation

$$(p + q)^3 = p^3 + q^3 + pq^2 + p^2q + pqp + qp^2 + qpq + q^2p$$

can be done with the command

```
# QUAT_POLY_CONV `(x + y) pow 3`;;
val it : thm =
  |- (p + q) pow 3 =
      p pow 3 + q pow 3 + p * q pow 2 + p pow 2 * q +
      p * q * p + q * p pow 2 + q * p * q + q pow 2 * p
```

3.3   The Geometry of Quaternions

One simple fact, which makes quaternions useful in several physical and geometrical applications, is that the quaternionic product encodes both the scalar and the vector product. More precisely, if p and q are purely imaginary quaternions then we have

$$pq = - \underbrace{\langle p, q \rangle}_{\substack{\text{scalar} \\ \text{product}}} + \underbrace{p \wedge q}_{\substack{\text{vector} \\ \text{product}}} \in \mathbb{R} \oplus \mathbb{I},$$

which can be easily verified by direct computation.

Moreover, it is possible to use quaternions to encode orthogonal transformations. We briefly recall the essential mathematical construction. For $q \neq 0$, the conjugation map is defined as

$$c_q : \mathbb{H} \longrightarrow \mathbb{H}$$
$$c_q(x) := q^{-1} x q$$

and we have

$$c_{q_1} \circ c_{q_2} = c_{q_1 q_2}, \qquad c_q^{-1} = c_{q^{-1}}.$$

One important special case is when q is unitary, i.e., $\|q\| = 1$ for which we have $q^{-1} = \bar{q}$ (the conjugate) and thus $c_q(x) = \bar{q} x q$.

Now, we are ready to state some basic results, which we have formalized in our framework (see file Quaternions/qisom.hl in the HOL Light distribution).

Proposition 1. *If v is a non-zero purely imaginary quaternion, then $-c_v \colon \mathbb{R}^3 \to \mathbb{R}^3$ is the reflection with respect to the subspace of $\mathbb{R}^3$ orthogonal to v.*

Here is the corresponding statement proved in HOL Light

```
REFLECT_ALONG_EQ_QUAT_CONJUGATION
|- !v. ~(v = vec 0)
        ==> reflect_along v = \x. --HIm (inv (Hv v) * Hv x * Hv v)
```

The theorem of Cartan-Dieudonné asserts that any orthogonal transformation $f \colon \mathbb{R}^n \longrightarrow \mathbb{R}^n$ is the composition of at most n reflections. Using this and the previous proposition we get the following result.

Proposition 2. *Any orthogonal transformation $f \colon \mathbb{R}^3 \longrightarrow \mathbb{R}^3$ is of the form*

$$f = c_q \quad or \quad f = -c_q, \qquad \|q\| = 1.$$

The corresponding formalization is the following

```
ORTHOGONAL_TRANSFORMATION_AS_QUAT_CONJUGATION
|- !f. orthogonal_transformation f
        ==> (?q. norm q = &1 /\
             ((!x. f x = HIm (inv q * Hv x * q)) \/
              (!x. f x = --HIm (inv q * Hv x * q))))
```

3.4   Elementary Quaternionic Analysis

Passing from algebra to analysis, we need to prove a series of technical results about the analytical behaviour of the algebraic operations. Following the HOL Light practice, we use the formalism of net topology to express limits and continuity. To give an idea, here we report the theorem that states the uniform continuity of the quaternionic inverse $q \mapsto q^{-1}$

```
UNIFORM_LIM_QUAT_INV
|- !net P f l b.
     (!e. &0 < e
          ==> eventually (\x. !n. P n ==> norm (f n x - l n) < e)
                         net) /\
     &0 < b /\ eventually (\x. !n. P n ==> b <= norm (l n)) net
   ==> (!e. &0 < e
             ==> eventually
                 (\x. !n. P n
                          ==> norm (inv (f n x) - inv (l n)) < e)
             net)
```

We conducted a systematic formalization of the behaviour of algebraic operations from the point of view of limits and continuity, which brought us to prove more than fifty such theorems overall. Some of them are indeed trivial. For instance, the uniform continuity of the product is a direct consequence of a more general result already available on bilinear maps. Some are less immediate and forced us to dive into a technical $\epsilon\delta$-reasoning.

Next, we considered the differential structure. Given a function $f : \mathbb{R}^n \to \mathbb{R}^m$ we denote by $\mathrm{D}f_{x_0}(v)$ or $\frac{\mathrm{d}}{\mathrm{d}x}f(x)|_{x_0}(v)$ the (Frechét) *derivative* of f at x_0 applied to the vector v. When the derivative exists, it is the linear function from $\mathbb{R}^n$ to $\mathbb{R}^m$ that "best" approximates the variation of f in a neighborhood of x_0, i.e.,

$$f(x) - f(x_0) \approx \mathrm{D}f_{x_0}(x - x_0).$$

In HOL Light, the ternary predicate (f has_derivative f') (at x0) is used to assert that f is (Frechét) differentiable at x_0 and $f' = \mathrm{D}f_{x_0}$

We compute the derivative of the basic quaternionic operations. Notice that, if f is a quaternionic valued function, the derivative $\mathrm{D}f_{q_0}(x)$ is a quaternion (in the modern language of Differential Geometry this is the natural identification of the tangent space $T_{f(q_0)}\mathbb{H} \simeq \mathbb{H}$). For instance, given two differentiable functions $f(q)$ and $g(q)$, the derivative of their product at q_0 is

$$\frac{\mathrm{d}\left(f(q)g(q)\right)}{\mathrm{d}q}|_{q_0}(x) = f(q_0)\mathrm{D}g_{q_0}(x) + \mathrm{D}f_{q_0}(x)g(q_0).$$

In our formalism, the previous formula becomes the following theorem:

```
QUAT_HAS_DERIVATIVE_MUL_AT
|- !f f' g g' q.
    (f has_derivative f') (at q) /\ (g has_derivative g') (at q)
    ==> ((\x. f x * g x) has_derivative
        (\x. f q * g' x + f' x * g q)) (at q)
```

Another consequence that will be useful later, is the following formula for the power:

$$\frac{\mathrm{d}q^n}{\mathrm{d}q}|_{q_0}(x) = \sum_{i=1}^{n} q_0^{n-i}xq_0^{i-1},$$

that is, the HOL theorem

```
QUAT_HAS_DERIVATIVE_POW
|- !q0 n.
    ((\q. q pow n) has_derivative
    (\h. vsum (1..n) (\i. q0 pow (n - i) * h * q0 pow (i - 1))))
    (at q0)
```

which is easily proven by induction using the derivative of the product.

4   Slice Regular Functions

Complex holomorphic functions play a central role in mathematics. Given the deep link and the evident analogy between complex numbers and quaternions, it is natural to seek for a theory of *quaternionic holomorphic functions*. A more careful investigation shows that the situation is less simple than expected. Naive attempts to generalize the complex theory to the quaternionic case fail because they lead to conditions which are either too strong or too weak and do not produce *interesting* classes of functions.

Fueter in the 1920s, proposed a definition of *regular* quaternionic function which is now well-known to the experts and has been extensively studied and developed. Fueter's regular functions have significant applications to physics and engineering, but present also some undesirable aspects. For instance, the identity function and the polynomials $P(q) = a_0 + a_1 q + \ldots + a_n q^n$, $a_i \in \mathbb{H}$ are not Fueter-regular. A more detailed discussion on this subject is far beyond the goal of the present work. To the interested reader we recommend Sudbery's excellent survey [11].

In this setting, a novel promising approach has been recently proposed by Gentili and Struppa in their seminal paper in 2006 [6], where they introduce the definition of *slice regular* functions and prove that they expand into power series near the origin. Slice regular functions are now a stimulating and active subject of research for several mathematicians worldwide. A comprehensive introduction on the foundation of this new theory can be found in the book of Gentili et al. [5].

In this section, we use our quaternionic framework presented in the previous section to formalize the basics of the theory of slice regular functions.

4.1   The Definition of Slice Regular Function

A real 2-dimensional subspace $L \subset \mathbb{H}$ containing the real line is called a *slice* (or *Cullen slice*) of $\mathbb{H}$. The key fact is that the quaternionic product becomes commutative when it is restricted on a slice, that is if p, q are in the same slice L, then $pq = qp$. In other terms, each slice L can be seen as a copy of the complex field $\mathbb{C}$. More precisely, if I is a quaternionic imaginary unit (i.e., an unitary imaginary quaternion), then $L_I = \mathrm{Span}\{1, I\} = \mathbb{R} \oplus \mathbb{R}I$ is a slice and the injection $j_I : \mathbb{C} \to L_I \subset \mathbb{H}$, defined by

$$j_I : x + y\mathbf{i} \mapsto x + yI,$$

is a field homomorphism. Its formal counterpart is

```
let cullen_inc = new_definition
  'cullen_inc i z = Hx(Re z) + Hx(Im z) * i';;
```

We can now introduce the definition of Gentili and Struppa:

Definition 1 (Slice regular function). *Given a domain (i.e., an open, connected set) $\Omega \subset \mathbb{H}$ a function $f : \Omega \to \mathbb{H}$ is slice regular if it is holomorphic (in the complex sense) on each slice, that is, the restricted function $f_{L_I} : \Omega \cap L_I \to \mathbb{H}$ satisfies the condition*

$$\frac{1}{2} \left(\frac{\partial}{\partial x} + I \frac{\partial}{\partial y} \right) f_{L_I}(x + yI) = 0$$

for each $q = x + yI$ in $\Omega \cap L_I$, for every imaginary unit I. In that case, we define the slice derivative *of f in q to be the quaternion*

$$f'(q) = \frac{1}{2}\left(\frac{\partial}{\partial x} - I\frac{\partial}{\partial y}\right) f_{L_I}(x + yI).$$

Our first task is to code the previous definition within our formalism. One problem is the notation for partial derivatives, which is notorious for being occasionally opaque and potentially misleading. When it has to be rendered in a formal language, its translation might be tricky or at least cumbersome. This is essentially due to the fact that it is a convention that induces us to use the same name for different functions, depending on the name of the arguments.[2]

We decided that the best way to avoid potential problems in our development was to systematically replace partial derivatives with (Frechét) derivatives. This leads to an alternative, and equivalent, definition of slice regular function which could be interesting in its own.

The basic idea is the following. A complex function f is holomorphic in z_0 precisely when its derivative Df_{z_0} is $\mathbb{C}$-linear. Hence, by analogy, a quaternionic function should be slice regular if its derivative is $\mathbb{H}$-linear on slices in a suitable sense. This is indeed the case: consider $f : \Omega \to \mathbb{H}$ as before and a quaternion $q_0 \in \Omega$. Let L be a slice containing q_0 and denote by f_L the restriction of f to $\Omega \cap L$. Then we have

Proposition 3. *The function f is slice regular in q_0 if and only if the derivative of f_L is right-$\mathbb{H}$-linear, that is, there exists a quaternion c such that*

$$D(f_L)_{q_0}(p) = pc.$$

In that case, c is the slice derivative $f'(q_0)$.

We then take the alternative formulation given by the above Proposition as the definition of slice regular function in our developement. The resulting formalization in HOL is the following

```
let has_slice_derivative = new_definition
  '!f (f':quat) net.
    (f has_slice_derivative f') net <=>
    (!l. subspace l /\ dim l = 2 /\ Hx(&1) IN l /\
        netlimit net IN l
        ==> (f has_derivative (\q. q * f')) (net within l))';;
```

---

[2] Spivak, in his book *Calculus on manifolds* [10, p. 65], notices that if $f(u, v)$ is a function and $u = g(x, y)$ and $v = h(x, y)$, then the chain rule is often written

$$\frac{\partial f}{\partial x} = \frac{\partial f}{\partial u}\frac{\partial u}{\partial x} + \frac{\partial f}{\partial v}\frac{\partial v}{\partial x},$$

where f denotes two different functions on the left- and right-hand of the equation.

Notice that the predicate `has_slice_derivative` formalizes at the same time the notion of slice regular function and the notion of slice derivative. The domain Ω does not appear in the definition because functions in HOL are total and, in any case, the notion of slice derivative is local.

Our formalization of slice derivative is slightly more general than the one of Proposition 3 for the fact that we use HOL nets. The reader who is not accustomed to the use of nets can simply think the variable `net` as a denoting the limit $q \to q_0$ and `netlimit net` as the limit point q_0. Other than that, these details about nets are largely irrelevant in the rest of the paper.

We formally proved Proposition 3 in HOL Light. Here is the statement for the case when $q_0 = x + yI$ is not real.

```
HAS_SLICE_DERIVATIVE
|- !f f' i x y.
     i pow 2 = -- Hx(&1) /\ ~(y = &0) /\
     f differentiable at (Hx x + Hx y * i)
     ==> ((f has_slice_derivative f') (at (Hx x + Hx y * i)) <=>
         (?fx fy.
            ((\a. f(Hx(drop a) + Hx y*i)) has_vector_derivative fx)
            (at(lift x)) /\
            ((\b. f(Hx x + Hx(drop b)*i)) has_vector_derivative fy)
            (at(lift y)) /\
            fx + i * fy = Hx(&0) /\ f' = fx /\ f' = --(i * fy)))
```

Since any slice L can be obtained as the image of j_I for any imaginary unit $I \in L$, then we also have the following useful reformulation

```
HAS_SLICE_DERIVATIVE_CULLEN_INC
|- !i f f' z0.
     i pow 2 = --Hx(&1)
     ==> ((f has_slice_derivative f')
         (at (cullen_inc i z) within cullen_slice i) <=>
         (f o cullen_inc i has_derivative
         (\z. cullen_inc i z * f')) (at z0))
```

After the definition, we provided a series of lemmas that allow us to compute the slice derivative of algebraic expressions. In particular, the powers q^n are slice regular and, if $f(q)$ and $g(q)$ are slice regular functions and c is a quaternion, then $f(q) + g(q)$ and $f(q)c$ are slice regular. It follows that *right* polynomials (i.e., polynomials with coefficients on the right)

$$c_0 + qc_1 + q^2c_2 + \cdots + q^nc_n$$

are all slice regular functions. Most of these results are easy consequences of those discussed in Sect. 3.4.

We should stress that the product $f(q)g(q)$, including left multiplication $cf(q)$ and arbitrary polynomials of the form

$$c_0 + c_{1,1}q + c_{2,0}qc_{2,1}qc_{2,2} + c_{3,0}qc_{3,1}qc_{3,2}qc_{3,3} + \cdots,$$

is not slice regular in general.

A more explicit link between slice regular functions and complex holomorphic functions is given by the *splitting lemma*, which is a fundamental tool for several subsequent results. Given two imaginary units I, J orthogonal to one other, every quaternion can be *split*, in an unique way, into a sum $q = z + wJ$ with $z, w \in L_I$. Now, given a function $f \colon \Omega \to \mathbb{H}$ we can split its restriction f_{L_I} as

$$f_{L_I}(z) = F(z) + G(z)J$$

with $F, G \colon \Omega \cap L_I \to L_I$. Then we have

Lemma 1 (Splitting Lemma). *The function f is slice regular at $q_0 \in L_I$ if and only if the functions F and G are holomorphic at q_0.*

Notice that, in the above statement, the two functions F, G are 'complex holomorphic' with respect to the implicit identification $\mathbb{C} \simeq L_I$ given by $j_I \colon x + yi \mapsto x + yI$. This has been made explicit in the following formal statement using our injection cullen_inc:

```
QUAT_SPLITTING_LEMMA
|- !f s i j.
     open s /\ i pow 2 = --Hx (&1) /\ j pow 2 = --Hx (&1) /\
     orthogonal i j
     ==> (?g h.
           (!z. f (cullen_inc i z) =
                cullen_inc i (g z) + cullen_inc i (h z) * j) /\
           (!g' h' z.
              z IN s
              ==> ((g has_complex_derivative g') (at z) /\
                   (h has_complex_derivative h') (at z) <=>
                   (f o cullen_inc i has_derivative
                     (\z. cullen_inc i z *
                           (cullen_inc i g' + cullen_inc i h' * j)))
                    (at z))) /\
           (g holomorphic_on s /\ h holomorphic_on s <=>
            (f slice_regular_on s) i))
```

4.2 Power Expansions of Slice Regular Functions

We now approach power series expansions of slice regular functions at the origin, which is one of the corner stone for the development of the whole theory. While the HOL Light library has a rather complete support for sequences and series in general, at the beginning of our work it was still lacking the proof of various theorems that were important prerequisites for our task.

We undertake a systematic formalization of the missing theory, including

1. the definition of limit superior and inferior and their basic properties;
2. the root test for series;
3. the Cauchy-Hadamard formula for the radius of convergence.

We avoid discussing this part of the work in detail in this paper. All these preliminaries have been recently included in the HOL Light standard library.[3]

Theorem 1 (Abel's Theorem for slice regular functions). *The quaternionic power series*

$$\sum_{n\in\mathbb{N}} q^n a_n \tag{1}$$

is absolutely convergent in the ball $B = B\left(0, 1/\limsup_{n\to+\infty} \sqrt[n]{|a_n|}\right)$ *and uniformly convergent on any compact subset of* B. *Moreover, its sum defines a slice regular function on* B.

The corresponding formalization is split into several theorems. As for the convergence, we have three statements, one for each kind of convergence (pointwise, absolute, uniform). As an example, we include the statement for the uniform convergence:

```
QUAT_UNIFORM_CONV_POWER_SERIES
|- !a b s k.
     ((\n. root n (norm (a n))) has_limsup b)
     (sequentially within k) /\
     compact s /\
     (!q. q IN s ==> b * norm q < &1)
     ==> ((\i q. q pow i * a i) uniformly_summable_on s) k
```

The predicate 'uniformly_summable_on' is a compact notation for uniform convergence for series. Note that the hypothesis 'b * norm q < &1' allows a correct representation of the domain of convergence also in the case of infinite radius (case $b = 0$).

With a little extra effort we proved the same results for the formal derivative of the series (1).

Finally, from the previous results, and the fact that derivative distributes over uniformly convergent series, we proved that right quaternionic power series are slice regular functions on any compact subsets of their domain of convergence

```
QUAT_HAS_SLICE_DERIVATIVE_POWER_SERIES_COMPACT
|- !a b k q0 s.
     ((\n. root n (norm (a n))) has_limsup b)
     (sequentially within k) /\
     compact s /\ s SUBSET {q | b * norm q < &1} /\
     ~(s = {}) /\ q0 IN s
     ==> ((\q. infsum k (\n. q pow n * a n)) has_slice_derivative
          infsum k (\n. q0 pow (n - 1) * Hx (&n) * a n)) (at q0)
```

which completes the formalization of Theorem 1.

Next, from the Splitting Lemma 1, we can derive the existence of the power series expansion of a slice regular function f from the analyticity of its holomorphic components F and G.

---

[3] Commit on Apr 10, 2017, HOL Light GitHub repository.

Theorem 2. *Let $f : B(0, R) \to \mathbb{H}$ be a slice regular function. Then*

$$f(q) = \sum_{n \in \mathbb{N}} q^n \frac{1}{n!} f^{(n)}(0),$$

where $f^{(n)}$ is the n-th slice derivative of f.

The resulting formalization is the following

```
SLICE_REGULAR_SERIES_EXPANSION
|- !r q f.
    &0 < r /\ q IN ball (Hx(&0),r) /\
    (!i. (f slice_regular_on ball (Cx(&0),r)) i)
    ==> (?z i.
            i pow 2 = --Hx(&1) /\ q = cullen_inc i z /\
            f q =
            infsum (:num)
            (\n. cullen_inc i z pow n *
                 cullen_inc i (inv (Cx(&(FACT n)))) *
                 higher_slice_derivative i n f (Hx(&0))))
```

5   Pythagorean-Hodograph Curves

The *hodograph* of a parametric curve $\mathbf{r}(t)$ in $\mathbb{R}^n$ is just its derivative $\mathbf{r}'(t)$, regarded as a parametric curve in its own right. A parametric polynomial curve $\mathbf{r}(t)$ is said to be a *Pythagorean-Hodograph* curve if it satisfies the *Pythagorean condition*, i.e., there exists a polynomial $\sigma(t)$ such that

$$\|\mathbf{r}'(t)\|^2 = x_1^2(t) + \cdots + x_n^2(t) = \sigma^2(t), \tag{2}$$

that is, the parametric speed $\|\mathbf{r}'(t)\|$ is polynomial.

Pythagorean-Hodograph curves (PH curves) were introduced by Farouki and Sakkalis in 1990. They have significant computational advantages when used for computer-aided design (CAD) and robotics applications since, among other things, their arc length can be computed precisely, i.e., without numerical quadrature, and their offsets are rational curves. Farouki's book [3] offers a fairly complete and self-contained exposition of this theory.

5.1   Formalization of PH Curves and Hermite Interpolation Problem

The formal definition of PH curve in HOL Light is straightforward:

```
let pythagorean_hodograph = new_definition
 'pythagorean_hodograph r <=>
  vector_polynomial_function r /\
  real_polynomial_function
    (\t. norm (vector_derivative r (at t)))';;
```

In our work, we deal with spacial PH curves which can be succinctly and profitably expressed in terms of the algebra of quaternions, and thus, are a natural application of our formalization of quaternionic algebra.

It turns out that, regarding $\mathbf{r}(t) = x(t)\mathbf{i} + y(t)\mathbf{j} + z(t)\mathbf{k}$ as a pure vector in $\mathbb{R}^3 \subset \mathbb{H}$, condition (2) holds if and only if exists a quaternionic polynomial $A(t)$ such that

$$\mathbf{r}'(t) = A(t)\mathbf{u}\bar{A}(t) \tag{3}$$

where $\mathbf{u}$ is any fixed unit vector and $\bar{A}(t)$ is the usual quaternionic conjugate of $A(t)$. We proved formally that the definition (2) follows from the previous condition.

```
QUAT_PH_CURVE
|- !r A u.
    u pow 2 = --Hx (&1) /\
    vector_polynomial_function A /\
    (!t. (r has_vector_derivative A t * u * cnj (A t)) (at t))
    ==> pythagorean_hodograph r
```

One basic question, with many practical applications, is whether there exists a PH curve with prescribed conditions on its endpoints.

Property 1 (Hermite Interpolation Problem). Given the initial and final point $\{\mathbf{p}_i, \mathbf{p}_f\}$ and derivatives $\{\mathbf{d}_i, \mathbf{d}_f\}$, find a PH interpolation for this data set.

Following the work of Farouki et al. [2], here we treat only the case of cubic and quintic solutions of the above problem.

From condition (3) the problem can be reduced to finding a quaternionic polynomial $A(t)$, of degree 1 (for cubics) or 2 (for quintics), such that the curve $\mathbf{r}(t)$ obtained by integrating (3) satisfies $\mathbf{r}(0) = \mathbf{p}_i$, $\mathbf{r}(1) = \mathbf{p}_f$ and $\mathbf{r}'(0) = \mathbf{d}_i$, $\mathbf{r}'(1) = \mathbf{d}_f$.

5.2   PH Cubic and Quintic Interpolant

As is well-known, for a given initial data set $\{\mathbf{p}_i, \mathbf{p}_f, \mathbf{d}_i, \mathbf{d}_f\}$, there is a unique "ordinary" cubic interpolant [3]. It turns out that such a curve is PH if and only if the data set satisfies specific conditions [2, Sect. 5], namely:

$$\mathbf{w} \cdot (\delta_i - \delta_f) = 0$$

$$\left(\mathbf{w} \cdot \frac{\delta_i + \delta_f}{|\delta_i + \delta_f|}\right)^2 + \frac{(\mathbf{w} \cdot \mathbf{z})^2}{|\mathbf{z}|^4} = |\mathbf{d}_i||\mathbf{d}_f|$$

where $\mathbf{w} = 3(\mathbf{p}_f - \mathbf{p}_i) - (\mathbf{d}_i + \mathbf{d}_f)$, $\delta_i = \frac{\mathbf{d}_i}{|\mathbf{d}_i|}$, $\delta_f = \frac{\mathbf{d}_f}{|\mathbf{d}_f|}$ and $\mathbf{z} = \frac{\delta_i \wedge \delta_f}{|\delta_i \wedge \delta_f|}$.

We formalized only one implication of this result, i.e., the sufficient condition for the "ordinary" cubic interpolant to be PH. The HOL theorem is

```
PH_CUBIC_INTERPOLANT_EXISTS
|- !Pf Pi di df:quat.
      let w = Hx(&3) * (Pf - Pi) - (di + df) in
      let n = \v. Hx(inv(norm v)) * v in
      let z = Hx(inv (norm (n di + n df))) *
                  Hv(HIm(n di) cross HIm(n df)) in
      let r = \t. bernstein 3 0 (drop t) % Pi +
                    bernstein 3 1 (drop t) % (Pi + Hx(&1 / &3) * di) +
                    bernstein 3 2 (drop t) % (Pf - Hx(&1 / &3) * df) +
                    bernstein 3 3 (drop t) % Pf in
      Re Pf = &0 /\ Re Pi = &0 /\ Re di = &0 /\ Re df = &0 /\
      ~(Hx(&0) = di) /\ ~(Hx(&0) = df) /\ (!a. ~(n di = Hx a * df))
      ==>
      pathstart r = Pi /\ pathfinish r = Pf /\
      pathstart (\t. vector_derivative r (at t)) = di /\
      pathfinish (\t. vector_derivative r (at t)) = df /\
      (w dot (n di - n df) = &0 /\
      (w dot (n (n di + n df))) pow 2 +
      inv(norm z) pow 4 * (w dot z) pow 2 =
      norm di * norm df
      ==> pythagorean_hodograph r)`
```

where the curve $\mathbf{r}(t)$ is expressed in the Bernstein form.

We also formalized the analogous result for quintics. In this case the theory shows several differences, since, for instance, an Hermite PH quintic interpolant can be found for every initial data set. Actually, there is a two-parameter family of such interpolants [2, Sect. 6] and the algebraic expression of $\mathbf{r}(t)$ is substantially more complex with respect to the case of cubics. The formal statement of the theorem is about 40 lines of code and thus cannot be included here for lack of space (see theorem PH_QUINTIC_INTERPOLANT in file ph_curve.hl in our online repository).

Both of the aforementioned proofs consist essentially in algebraic manipulation on quaternions, so our formal framework has been very useful to automate many calculations that were implicit in the original paper [2, Sect. 6].

6 Conclusions

We laid the foundations for quaternionic calculus in the HOL Light theorem prover, which might be of general interest for developing further formalization in pure mathematics, physics, and for several possible applications in formal methods.

We also presented two applications. First, a formalization of quaternionic analysis with a focus on the theory of slice regular functions, as proposed by Gentili and Struppa. Secondly, the computer verified solutions to the Hermite interpolation problem for cubic and quintic PH curves.

Along the way, we provided a few extensions of the HOL Light library about multivariate and complex analysis, comprising limit superior and inferior, root test for series, Cauchy-Hadamard formula for the radius of convergence and some basic theorems about derivatives.

Overall, our contribution takes about 10,000 lines of code and consists in about 600 theorems, of which more than 350 have been included in the HOL Light library.

This work is open to a wide range of possible improvements and extensions. The most obvious line of developement would be to formalize further mathematical results about quaternions; there is an endless list of potential interesting candidates within reach from the present state of art.

For the core formalization of quaternions, we only provided basic procedures for algebraic simplification. They were somehow sufficient for automating several computations occurring in our development, but it surely would be interesting to implement more powerful decision procedures. Some of them would probably involve advanced techniques from non-commutative algebra.

References

1. Ciolli, G., Gentili, G., Maggesi, M.: A certified proof of the Cartan fixed point theorems. J. Automated Reason. **47**(3), 319–336 (2011). https://doi.org/10.1007/s10817-010-9198-6
2. Farouki, R.T., Giannelli, C., Manni, C., Sestini, A.: Identification of spatial PH quintic hermite interpolants with near-optimal shape measures. Comput. Aided Geom. Des. **25**(4), 274–297 (2008)
3. Farouki, R.: Pythagorean-Hodograph Curves: Algebra and Geometry Inseparable. Geometry and Computing. Springer, Heidelberg (2009)
4. Fuchs, L., Théry, L.: Implementing geometric algebra products with binary trees. Adv. Appl. Clifford Algebras **24**(2), 589–611 (2014)
5. Gentili, G., Stoppato, C., Struppa, D.: Regular Functions of a Quaternionic Variable. Springer Monographs in Mathematics. Springer, Heidelberg (2013)
6. Gentili, G., Struppa, D.C.: A new approach to Cullen-regular functions of a quaternionic variable. Comptes Rendus Math. **342**(10), 741–744 (2006)
7. Harrison, J.: A HOL theory of euclidean space. In: Hurd, J., Melham, T. (eds.) TPHOLs 2005. LNCS, vol. 3603, pp. 114–129. Springer, Heidelberg (2005). doi:10. 1007/11541868_8
8. Harrison, J.: Formalizing basic complex analysis. In: Matuszewski, R., Zalewska, A. (eds.) From Insight to Proof: Festschrift in Honour of Andrzej Trybulec. Studies in Logic, Grammar and Rhetoric, vol. 10, no. 23, pp. 151–165. University of Białystok (2007). http://mizar.org/trybulec65/
9. Ma, S., Shi, Z., Shao, Z., Guan, Y., Li, L., Li, Y.: Higher-order logic formalization of conformal geometric algebra and its application in verifying a robotic manipulation algorithm. Adv. Appl. Clifford Algebras **26**(4), 1305–1330 (2016)
10. Spivak, M.: Calculus on Manifolds: A Modern Approach to Classical Theorems of Advanced Calculus. Advanced Book Program. Avalon Publishing, New York (1965)
11. Sudbery, A.: Quaternionic analysis. Math. Proc. Cambridge Philos. Soc. **85**(02), 199–225 (1979)

A Formalized General Theory of Syntax with Bindings

Lorenzo Gheri[1] and Andrei Popescu[1,2(✉)]

[1] Department of Computer Science, Middlesex University, London, UK
uuomul@yahoo.com
[2] Institute of Mathematics Simion Stoilow of the Romanian Academy,
Bucharest, Romania

Abstract. We present the formalization of a theory of syntax with bindings that has been developed and refined over the last decade to support several large formalization efforts. Terms are defined for an arbitrary number of constructors of varying numbers of inputs, quotiented to alpha-equivalence and sorted according to a binding signature. The theory includes a rich collection of properties of the standard operators on terms, such as substitution and freshness. It also includes induction and recursion principles and support for semantic interpretation, all tailored for smooth interaction with the bindings and the standard operators.

1 Introduction

Syntax with bindings is an essential ingredient in the formal specification and implementation of logics and programming languages. However, correctly and formally specifying, assigning semantics to, and reasoning about bindings is notoriously difficult and error-prone. This fact is widely recognized in the formal verification community and is reflected in manifestos and benchmarks such as the influential POPLmark challenge [1].

In the past decade, in a framework developed intermittently starting with the second author's PhD [42] and moving into the first author's ongoing PhD, a series of results in logic and λ-calculus have been formalized in Isabelle/HOL [35,37]. These include classic results (e.g., FOL completeness and soundness of Skolemization [7,13,15], λ-calculus standardization and Church-Rosser theorems [42,44], System F strong normalization [45]), as well as the meta-theory of Isabelle's Sledgehammer tool [7,8].

In this paper, we present the Isabelle/HOL formalization of the framework itself (made available from the paper's website [21]). While concrete system syntaxes differ in their details, there are some fundamental phenomena concerning bindings that follow the same generic principles. It is these fundamental phenomena that our framework aims to capture, by mechanizing a form of universal algebra for bindings. The framework has evolved over the years through feedback from concrete application challenges: Each time a tedious, seemingly routine construction was encountered, a question arose as to whether this could be performed once and for all in a syntax-agnostic fashion.

M. Ayala-Rincón and C.A. Muñoz (Eds.): ITP 2017, LNCS 10499, pp. 241–261, 2017.
DOI: 10.1007/978-3-319-66107-0_16

The paper is structured as follows. We start with an example-driven overview of our design decisions (Sect. 2). Then we present the general theory: terms as alpha-equivalence classes of "quasiterms," standard operators on terms and their basic properties (Sect. 3), custom induction and recursion schemes (Sect. 4), including support for the semantic interpretation of syntax, and the sorting of terms according to a signature (Sect. 5). Within the large body of formalizations in the area (Sect. 6), distinguishing features of our work are the general setting (many-sorted signature, possibly infinitary syntax), a rich theory of the standard operators, and operator-aware recursion. More details on this paper's results can be found in an extended technical report [22].

2   Design Decisions

In this section, we use some examples to motivate our design choices for the theory. We also introduce conventions and notations that will be relevant throughout the paper.

The paradigmatic example of syntax with bindings is that of the λ-calculus [4]. We assume an infinite supply of variables, $x \in$ **var**. The λ-terms, $X, Y \in$ **term**$_\lambda$, are defined by the following BNF grammar:

$$X ::= \mathsf{Var}\ x \mid \mathsf{App}\ X\ Y \mid \mathsf{Lm}\ x\ X$$

Thus, a λ-term is either a variable, or an application, or a λ-abstraction. This grammar specification, while sufficient for first-order abstract syntax, is incomplete when it comes to syntax with bindings—we also need to indicate which operators introduce bindings and in which of their arguments. Here, Lm is the only binding operator: When applied to the variable x and the term X, it binds x in X. After knowing the binders, the usual convention is to *identify terms modulo alpha-equivalence*, i.e., to treat as equal terms that only differ in the names of bound variables, such as, e.g., $\mathsf{Lm}\ x\ (\mathsf{App}\ (\mathsf{Var}\ x)\ (\mathsf{Var}\ y))$ and $\mathsf{Lm}\ z\ (\mathsf{App}\ (\mathsf{Var}\ z)\ (\mathsf{Var}\ y))$. The end results of our theory will involve terms modulo alpha. We will call the raw terms "quasiterms," reserving the word "term" for alpha-equivalence classes.

2.1   Standalone Abstractions

To make the binding structure manifest, we will "quarantine" the bindings and their associated intricacies into the notion of *abstraction*, which is a pairing of a variable and a term, again modulo alpha. For example, for the λ-calculus we will have

$$X ::= \mathsf{Var}\ x \mid \mathsf{App}\ X\ Y \mid \mathsf{Lam}\ A \qquad A ::= \mathsf{Abs}\ x\ X$$

where X are terms and A abstractions. Within $\mathsf{Abs}\ x\ X$, we assume that x is bound in X. The λ-abstractions $\mathsf{Lm}\ x\ X$ of the the original syntax are now written $\mathsf{Lam}\ (\mathsf{Abs}\ x\ X)$.

2.2 Freshness and Substitution

The two most fundamental and most standard operators on λ-terms are:

- the freshness predicate, fresh : $\mathbf{var} \rightarrow \mathbf{term}_\lambda \rightarrow \mathbf{bool}$, where fresh x X states that x is fresh for (i.e., does not occur free in) X; for example, it holds that fresh x (Lam (Abs x (Var x))) and fresh x (Var y) (when $x \neq y$), but not that fresh x (Var x).
- the substitution operator, $\_[\_/\_]$: $\mathbf{term}_\lambda \rightarrow \mathbf{term}_\lambda \rightarrow \mathbf{var} \rightarrow \mathbf{term}_\lambda$, where $Y[X/x]$ denotes the (capture-free) substitution of term X for (all free occurrences of) variable x in term Y; e.g., if Y is Lam (Abs x (App (Var x) (Var y))) and $x \notin \{y, z\}$, then:
  - $Y[(\mathsf{Var}\ z)/y] = $ Lam (Abs x (App (Var x) (Var z)))
  - $Y[(\mathsf{Var}\ z)/x] = Y$ (since bound occurrences like those of x in Y are not affected)

And there are corresponding operators for abstractions—e.g., freshAbs x (Abs x (Var x)) holds. Freshness and substitution are pervasive in the meta-theory of λ-calculus, as well as in most logical systems and formal semantics of programming languages. The basic properties of these operators lay at the core of important meta-theoretic results in these fields—our formalized theory aims at the exhaustive coverage of these basic properties.

2.3 Advantages and Obligations from Working with Terms Modulo Alpha

In our theory, we start with defining quasiterms and quasiabstractions and their alpha-equivalence. Then, after proving all the syntactic constructors and standard operators to be compatible with alpha, we quotient to alpha, obtaining what we call terms and abstractions, and define the versions of these operators on quotiented items. For example, let $\mathbf{qterm}_\lambda$ and $\mathbf{qabs}_\lambda$ be the types of quasiterms and quasiabstractions in λ-calculus. Here, the quasiabstraction constructor, qAbs : $\mathbf{var} \rightarrow \mathbf{qterm}_\lambda \rightarrow \mathbf{qabs}_\lambda$, is a free constructor, of the kind produced by standard datatype specifications [6, 10]. The types $\mathbf{term}_\lambda$ and $\mathbf{abs}_\lambda$ are $\mathbf{qterm}_\lambda$ and $\mathbf{qabs}_\lambda$ quotiented to alpha. We prove compatibility of qAbs with alpha and then define Abs : $\mathbf{var} \rightarrow \mathbf{term}_\lambda \rightarrow \mathbf{abs}_\lambda$ by lifting qAbs to quotients.

The decisive advantages of working with quasiterms and quasiabstractions modulo alpha, i.e., with terms and abstractions, are that (1) substitution behaves well (e.g., is compositional) and (2) Barendregt's variable convention [4] (of assuming, w.l.o.g., the bound variables fresh for the parameters) can be invoked in proofs.

However, this choice brings the obligation to prove that all concepts on terms are compatible with alpha. Without employing suitable abstractions, this can become quite difficult even in the most "banal" contexts. Due to nonfreeness, primitive recursion on terms requires a proof that the definition is well formed, i.e., that the overlapping cases lead to the same result. As for Barendregt's convention, its rigorous usage in proofs needs a principle that goes beyond the usual structural induction for free datatypes.

A framework that deals gracefully with these obligations can make an important difference in applications—enabling the formalizer to quickly leave behind low-level "bootstrapping" issues and move to the interesting core of the results. To address these obligations, we formalize state-of-the-art techniques from the literature [41, 44, 57].

2.4   Many-Sortedness

While λ-calculus has only one syntactic category of terms (to which we added that of abstractions for convenience), this is often not the case. FOL has two: terms and formulas. The Edinburgh Logical Framework (LF) [25] has three: object families, type families and kinds. More complex calculi can have many syntactic categories.

Our framework will capture these phenomena. We will call the syntactic categories *sorts*. We will distinguish syntactic categories for terms (the sorts) from those for variables (the *varsorts*). Indeed, e.g., in FOL we do not have variables ranging over formulas, in the π-calculus [34] we have channel names but no process variables, etc.

Sortedness is important, but formally quite heavy. In our formalization, we postpone dealing with it for as long as possible. We introduce an intermediate notion of *good* term, for which we are able to build the bulk of the theory—only as the very last step we introduce many-sorted signatures and transit from "good" to "sorted".

2.5   Possibly Infinite Branching

Nominal Logic's [40, 57] notion of finite support has become central in state-of-the-art techniques for reasoning about bindings. Occasionally, however, important developments step outside finite support. For example, (a simplified) CCS [33] has the following syntactic categories of data expressions $E \in \mathbf{exp}$ and processes $P \in \mathbf{proc}$:

$$E ::= \mathsf{Var}\; x \mid 0 \mid E + E \qquad\qquad P ::= \mathsf{Inp}\; c\; x\; P \mid \mathsf{Out}\; c\; e\; P \mid \sum_{i \in I} P_i$$

Above, $\mathsf{Inp}\; c\; x\; P$, usually written $c(x).P$, is an input prefix $c(x)$ followed by a continuation process P, with c being a channel and x a variable which is bound in P. Dually, $\mathsf{Out}\; c\; E\; P$, usually written $c\overline{E}.P$, is an output-prefixed process with E an expression. The exotic constructor here is the sum $\sum$, which models nondeterministic choice from a collection $(P_i)_{i \in I}$ of alternatives indexed by a set I. It is important that I is allowed to be infinite, for modeling different decisions based on different received inputs. But then process terms may use infinitely many variables, i.e., may not be finitely supported. Similar issues arise in infinitary FOL [29] and Hennessey-Milner logic [26]. In our theory, we cover such infinitely branching syntaxes.

3   General Terms with Bindings

We start the presentation of our formalized theory, in its journey from quasiterms (Sect. 3.1) to terms via alpha-equivalence (Sect. 3.2). The journey is fueled by the availability of fresh variables, ensured by cardinality assumptions on constructor branching and variables (Sect. 3.3). It culminates with a systematic study of the standard term operators (Sect. 3.4).

3.1   Quasiterms

The types **qterm** and **qabs**, of quasiterms and quasiabstractions, are defined as mutually recursive datatypes polymorphic in the following type variables: **index** and **bindex**, of indexes for free and bound arguments, **varsort**, of varsorts, i.e., sorts of variables, and **opsym**, of (constructor) operation symbols. For readability, below we omit the occurrences of these type variables as parameters to **qterm** and **qabs**:

datatype qterm = qVar **varsort var** |
$\qquad\qquad\qquad$ qOp **opsym** ((index, qterm) input) ((bindex, qabs) input)
$\quad$ and $\quad$ qabs = qAbs **varsort var qterm**

Thus, any quasiabstraction has the form qAbs xs x X, putting together the variable x of varsort xs with the quasiterm X, indicating the binding of x in X. On the other hand, a quasiterm is either an injection qVar xs x, of a variable x of varsort xs, or has the form qOp δ inp $binp$, i.e., consists of an operation symbol applied to some inputs that can be either free, inp, or bound, $binp$.

We use (α, β) **input** as a type synonym for $\alpha \to \beta$ **option**, the type of partial functions from α to β; such a function returns either None (representing "undefined") or Some b for $b : \beta$. This type models inputs to the quasiterm constructors of varying number of arguments. An operation symbol δ : **opsym** can be applied, via qOp, to: (1) a varying number of free inputs, i.e., families of quasiterms modeled as members of (**index, qterm**) **input** and (2) a varying number of bound inputs, i.e., families of quasiabstractions modeled as members of (**index, qabs**) **input**. For example, taking **index** to be **nat** we capture n-ary operations for any n (passing to qOp δ inputs defined only on $\{0, \ldots, n-1\}$), as well as as countably-infinitary operations (passing to qOp δ inputs defined on the whole **nat**).

Note that, so far, we consider sorts of variables but not sorts of terms. The latter will come much later, in Sect. 5, when we introduce signatures. Then, we will gain control (1) on which varsorts should be embedded in which term sorts and (2) on which operation symbols are allowed to be applied to which sorts of terms. But, until then, we will develop the interesting part of the theory of bindings without sorting the terms.

On quasiterms, we define freshness, qFresh : **varsort** → **var** → **qterm** → **bool**, substitution, _[_/_]_ : **qterm** → **qterm** → **var** → **varsort** → **qterm**, parallel substitution, _[_] : **qterm** → (**varsort** → **var** → **qterm option**) → **qterm**, swapping, $\_[\_ \wedge \_]\_$: **qterm** $\rightarrow$ **var** $\rightarrow$ **var** $\rightarrow$ **varsort** $\rightarrow$ **qterm**, and alpha-equivalence, alpha : **qterm** $\rightarrow$ **qterm** $\rightarrow$ **bool**—and corresponding operators on quasiabstractions: qFreshAbs, alphaAbs, etc.

The definitions proceed as expected, with picking suitable fresh variables in the case of substitutions and alpha. For parallel substitution, given a (partial) variable-to-quasiterm assignment ρ : **varsort** $\rightarrow$ **var** $\rightarrow$ **qterm option**, the quasiterm $X[\rho]$ is obtained by substituting, for each free variable x of sort xs in X for which ρ is defined, the quasiterm Y where ρ xs $x =$ Some Y. We only show the formal definition of alpha.

3.2 Alpha-Equivalence

We define the predicates alpha (on quasiterms) and alphaAbs (on quasiabstractions) mutually recursively, as shown in Fig. 1. For variable quasiterms, we require equality on both the variables and their sorts. For qOp quasiterms, we recurse through the components, inp and $binp$. Given any predicate P : β^2 $\rightarrow$ **bool**, we write $\uparrow P$ for its lifting to (α, β) **input**^{2} $\rightarrow$ **bool**, defined as $\uparrow P$ inp inp' $\iff$ $\forall i.$ case $(inp$ i, inp' $i)$ of (None, None) $\Rightarrow$ True | (Some b, Some b') $\Rightarrow$ P b b' | $\_ \Rightarrow$ False. Thus, $\uparrow P$ relates two inputs just in case they have the same domain and their results are componentwise related.

$$
\begin{aligned}
\text{alpha } (\text{qVar } xs\ x)\ (\text{qVar } xs'\ x') &\iff xs = xs' \wedge x = x' \\
\text{alpha } (\text{qOp } \delta\ inp\ binp)\ (\text{qOp } \delta'\ inp'\ binp') &\iff \delta = \delta' \wedge \uparrow\text{alpha } inp\ inp' \wedge \uparrow\text{alphaAbs } binp\ binp' \\
\text{alpha } (\text{qVar } xs\ x)\ (\text{qOp } \delta'\ inp'\ binp') &\iff \text{False} \\
\text{alpha } (\text{qOp } \delta\ inp\ binp)\ (\text{qVar } xs'\ x') &\iff \text{False} \\
\text{alphaAbs } (\text{qAbs } xs\ x\ X)\ (\text{qAbs } xs'\ x'\ X') &\iff xs = xs' \wedge (\exists y \notin \{x, x'\}.\ \text{qFresh } xs\ y\ X\ \wedge \\
&\qquad \text{qFresh } xs\ y\ X' \wedge \text{alpha } (X[y \wedge x]_{xs})\ (X'[y \wedge x']_{xs}))
\end{aligned}
$$

Fig. 1. Alpha-equivalence

Convention 1. Throughout this paper, we write $\uparrow$ for the natural lifting of the various operators from terms and abstractions to free or bound inputs.

In Fig. 1's clause for quasiabstractions, we require that the bound variables are of the same sort and there exists some fresh y such that alpha holds for the terms where y is swapped with the bound variable. Following Nominal Logic, we prefer to use swapping instead of substitution in alpha-equivalence, since this leads to simpler proofs [41].

3.3   Good Quasiterms and Regularity of Variables

In general, alpha will not be an equivalence, namely, will not be transitive: Due to the arbitrarily wide branching of the constructors, we may not always have fresh variables y available in an attempt to prove transitivity by induction. To remedy this, we restrict ourselves to "good" quasiterms, whose constructors do not branch beyond the cardinality of **var**. Goodness is defined as the mutually recursive predicates qGood and qGoodAbs:

$$\text{qGood (qVar } xs \ x) \iff \text{True}$$
$$\text{qGood (qOp } \delta \ inp \ binp) \iff \uparrow \text{qGood } inp \ \wedge \ \uparrow \text{qGoodAbs } binp \ \wedge$$
$$|\text{dom } inp| < |\textbf{var}| \ \wedge \ |\text{dom } binp| < |\textbf{var}|$$
$$\text{qGoodAbs (qAbs } xs \ x \ X) \iff \text{qGood } X$$

where, given a partial function f, we write dom f for its domain.

Thus, for good items, we hope to always have a supply of fresh variables. Namely, we hope to prove qGood $X \implies \forall xs. \ \exists x. \ \text{qFresh } xs \ x \ X$. But goodness is not enough. We also need a special property for the type **var** of variables. In the case of finitary syntax, it suffices to take **var** to be countably infinite, since a finitely branching term will contain fewer than $|\textbf{var}|$ variables (here, meaning a finite number of them)—this can be proved by induction on terms, using the fact that a finite union of finite sets is finite.

So let us attempt to prove the same in our general case. In the inductive qOp case, we know from goodness that the branching is smaller than $|\textbf{var}|$, so to conclude we would need the following: *A union of sets smaller than* $|\textbf{var}|$ *indexed by a set smaller than* $|\textbf{var}|$ *stays smaller than* $|\textbf{var}|$. It turns out that this is a well-studied property of cardinals, called *regularity*—with $|\textbf{nat}|$ being the smallest regular cardinal. Thus, the desirable generalization of countability is regularity (which is available from Isabelle's cardinal library [12]). Henceforth, we will assume:

Assumption 2. $|\textbf{var}|$ is a regular cardinal.

We will thus have not only one, but a $|\textbf{var}|$ number of fresh variables:

Prop 3. qGood $X \implies \forall xs. \ |\{x. \ \text{qFresh } xs \ x \ X\}| = |\textbf{var}|$

Now we can prove, for good items, the properties of alpha familiar from the λ-calculus, including it being an equivalence and an alternative formulation of the abstraction case, where "there exists a fresh y" is replaced with "for all fresh y." While the "exists" variant is useful when proving that two terms are alpha-equivalent, the "forall" variant gives stronger inversion and induction rules for proving implications from alpha. (Such fruitful "exsist-fresh/forall-fresh," or "some-any" dychotomies have been previously discussed in the context of bindings, e.g., in [3, 32, 39].)

Prop 4. The following hold:

(1) alpha and alphaAbs are equivalences on good quasiterms and quasiabstractions
(2) The predicates defined by replacing, in Fig. 1's definition, the abstraction case with

$$\text{alphaAbs (qAbs } xs \ x \ X) \ (\text{qAbs } xs' \ x' \ X') \iff$$
$$xs = xs' \wedge (\forall y \notin \{x, x'\}. \ \text{qFresh } xs \ y \ X \wedge \text{qFresh } xs \ y \ X' \implies \text{alpha}(X[y \wedge x]_{xs})(X'[y \wedge x']_{xs}))$$

coincide with alpha and alphaAbs.

3.4 Terms and Their Properties

We define **term** and **abs** as collections of alpha- and alphaAbs- equivalence classes of qterm and qabs. Since qGood and qGoodAbs are compatible with alpha and alphaAbs, we lift them to corresponding predicates on terms and abstractions, good and goodAbs.

We also prove that all constructors and operators are alpha-compatible, which allows lifting them to terms: Var : **varsort** → **var** → **term**, Op : **opsym** → (**index**, **term**)**input** → (**bindex**, **abs**)**input** → **term**, Abs : **varsort** → **var** → **term** → **abs**, fresh : **varsort** → **term** → **bool**, _[_/_]_ : **term** → **term** → **var** → **varsort** → **term**, etc.

To establish an abstraction barrier that sets terms free from their quasiterm origin, we prove that the syntactic constructors mostly behave like free constructors, in that Var, Op and Abs are exhaustive and Var and Op are injective and nonoverlapping. True to the quarantine principle expressed in Sect. 2.1, the only nonfreeness incident occurs for Abs. Its equality behavior is regulated by the "exists fresh" and "forall fresh" properties inferred from the definition of alphaAbs and Prop. 4(2), respectively:

Prop 5. Assume good X and good X'. Then the following are equivalent:

(1) Abs xs x X = Abs xs' x' X'
(2) $xs = xs' \wedge (\exists y \notin \{x, x'\}.$ fresh xs y $X \wedge$ fresh xs y $X' \wedge X[y \wedge x]_{xs} = X'[y \wedge x']_{xs})$
(3) $xs = xs' \wedge (\forall y \notin \{x, x'\}.$ fresh xs y $X \wedge$ fresh xs y $X' \implies X[y \wedge x]_{xs} = X'[y \wedge x']_{xs})$

Useful rules for abstraction equality also hold with substitution:

Prop 6. Assume good X and good X'. Then the following hold:

(1) $y \notin \{x, x'\} \wedge$ fresh xs y $X \wedge$ fresh xs y $X' \wedge X[(\text{Var } xs$ $y) / x]_{xs} = X'[(\text{Var } xs$ $y) / x']_{xs} \implies$ Abs xs x X = Abs xs x' X'
(2) fresh xs y $X \implies$ Abs xs x X = Abs xs y $(X[(\text{Var } xs$ $y) / x]_{xs})$

To completely seal the abstraction barrier, for all the standard operators we prove simplification rules regarding their interaction with the constructors, which makes the former behave as if they had been defined in terms of the latter. For example, the following facts resemble an inductive definition of freshness (as a predicate):

Prop 7. Assume good X, ↑good inp, ↑good $binp$, $|\text{dom } inp| < |\textbf{var}|$ and $|\text{dom } binp| < |\textbf{var}|$. The following hold:

(1) $(ys, y) \neq (xs, x) \implies$ fresh ys y $(\text{Var } xs$ $x)$
(2) ↑(fresh ys y) $inp \wedge$ ↑(freshAbs ys y) $binp \implies$ fresh ys y $(\text{Op } \delta$ inp $binp)$
(3) $(ys, y) = (xs, x) \vee$ fresh ys y $X \implies$ freshAbs ys y $(\text{Abs } xs$ x $X)$

Here and elsewhere, when dealing with Op, we make cardinality assumptions on the domains of the inputs to make sure the terms Op δ inp $binp$ are good.

We can further improve on Prop. 7, obtaining "iff" facts that resemble a primitively recursive definition of freshness (as a function):

Prop 8. Prop. 7 stays true if the implications are replaced by equivalences ($\Longleftrightarrow$).

For substitution, we prove facts with a similarly primitive recursion flavor:

Prop 9. Assume good X, good Y, $\uparrow$ good inp, $\uparrow$ good $binp$, $|\text{dom } inp| < |\textbf{var}|$ and $|\text{dom } binp| < |\textbf{var}|$. The following hold:

(1) $(\text{Var } xs\ x)\,[Y/y]_{ys} = (\text{if } (xs, x) = (ys, y)\text{then } Y \text{ else Var } xs\ x)$
(2) $(\text{Op } \delta\ inp\ binp)\,[Y/y]_{ys} = \text{Op } \delta\ (\uparrow (-[Y/y]_{ys})\ inp)\ (\uparrow (-[Y/y]_{ys})\ binp)$
(3) $(xs, x) \neq (ys, y) \wedge \text{ fresh } xs\ x\ Y \implies (\text{Abs } xs\ x\ X)\,[Y/y]_{ys} = \text{Abs } xs\ x\ (X\,[Y/y]_{ys})$

We also prove generalizations of Prop. 9's facts for parallel substitution, for example, $\uparrow (\text{fresh } xs\ x)\ \rho \implies (\text{Abs } xs\ x\ X)\,[\rho] = \text{Abs } xs\ x\ (X\,[\rho])$.

Note that, for properties involving Abs, the simplification rules require freshness of the bound variable: $\text{freshAbs } ys\ y\ (\text{Abs } xs\ x\ X)$ is reducible to $\text{fresh } ys\ y\ X$ only if (xs, x) is distinct from (ys, y), $(\text{Abs } xs\ x\ X)\,[Y/y]_{ys}$ is expressible in terms of $X\,[Y/y]_{ys}$ only if (xs, x) is distinct from (ys, y) and fresh for Y, etc.

Finally, we prove lemmas that regulate the interaction between the standard operators, in all possible combinations: freshness versus swapping, freshness versus substitution, substitution versus substitution, etc. Here are a few samples:

Prop 10. If the terms X, Y, Y_1, Y_2, Z are good and the assignments ρ, ρ' are $\uparrow$ good, then:

(1) Swapping distributes over all operators, including, e.g., substitution:
$$Y\,[X/x]_{xs}\,[z_1 \wedge z_2]_{zs} = (Y\,[z_1 \wedge z_2]_{zs})\,[(X\,[z_1 \wedge z_2]_{zs})\,/\,(x[z_1 \wedge z_2]_{xs,zs})]_{xs}$$
where $x[z_1 \wedge z_2]_{xs,zs} = (\text{if } xs = zs \text{ then } x[z_1 \wedge z_2]\text{else}x)$
(2) Substitution of the same variable (and of the same varsort) distributes over itself:
$$X\,[Y_1/y]_{ys}\,[Y_2/y]_{ys} - X\,[(Y_1\,[Y_2/y]_{ys})/y]_{ys}$$
(3) Substitution of different variables distributes over itself, assuming freshness:
$$(ys \neq zs \vee y \neq z) \wedge \text{ fresh } ys\ y\ Z \implies X\,[Y/y]_{ys}\,[Z/z]_{zs} = (X\,[Z/z]_{zs})\,[(Y\,[Z/z]_{zs})/y]_{ys}$$

(4) Freshness for a substitution decomposes into freshness for its participants:
$$\text{fresh } zs\ z\ (X[Y/y]_{ys}) \Longleftrightarrow ((zs, z) = (ys, y) \vee \text{fresh } zs\ z\ X) \wedge (\text{fresh } ys\ y\ X \vee \text{fresh } zs\ z\ Y)$$

(5) Parallel substitution is compositional:
$$X\,[\rho]\,[\rho'] = X\,[\rho \bullet \rho']$$
where $\rho \bullet \rho'$ is the monadic composition of ρ and ρ', defined as
$$(\rho \bullet \rho')\ xs\ x = \text{ case } \rho\ xs\ x \text{ of None} \Rightarrow \rho'\ xs\ x \mid \text{Some } X \Rightarrow X[\rho']$$

In summary, we have formalized quite exhaustively the general-purpose properties of all syntactic constructors and standard operators. Some of these properties are subtle. In formalization of concrete results for particular syntaxes, they are likely to require a lot of time to even formulate them correctly, let alone prove them—which would be wasteful, since they are independent on the particular syntax.

4 Reasoning and Definition Principles

We formalize schemes for induction (Sect. 4.1), recursion and semantic interpretation (Sect. 4.2) that realize the Barendregt convention and are compatible with the standard operators.

4.1 Fresh Induction

We introduce fresh induction by an example. To prove Prop. 10(4), we use (mutual) structural induction over terms and abstractions, proving the statement together with the corresponding statement for abstractions, freshAbs zs z $(A[Y/y]_{ys}) \Longleftrightarrow ((zs, z) = (ys, y) \lor$ freshAbs zs z $A) \land ($freshAbs ys y $A \lor$ fresh zs z $Y)$. The proof's only interesting case is the Abs case, say, for abstractions of the form Abs xs x X. However, if we were able to assume freshness of (xs, x) for all the statement's parameters, namely Y, (ys, y) and (zs, z), this case would also become "uninteresting," following automatically from the induction hypothesis by mere simplification, as shown below (with the freshness assumptions highlighted):

freshAbs zs z $((\text{Abs } xs \ x \ X) [Y/y]_{ys})$

$\updownarrow$ (by Prop. 9(3), since $(xs, x) \neq (ys, y)$ and fresh xs x Y)

freshAbs zs z $(\text{Abs } xs \ x \ (X [Y/y]_{ys}))$

$\updownarrow$ (by Prop. 8(3), since $(xs, x) \neq (zs, z)$)

fresh zs z $(X [Y/y]_{ys})$

$\updownarrow$ (by Induction Hypothesis)

$((zs, z) = (ys, y) \lor$ fresh zs z $X) \land ($fresh ys y $X \lor$ fresh zs z $Y)$

$\updownarrow$ (by Prop. 8(3) applied twice, since $(xs, x) \neq (zs, z)$ and $(xs, x) \neq (ys, y)$)

$((zs, z) = (ys, y) \lor$ freshAbs zs z $(\text{Abs } xs \ x \ X)) \land ($freshAbs ys y $(\text{Abs } xs \ x \ X) \lor$ fresh zs z $Y)$

The practice of assuming freshness, known in the literature as the Barendregt convention, is a hallmark in informal reasoning about bindings. Thanks to insight from Nominal Logic [41,55,57], we also know how to apply this morally correct convention fully rigorously. To capture it in our formalization, we model parameters p : **param** as anything that allows for a notion of freshness, or, alternatively, provides a set of (free) variables for each varsort, varsOf : **param** $\rightarrow$ **varsort** $\rightarrow$ **var set**. With this, a "fresh induction" principle can be formulated, if all parameters have fewer variables than $|\textbf{var}|$ (in particular, if they have only finitely many).

Theorem 11. Let φ : **term** $\rightarrow$ **param** $\rightarrow$ **bool** and φAbs : **abs** $\rightarrow$ **param** $\rightarrow$ **bool**. Assume:

(1) $\forall xs, p.$ $|\text{varsOf } xs \ p| < |\textbf{var}|$

(2) $\forall xs, x, p.$ φ $(\text{Var } xs \ x)$ p

(3) $\forall \delta, inp, binp, p.$ $|\text{dom } inp| < |\textbf{var}| \land |\text{dom } binp| < |\textbf{var}| \land \uparrow (\lambda X. \text{ good } X \land (\forall q. \varphi \ X \ q))$ inp $\land \uparrow (\lambda A. \text{ goodAbs } A \land (\forall q. \varphi Abs \ A \ q))$ $binp \Longrightarrow$ φ $(\text{Op } \delta \ inp \ binp)$ p

(4) $\forall xs, x, X, p.$ good $X \land \varphi \ X \ p \land x \notin \text{varsOf } xs \ p \Longrightarrow \varphi Abs \ (\text{Abs } xs \ x \ X)$

Then $\forall X, p.$ good $X \Longrightarrow \varphi X p$ and $\forall A, p.$ goodAbs $A \Longrightarrow \varphi Abs A p$.

Highlighted is the essential difference from the usual structural induction: The bound variable x can be assumed fresh for the parameter p (on its varsort, xs). Note also that, in the Op case, we lift to inputs the predicate as quantified universally over all parameters.

Back to Prop. 10(4), this follows automatically by fresh induction (plus the shown simplifications), after recognizing as parameters the variables (ys, y) and (zs, z) and the term Y—formally, taking **param** = (**varsort** × **var**)^2 × **term** and varsOf $xs ((ys, y), (zs, z), Y) = \{y \mid xs = ys\} \cup \{z \mid xs = zs\} \cup \{x \mid \neg \text{fresh } xs \, x \, Y\}$.

4.2 Freshness- and Substitution- Sensitive Recursion

A *freshness-substitution (FS) model* consists of two collections of elements endowed with term- and abstraction- like operators satisfying some characteristic properties of terms. More precisely, it consists of:

- two types, **T** and **A**
- operations corresponding to the constructors: VAR : **varsort** → **var** → **T**, OP : **opsym** → (**index**, **T**) **input** → (**bindex**, **A**) **input** → **T**, ABS : **varsort** → **var** → **T** → **A**
- operations corresponding to freshness and substitution: FRESH : **varsort** → **var** → **T** → **bool**, FRESHABS : **varsort** → **var** → **A** → **bool**, _[_/_]_ : **T** → **T** → **var** → **varsort** → **T** and _[_/_]_ : **A** → **T** → **var** → **varsort** → **A**

and it is required to satisfy the analogues of:

- the implicational simplification rules for fresh from Prop. 7
  (for example, $(ys, y) \neq (xs, x) \Longrightarrow$ FRESH $ys \, y$ (VAR $xs \, x$))
- the simplification rules for substitution from Prop. 9
- the substitution-based abstraction equality rules from Prop. 6.

Theorem 12. The good terms and abstractions form the initial FS model. Namely, for any FS model as above, there exist the functions $f :$ **term** → **T** and $fAbs :$ **abs** → **A** that commute, on good terms, with the constructors and with substitution and preserve freshness:

$f(\text{Var } xs \, x) = \text{VAR } xs \, x$ $\qquad$ $f(\text{Op } \delta \, inp \, binp) = \text{OP } \delta \, (\uparrow f \, inp) \, (\uparrow fAbs \, binp)$
$fAbs(\text{Abs } xs \, x \, X) = \text{ABS } xs \, x \, (f \, X)$
$f(X [Y/y]_{ys}) = (f X) [(f Y)/y]_{ys}$ $\qquad$ $fAbs(A [Y/y]_{ys}) = (fAbs A) [(f Y)/y]_{ys}$
fresh $xs \, x \, X \Longrightarrow$ FRESH $xs \, x \, (f X)$ $\qquad$ freshAbs $xs \, x \, A \Longrightarrow$ FRESHABS $xs \, x \, (fAbs A)$

In addition, the two functions are uniquely determined on good terms and abstractions, in that, for all other functions $g :$ **term** → **T** and $gAbs :$ **abs** → **A** satisfying the same commutation and preservation properties, it holds that f and g are equal on good terms and $fAbs$ and $gAbs$ are equal on good abstractions.

Like any initiality property, this theorem represents a primitive recursion principle. Consider first the simpler case of lists over a type **G**, with constructors

Nil : **G list** and Cons : **G** → **G list** → **G list**. To define, by primitive recursion, a function from lists, say, length : **G list** → **nat**, we need to indicate what is Nil mapped to, here length Nil = 0, and, recursively, what is Cons mapped to, here length (Cons a as) = 1 + length as. We can rephrase this by saying: If we define "list-like" operators on the target domain— here, taking NIL : **nat** to be 0 and CONS : **G** → **nat** → **nat** to be $\lambda g, n.\ 1 + n$—then the recursion principle offers us a function length that commutes with the constructors: length Nil = NIL = 0 and length (Cons a as) = CONS a (length as) = 1 + length as. For terms, we have a similar situation, except that (1) substitution and freshness are considered in addition to the constructors and (2) paying the price for lack of freeness, some conditions need to be verified to deem the operations "term-like".

This recursion principle was discussed in [44] for the syntax of λ-calculus and shown to have many useful applications. Perhaps the most useful one is the seamless interpretation of syntax in semantic domains, in a manner that is guranteed to be compatible with alpha, substitution and freshness. We formalize this in our general setting:

A *semantic domain* consists of two collections of elements endowed with interpretations of the Op and Abs constructors, the latter in a higher-order fashion—interpreting variable binding as (meta-level) functional binding. Namely, it consists of:

– two types, **Dt** and **Da**
– a function op : **opsym** → (**index, Dt**) **input** → (**bindex, Da**) **input** → **Dt**
– a function abs : **varsort** → (**Dt** → **Dt**) → **Da**

Theorem 13. The terms and abstractions are interpretable in any semantic domain. Namely, if **val** is the type of valuations of variables in the domain, **varsort** → **var** → **Dt**, there exist the functions sem : **term** → **val** → **Dt** and semAbs : **abs** → **val** → **Da** such that:

– sem (Var xs x) $\rho = \rho$ xs x
– sem (Op δ inp $binp$) ρ = op δ ($\uparrow$ ($\lambda X.$ sem X ρ) inp) ($\uparrow$ ($\lambda A.$ semAbs A ρ) $binp$)
– semAbs (Abs xs x X) ρ = abs xs ($\lambda d.$ sem X ($\rho[(xs, x) \leftarrow d]$))

In addition, the interpretation functions map syntactic substitution and freshness to semantic versions of the concepts:

– sem ($X[Y/y]_{ys}$) ρ = sem X ($\rho[(ys, y) \leftarrow$ sem Y $\rho]$)
– fresh xs x X $\implies$ ($\forall \rho, \rho'.\ \rho =_{(xs,x)} \rho' \implies$ sem X ρ = sem X ρ'),
  where "$=_{(xs,x)}$" means equal everywhere but on (xs, x)

Theorem 13 is the foundation for many particular semantic interpretations, including that of λ-terms in Henkin models and that of FOL terms and formulas in FOL models. It guarantees compatibility with alpha and proves, as bonuses, a freshness and a substitution property. The freshness property is nothing but the notion that the interpretation only depends on the free variables, whereas the substitution property generalizes what is usually called *the substitution lemma*, stating that interpreting a substituted term is the same as interpreting the original term in a "substituted" environment.

This theorem follows by an instantiation of the recursion Theorem 12: taking **T** and **A** to be **val** → **Dt** and **val** → **Da** and taking the term/abstraction-like operations as prescribed by the desired clauses for sem and semAbs—e.g., VAR xs x is $\lambda\rho.\ \rho\ xs\ x$.

5 Sorting the Terms

So far, we have a framework where the operations take as free and bound inputs partial families of terms and abstractions. All theorems refer to good (i.e., sufficiently low-branching) terms and abstractions. However, we promised a theory that is applicable to terms over many-sorted binding signatures. Thanks to the choice of a flexible notion of input, it is not hard to cast our results into such a many-sorted setting. Given a suitable notion of signature Sect. 5.1, we classify terms according to sorts Sect. 5.2 and prove that well-sorted terms are good Sect. 5.3—this gives us sorted versions of all theorems Sect. 5.4.

5.1 Binding Signatures

A *(binding) signature* is a tuple (**index, bindex, varsort, sort, opsym,** asSort, stOf, arOf, barOf), where **index**, **bindex**, **varsort** and **opsym** are types (with the previously discussed intuitions) and **sort** is a new type, of sorts for terms. Moreover:

- asSort : **varsort** → **sort** is an injective map, embedding varsorts into sorts
- stOf : **opsym** → **sort**, read "the (result) sort of"
- arOf : **opsym** → (**index, sort**) **input**, read "the (free) arity of"
- barOf : **opsym** → (**bindex, varsort** × **sort**) **input**, read "the bound arity of"

Thus, a signature prescribes which varsorts correspond to which sorts (as discussed in Sect. 2.4) and, for each operation symbol, which are the sorts of its free inputs (the arity), of its bound (abstraction) inputs (the bound arity), and of its result.

When we give examples for our concrete syntaxes in Sect. 2, we will write $(i_1 \mapsto a_1, \ldots, i_n \mapsto a_n)$ for the partial function that sends each i_k to Some a_k and everything else to None. In particular, () denotes the totally undefined function.

For the λ-calculus syntax, we take **index** = **bindex** = **nat**, **varsort** = **sort** = {lam} (a singleton datatype), **opsym** = {App, Lam}, asSort to be the identity and stOf to be the unique function to {lam}. Since App has two free inputs and no bound input, we use the first two elements of **nat** as free arity and nothing for the bound arity: arOf App = $(0 \mapsto$ lam, $1 \mapsto$ lam), barOf App = (). By contrast, since Lam has no free input and one bound input, we use nothing for the free arity, and the first element of **nat** for the bound arity: arOf Lam = (), barOf Lam = $(0 \mapsto$ (lam, lam)).

For the CCS example in Sect. 2.5, we fix a type **chan** of channels. We choose a cardinal upper bound κ for the branching of sum ($\sum$), and choose a type **index** of cardinality κ. For **bindex**, we do not need anything special, so we take it to be **nat**. We have two sorts, of expressions and processes, so we take **sort** = {exp, proc}. Since we have expression variables but no process variables, we take **varsort** = {varexp} and asSort to send varexp to exp. We define **opsym** as the following datatype: **opsym** = Zero | Plus | Inp **chan** | Out **chan** | $\sum$ (**index set**). The free and bound arities and sorts of the operation symbols are as expected. For example, Inp c acts similarly to λ-abstraction, but binds, in proc terms, variables of a different sort, varexp: arOf (Inp c) = (), barOf (Inp c) = $(0 \mapsto$ (varexp, proc)). For $\sum I$ with I : **index set**, the arity is only defined for elements of I, namely arOf ($\sum I$) = $((i \in I) \mapsto$ proc).

5.2 Well-Sorted Terms over a Signature

Based on the information from a signature, we can distinguish our terms of interest, namely those that are well-sorted in the sense that:

– all variables are embedded into terms of sorts compatible with their varsorts
– all operation symbols are applied according their free and bound arities

This is modeled by well-sortedness predicates wls : **sort** → **term** → **bool** and wlsAbs : **varsort** → **sort** → **abs** → **bool**, where wls s X states that X is a well-sorted term of sort s and wlsAbs (xs, s) A states that A is a well-sorted abstraction binding an xs-variable in an s-term. They are defined mutually inductively by the following clauses:

$$\text{wls (asSort } xs) \text{ (Var } xs \ x)$$
$$\uparrow \text{wls (arOf } \delta) \ inp \ \wedge \ \uparrow \text{wlsAbs (barOf } \delta) \ binp \implies \text{wls (stOf } \delta) \text{ (Op } \delta \ inp \ binp)$$
$$\text{isInBar } (xs, s) \ \wedge \ \text{wls } s \ X \implies \text{wlsAbs } (xs, s) \text{ (Abs } xs \ x \ X)$$

where isInBar (xs, s) states that the pair (xs, s) is in the bound arity of at least one operation symbol δ, i.e., barOf δ $i = (xs, s)$ for some i— this rules out unneeded abstractions.

Let us illustrate sorting for our running examples. In the λ-calculus syntax, let X = Var lam x, A = Abs lam x X, and Y = Op Lam () $(0 \mapsto A)$. These correspond to what, in the unsorted BNF notation from Sect. 2.1, we would write Var x, Abs x X and Lam (Abs x X). In our sorting system, X and Y are both well-sorted terms at sort lam (written wls lam X and wls lam Y) and A is a well-sorted abstraction at sort (lam, lam) (written wlsAbs (lam, lam) A).

For CCS, we have that E = Op Zero () () and F = Op Plus $(0 \mapsto E, 1 \mapsto E)$ () are well-sorted terms of sort exp. Moreover, P = Op $(\sum \emptyset)$ () () and Q = Op (Out c) $(0 \mapsto F, 1 \mapsto P)$ () are well-sorted terms of sort proc. (Note that P is a sum over the empty set of choices, i.e., the null process, whereas Q represents a process that outputs the value of $0 + 0$ on channel c and then stops.) If, e.g., we swap the arguments of Out c in Q, we obtain Op (Out c) $(0 \mapsto P, 1 \mapsto F)$ (), which is not well-sorted: In the inductive clause for wls, the input $(0 \mapsto P, 1 \mapsto F)$ fails to match the arity of Out c, $(0 \mapsto$ exp, $1 \mapsto$ proc).

5.3   From Good to Well-Sorted

Recall that goodness means "does not branch beyond $|\mathbf{var}|$." On the other hand, well-sortedness imposes that, for each applied operation symbol δ, its inputs have same domains, i.e., *only branch as much*, as the arities of δ. Thus, it suffices to assume the arity domains smaller than $|\mathbf{var}|$. We will more strongly assume that the types of sorts and indexes (the latter subsuming the arity domains) are all smaller than $|\mathbf{var}|$:

Assumption 14. $|\mathbf{sort}| < |\mathbf{var}| \ \wedge \ |\mathbf{index}| < |\mathbf{var}| \ \wedge \ |\mathbf{bindex}| < |\mathbf{var}|$

Now we can prove:

Prop 15. $(\mathsf{wls}\ s\ X \Longrightarrow \mathsf{good}\ X) \ \wedge \ (\mathsf{wls}\ (xs, s)\ A \Longrightarrow \mathsf{goodAbs}\ A)$

In addition, we prove that all the standard operators preserve well-sortedness. For example, we prove that if we substitute, in the well-sorted term X of sort s, for the variable y of varsort ys, the well-sorted term Y of sort corresponding to ys, then we obtain a well-sorted term of sort s: $\mathsf{wls}\ s\ X \ \wedge \ \mathsf{wls}\ (\mathsf{asSort}\ ys)\ Y \Longrightarrow \mathsf{wls}\ s\ (X\,[Y/y]_{ys})$.

Using the preservation properties and Prop. 15, we transfer the entire theory of Sects. 3.4 and 4 from good terms to well-sorted terms—e.g., Prop. 10(2) becomes:

$\mathsf{wls}\ s\ X \ \wedge \ \mathsf{wls}\ (\mathsf{asSort}\ ys)\ Y_1 \ \wedge \ \mathsf{wls}\ (\mathsf{asSort}\ ys)\ Y_2 \Longrightarrow X\,[Y_1/y]_{ys}\,[Y_2/y]_{ys} = \ldots$

The transfer is mostly straightforward for all facts, including the induction theorem. (For stating the well-sorted version of the recursion and semantic interpretation theorems, there is some additional bureaucracy since we also need sorting predicates on the target domain —the extended technical report [?] gives details.)

There is an important remaining question: Are our two Assumptions (2 and 14) satisfiable? That is, can we find, for any types **sort**, **index** and **bindex**, a type **var** larger than these such that $|\mathbf{var}|$ is regular? Fortunately, the theory of cardinals again provides us with a positive answer: Let $\mathbf{G} = \mathbf{nat} + \mathbf{sort} + \mathbf{index} + \mathbf{bindex}$. Since any successor of an infinite cardinal is regular, we can take **var** to have the same cardinality as the successor of $|\mathbf{G}|$, by defining **var** as a suitable subtype of $\mathbf{G}$ set. In the case of all operation symbols being finitary, i.e., with their arities having finite domains, we do not need the above fancy construction, but can simply take **var** to be a copy of **nat**.

5.4   End Product

All in all, our formalization provides a theory of syntax with bindings over an arbitrary many-sorted signature. The signature is formalized as an Isabelle locale [28] that fixes the types **var**, **sort**, **varsort**, **index**, **bindex** and **opsym** and the constants asSort, arOf and barOf and assumes the injectivity of asSort and the **var** properties (Assumptions 2 and 14). All end-product theorems are placed in this locale.

The whole formalization consists of 22700 lines of code (LOC). Of these, 3300 LOC are dedicated to quasiterms, their standard operators and alpha-equivalence. 3700 LOC are dedicated to the definition of terms and the lifting of results from quasiterms. Of the latter, the properties of substitution were the most extensive—2500 LOC out of the whole 3700—since substitution, unlike freshness and swapping, requires heavy variable renaming, which complicates the proofs.

The induction and recursion schemes presented in Sect. 4 are not the only schemes we formalized (but are the most useful ones). We also proved a variety of lower-level induction schemes based on the skeleton of the terms (a generalization of depth for possibly infinitely branching terms) and schemes that are easier to instantiate—e.g., by pre-instantiating Theorem 11 with commonly used parameters such as variables, terms and environments. As for the recursion Theorem 12, we additionally proved a more flexible scheme that allows the recursive argument, and not only the recursive result, to be referred—this is *full-fledged primitive recursion*, whereas Theorem 12 only implements *iteration*. Also, we proved schemes for recursion that factor swapping [38] instead of and in addition to substitution. All together, these constitute 8000 LOC.

The remaining 7700 LOC of the formalization are dedicated to transiting from good terms to sorted terms. Of these, 3500 LOC are taken by the sheer statement of our many end-product theorems. Another fairly large part, 2000 LOC, is dedicated to transferring all the variants of the recursion Theorem 12 and the interpretation Theorem 13, which require conceptually straightforward but technically tedious moves back and forth between sorted terms and sorted elements of the target domain.

6   Discussion, Related Work and Future Work

There is a large amount of literature on formal approaches to syntax with bindings. (See [1, Sect. 2], [18, Sect. 6] and [42, Sect. 2.10, Sect. 3.7] for overviews.) Our work, nevertheless, fills a gap in the literature: It is the first theory of binding syntax mechanized in a universal algebra fashion, i.e., with sorts and many-sorted term constructors specified by a binding signature, as employed in several theoretical developments, e.g., [19,41,48,52]. The universal algebra aspects of our approach are the consideration of an *arbitrary signature* and the singling out of the collection of terms and the operations on them as an *initial object in a category of models/algebras* (which yields a recursion principle). We do not consider arbitrary equational varieties (like in [52]), but only focus on selected equations and Horn clauses that characterize the term models (like in [41]).

Alternatives to Universal Algebra. A popular alternative to our universal algebra approach is higher-order abstract syntax (HOAS) [16–18,24,25]: the reduction of all bindings to a single binding—that of a fixed λ-calculus. Compared to universal algebra, HOAS's advantage is lighter formalizations, whereas the disadvantage is the need to prove the representation's adequacy (which involves reasoning about substitution) and, in some frameworks, the need to rule out the exotic terms.

Another alternative, very successfully used in HOL-based provers such as HOL4 [51] and Isabelle/HOL, is the "package" approach: Instead of deeply embedding sorts and operation symbols like we do, packages take a user specification of the desired types and operations and prove all the theorems for that instance (on a dynamic basis). Nominal Isabelle [54, 56] is a popular such package, which implements terms with bindings for Isabelle/HOL. From a theoretical perspective, a universal algebra theory has a wider appeal, as it models "statically" the meta-theory in its whole generality. However, a package is more practical, since most proof assistant users only care about the particular instance syntax used in their development. In this respect, simply instantiating our signature with the particular syntax is not entirely satisfactory, since it is not sufficiently "shallow"—e.g., one would like to have actual operations such as Lam instead of applications of Op to a Lam operation symbol, and would like to have actual types, such as **exp** and **proc**, instead of the well-sortedness predicate applied to sorts, wls exp and wls proc. For our applications, so far we have manually transited from our "deep" signature instances to the more usable shallow version sketched above. In the future, we plan to have this transit process automated, obtaining the best of both worlds, namely a universal algebra theory that also acts as a *statically certified* package. (This approach has been prototyped for a smaller theory: that of nonfree equational datatypes [49]).

Theory of Substitution and Semantic Interpretation. The main goal of our work was the development of as much as possible from the theory of syntax for an arbitrary syntax. To our knowledge, none of the existing frameworks provides support for substitution and the interpretation of terms in semantic domains at this level of generality. Consequently, formalizations for concrete syntaxes, even those based on sophisticated packages such as Nominal Isabelle or the similar tools and formalizations in Coq [2, 3, 27], have to redefine these standard concepts and prove their properties over and over again—an unnecessary consumption of time and brain power.

Induction and Recursion Principles. There is a rich literature on these topics, which are connected to the quest, pioneered by Gordon and Melham [23], of understanding terms with bindings modulo alpha as an abstract datatype. We formalized the Nominal structural induction principle from [41], which is also implemented in Nominal Isabelle. By contrast, we did not go after the Nominal recursion principle. Instead, we chose to stay more faithful to the abstract datatype desideratum, generalizing to an arbitrary syntax our own schema for substitution-aware recursion [44] and Michael Norrish's schema for swapping-aware recursion [38]—both of which can be read as stating that terms with bindings are Horn-abstract datatypes, i.e., are initial models of certain Horn theories [44, Sect. 3, Sect. 8].

Generality of the Framework. Our constructors are restricted to binding at most one variable in each input—a limitation that makes our framework far from ideal for representing complex binders such as the let patterns of POPLmark's Challenge 2B. In contrast, the specification language Ott [50] and Isabelle's Nominal2 package [56] were specifically designed to address such complex, possibly recursive binders. Incidentally, the Nominal2 package also separates abstractions from terms, like we do, but their abstractions are significantly more expressive; their terms are also quotiented to alpha-equivalence, which is defined via flattening the binders into finite sets or lists of variables (atoms).

On the other hand, to the best of our knowledge, our formalization is the first to capture infinitely branching terms and our foundation of alpha equivalence on the regularity of |var| is also a theoretical novelty—constituting a less exotic alternative to Gabbay's work on infinitely supported objects in nonstandard set theory [20]. This flexibility would be needed to formalize calculi such as infinite-choice process algebra, for which infinitary structures have been previously employed to give semantics [31].

Future Generalizations and Integrations. Our theory currently addresses mostly *structural* aspects of terms. A next step would be to cover *behavioral* aspects, such as formats for SOS rules and their interplay with binders, perhaps building on existing Isabelle formalizations of process algebras and programming languages (e.g., [5, 30, 36, 43, 46, 47]).

Another exciting prospect is the integration of our framework with Isabelle's recent package for inductive and coinductive datatypes [10] based on bounded natural functors (BNFs), which follows a compositional design [53] and provides flexible ways to nest types [11] and mix recursion with corecursion [9,14], but does not yet cover terms with bindings. Achieving compositionality in the presence of bindings will require a substantial refinement of the notion of BNF (since terms with bindings form only partial functors w.r.t. their sets of free variables).

Acknowledgment. We thank the anonymous reviewers for suggesting textual improvements. Popescu has received funding from UK's Engineering and Physical Sciences Research Council (EPSRC) via the grant EP/N019547/1, Verification of Web-based Systems (VOWS).

References

1. The POPLmark Challenge (2009). http://fling-l.seas.upenn.edu/plclub/cgi-bin/poplmark/
2. Aydemir, B.E., Bohannon, A., Weirich, S.: Nominal reasoning techniques in Coq: (extended abstract). Electron. Notes Theor. Comput. Sci. **174**(5), 69–77 (2007)
3. Aydemir, B.E., Charguéraud, A., Pierce, B.C., Pollack, R., Weirich, S.: Engineering formal metatheory. In: POPL 2008, pp. 3–15 (2008)
4. Barendregt, H.P.: The Lambda Calculus. North-Holland, Amsterdam (1984)
5. Bengtson, J., Parrow, J., Weber, T.: Psi-calculi in Isabelle. J. Autom. Reason. **56**(1), 1–47 (2016)
6. Berghofer, S., Wenzel, M.: Inductive datatypes in HOL — lessons learned in formallogic engineering. In: Bertot, Y., Dowek, G., Théry, L., Hirschowitz, A., Paulin, C. (eds.) TPHOLs 1999. LNCS, vol. 1690, pp. 19–36. Springer, Heidelberg (1999). doi:10.1007/3-540-48256-3_3

7. Blanchette, J.C., Popescu, A.: Mechanizing the metatheory of sledgehammer. In: Fontaine, P., Ringeissen, C., Schmidt, R.A. (eds.) FroCoS 2013. LNCS (LNAI), vol. 8152, pp. 245–260. Springer, Heidelberg (2013). doi:10.1007/978-3-642-40885-4_17

8. Blanchette, J.C., Böhme, S., Popescu, A., Smallbone, N.: Encoding monomorphic and polymorphic types. In: Piterman, N., Smolka, S.A. (eds.) TACAS 2013. LNCS, vol. 7795, pp. 493–507. Springer, Heidelberg (2013). doi:10.1007/978-3-642-36742-7_34

9. Blanchette, J.C., Bouzy, A., Lochbihler, A., Popescu, A., Traytel, D.: Friends with benefits - implementing corecursion in foundational proof assistants. In: Yang, H. (ed.) ESOP 2017. LNCS, vol. 10201, pp. 111–140. Springer, Heidelberg (2017). doi:10.1007/978-3-662-54434-1_5

10. Blanchette, J.C., Hölzl, J., Lochbihler, A., Panny, L., Popescu, A., Traytel, D.: Truly modular (co)datatypes for Isabelle/HOL. In: Klein, G., Gamboa, R. (eds.) ITP 2014. LNCS, vol. 8558, pp. 93–110. Springer, Cham (2014). doi:10.1007/978-3-319-08970-6_7

11. Blanchette, J.C., Meier, F., Popescu, A., Traytel, D.: Foundational nonuniform (co)datatypes for higher-order logic. In: LICS. IEEE (2017)

12. Blanchette, J.C., Popescu, A., Traytel, D.: Cardinals in Isabelle/HOL. In: Klein, G., Gamboa, R. (eds.) ITP 2014. LNCS, vol. 8558, pp. 111–127. Springer, Cham (2014). doi:10.1007/978-3-319-08970-6_8

13. Blanchette, J.C., Popescu, A., Traytel, D.: Unified classical logic completeness—a coinductive pearl. In: Demri, S., Kapur, D., Weidenbach, C. (eds.) IJCAR 2014. LNCS (LNAI), vol. 8562, pp. 46–60. Springer, Cham (2014). doi:10.1007/978-3-319-08587-6_4

14. Blanchette, J.C., Popescu, A., Traytel, D.: Foundational extensible corecursion: a proof assistant perspective. In: ICFP, pp. 192–204 (2015)

15. Blanchette, J.C., Popescu, A., Traytel, D.: Soundness and completeness proofs by coinductive methods. J. Autom. Reason. 58(1), 149–179 (2017)

16. Chlipala, A.J.: Parametric higher-order abstract syntax for mechanized semantics. In: ICFP, pp. 143–156 (2008)

17. Despeyroux, J., Felty, A., Hirschowitz, A.: Higher-order abstract syntax in Coq. In: Dezani-Ciancaglini, M., Plotkin, G. (eds.) TLCA 1995. LNCS, vol. 902, pp. 124–138. Springer, Heidelberg (1995). doi:10.1007/BFb0014049

18. Felty, A.P., Momigliano, A.: Hybrid - a definitional two-level approach to reasoning with higher-order abstract syntax. J. Autom. Reason. 48(1), 43–105 (2012)

19. Fiore, M., Plotkin, G., Turi, D.: Abstract syntax and variable binding (extended abstract). In: LICS 1999, pp. 193–202 (1999)

20. Gabbay, M.J.: A general mathematics of names. Inf. Comput. 205(7), 982–1011 (2007)

21. Gheri, L., Popescu, A.: This Paper's Homepage. http://andreipopescu.uk/papers/BindingTheory.html

22. Gheri, L., Popescu, A.: A formalized general theory of syntax with bindings. CoRR (2017)

23. Gordon, A.D., Melham, T.: Five axioms of alpha-conversion. In: Goos, G., Hartmanis, J., Leeuwen, J., Wright, J., Grundy, J., Harrison, J. (eds.) TPHOLs 1996. LNCS, vol. 1125, pp. 173–190. Springer, Heidelberg (1996). doi:10.1007/BFb0105404

24. Gunter, E.L., Osborn, C.J., Popescu, A.: Theory support for weak higher order abstract syntax in Isabelle/HOL. In: LFMTP, pp. 12–20 (2009)

25. Harper, R., Honsell, F., Plotkin, G.: A framework for defining logics. In: LICS 1987, pp. 194–204. IEEE Computer Society Press (1987)

26. Hennessy, M., Milner, R.: On observing nondeterminism and concurrency. In: Bakker, J., Leeuwen, J. (eds.) ICALP 1980. LNCS, vol. 85, pp. 299–309. Springer, Heidelberg (1980). doi:10.1007/3-540-10003-2_79
27. Hirschowitz, A., Maggesi, M.: Nested abstract syntax in Coq. J. Autom. Reason. 49(3), 409–426 (2012)
28. Kammüller, F., Wenzel, M., Paulson, L.C.: Locales a sectioning concept for Isabelle. In: Bertot, Y., Dowek, G., Théry, L., Hirschowitz, A., Paulin, C. (eds.) TPHOLs 1999. LNCS, vol. 1690, pp. 149–165. Springer, Heidelberg (1999). doi:10.1007/3-540-48256-3_11
29. Keisler, H.J.: Model Theory for Infinitary Logic. North-Holland, Amsterdam (1971)
30. Lochbihler, A.: Java and the Java memory model — a unified, machine-checked formalisation. In: Seidl, H. (ed.) ESOP 2012. LNCS, vol. 7211, pp. 497–517. Springer, Heidelberg (2012). doi:10.1007/978-3-642-28869-2_25
31. Luttik, B.: Choice quantification in process algebra. Ph.D. thesis, University of Amsterdam, April 2002
32. Miller, D., Tiu, A.: A proof theory for generic judgments. ACM Trans. Comput. Logic 6(4), 749–783 (2005)
33. Milner, R.: Communication and Concurrency. Prentice Hall, Upper Saddle River (1989)
34. Milner, R.: Communicating and Mobile Systems: The π-Calculus. Cambridge University Press, Cambridge (2001)
35. Nipkow, T., Klein, G.: Concrete Semantics: With Isabelle/HOL. Springer, Heidelberg (2014). doi:10.1007/978-3-319-10542-0
36. Nipkow, T., von Oheimb, D.: Java$_{light}$ is type-safe - definitely. In: POPL, pp. 161–170 (1998)
37. Nipkow, T., Paulson, L.C., Wenzel, M.: Isabelle/HOL: A Proof Assistant for Higher-Order Logic. Springer, Heidelberg (2002). doi:10.1007/3-540-45949-9
38. Norrish, M.: Mechanising lambda-calculus using a classical first order theory of terms with permutations. High.-Order Symb. Comput. 19(2–3), 169–195 (2006)
39. Norrish, M., Vestergaard, R.: Proof pearl: de bruijn terms really do work. In: Schneider, K., Brandt, J. (eds.) TPHOLs 2007. LNCS, vol. 4732, pp. 207–222. Springer, Heidelberg (2007). doi:10.1007/978-3-540-74591-4_16
40. Pitts, A.M.: Nominal logic: a first order theory of names and binding. In: Kobayashi, N., Pierce, B.C. (eds.) TACS 2001. LNCS, vol. 2215, pp. 219–242. Springer, Heidelberg (2001). doi:10.1007/3-540-45500-0_11
41. Pitts, A.M.: Alpha-structural recursion and induction. J. ACM 53(3), 459–506 (2006)
42. Popescu, A.: Contributions to the theory of syntax with bindings and to process algebra. Ph.D. thesis, University of Illinois (2010). andreipopescu.uk/thesis.pdf
43. Popescu, A., Gunter, E.L.: Incremental pattern-based coinduction for process algebra and its isabelle formalization. In: Ong, L. (ed.) FoSSaCS 2010. LNCS, vol. 6014, pp. 109–127. Springer, Heidelberg (2010). doi:10.1007/978-3-642-12032-9_9
44. Popescu, A., Gunter, E.L.: Recursion principles for syntax with bindings and substitution. In: ICFP, pp. 346–358 (2011)
45. Popescu, A., Gunter, E.L., Osborn, C.J.: Strong normalization of system F by HOAS on top of FOAS. In: LICS, pp. 31–40 (2010)
46. Popescu, A., Hölzl, J., Nipkow, T.: Proving concurrent noninterference. In: Hawblitzel, C., Miller, D. (eds.) CPP 2012. LNCS, vol. 7679, pp. 109–125. Springer, Heidelberg (2012). doi:10.1007/978-3-642-35308-6_11

47. Popescu, A., Hölzl, J., Nipkow, T.: Formalizing probabilistic noninterference. In: Gonthier, G., Norrish, M. (eds.) CPP 2013. LNCS, vol. 8307, pp. 259–275. Springer, Cham (2013). doi:10.1007/978-3-319-03545-1_17
48. Popescu, A., Rosu, G.: Term-generic logic. Theor. Comput. Sci. **577**, 1–24 (2015)
49. Schropp, A., Popescu, A.: Nonfree datatypes in Isabelle/HOL. In: Gonthier, G., Norrish, M. (eds.) CPP 2013. LNCS, vol. 8307, pp. 114–130. Springer, Cham (2013). doi:10.1007/978-3-319-03545-1_8
50. Sewell, P., Nardelli, F.Z., Owens, S., Peskine, G., Ridge, T., Sarkar, S., Strnisa, R.: Ott: effective tool support for the working semanticist. J. Funct. Program. **20**(1), 71–122 (2010)
51. Slind, K., Norrish, M.: A brief overview of HOL4. In: Mohamed, O.A., Muñoz, C., Tahar, S. (eds.) TPHOLs 2008. LNCS, vol. 5170, pp. 28–32. Springer, Heidelberg (2008). doi:10.1007/978-3-540-71067-7_6
52. Sun, Y.: An algebraic generalization of frege structures–binding algebras. Theor. Comput. Sci. **211**(1–2), 189–232 (1999)
53. Traytel, D., Popescu, A., Blanchette, J.C.: Foundational, compositional (co)datatypes for higher-order logic: Category theory applied to theorem proving. In: LICS 2012, pp. 596–605. IEEE (2012)
54. Urban, C.: Nominal techniques in Isabelle/HOL. J. Autom. Reason. **40**(4), 327–356 (2008)
55. Urban, C., Berghofer, S., Norrish, M.: Barendregt's variable convention in rule inductions. In: Pfenning, F. (ed.) CADE 2007. LNCS, vol. 4603, pp. 35–50. Springer, Heidelberg (2007). doi:10.1007/978-3-540-73595-3_4
56. Urban, C., Kaliszyk, C.: General bindings and alpha-equivalence in nominal Isabelle. In: Barthe, G. (ed.) ESOP 2011. LNCS, vol. 6602, pp. 480–500. Springer, Heidelberg (2011). doi:10.1007/978-3-642-19718-5_25
57. Urban, C., Tasson, C.: Nominal techniques in Isabelle/HOL. In: Nieuwenhuis, R. (ed.) CADE 2005. LNCS (LNAI), vol. 3632, pp. 38–53. Springer, Heidelberg (2005). doi:10.1007/11532231_4

Proof Certificates in PVS

Frédéric Gilbert[✉]

École des Ponts ParisTech, Inria, CEA LIST, Paris, France
frederic.a.gilbert@inria.fr

Abstract. The purpose of this work is to allow the proof system PVS to export proof certificates that can be checked externally. This is done through the instrumentation of PVS to record detailed proofs step by step during the proof search process. At the current stage of this work, proofs can be built for any PVS theory. However, some reasoning steps rely on unverified assumptions. For a restricted fragment of PVS, the proofs are exported to the universal proof checker Dedukti, and the unverified assumptions are proved externally using the automated theorem prover MetiTarski.

1 Introduction

Given the complexity of proof assistants such as PVS, external verifications become necessary to reach the highest levels of trust in its results. A possible way to this end is to require the system to export certificates that can be checked using third-party tools. The purpose of this work is to instrument PVS to export certificates that can be verified externally.

This approach is comparable to the OpenTheory project [3], in which the higher order logic theorem provers HOL Light, HOL4, and ProofPower are instrumented to export verifiable certificates in a shared format. In HOL Light, HOL4, and ProofPower, the detail of each reasoning step is expressed using a small number of simple logical rules, which are used as a starting point to the generation of OpenTheory certificates. As this is not the case in PVS, the whole proof system needs to be instrumented to generate complete certificates. At the current stage of this work, this instrumentation is not complete, leading to the presence of unverified assumptions in the generated certificates. For a restricted fragment of PVS, the proof certificates are exported to the universal proof checker Dedukti [5], and the unverified assumptions are proved externally using the automated theorem prover MetiTarski [1].

In PVS [4], the proof process is decomposed into a succession of proof steps. These proof steps are recorded into a proof trace format, the .prf files. These proof traces can be used to rerun and verify a proof, but only internally. In order to check these proof traces externally, one would have to reimplement PVS proof mechanisms almost entirely.

F. Gilbert—This work has been completed as part of two visits to the National Institute of Aerospace (NIA) and the NASA Langley Research Center under NASA/NIA Research Cooperative Agreement No. NNL09AA00A.

M. Ayala-Rincón and C.A. Muñoz (Eds.): ITP 2017, LNCS 10499, pp. 262–268, 2017.
DOI: 10.1007/978-3-319-66107-0_17

The purpose of the proof certificates presented in this work is to check PVS proofs externally using small systems. To this end, we present a decomposition of PVS proof steps into a small number of atomic rules, which are easier to encode into a third-party system than the original proof steps. The proof certificates are built on these atomic rules, and can be checked without having to reimplement PVS proof steps.

These atomic rules are defined as a refinement of an intermediate decomposition of proof steps which is already present in PVS. This intermediate decomposition is based on a specific subset of proof steps, the primitive rules. In PVS, every proof step, including defined rules and strategies, can be decomposed as a sequence of primitive rules. As any primitive step is a proof step, this intermediate level of decomposition can be formalized in the original format of .prf proof traces. In fact, such a decomposition can be performed using the PVS package Manip [2], in which the instruction `expand-strategy-steps` allows one to decompose every proof step into a succession of primitive rules.

However, this intermediate decomposition is not sufficient to make proof traces verifiable externally using small systems. Indeed, the complexity of PVS proof mechanisms lies for the largest part in the primitive rules themselves. In particular, the implementation of primitive rules is one order of magnitude larger than the implementation of strategies. For instance, the primitive rule `simplify` hides advanced reasoning techniques, including simplifications, rewritings, and Shostak's decision procedures.

In order to provide a refinement of the primitive rule decomposition, we modify PVS directly to record reasoning at a higher level of precision. The main part of this modification is done in the code of the primitive rules themselves. This instrumentation doesn't affect the reasoning in any way besides some slowdown due to the recording of proofs. In particular, it doesn't affect the emission of .prf proof traces, which continue to be used internally to rerun proofs as in the original system.

The coherence of a PVS theory is based on both reasoning and typing. At the current stage of this work, the proof certificates are limited to reasoning. Moreover, primitive rules are not entirely instrumented, and the corresponding gaps in reasoning are completed with unverified assumptions.

In the next section, we present the formalization of proof certificates in PVS. Then, we present a first attempt to export these proofs to the universal proof checker Dedukti [5], and to export their unverified assumptions to the theorem prover MetiTarski [1].

2   Proofs Certificates in PVS

2.1   Expressions and Conversion

Proof are added as a new layer of abstract syntax, on top of the existing layers of PVS expressions and PVS sequents. For readability, we will denote PVS expressions as they are printed in PVS. We stress the fact that this denotation is not faithful, as several components of PVS expressions, such as types and resolutions, are erased through PVS printing.

As several other proof systems, Dedukti is equipped with a notion of conversion, which includes, among others, β-conversion and constant definitions, which will be referred to as δ-conversion. As a consequence, it is not necessary to record the expansion of a definition or the reduction of a β-redex in Dedukti, which allows us to keep proofs compact.

Following this idea, we equip PVS expressions with a conversion, denoted $\equiv$. This conversion includes β-conversion, and non-recursive definitions, expressed as δ-rules. However, δ rules are not used for recursive definitions as this would lead to infinite reductions: instead, the expansions or contractions of recursive definitions are kept as explicit reasoning steps.

2.2 Reasoning

In PVS, internally, the formulas appearing on both sides of a sequents are recorded in a single list, where all formulas belonging to the left hand side appear under a negation. For instance, a sequent appearing as NOT A, B ⊢ C is recorded internally as the list NOT NOT A, NOT B, C. Denoting Γ the union of this list together with the list of hidden formulas, the corresponding sequent will be denoted ⊢ Γ.

We equip sequents with the identification modulo permutation. In this setting, sequents correspond to multisets, and we don't need to record any exchange rule, which makes proofs more compact. On top of this layer of sequents, we use the following rules, which are presented modulo conversion $\equiv$.

Structural Rules

$$\frac{}{\vdash \Gamma, \text{A}, \text{ NOT A}} \qquad \frac{\vdash \Gamma, \text{A} \qquad \vdash \Gamma, \text{NOT A}}{\vdash \Gamma} \qquad \frac{\vdash \Gamma}{\vdash \Gamma, \text{A}} \qquad \frac{\vdash \Gamma, \text{A}, \text{ A}}{\vdash \Gamma, \text{A}}$$

Propositional Rules

$$\frac{}{\vdash \Gamma, \text{TRUE}} \qquad \frac{\vdash \Gamma, \text{NOT TRUE}}{\vdash \Gamma} \qquad \frac{\vdash \Gamma, \text{FALSE}}{\vdash \Gamma} \qquad \frac{}{\vdash \Gamma, \text{NOT FALSE}}$$

$$\frac{\vdash \Gamma, \text{A} \qquad \vdash \Gamma, \text{B}}{\vdash \Gamma, \text{A AND B}} \qquad \frac{\vdash \Gamma, \text{NOT A}, \text{ NOT B}}{\vdash \Gamma, \text{NOT (A AND B)}}$$

$$\frac{\vdash \Gamma, \text{A}, \text{ B}}{\vdash \Gamma, \text{A OR B}} \qquad \frac{\vdash \Gamma, \text{NOT A} \qquad \vdash \Gamma, \text{NOT B}}{\vdash \Gamma, \text{NOT (A OR B)}}$$

$$\frac{\vdash \Gamma, \text{NOT A}, \text{ B}}{\vdash \Gamma, \text{A IMPLIES B}} \qquad \frac{\vdash \Gamma, \text{NOT B} \qquad \vdash \Gamma, \text{A}}{\vdash \Gamma, \text{NOT (A IMPLIES B)}} \qquad \frac{\vdash \Gamma, \text{A}}{\vdash \Gamma, \text{NOT NOT A}}$$

$$\frac{\vdash \Gamma, \text{A IMPLIES B} \qquad \vdash \Gamma, \text{B IMPLIES A}}{\vdash \Gamma, \text{A IFF B}}$$

$$\frac{\vdash \Gamma, \text{NOT (A IMPLIES B)}, \text{ NOT (B IMPLIES A)}}{\vdash \Gamma, \text{NOT (A IFF B)}}$$

$$\frac{\vdash \Gamma, \text{A IMPLIES B} \qquad \vdash \Gamma, \text{NOT A IMPLIES C}}{\vdash \Gamma, \text{IF(A, B, C)}}$$

$$\frac{\vdash \Gamma, \text{NOT (A AND B)} \qquad \vdash \Gamma, \text{NOT (NOT A AND C)}}{\vdash \Gamma, \text{NOT IF(A, B, C)}}$$

Quantification Rules

$$\frac{\vdash \Gamma, A}{\vdash \Gamma, \text{FORALL (x : T) : A}} \qquad \frac{\vdash \Gamma, \text{NOT A[t/x]}}{\vdash \Gamma, \text{NOT FORALL (x : T) : A}}$$

$$\frac{\vdash \Gamma, A[t/x]}{\vdash \Gamma, \text{EXISTS (x : T) : A}} \qquad \frac{\vdash \Gamma, \text{NOT A}}{\vdash \Gamma, \text{NOT EXISTS (x : T) : A}}$$

Equality Rules

$$\frac{}{\vdash \Gamma, t = t} \qquad \frac{\vdash \Gamma, t = u \qquad \vdash \Gamma, u = v}{\vdash \Gamma, t = v}$$

$$\frac{\vdash \Gamma, A(t) \qquad \vdash \Gamma, t = u}{\vdash \Gamma, A(u)} \qquad \frac{\vdash \Gamma, u = v}{\vdash \Gamma, f(u) = f(v)}$$

$$\frac{\vdash \Gamma, \text{NOT A, u = v}}{\vdash \Gamma, \text{IF(A, u, t) = IF(A, v, t)}} \qquad \frac{\vdash \Gamma, A, u = v}{\vdash \Gamma, \text{IF(A, t, u) = IF(A, t, v)}}$$

Extensionality Rules

$$\frac{\vdash \Gamma, A \text{ IFF } B}{\vdash \Gamma, A = B} \qquad \frac{\vdash \Gamma, t = u}{\vdash \Gamma, \text{LAMBDA (x : T) : t = LAMBDA (x : T) : u}}$$

$$\frac{\vdash \Gamma, t = u}{\vdash \Gamma, \text{FORALL (x : T) : t = FORALL (x : T) : u}}$$

$$\frac{\vdash \Gamma, t = u}{\vdash \Gamma, \text{EXISTS (x : T) : t = EXISTS (x : T) : u}}$$

Extra Rules

$$\frac{\vdash \Gamma, \Delta \qquad \vdash \Gamma, \Delta_1 \qquad \cdots \qquad \vdash \Gamma, \Delta_n}{\vdash \Gamma, \Delta} \; TCC$$

$$\frac{\vdash \Gamma, \Delta_1 \qquad \cdots \qquad \vdash \Gamma, \Delta_n}{\vdash \Gamma, \Delta} \; Assumption$$

Only the two last rules, *TCC* and *Assumption*, are specific to this system. The first one is due to the appearance of type-checking conditions during the proof run, for instance after giving an instantiation for an existential proposition. As typing is not checked in such proofs, this condition is not necessary, but this rule allows us to ensure that all steps of reasoning are recorded in proofs, included the reasoning steps ensuring typing constraints.

The second one, *Assumption*, is generated from all reasoning steps in PVS which haven't been instrumented yet. In practice, the use of *Assumption* doesn't imply that the corresponding reasoning gap cannot be described using the other rules. For instance, the primitive rule *bddsimp*, which calls a function outside the PVS kernel, was not instrumented. Yet, the corresponding reasoning steps could be justified using structural and propositional rules. On the other hand, the strategy *prop*, which has the same role, doesn't generate any *Assumption* rule, as the underlying primitive rules *flatten* and *split* are both instrumented.

2.3 Proof Objects

In order to record lightweight proofs, we record only the rules used in the proofs, provided with a sufficient amount of rule parameters.

For instance, the proof

$$
\cfrac{\cfrac{\cfrac{\vdash \texttt{NOT A, NOT B, A}}{\vdash \texttt{NOT A, NOT B, NOT NOT A}}}{\vdash \texttt{NOT (A AND B), NOT NOT A}}}{\vdash \texttt{(A AND B) IMPLIES NOT NOT A}}
$$

will be recorded as

```
RImplies(A AND B, NOT NOT A,
RNotAnd(A, B,
  RNotNot(A,
    RAxiom(A))))
```

where `RImplies`, `RNotAnd`, `RNotNot`, and `RAxiom` denote the rules used in the proof, and accept as argument a list of parameters followed by a (possibly empty) list of subproofs.

3 Checking PVS Proofs Using Dedukti and Metitarski

This part of the work is only at the stage of a first prototype. The universal proof checker Dedukti is used to verify the proof certificates. As these certificates contain unverified assumptions, the automated theorem prover MetiTarski is used to prove them externally.

3.1 Translating Proofs to Dedukti

Dedukti is a dependently typed language. However, as we only record reasoning in this work, we use a translation which doesn't make PVS types appear. We declare one universal type `type` for all PVS expressions. In order to translate applications, we use a constant `apply : type -> type -> type`. Conversely, we use a constants `lambda : (type -> type) -> type` to translate lambda expressions.

A similar technique is used to translate the other constructions appearing in the rules, such as `FORALL`.

The translation from PVS proofs to Dedukti is a translation from sequent calculus to natural deduction. The use of Dedukti being based on the Curry-Howard isomorphism, a proof of a proposition `A` is expected as a term of type `A`. The main translation function takes a proof of a sequent $\vdash A_1, ..., A_n$ and a list of proof variables $h_1, ..., h_n$ to a produce a term p which has the type `FALSE` in the context $h_1 : \texttt{NOT } A_1, ..., h_n : \texttt{NOT } A_n$. This translation is based on the declaration of the rules as constants in Dedukti.

Using this main translation function, for any proposition A proved in PVS, and for any proof variable h, we build a proof p of type FALSE in the context h : NOT A. Then, using a rule of negation introduction together with a rule of double negation elimination, we get a proof term of type A in the empty context, as expected.

3.2 Checking Assumptions with MetiTarski

Every rule except Assumption is valid in classical higher-order logic. In order to check the assumption rules as well, we use an automated theorem prover. We chose the first-order theorem prover MetiTarski for this purpose.

Using conjunctions, disjunctions and implications, every assumption rule is translated into a single proposition, which in turn is translated to the TPTP [6] format. The main issue in this translation is the presence of higher-order expressions, such as lambda terms of if-then-else expressions for instance. In this work, these terms are translated as constant symbols: the obtained expressions are correct TPTP expressions, and their validity in first-order logic ensures the validity of the original expression in higher-order logic.

4   Results

The instrumentation of PVS to build proof certificates is not restricted to any fragment of PVS. It has been tested using the arithmetic theories (ints) of the NASA Library nasalib. The generation of all certificates for the whole (ints) library (32 files, 268 proofs) was performed in one hour.

The exportation to Dedukti and MetiTarski has been tested on the following example:

```
induction : THEORY
   BEGIN
   f : [nat -> nat]
   nat_sum : LEMMA
       (f(0) = 0 AND (FORALL (n:nat): f(n+1) = f(n) + n + 1))
             IMPLIES FORALL (n:nat): 2 * f(n) = n * (n + 1)
END induction
```

This theorem was proved in two steps: flatten, and induct-and-simplify. The Dedukti file generated has been successfully checked by Dedukti. It contained 19 unverified assumptions. All of them have been successfully proved using MetiTarski.

References

1. Akbarpour, B., Paulson, L.C.: Metitarski: an automatic theorem prover for real-valued special functions. J. Autom. Reason. **44**(3), 175–205 (2010)
2. Di Vito, B.L.: Manip user's guide, version 1.3 (2011)
3. Hurd, J.: The opentheory standard theory library. In: Bobaru, M., Havelund, K., Holzmann, G.J., Joshi, R. (eds.) NFM 2011. LNCS, vol. 6617, pp. 177–191. Springer, Heidelberg (2011). doi:10.1007/978-3-642-20398-5_14
4. Owre, S., Rushby, J.M., Shankar, N.: PVS: a prototype verification system. In: Kapur, D. (ed.) CADE 1992. LNCS, vol. 607, pp. 748–752. Springer, Heidelberg (1992). doi:10.1007/3-540-55602-8_217
5. Saillard, R.: Dedukti: a universal proof checker. In: Foundation of Mathematics for Computer-Aided Formalization Workshop (2013)
6. Sutcliffe, G.: The TPTP problem library and associated infrastructure. J. Autom. Reason. **43**(4), 337 (2009)

Efficient, Verified Checking
of Propositional Proofs

Marijn Heule[1(✉)], Warren Hunt Jr.[1], Matt Kaufmann[1], and Nathan Wetzler[2]

[1] The University of Texas at Austin, Austin, TX, USA
[2] Intel Corporation, Hillsboro, OR, USA
{marijn,hunt,kaufmann}@cs.utexas.edu, nathan.wetzler@gmail.com

Abstract. Satisfiability (SAT) solvers—and software in general—sometimes have serious bugs. We mitigate these effects by validating the results. Today's SAT solvers emit proofs that can be checked with reasonable efficiency. However, these checkers are not trivial and can have bugs as well. We propose to check proofs using a formally verified program that adds little overhead to the overall process of proof validation. We have implemented a sequence of increasingly efficient, verified checkers using the ACL2 theorem proving system, and we discuss lessons from this effort. This work is already being used in industry and is slated for use in the next SAT competition.

1 Introduction

This paper presents a formally verified application, a SAT proof-checker, that has sufficient efficiency to support its practical use. Our checker, developed using the ACL2 theorem-proving system [12,15], validates the results of SAT solvers by checking the emitted proofs. Our intention here is to provide some useful lessons from the development of an efficient, formally verified application using ACL2. We therefore avoid lower-level details of algorithms, mathematics, and proof development.

The Problem. Boolean satisfiability (SAT) solving has become a key technology for formal verification. Users of SAT solvers increasingly seek confidence in claims that given formulas are unsatisfiable[1]. Contemporary SAT solvers therefore emit proofs [10] that can be validated by *SAT proof-checkers* [9,27]. Such a proof is a sequence of *steps*, each of which is interpreted as transforming a formula to a new formula. Checking the proof is just the result of iterating through the steps; for each step, the checker performs a validation intended to guarantee that if the current formula (initially the input formula) is satisfiable, then so

---

This work was supported by NSF under Grant No. CCF-1526760. We thank the reviewers for useful feedback.
[1] Checking a claim of satisfiability is easy.

M. Ayala-Rincón and C.A. Muñoz (Eds.): ITP 2017, LNCS 10499, pp. 269–284, 2017.
DOI: 10.1007/978-3-319-66107-0_18

is the transformed formula. Typically the final formula is clearly unsatisfiable; then the validation process guarantees that the input formula is unsatisfiable.

How can we trust SAT proof-checkers? Although they are usually much simpler than SAT solvers, they are not trivial, and any software is susceptible to bugs. We implemented a verified SAT proof-checker in ACL2 [26], but this checker was not intended be *efficient*. For example, a specific proof that was validated in about 1.5 s by the unverified checker DRAT-trim [27] took about a week to validate using this verified checker. Several reasons explain the slowdown. The verified checker used list-based data structures, providing linear-time accesses, while the unverified checker used arrays and various low-level optimizations. Additionally, proofs of unsatisfiability usually contain many *deletion steps*, while deletion is not supported by that verified checker. The size of the formula is important because a key procedure, the *RAT* check [11], may need to consider every clause in the formula. Finally, RAT checking is based on a procedure, *unit propagation*, that can require expensive search. (These two aspects of RAT checks—some checks that are linear in the size of the formula, and search—are all we need to know about RAT checks for this paper.)

Alternatively one could verify the correctness of the solver in a theorem prover. That approach does not require proof logging and validation. However, SAT solvers are complicated and frequently improved, thereby making the verification task hard. Moreover, verified SAT solvers tend to be orders of magnitude slower compared to unverified solvers [1]. That said, verification of SAT solvers has been studied by various authors in the last decade. The DPLL [4,5] algorithm, which was the core algorithm of solvers until the late 90's, has been formalized and verified by Lescuyer and Conchon [17] in Coq and by Shankar and Vaucher [23] in PVS. The conflict driven clause-learning paradigm of modern SAT solvers [20] was verified by Marić [18,19] in Isabelle/HOL (2010), by Oe et al. [22] in Guru (2012), and by Blanchette et al. [1] in Isabelle/HOL (2016).

Towards a Solution. At least three parallel efforts have attempted to produce efficient, formally verified SAT checkers [3,16]. A key idea was to avoid all search (all of which results from unit propagation) by adding certain "hints" to each proof step, resulting in a new proof format, LRAT (Linear RAT) [3]. In this paper we discuss one of those three efforts: an LRAT proof-checker developed in ACL2 (the others being checkers in Coq [3] and Isabelle/HOL [16]). The SAT proof mentioned above that took a week to check now takes under 3 s to check with the new ACL2-based checker. As suggested by some data reported below, our checker may run sufficiently fast so that it adds relatively little overhead beyond using a fast C-based checker. This work is already used in industry at Centaur Technology [25], and we expect it to be used in the 2017 SAT competition. In this system, one does not need to reason about the original proof produced by a solver or the proof conversion process of DRAT-trim: if our verified checker validates the final optimized proof, then the input formula is unsatisfiable.

This paper is not intended to provide proof details, but rather, to extract some lessons in the effective use of one proof assistant (ACL2). This paper assumes no knowledge of ACL2, SAT solving, or SAT proof-checking (such as RAT and LRAT); all necessary background is provided above or as needed below.

We begin with a few ACL2 preliminaries. Then in Sect. 3 we describe a sequence of increasingly efficient checkers. That description provides background for discussion in Sect. 4 of the ACL2 soundness proofs done for each of these checkers. Section 5 concludes with remarks that summarize our findings.

2 ACL2 Preliminaries

The ACL2[2] theorem-proving system [12,15] includes a programming language based on an applicative subset of Common Lisp [24]. Lisp is one of the oldest programming languages [21] and is supported by several efficient compilers, both commercial and free. Moreover, ACL2 was designed with efficient execution in mind [7,28]; indeed, efficiency is important since the ACL2 theorem prover is mostly written in its own language. Thus, ACL2 provides a platform where one can write programs that execute efficiently and also prove programs correct.

We focus below on three ACL2 features that support efficient execution of our SAT proof-checker: *guards*, *stobjs*, and *fast-alists*. Then we close this section by explaining ACL2 notions used in the rest of this paper.

Guards. The ACL2 logic is an untyped first-order logic of total functions. The expression (first 3) denotes the application of a function, first to a single parameter, 3. Thus, even a "bad" expression like (first 3)—first is intended to be applied to a list (to return its first element), not a number—are logically well-formed. Indeed, ACL2 can prove that (first 3) is equal to (first 4), and ACL2 provides a way to evaluate (first 3) without error. On the other hand, Common Lisp signals an error when evaluating this expression. It would be wrong for ACL2 to use Common Lisp to do all of its evaluation, while taking advantage of modern Common Lisp compilers is exactly what we want to do.

A solution is provided by *guards*. The ACL2 guard for a function is an expression whose variables are all formal parameters of that function. Guards can be viewed as analogous of types, in that they are preconditions on the arguments of a function. In contrast with most type systems, however, a guard can be any expression involving any subset of the formal parameters of a function. For example, the guard for first, with formal parameter x, is that x is a list.[3]

ACL2 relies on Common Lisp to evaluate using the definitions provided, but only after *guard verification* is performed on those definitions: proving formulas guaranteeing that for every function call during evaluation, the arguments of that call satisfy its function's guard. Guard verification was an important part of our proof effort (see Sect. 4.3), resulting in a verified checker that executes efficiently in Common Lisp.

---

[2] "A Computational Logic for Applicative Common Lisp".

[3] More accurately, first is a macro expanding to a corresponding call of the function car, whose guard specifies that the argument is a pair or the empty list.

Stobjs. Single-threaded objects, or *stobjs* [2], are mutable objects that support fast execution in ACL2.[4] A stobj *s* is introduced as a record with fields, some of which may be arrays. Henceforth, *s* may be an argument to a function, but ACL2 enforces syntactic requirements, in particular: if *s* is modified by a function then it must be returned, and its use must be single-threaded. Such restrictions guarantee that there is only one instance of *s* present at any time during evaluation, and therefore it is sound to modify *s* in place, which can boost speed significantly since it avoids allocating new structures.

Fast-Alists. In Lisp parlance, an *alist* (or *association list*) is a representation of a finite function as a list of ordered pairs $\langle i, j \rangle$ for which the *key*, *i*, is mapped to the *value*, *j*. ACL2 supports so-called *fast-alists*, sometimes called *applicative hash tables*. For any fast-alist, the implementation provides a corresponding hash table so that the function hons-get obtains the value for a given key in essentially constant time—provided a certain single-threaded discipline is maintained. Unlike stobjs, the discipline is not enforced at definition time; instead, a runtime warning is printed when it is violated, in which case the alist is searched linearly until a pair $\langle i, j \rangle$ is found for a given key, *i*. In practice, it is straightforward for ACL2 programmers to use fast-alists so that the discipline is maintained.

Other Preliminaries. We mention a few other aspects of ACL2, towards making this paper self-contained. The ACL2 prover is extensively discussed in its documentation[5] and in other places [12,15]. While automated induction is certainly helpful, the "workhorse" of the prover is rewriting. Definitions and (by default) theorems are stored as rewrite rules. It is often helpful to *disable* (turn off) some rules either to speed up the prover or to implement some rewriting strategy. A *book* is an ACL2 input file, typically containing definitions and theorems. Finally, symbols are case-insensitive and in particular, Boolean values are represented by the symbols T (true) and NIL (false).

3   SAT Proof-Checker Code

Our most efficient SAT proof-checker is the last in a sequence of verified SAT proof-checkers developed in ACL2. Section 3.1 enumerates these checkers, providing a name and some explanation for each. The statistics provided in Sect. 3.2 demonstrate improved performance offered by each successive checker. All supporting materials for the checkers listed below, including proofs, may be found in the projects/sat/lrat/ directory within the ACL2 community books[6]; see its README file.

---

[4] Thus, stobjs play a role somewhat like monads in higher-order functional languages.
[5] http://www.cs.utexas.edu/users/moore/acl2/current/manual/.
[6] https://github.com/acl2/acl2/tree/master/books/.

3.1 A Sequence of Checkers

[rat] **A Verified RAT Checker** [26]. A formula is a list of clauses, implicitly conjoined (hence, in what is typically called *conjunctive normal form*). A proof designates an ordered sequence of clauses, each of which is to be added to the formula, in order. The RAT check is intended to ensure that when a clause C in the proof is added to the current formula F: if F is satisfiable, then F remains satisfiable after adding C. The RAT check is proved sound: if the proof passes that check and contains the empty clause, then the original formula is unsatisfiable.

[drat] **A Verified DRAT Checker.** Our first proof effort was to extend the verified RAT checker to handle deletion—the "D" in "DRAT"—of clauses from a formula. Thus a proof step became a pair consisting of a Boolean flag and a clause, where: a flag of T indicates that the clause is to be added, as before; but a flag of NIL indicates that the clause is to be removed. Since deletion obviously preserves satisfiability, we quite easily modified the [rat] soundness proof to accommodate this enhanced notion of SAT proof.

Only modest benefit might accrue from extending the initial checker in a straightforward way with deletion: on the easiest problem in our test suite, [rat] requires 20 s, while [drat] took 9 s. All [lrat-*] checkers can verify the proof of the same problem in a fraction of a second. However, it is well established that without deletion, high-performance checkers will suffer greatly [9]. Thus, incorporating deletion was an important first step.

[lrat-1] **A Verified LRAT Checker Using Fast-Alists.** In order to speed up SAT proof-checking, we wanted to exploit proof hints recently provided by the *LRAT* format [3], which facilitate fast lookup of clauses in formulas. So we developed an ACL2 checker that represents formulas using fast-alists, which provide a Lisp hash-table for nearly constant-time lookup. Our fast-alists contain pairs of the form $\langle i, c \rangle$, where the key, i, is a positive integer that denotes the index of the associated clause, c. But a formula can also contain pairs $\langle i, D \rangle$ where D is a special deletion indicator, meaning that the clause with index i has been deleted from the formula. A deletion proof step provides an index i to delete, and is processed by updating the formula's fast-alist with a new pair $\langle i, D \rangle$.

For this checker, a formula is actually an ordered pair $\langle m, a \rangle$, where a is a fast-alist as described above and m is the maximum index in that alist. That value is passed to the function that may be called to perform a *full RAT check*, which recurs through the entire formula starting with index $i = m$. Each step in that recursion looks up i in the fast-alist to find either a clause that is checked, or the deletion indicator, D. The repeated use of the lookup function, hons-get, on clause indices turned out to be somewhat expensive, in spite of its use of a Lisp hash-table. That expense is addressed with improvements discussed below.

[lrat-2] **A Faster Verified LRAT Checker that Shrinks Fast-Alists.** ACL2 supports profiling, which we used on the [lrat-1] proof-checker. We found that 69% of the time was spent performing lookup with hons-get. On reflection,

this was not a surprise: the full RAT check walks through the entire (fast-)alist, which grows with every proof step that adds a clause. This quadratic behavior would not be present if fast-alists were nothing more than mutable hash tables; but in ACL2 they are also alists, which grow with each update. Note the [lrat-1] checker applies `hons-get` at every step of the full RAT check: the ordered pair $\langle i, c \rangle$ seems to suggest that a suitable check needs to be done on the clause c, but this pair may be overridden by a pair $\langle i, \mathrm{D} \rangle$ in the formula indicating that the clause c has actually been deleted, and thus should not be checked.

This checker (also those that follow) heuristically chooses when to shrink the formula's fast-alist, by removing from it all traces of deleted clauses. This happens immediately before checking any proof step's addition of a clause, whenever the number of deleted clauses in the formula exceeds the number of *active* (not deleted) clauses by at least a certain factor. Based on some experimentation, that factor is set to 1/3 when about to do the full RAT check, which as mentioned above must consider every clause in the formula; otherwise, the factor is set to 10. The function `shrink-formula-fal` creates a smaller formula, equivalent to its input, by removing pairs that represent deletion. It does this by first using an ACL2 primitive that exploits the underlying hash table to remove, very efficiently, all pairs $\langle i, c \rangle$ that are overridden by deletion pairs $\langle i, \mathrm{D} \rangle$; a linear walk removing all deletion pairs, followed by creation of a new fast-alist, then finishes the job.

[lrat-3] A Verified LRAT Checker with a Simpler Representation of Formulas. The previous version still represents a formula as a pair $\langle m, a \rangle$, where a is a fast-alist and m is its maximum index. The [lrat-3] checker represents a formula simply as a fast-alist, since starting with [lrat-2], the full RAT check recurs through the fast-alist without needing the maximum index in advance. Other improvements (all small) include better error messages.

[lrat-4] A Verified LRAT Checker with Assignments Based on Single-Threaded Objects. The previous versions all represent an assignment as a list of (true) literals. Our next change was to represent assignments using single-threaded objects in order to improve performance. Profiling showed that most of the time in [lrat-3] was being spent evaluating clauses and literals. Evidently, the linear lookup into a long assignment (list of literals) can be expensive. Using a stobj avoids memory allocation for assignments, but probably much more important, it supports constant-time evaluation of literals.

Our stobj, `a$`, contains the three fields below. It uses standard representations: of propositional variables as natural numbers, of literals as non-zero integers, and of logical negation as arithmetic negation (-5 represents "not 5").

- `a$arr`: an array whose ith value is T, NIL, or 0, according to whether variable i is true, false, or of unknown value
- `a$stk`: a stack of variables, implemented as an array
- `a$ptr`: a natural number indicating the top of the stack

Returning to the `a$` stobj, we observe that the `a$arr` field alone does not provide direct support for reverting an assignment after having extended it.

We use a standard "trail" [6] approach to address this, by creating a stack of variables that have been assigned, such that whenever a Boolean value is written at position V of a$arr, V is also pushed onto the stack, by writing V at position a$ptr of a$stk and then incrementing a$ptr. That extension is undone by way of an inverse operation: the variable at the top of the stack serves as an index into a$arr at which to write 0 ("unassigned"), and then the stack pointer a$ptr is decremented.

[lrat-5] Compression and Incremental Reading. SAT proofs have grown to the point where the proof files that need to be certified are gigabytes in size. To help manage the sheer size of these proofs, we developed a lightweight procedure to compress LRAT files into CLRAT (Compressed LRAT) files, using techniques similar to those used for compression of DRAT files [8]. Our compression results in files about 40% the size of the original. Our CLRAT proof-file reader is guard-verified, both to support efficient execution and to increase confidence that we are parsing the input in a manner consistent with its specified syntax.

Compressed files reduce the size of proof files, but they do not reduce the number of proof steps that must be processed. Our earlier SAT proof-checkers read an entire proof file (into memory) before checking the veracity of every proof step, but given the ever increasing size of proof files, this approach is no longer tenable. We can now read SAT proofs in sections, for example of a few megabytes each; thus, we read (some of a proof file), then check (part of a proof), then read some more, then check some more, and so on, thus supporting proof files of arbitrary length. This checker has the highest performance of all of our verified SAT proof-checkers.

To provide for incrementally reading a large file, we extended the ACL2 function read-file-into-string so that it could read successive segments of the file, as specified by the user. Our correctness proof confirms that performing the interleaved file-reading and proof-checking is sound. The main advantage of interleaving proof reading and proof validation is that we can avoid having the entire proof in memory, which significantly reduces the memory footprint of the checker.

3.2   Performance

Table 1 compares performance for the checkers discussed above[7]. All runtimes are in seconds and include both parsing and checking time, and each is labeled by the proof file for the run. Each column header indicates one of the checkers discussed above, with a reminder of how it differs from the preceding checker. The [lrat-5] times do not include the use of diff described in Sect. 4.4, although that was done and was measured at under 1/50 s in each case. We omit columns for [lrat-2] (which is similar to [lrat-3] and for the early checkers that were much less efficient (for example, roughly one week for [rat] on R_4_4_18).

---

[7] We used ACL2 GitHub commit 639ef8760d30a63e2f21e160cdf02b75e1154fcc and SBCL Version 1.3.15, on a 3.5 GHz Intel(R) Xeon(R) with 32 GB of memory running on Ubuntu Linux.

Notice that an improvement can make much more of a difference for some tests than for others. In particular, as we move down through the last two columns we see that the list-based [lrat-3] checker compares well with the stobj-based [lrat-4] checker, until we get to a hard benchmark from the SAT 2016 competition, "Schur_161_5_d43.cnf", with a 5.6 GB proof (a rather typical size). Profiling showed that most of the time for [lrat-2] and [lrat-3] is in evaluating clauses and literals with respect to assignments. Since an [lrat-3] assignment is a linear list (of all true literals), it makes sense that the constant-time array access provided by an [lrat-4] stobj can reduce the time considerably. The [lrat-5] time of just over 4 min adds less than 25% to the 20 min it takes for the DRAT-trim checker to process a DRAT proof into an LRAT proof, which bodes well for using [lrat-5] in SAT competitions.

Table 1. Times in seconds when running checkers on various inputs

benchmark	[lrat-1] (fast-alist)	[lrat-3] (shrink)	[lrat-4] (stobjs)	[lrat-5] (incremental)
uuf-100-3	0.09	0.03	0.05	0.01
tph6[-dd]	3.08	0.57	0.33	0.33
R_4_4_18	164.74	5.13	2.23	2.24
transform	25.63	6.16	5.81	5.82
Schur_161_5_d43	5341.69	2355.26	840.04	259.82

We also produced RAT proofs of all application benchmarks of the SAT 2016 Competition that CryptoMinisat 5.0[8] could solve in 5000 s. We choose CryptoMinisat as it produces proofs with the most RAT clauses among those solvers that participated in the SAT 2016 Competition. CryptoMinisat solved 95 unsatisfiable benchmarks within the time limit. On 5 problems we ran into memory issues when converting the DRAT proof produced by CryptoMinisat into CLRAT proofs. One benchmark used duplicate literals, which is not allowed in our formalization. Figure 1 shows the results on the 89 validated proofs. For benchmarks that can be solved within 20 s, solving, DRAT to CLRAT conversion, and verified CLRAT checking are similar. For hard problems, solving takes about one third the time compared to DRAT to CLRAT conversion, while verified CLRAT checking takes about one third the time compared to solving. Hence, verified CLRAT checking adds relatively small overhead to the tool chain.

4   Correctness Proofs

We next consider, in order, each checker of the preceding section except the first, [rat], explaining some key high-level approaches to its correctness proof. Our focus is not on proving the basic algorithm correct, as this was done previously for the [rat] checker [26], including an analogue of the key inductive step (called

---

[8] https://github.com/msoos/cryptominisat.

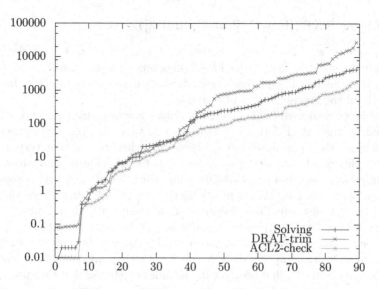

Fig. 1. Cactus plot of the solving times (including DRAT proof logging) of benchmarks for the SAT 2016 competition application benchmarks using CryptoMinisat, the validation times (including CLRAT proof logging) of DRAT-trim, and checking CLRAT proofs using ACL2-check.

`satisfiable-add-proof-clause` in Sect. 4.2): it preserves satisfiability to add a validated clause from an alleged proof. Rather, we discuss the steps taken in order to yield proofs for increasingly efficient code.

All of the soundness theorems for [rat] up through [lrat-3] have essentially the form displayed below: given a formula (list of clauses) and a valid refutation of it, then that formula is unsatisfiable. We will see a small variant for [lrat-4] and a major improvement for [lrat-5].

Soundness.

```
(implies (and (formula-p formula)
              (refutation-p proof formula))
         (not (satisfiable formula))))
```

4.1   Deletion ([drat])

Our first checker is a replacement for the initial checker [26]. A comparison of the two books shows that the original structure was preserved, the key difference being in the notion of a proof step: instead of a clause, it is a pair consisting of a flag and a clause, where the flag indicates whether the clause is to be added to the formula or deleted from it. Conceptually, deletion is trivially correct: if a formula is satisfiable, then it is still satisfiable after deleting one of its clauses. Our soundness proof effort took advantage of the automation provided by ACL2, in particular conditional rewriting: most lemmas were still proved automatically when we modified the checker, and the rest were straightforward to fix.

4.2   Linear RAT ([lrat-1], [lrat-2], [lrat-3])

In this section we discuss some lessons that can be learned from the proofs
of soundness for the first three LRAT checkers [lrat-1], [lrat-2], and [lrat-3]
described in Sect. 3. Recall that these checkers departed from [drat] by using
fast-alists in the representation of formulas.

In order to deal with the new formula data structure and the new proof hints
provided by the LRAT format, we chose to develop the soundness proof from
scratch, since the main developer for these new checkers was not very familiar
with the [rat] development. An early step was to write out a hand proof, so as
to avoid getting lost in a proof of this complexity. We started with a top-down
approach, supported by ACL2 utility skip-proofs [14]: first prove the main
result from the key lemmas (whose proofs are skipped), then similarly prove
each key lemma from its (proofs skipped) key sublemmas, and so on.[9]

To see this top-down style in action, consider the [lrat-1] book satisfiable-
add-proof-clause.lisp. As displayed below (with comments added, each following a semicolon (;)), that book locally includes two books that each prove a key
lemma in order to export those lemmas (not the other contents of the books) from
its scope, which are then used to prove the desired theorem.

```
(local ; Do not export the following outside this book.
 (encapsulate () ; Introduce a scope
   ; Load the two indicated books.
   (local (include-book"satisfiable-add-proof-clause-rup"))
   (local (include-book "satisfiable-add-proof-clause-drat"))
   ; Export two key lemmas outside the encapsulate scope.
   (defthm satisfiable-add-proof-clause-rup ...)
   (defthm satisfiable-add-proof-clause-drat ...)))
; Prove the main theorem of this book.
(defthm satisfiable-add-proof-clause
   ; Theorem statement is omitted in this display.
   :hints (("Goal"
            ; Prove that the two key lemmas imply this theorem.
            :use (satisfiable-add-proof-clause-rup
                  satisfiable-add-proof-clause-drat)
            ; Disabling most rules improves reliability and speed.
            :in-theory (union-theories '(verify-clause)
                                       (theory 'minimal-theory)))))
```

The [lrat-1] book sat-drat-claim-2-3.lisp also follows our hand proof.

The correctness proof for [lrat-1] was tedious, but presented no surprises. One
key proof technique, found in the [lrat-1] book soundness.lisp, is to define a

---

[9] The hand proof may be found in a comment near the top of the book
satisfiable-add-proof-clause.lisp (see for example community books directory
projects/sat/list-based/). That informal proof is annotated with names of lemmas from the actual proof script.

function `extend-with-proof` that recurs much like the checker, except instead of returning a Boolean, it returns the formula produced by applying the proof steps in sequence, starting with the original formula, with each step deleting or adding a clause. The following lemma is then key; with enough lemmas in place, it is proved automatically by induction using the recursion scheme for that function.

```
(defthm proof-contradiction-p-implies-false
  (implies (and (formula-p formula)
                (proofp proof)
                (proof-contradiction-p proof))
           (equal (evaluate-formula (extend-with-proof formula proof)
                                    assignment)
                  nil)))
```

Of course, the phrase "enough lemmas in place" above hides all the real work in the proof, for example in proving that the RAT check suffices for concluding that the addition of a clause preserves satisfiability.

With the proof of [lrat-1] complete, the next step was to improve efficiency by shrinking the formula from time to time, as explained in the description of [lrat-2] in Sect. 3. The [lrat-2] code was thus structurally similar but incorporated this shrinking. By keeping the top-level shrinking function disabled, it was reasonably straightforward to update the proof. Our process was to see where the former proof failed: when an ACL2 proof fails, it prints *key checkpoints*, which are formulas that can no longer be simplified. They often provide good clues for lemmas to formulate and prove.

The migration from [lrat-2] to [lrat-3] was very easy, including modifying the soundness proof. The key change was to avoid storing a maximum index field in the formula, so that the formula became exactly its fast-alist. This change had little effect on efficiency, though it did avoid some memory allocation (from building `cons` pairs). Rather, the point was to simplify the proof development, in preparation for our final step.

4.3 Using Stobjs ([lrat-4])

The introduction of stobjs for assignments presented the possibility of modifying the existing soundness proof. However, that seemed potentially difficult, given the disparity in the two representations of assignments: in the list version, assignments are extended using `cons` and retracted by going out of the scope of a LET binding; by contrast, the stobj version modifies assignments by updating array entries and stack pointers.

So instead of modifying the proof of the [lrat-3] soundness theorem, we decided to *apply* that theorem by relating the [lrat-3] list-based checker and the [lrat-4] stobj-based checker. A summary of that approach is presented below, followed by some deeper exploration. See the [lrat-4] (stobj-based) book `equiv.lisp` for the ACL2 theorems that relate the two checkers.

Applying [lrat-3] Correctness Using a Correspondence Theorem. The [lrat-3] and [lrat-4] checkers are connected using the correspondence

theorem below, `refutation-p-equiv`[10]. It is formulated using a function
`refutation-p$`, defined for the stobj-based checker in analogy to the list-based
recognizer function `refutation-p` for valid refutations, but using a so-called
local stobj.

```
(defthm refutation-p-equiv
  (implies (and (formula-p formula)
                (refutation-p$ proof formula))
           (refutation-p proof formula)))
```

That correspondence theorem trivially combines with the list-based checker's
soundness theorem (stated near the beginning of Sect. 4) to yield soundness for
the stobj-based checker.

```
(defthm main-theorem-stobj-based
  (implies (and (formula-p formula)
                (refutation-p$ proof formula))
           (not (satisfiable formula))))
```

Guard Verification and a Stobj Invariant. Our first step was to
verify guards for the stobj-based checker definitions, to support high-
performance execution. This step was undertaken before starting the proof of
`refutation-p-equiv` or its supporting lemmas, so that useful insights and lemmas developed during guard verification could be reused when developing the
correspondence proofs. In particular, it was clear that guard verification would
require developing an invariant on the stobj—e.g., to guarantee that extending
an assignment never writes to the stack at an out-of-bounds index—and that
proving invariance could be useful when proving the correspondence theorems.

The stobj invariant, `a$p`, is defined in terms of several recursively-defined properties. Informally, it says that the stack and array of `a$` correspond nicely: the
stack has no duplicates, and the variables below the top of the stack are exactly the
variables with an assigned value of true (`T`) or false (`NIL`) in the array, as opposed
to being undefined (value 0). It was rather challenging to complete all of the guard
verification, but then perhaps more straightforward to prove the correspondence
theorems, culminating in the theorem `refutation-p-equiv` shown above.

A Challenge in Proving Correspondence. A glitch arose while attempting
the correspondence proofs. Consider the following correspondence theorem.

```
(defthm negate-clause-equiv-1
  (implies (and (a$p a$)
                (= (a$ptr a$) 0)
```

---

[10] The subsidiary correspondence theorems all state equivalences, so the suffix "`-equiv`"
was used in the names of correspondence theorems, even though the top-level theorem, `refutation-p-equiv`, is actually an implication.

```
(clausep$ clause a$))
(equal (list-assignment (mv-nth 1 (negate-clause clause a$)))
       (negate-clause-or-assignment clause)))
```

The call of `negate-clause` on the left-hand side of the equality pushes each literal of the clause onto the stack, and then the function `list-assignment` extracts a list-based assignment from the resulting stack. However, the function `negate-clause-or-assignment` (defined for [lrat-3]) simply mapped negation over the clause, for example transforming the clause (3 -4 5) to (-3 4 -5)— whereas the left-hand side produces (-5 4 -3)—reversed! Fortunately, this was the only case in which the list-based and corresponding stobj-based function didn't match up.

By the time this issue surfaced, soundness had been established for the [lrat-3] (list-based) checker, guards had been verified for the [lrat-4] (stobj-based) checker, and some of the equivalence proofs had been completed. So we followed the steps below to modify the [lrat-3] checker to support the remaining equivalence proofs and avoid excessive re-work; after these steps, we completed the remaining correspondence proofs without undue difficulty.

1. We modified [lrat-3] by disabling `negate-clause-or-assignment` and attempting the proofs, expecting them to fail since that definition was no longer available.
2. We fixed the failed proofs—there were only a few—by providing them with hints to re-enable `negate-clause-or-assignment`.
3. We redefined `negate-clause-or-assignment` as a call to a tail-recursive function that reversed the order. Because of the steps above, the only proofs that failed were those explicitly enabling `negate-clause-or-assignment`.
4. With relatively modest effort we fixed all failed proofs.

4.4   The [lrat-5] Proof

Our [lrat-4] and [lrat-5] code were essentially the same except for the highest-level functions. It was thus straightforward.to work through the proof in a top-down style, reusing previous lemmas once we worked our way down to reasoning about functions that had not changed.

We improved the soundness theorem. Previous versions simply stated that every formula with a refutation is unsatisfiable. To see why that statement is insufficient, imagine an "evil" parser that always returns the trivial formula, containing only the empty clause. Then when the checker validates a proof, such a soundness theorem will only tell us that the empty clause is unsatisfiable! In principle a solution is to verify the parser, but that seems to us a difficult undertaking.

Instead we define a function, `proved-formula`, which takes two input files and various other parameters (such as how much of the proof file to read at each iteration). When a proof is successfully checked, this function returns the formula proved—essentially, what was read from the formula input file. This is the function that we actually run to check proofs. The following theorem states that if `proved-formula` is applied to a given formula file, `cnf-file`, and proof file, `clrat-file`, and it returns a formula F (rather than `nil`, which represents failure), then F is unsatisfiable.[11]

```
(defthm soundness
  (let ((formula (mv-nth 1 (proved-formula cnf-file clrat-file ...))))
    (implies formula
             (not (satisfiable formula)))))
```

For extra confidence, a very simple program[12], whose correctness can easily be ascertained by inspection, can print to a new file the formula returned by `proved-formula`. We have used this utility to compare the new file to the input formula using the `diff` utility, thereby providing confidence that the unsatisfiable formula returned by `proved-formula` truly represents the contents of the input formula file.

5   Conclusion

We now have an efficient, verified SAT checker that can rapidly check SAT proof files of many gigabytes. We expect that it will be used in applications of SAT solvers that demand validation, both in SAT competitions and in industry. Performance data on hard problems of the recent SAT competition suggest that the ACL2-based [lrat-5] checker generally adds less than 25% to the time spent by unverified proof-checking alone. The soundness proof for the stobj-based checker was split quite nicely into a sequence of proof efforts. Here are some reflections on those efforts, based on checker names introduced in Sect. 3.1.

1. We easily proved the soundness of [drat] by modifying the proof for [rat].
2. We developed the soundness proof for [lrat-1] essentially from scratch, starting with development of a hand proof. We believe that this helped us to deal with proof fallout from changes to the code from [drat] to [lrat-1].
3. We modified the proof for [lrat-1] in a modular way to produce a proof for [lrat-2], which shrinks the formula's fast-alist heuristically to boost performance significantly. This step (and others) benefited from ACL2's automation, in particular its display of key checkpoints upon proof failure. We believe that the structuring of the [lrat-1] soundness proof to follow a hand proof helped us to be efficient, by adding clarity to what we were trying to do.
4. The change from [lrat-2] to [lrat-3] was quite easy. The simplified notion of formula was expected to be useful for the next step, and we believe it was.
5. The change from [lrat-3] to [lrat-4] introduced stobj-based code. It seemed simplest to avoid trying to modify the soundness proof, instead deriving soundness as a corollary of a correspondence theorem that relates those two checkers. That worked out nicely, though it involved modifying a function in [lrat-3]. That modification was done in a modular way, in a succession of

---

[11] The `mv-nth` expression extracts the returned formula from a multiply-valued result.

[12] `projects/sat/lrat/incremental/print-formula.lisp` in the community books.

steps for which that function was disabled. Guard verification was challenging, but its supporting theorems and techniques helped with the soundness proof. Specifically, patterned-based congruences [13] developed for guard verification were also used in proving correspondence theorems.

6. The change from [lrat-4] to [lrat-5] caused us to extend the ACL2 system (and logical theory) with a utility for reading a portion of a file into a string. This utility supports efficient input from very large proof files.

Our software development approach used a form of refinement. We first specified and verified a very simple, but inefficient SAT proof checker. We then introduced another more efficient, but more complex, SAT proof-checker, that we then verified. We continued this process until we had a solution that was fast enough and verified to be correct. We believe that this stepwise approach was an effective, efficient way to develop a high-performance formally verified SAT proof-checker. This effort adds evidence one can build formally-verified production-class software.

References

1. Blanchette, J.C., Fleury, M., Weidenbach, C.: A verified SAT solver framework with learn, forget, restart, and incrementality. In: Olivetti, N., Tiwari, A. (eds.) IJCAR 2016. LNCS, vol. 9706, pp. 25–44. Springer, Cham (2016). doi:10.1007/978-3-319-40229-1_4
2. Boyer, R.S., Moore J S.: Single-threaded objects in ACL2. In: Krishnamurthi, S., Ramakrishnan, C.R. (eds.) PADL 2002. LNCS, vol. 2257, pp. 9–27. Springer, Heidelberg (2002). doi:10.1007/3-540-45587-6_3
3. Cruz-Filipe, L., Heule, M.J.H., Hunt Jr., W.A., Kaufmann, M., Schneider-Kamp, P.: Efficient certified RAT verification. In: de Moura, L. (ed.) CADE 2017. LNAI, vol. 10395, pp. 220–236. Springer, Cham (2017). doi:10.1007/978-3-319-63046-5_14
4. Davis, M., Logemann, G., Loveland, D.: A machine program for theorem-proving. Commun. ACM 5(7), 394–397 (1962)
5. Davis, M., Putnam, H.: A computing procedure for quantification theory. J. ACM (JACM) 7(3), 201–215 (1960)
6. Eén, N., Sörensson, N.: An extensible SAT-solver. In: Giunchiglia, E., Tacchella, A. (eds.) SAT 2003. LNCS, vol. 2919, pp. 502–518. Springer, Heidelberg (2004). doi:10.1007/978-3-540-24605-3_37
7. Greve, D.A., Kaufmann, M., Manolios, P., Moore J S., Ray, S., Ruiz-Reina, J.L., Sumners, R., Vroon, D., Wilding, M.: Efficient execution in an automated reasoning environment. J. Funct. Program. 18(1), 15–46 (2008)
8. Heule, M.J.H., Biere, A.: Clausal proof compression. In: 11th International Workshop on the Implementation of Logics. EPiC Series in Computing, vol. 40, pp. 21–26 (2016)
9. Heule, M.J.H., Hunt Jr., W.A., Wetzler, N.D.: Trimming while checking clausal proofs. In: Formal Methods in Computer-Aided Design, FMCAD 2013, Portland, OR, USA, 20–23 October 2013, pp. 181–188 (2013)
10. Heule, M.J.H., Hunt Jr., W.A., Wetzler, N.D.: Verifying refutations with extended resolution. In: Bonacina, M.P. (ed.) CADE 2013. LNCS (LNAI), vol. 7898, pp. 345–359. Springer, Heidelberg (2013). doi:10.1007/978-3-642-38574-2_24

11. Järvisalo, M., Heule, M.J.H., Biere, A.: Inprocessing rules. In: Gramlich, B., Miller, D., Sattler, U. (eds.) IJCAR 2012. LNCS (LNAI), vol. 7364, pp. 355–370. Springer, Heidelberg (2012). doi:10.1007/978-3-642-31365-3_28

12. Kaufmann, M., Manolios, P., Moore J S.: Computer-Aided Reasoning: An Approach. Kluwer Academic Press, Boston (2000)

13. Kaufmann, M., Moore J S.: Rough diamond: an extension of equivalence-based rewriting. In: Klein, G., Gamboa, R. (eds.) ITP 2014. LNAI, vol. 8558, pp. 537–542. Springer, Cham (2014). doi:10.1007/978-3-319-08970-6_35

14. Kaufmann, M.: Modular proof: the fundamental theorem of calculus. In: Kaufmann, M., Manolios, P., Moore J S. (eds.) Computer-Aided Reasoning: ACL2 Case Studies. Advances in Formal Methods, vol. 4, pp. 75–91. Springer, Boston (2000). doi:10.1007/978-1-4757-3188-0_6

15. Kaufmann, M., Moore J S.: ACL2 home page. http://www.cs.utexas.edu/users/moore/acl2. Accessed 2016

16. Lammich, P.: Efficient verified (UN)SAT certificate checking. In: de Moura, L. (ed.) CADE 2017. LNAI, vol. 10395, pp. 237–254. Springer, Cham (2017). doi:10.1007/978-3-319-63046-5_15

17. Lescuyer, S., Conchon, S.: A reflexive formalization of a SAT solver in Coq. In: International Conference on Theorem Proving in Higher Order Logics (TPHOLs) (2008)

18. Marić, F.: Formalization and implementation of modern SAT solvers. J. Autom. Reason. **43**(1), 81–119 (2009)

19. Marić, F.: Formal verification of a modern SAT solver by shallow embedding into Isabelle/HOL. Theor. Comput. Sci. **411**(50), 4333–4356 (2010)

20. Marques-Silva, J.P., Lynce, I., Malik, S.: Conflict-driven clause learning SAT solvers. In: Biere, A., Heule, M.J.H., van Maaren, H., Walsh, T. (eds.) Handbook of Satisfiability, chap. 4, pp. 131–153. IOS Press, Amsterdam (2009)

21. McCarthy, J.: Recursive functions of symbolic expressions and their computation by machine (part I). CACM **3**(4), 184–195 (1960)

22. Oe, D., Stump, A., Oliver, C., Clancy, K.: versat: a verified modern SAT solver. In: Kuncak, V., Rybalchenko, A. (eds.) VMCAI 2012. LNCS, vol. 7148, pp. 363–378. Springer, Heidelberg (2012). doi:10.1007/978-3-642-27940-9_24

23. Shankar, N., Vaucher, M.: The mechanical verification of a DPLL-based satisfiability solver. Electron. Notes Theor. Comput. Sci. **269**, 3–17 (2011)

24. Steele Jr., G.L.: Common Lisp the Language, 2nd edn. Digital Press, Burlington (1990)

25. Swords, S.: Private communication, March/April 2017

26. Wetzler, N.D., Heule, M.J.H., Hunt Jr., W.A.: Mechanical verification of SAT refutations with extended resolution. In: Blazy, S., Paulin-Mohring, C., Pichardie, D. (eds.) ITP 2013. LNCS, vol. 7998, pp. 229–244. Springer, Heidelberg (2013). doi:10.1007/978-3-642-39634-2_18

27. Wetzler, N.D., Heule, M.J.H., Hunt Jr., W.A.: DRAT-trim: efficient checking and trimming using expressive clausal proofs. In: Sinz, C., Egly, U. (eds.) SAT 2014. LNCS, vol. 8561, pp. 422–429. Springer, Cham (2014). doi:10.1007/978-3-319-09284-3_31

28. Wilding, M.: Design goals for ACL2. Tech. Rep. CLI Technical Report 101, Computational Logic, Inc., August 1994. https://www.cs.utexas.edu/users/moore/publications/km94.pdf

Proof Tactics for Assertions in Separation Logic

Zhé Hóu$^{(\boxtimes)}$, David Sanán, Alwen Tiu, and Yang Liu

Nanyang Technological University, Singapore, Singapore

Abstract. This paper presents tactics for reasoning about the assertions of separation logic. We formalise our proof methods in Isabelle/HOL based on Klein et al.'s separation algebra library. Our methods can also be used in other separation logic frameworks that are instances of the separation algebra of Calcagno et al. The first method, *separata*, is based on an embedding of a labelled sequent calculus for abstract separation logic (ASL) by Hóu et al. The second method, *starforce*, is a refinement of separata with specialised proof search strategies to deal with separating conjunction and magic wand. We also extend our tactics to handle pointers in the heap model, giving a third method *sepointer*. Our tactics can *automatically* prove many complex formulae. Finally, we give two case studies on the application of our tactics.

1 Introduction

Separation Logic (SL) is widely used to reason about programs with pointers and mutable data structures [34]. Many tools for separation logic have emerged since its inception and some of them have proven successful in real-life applications, such as the bi-abduction based techniques used in Infer [1]. Most tools for separation logic are built for small subsets of the assertion logic, notably the symbolic heap fragment [5], and applied to verify correctness and memory safety properties of computer programs. However, when verifying concurrent programs, often there is the need to use a larger fragment of the assertion language. For instance, the Separation Logic framework in Isabelle/HOL [28] and the Iris 3.0 framework [26] both use the full set of logical connectives, along with other features. Currently the frameworks that use larger fragments of the assertion language tend to focus more on the reasoning of Hoare triples than the assertions. An exception is the Iris 3.0 framework, in which the authors developed tactics for interactive proofs. Automated tools, however, are still beyond reach for larger fragments of SL and are the future work of the Iris project [26].

We are also motivated by our own project, which aims at verifying that an execution stack, including the processor architecture, micro-kernel, and applications, is correct and secure. Similar projects are NICTA's seL4 [30] and Yale's CertiKOS [19]. In particular, we are verifying the XtratuM hypervisor which runs on a multi-core LEON3 processor. Since concurrency is important in our project, it is useful to build formal models using techniques such as rely/guarantee and

© Springer International Publishing AG 2017
M. Ayala-Rincón and C.A. Muñoz (Eds.): ITP 2017, LNCS 10499, pp. 285–303, 2017.
DOI: 10.1007/978-3-319-66107-0_19

separation logic, and we will use the full assertion language because logical connectives such as the "magic wand" ($-\!*$) and "septraction" ($-\!\circledast$) are useful in rely/guarantee reasoning [38]. We aim to build a framework in Isabelle/HOL that can provide high confidence for the verification tasks. Automatic tactics in a proof assistant are therefore highly desirable because they can minimise the overhead of translating back and forth between the proof assistant and external provers, and it is easier to integrate them with other tactics.

This paper presents automatic proof tactics for reasoning about assertions in separation logic. Although frame inference is not in our scope, our tactics can be used to reason about assertions in frame inference. The tactics are independent of the separation logic framework and the choice of proof assistant, as long as the assertion logic is an instance of Calcagno et al.'s original definition of separation algebra [11]. For demonstration purposes and for the sake of our own project, we base our implementation on the work of Klein et al. [25], which formalises Calcagno et al.'s separation algebra and uses a shallow embedding of separation logic assertions into Isabelle/HOL formulae. At the core of our tactics lies the labelled sequent calculus LS_{PASL} of Hóu et al. [22], which is one of the few proof systems that have been shown successful in reasoning about the full language of assertions of separation logics with various flavours of semantics.

We first formalise each inference rule in LS_{PASL} as a lemma in a proof assistant, we then give a basic proof search procedure *separata* which can easily solve the formulae in previous BBI and PASL benchmarks [21,33]. To improve the performance and automation, we develop several more advanced tactics. The widely-used separating conjunction (denoted by $*$) behaves like linear conjunction in linear logic. It often creates difficulty in proof search because one has to find the correct splitting of resources to complete the proof. Effective 'resource management' in linear logic proof search is a well-known problem and it has been studied in the literature [12,20]. Unlike the case with linear logic, where resource is a multiset, we need to deal with a more complex structure capturing relations between heaps, and it is not clear how search techniques for linear logic [20] can be employed. We propose a new formula-driven algorithm to solve the heap partitioning problem. We also present a tactic to simplify the formula when it involves a combination of $*$ and $-\!*$ connectives. Finally, we extend the above tactics with inference rules [22] to handle pointers in the heap semantics.

We demonstrate that our tactics are able to prove many separation logic formulae automatically. These formulae are taken from benchmarks for BBI and abstract separation logic provers and the *sep_solve* method developed in seL4. We give a case study where we formalise Feng's semantics of actions in local rely/guarantee [17] using our extension of separation algebra, and prove some properties of the semantics using our tactics. Lastly, we show that our tactics can be easily implemented in other frameworks in case the user cannot directly use our implementation.

2   Preliminaries

This section gives an overview of Klein et al.'s formalization of Calcagno et al.'s separation algebra [25]. We extend their work with additional properties which are useful in applications. Then we briefly revisit the labelled sequent calculus LS_{PASL}.

2.1   Separation Algebra

A separation algebra [11] is a partial commutative monoid $(\Sigma, +, 0)$ where Σ is a non-empty set of elements (referred to as "worlds"), $+$ is a binary operator over worlds, and 0 is the unit for $+$. In Calcagno et al.'s definition, $+$ is a partial function, whereas Klein et al. defined it as a total function. For generality we shall assume that $+$ at least satisfies (in the sequel, if not stated otherwise, variables are implicitly universally quantified)

partial-functionality: if $x + y = z$ and $x + y = w$ then $z = w$.

Some formalizations of separation algebra also include a binary relation #, called "separateness" [11], over worlds. Two properties are given to the separateness relation: (1) $x \# 0$; and (2) $x \# y$ implies $y \# x$. The first one says every element is separated from the unit 0, the second one ensures the commutativity of #. As usual, the $+$ operator enjoys the following properties:

identity: $x + 0 = x$.
commutativity: $x \# y$ implies $x + y = y + x$.
associativity: if $x \# y$ and $y \# z$ and $x \# z$ then $(x + y) + z = x + (y + z)$.

Klein et al. then extend the above definitions with two more properties to obtain separation algebra: (1) if $x \# (y + z)$ and $y \# z$ then $x \# y$; (2) if $x \# (y + z)$ and $y \# z$ then $(x + y) \# z$. Finally, cancellative separation algebra extends the above with

cancellativity: if $x + z = y + z$ and $x \# z$ and $y \# z$ then $x = y$.

Assertions in separation algebra include the formulae of predicate calculus which are made from $\top, \bot, \neg, \rightarrow, \wedge, \vee$, and quantifiers $\exists, \forall$. In addition, there are multiplicative constant and connectives emp, $*$, and $-\!*$. In Isabelle/HOL, assertions can be encoded as predicates of type $'h \Rightarrow bool$ where $'h$ is the type of worlds in separation algebra. We write $\lfloor A \rfloor_w$ for the boolean formula resulting from applying the world w on the assertion A. The semantics of multiplicative assertions can be defined as:

$\lfloor emp \rfloor_w$ iff $w = 0$.
$\lfloor P * Q \rfloor_w$ iff there exists x, y such that $x \# y$ and $w = x + y$ and $\lfloor P \rfloor_x$ and $\lfloor Q \rfloor_y$.
$\lfloor P -\!* Q \rfloor_w$ iff for all x, if $w \# x$ and $\lfloor P \rfloor_x$ then $\lfloor Q \rfloor_{(w+x)}$.

2.2 Further Extension of Separation Algebra

We extend Klein et al.'s library with the following properties that hold in many applications such as heap model and named permissions, as discussed in [8,9,16]:

indivisible unit: if $x + y = 0$ and $x \# y$ then $x = 0$.
disjointness: $x \# x$ implies $x = 0$.
cross-split: if $a + b = w$, $c + d = w$, $a \# b$ and $c \# d$, then there exist e, f, g, h such that $e + f = a$, $g + h = b$, $e + g = c$, $f + h = d$, $e \# f$, $g \# h$, $e \# g$, and $f \# h$.

We call our extension *heap-sep-algebra* because our main application is the heap model. The following tactics also work for the algebra of Calcagno et al. if we remove from the tactics these extended properties.

2.3 The Labelled Sequent Calculus LS_{PASL}

The sequent calculus LS_{PASL} [21] for abstract separation logic is given in Fig. 1, where we omit the rules for classical connectives. A distinguishing feature of LS_{PASL} is that it has "structural rules" which manipulate ternary relational atoms. We define the ternary relation as: $(a, b \triangleright c) \equiv a \# b$ and $a + b = c$.

A sequent $\mathscr{G}; \Gamma \vdash \Delta$ contains a set $\mathscr{G}$ of ternary relational atoms, and sets Γ, Δ of labelled formulae of the form $h : A$, which corresponds to $\lfloor A \rfloor_h$ in the semantics.

Semicolon on the left hand side of $\vdash$ means classical conjunction and on the right means classical disjunction. The sequents on the top of a rule are premises, the one below is the conclusion. These inference rules are often used *backwards* in proof search. That is, to derive the conclusion, we need to derive the premises. The structural rules eq, u capture identity; e, a are for commutativity and associativity respectively; d for disjointness, which suffices to derive indivisible unit iu; p, c are for partial-functionality and cancellativity, and cs for cross-split.

3 Basic Proof Search

LS_{PASL} *Rules as Lemmas.* The first step towards developing automatic tactics in proof assistants based on the proof system LS_{PASL} is to translate each inference rule in LS_{PASL} to a lemma and prove that it is sound. Suppose a sequent takes the form

$$R_1; \ldots; R_l; s_1 : A_1; \ldots; s_m : A_m \vdash s_1' : B_1; \ldots; s_n' : B_n$$

where R_i are ternary relational atoms over labels/worlds, A_i and B_j are separation logic assertions, s_i and s_j' are labels denoting worlds. We translate the sequent to a formula

$$(R_1 \wedge \cdots \wedge R_l \wedge \lfloor A_1 \rfloor_{s_1} \wedge \cdots \wedge \lfloor A_m \rfloor_{s_m}) \rightarrow (\lfloor B_1 \rfloor_{s_1'} \vee \cdots \vee \lfloor B_n \rfloor_{s_n'}).$$

If a rule has premises P_1, P_2 and a conclusion C, we translate it to a lemma $(P_1 \wedge P_2) \rightarrow C$. If a rule has no premises, then we simply need to prove the conclusion. For instance, the rule *starr* in Fig. 1 is translated to the following lemma:

$$\frac{\mathscr{G};h=0;\Gamma\vdash\Delta}{\mathscr{G};\Gamma;h:emp\vdash\Delta}\ empl \qquad \frac{}{\mathscr{G};\Gamma\vdash 0:emp;\Delta}\ empr$$

$$\frac{(h_1,h_2\triangleright h_0);\mathscr{G};\Gamma;h_1:A;h_2:B\vdash\Delta}{\mathscr{G};\Gamma;h_0:A*B\vdash\Delta}\ starl \qquad \frac{(h_1,h_0\triangleright h_2);\mathscr{G};\Gamma;h_1:A\vdash h_2:B;\Delta}{\mathscr{G};\Gamma\vdash h_0:A-\!\!*B;\Delta}\ magicr$$

$$\frac{(h_1,h_2\triangleright h_0);\mathscr{G};\Gamma\vdash h_1:A;h_0:A*B;\Delta \quad (h_1,h_2\triangleright h_0);\mathscr{G};\Gamma\vdash h_2:B;h_0:A*B;\Delta}{(h_1,h_2\triangleright h_0);\mathscr{G};\Gamma\vdash h_0:A*B;\Delta}\ starr$$

$$\frac{(h_1,h_0\triangleright h_2);\mathscr{G};\Gamma;h_0:A-\!\!*B\vdash h_1:A;\Delta \quad (h_1,h_0\triangleright h_2);\mathscr{G};\Gamma;h_0:A-\!\!*B;h_2:B\vdash\Delta}{(h_1,h_0\triangleright h_2);\mathscr{G};\Gamma;h_0:A-\!\!*B\vdash\Delta}\ magicl$$

$$\frac{(0,h_2\triangleright h_2);\mathscr{G};h_1=h_2;\Gamma\vdash\Delta}{(0,h_1\triangleright h_2);\mathscr{G};\Gamma\vdash\Delta}\ eq \qquad \frac{(h_2,h_1\triangleright h_0);(h_1,h_2\triangleright h_0);\mathscr{G};\Gamma\vdash\Delta}{(h_1,h_2\triangleright h_0);\mathscr{G};\Gamma\vdash\Delta}\ e$$

$$\frac{(0,h_2\triangleright 0);h_1=0;\mathscr{G};\Gamma\vdash\Delta}{(h_1,h_2\triangleright 0);\mathscr{G};\Gamma\vdash\Delta}\ iu \qquad \frac{(h_1,h_1\triangleright h_2);h_1=0;\mathscr{G};\Gamma\vdash\Delta}{(h_1,h_1\triangleright h_2);\mathscr{G};\Gamma\vdash\Delta}\ d \qquad \frac{(h,0\triangleright h);\mathscr{G};\Gamma\vdash\Delta}{\mathscr{G};\Gamma\vdash\Delta}\ u$$

$$\frac{(h_3,h_5\triangleright h_0);(h_2,h_4\triangleright h_5);(h_1,h_2\triangleright h_0);(h_3,h_4\triangleright h_1);\mathscr{G};\Gamma\vdash\Delta}{(h_1,h_2\triangleright h_0);(h_3,h_4\triangleright h_1);\mathscr{G};\Gamma\vdash\Delta}\ a$$

$$\frac{(h_1,h_2\triangleright h_0);h_0=h_3;\mathscr{G};\Gamma\vdash\Delta}{(h_1,h_2\triangleright h_0);(h_1,h_2\triangleright h_3);\mathscr{G};\Gamma\vdash\Delta}\ p \qquad \frac{(h_1,h_2\triangleright h_0);h_2=h_3;\mathscr{G};\Gamma\vdash\Delta}{(h_1,h_2\triangleright h_0);(h_1,h_3\triangleright h_0);\mathscr{G};\Gamma\vdash\Delta}\ c$$

$$\frac{(h_5,h_6\triangleright h_1);(h_7,h_8\triangleright h_2);(h_5,h_7\triangleright h_3);(h_6,h_8\triangleright h_4);(h_1,h_2\triangleright h_0);(h_3,h_4\triangleright h_0);\mathscr{G};\Gamma\vdash\Delta}{(h_1,h_2\triangleright h_0);(h_3,h_4\triangleright h_0);\mathscr{G};\Gamma\vdash\Delta}\ cs$$

Side conditions:
In *starl* and *magicr*, the labels h_1 and h_2 do not occur in the conclusion.
In *a*, the label h_5 does not occur in the conclusion.
In *cs*, the labels h_5, h_6, h_7, h_8 do not occur in the conclusion.

Fig. 1. The inference rules for multiplicative connectives and structural rules in LS_{PASL}.

Lemma (lspasl-starr). $(((h_1, h_2 \triangleright h_0) \wedge \Gamma \to \lfloor A \rfloor_{h_1} \vee \lfloor A*B \rfloor_{h_0} \vee \Delta) \wedge$
$((h_1, h_2 \triangleright h_0) \wedge \Gamma \to \lfloor B \rfloor_{h_2} \vee \lfloor A*B \rfloor_{h_0} \vee \Delta)) \to ((h_1, h_2 \triangleright h_0) \wedge \Gamma \to \lfloor A*B \rfloor_{h_0} \vee \Delta)$

Note that we combine $\mathscr{G}$ and Γ in the rule into Γ in the lemma because in proof assistants Γ is an arbitrary formula which can be used to represent both. The above lemma is thus stronger than the soundness of a direct translation of sequents.

For each inference rule r in Fig. 1, we prove a corresponding lemma *lspasl-r* to show the soundness of the rule in Calcagno et al.'s separation algebra. In the sequel we may loosely refer to an inference rule as its corresponding lemma. We have also proved the inverted versions of the those lemmas which show that all the rules in LS_{PASL} are *invertible*. That is, if the conclusion is derivable, so are the premises. Completeness for Klein et al.'s formalisation is beyond this work because the semantics that LS_{PASL} is complete for, which is also widely used in the literature [9,16], does not consider the "separateness" relation, thus LS_{PASL} itself lacks the treatment of this relation.

Theorem (Soundness). LS_{PASL} *is sound with respect to* heap-sep-algebra.

Lemma (Invertibility). *The inference rules in LS_{PASL} are invertible.*

Proof Search Using LS_{PASL}. Proof assistants such as Isabelle/HOL can automatically deal with first-order connectives such as $\top, \bot, \wedge, \vee, \neg, \rightarrow, \exists$ and $\forall$, so we do not have to integrate the rule applications for these connectives in proof search. We divide the other inference rules in two groups: those that are truly invertible, and those that are only invertible because we "copy" the conclusion to the premises. The intuition is as follows: "invertible" rules are those that can be applied whenever possible without increasing the search space unnecessarily. The types of inference rules are summarised in Table 1.

Table 1. The types of inferences rules in LS_{PASL}.

Type		Rules
Invertible		lspasl-empl, lspasl-empr, lspasl-starl, lspasl-magicr
		lspasl-eq, lspasl-p, lspasl-c, lspasl-iu, lspasl-d
Quasi-invertible	Logical	lspasl-starr, lspasl-magicl
	Structural	lspasl-u, lspasl-e, lspasl-a, lspasl-cs

We analyse each rule *lspasl-r* in Table 1 and prove a lemma *lspasl-r-der* for a form of backward derivation. Such lemmas will be directly used in the tactics. Quasi-invertible rules such as lspasl-starr and lspasl-magicl need to be used with care because they may generate useless information and add unnecessary subgoals. Continuing with the example of the rule lspasl-starr, reading it backwards yields the following lemma:

Lemma (lspasl-starr-der). *If $(h_1, h_2 \triangleright h_0)$ and $\neg \lfloor A*B \rfloor_{h_0}$, then*
$((h_1, h_2 \triangleright h_0) \wedge \neg(\lfloor A \rfloor_{h_1} \vee \lfloor A*B \rfloor_{h_0}) \wedge (starr\_applied\ h_1\ h_2\ h_0\ (A*B)))$ *or*
$((h_1, h_2 \triangleright h_0) \wedge \neg(\lfloor B \rfloor_{h_2} \vee \lfloor A*B \rfloor_{h_0}) \wedge (starr\_applied\ h_1\ h_2\ h_0\ (A*B)))$

We include the assumptions in each disjunct so that contraction is admissible. We also include a dummy predicate "starr_applied" on each disjunct to record this rule application. This predicate is defined as $starr\_applied\ h_1\ h_2\ h_0\ F \equiv (h_1, h_2 \triangleright h_0) \wedge \neg\lfloor F \rfloor_{h_0}$.

We use three tactics to reduce search space when lspasl-starr or lspasl-magicl is applied. The first tactic is commonly used in provers for BBI and abstract separation logics [21,33]. For example, we forbid applications of lspasl-starr on the same pair of labelled formula and ternary relational atom more than once, because repeating applications on the same pair will not advance the proof search. To realise this, we generate the predicate "starr_applied" in proof search *only* when the corresponding pair is used in a rule application. We can then check if this predicate is generated during proof search, and avoid applying the rule on the same pair again.

The second tactic applies Lemma lspasl-starr-der2, which is an alternative of the above lemma that applies lspasl-starr on $\neg\lfloor A*B\rfloor_{h_0}$ and $(h_2, h_1 \triangleright h_0)$:

Lemma (lspasl-starr-der2). *If $(h_1, h_2 \triangleright h_0)$ and $\neg\lfloor A*B\rfloor_{h_0}$, then*
$((h_1, h_2 \triangleright h_0) \wedge \neg(\lfloor A\rfloor_{h_2} \vee \lfloor A*B\rfloor_{h_0}) \wedge (starr\_applied\ h_2\ h_1\ h_0\ (A*B)))$ *or*
$((h_1, h_2 \triangleright h_0) \wedge \neg(\lfloor B\rfloor_{h_1} \vee \lfloor A*B\rfloor_{h_0}) \wedge (starr\_applied\ h_2\ h_1\ h_0\ (A*B)))$

This is a crucial step because without it we will have to wait for the lspasl-e rule application to generate the commutative variant $(h_2, h_1 \triangleright h)$, and this particular rule application may be very late in proof search.

The third tactic is a look-ahead in the search: analyse each pair of $\neg\lfloor A*B\rfloor_{h_0}$ and $(h_1, h_2 \triangleright h_0)$ in the subgoal, and look for $\lfloor A\rfloor_{h_1}$ and $\lfloor B\rfloor_{h_2}$ in assumptions. If we can find at least one of them, then we can safely apply Lemma lspasl-starr-der and solve one subgoal immediately; thus the proof search space is not increased too much. We refer to the look-ahead tactics as *lspasl-starr-der-guided* (resp. *lspasl-magicl-der-guided*). Similar tactics are also developed for the rule lspasl-magicl. We apply lspasl-starr-der-guided and lspasl-magicl-der-guided whenever possible.

The structural rule lspasl-u requires more care, because it generates a new ternary relational atom out of nothing. A natural restriction is to forbid generating an atom if it already exists in the subgoal. Moreover, we only generate $(h, 0 \triangleright h)$ when (1) h occurs in some ternary relational atom (in the subgoal), or (2) h occurs in some labelled formula. We call these two applications *lspasl-u-der-tern* and *lspasl-u-der-form* respectively.

We develop two proof methods for the associativity rule lspasl-a. In the first method, *lspasl-a-der*, when we find the two assumptions $(h_1, h_2 \triangleright h_0)$ and $(h_3, h_4 \triangleright h_1)$, we only apply the rule lspsal-a when none of the following appear in the subgoal:

$$(0, h_2 \triangleright h_0), (h_1, 0 \triangleright h_0), (h_1, h_2 \triangleright 0), (0, h_4 \triangleright h_1), (h_3, 0 \triangleright h_1),$$
$$(\_, h_3 \triangleright h_0), (h_3, \_ \triangleright h_0), (h_2, h_4 \triangleright \_), (h_4, h_2 \triangleright \_).$$

In the first 5 cases, we can simplify the subgoal by unifying labels. For instance, the first case implies that $h_2 = h_0$, which can be derived by the rule lspasl-eq. The last 4 cases (_ means any label/world) indicate that one of the atoms to be generated *may* already exist in the subgoal, so we delay this rule application to the second method, *lspasl-a-der-full*, in which we generate all possible associative variants of the assumptions.

Real-world applications often involve reasoning of the form "if this assertion holds for all heaps, then ..." [30]. Hóu and Tiu's recent work included treatments for separation logic modalities with similar semantics [24]. For example, if the quantifier occurs on the left hand side of the sequent, they instantiate the quantified world to either an existing world or a fresh variable. This kind of reasoning often can be handled by existing lemmas in proof assistants, such as *meta-spec* in Isabelle/HOL. Therefore we do not detail the treatment for such quantifiers. We call the tactic for universal quantifiers on worlds *lsfasl-boxl-der* since it mimics the $\Box L$ rule in [24].

We are now ready to present the basic proof search. The first step is to "normalise" the subgoal from $P_1 \Longrightarrow P_2 \Longrightarrow \cdots \Longrightarrow P_n \Longrightarrow C$ to $P_1 \Longrightarrow P_2 \Longrightarrow \cdots \Longrightarrow P_n \Longrightarrow \neg C \Longrightarrow \bot$; otherwise, if C is some $A * B$, Lemma lspasl-starr-der will fail to apply on the subgoal. This preparation stage is called "prep".

Then we apply the "invertible" rules as much as possible, this is realised by a loop of applying the following lemmas until none are applicable: lspasl-empl-der, lspasl-starl-der, lspasl-magicr-der, lspasl-iu-der, lspasl-d-der, lspasl-eq-der, lspasl-p-der, lspasl-c-der, lspasl-starr-der-guided, lspasl-magicl-der-guided. This stage is called "invert".

The application of Lemma lsfasl-boxl-der follows, then come "quasi-invertible" rules. When applying the lemmas for structural rules lspasl-u-der-tern (identity), lspasl-e-der (commutativity), and lspasl-a-der (associativity), we apply them as much as possible based on existing ternary relational atoms in the (first) subgoal. We call this loop "non-inv-struct". We do not apply quasi-invertible logical rules as much as possible because that will produce too many subgoals. Thus in the "non-inv-logical" stage we apply one of lspasl-starr-der, lspasl-starr-der2, lspasl-magicl-der, lspasl-magicl-der2 only once.

Finally, we apply one of three rarely used lemmas in the end: lspasl-u-der-form, lspasl-cs-der, lspasl-a-der-full. We call this stage "rare".

The basic proof search procedure, named *separata*, is an infinite loop of the above stages until the subgoal is proven or none of the lemmas are applicable. We can express separata by the following regular expression, where "|" means "or" and "+" means one or more applications of the preceding element:

separata $\equiv$ (prep | (invert | lsfasl-boxl-der | non-inv-struct | non-inv-logical)+ | rare)+

4   Advanced Tactics for Proof Search

Although separata can handle all logical connectives, it is inefficient when the formula contains a complex structure with $*$ and $-\!\!*$. This section extends separata with specialised tactics for $*$ and $-\!\!*$, which pose the main difficulties in reasoning with separation logic. The former connective is pervasive in program verification, and the latter connective is important when reasoning about concurrent programs with rely/guarantee techniques [38]. We also integrate proof methods for pointers in the heap model.

4.1   Formula-Driven Tactics for the $*$ Connective

One of the hardest problems in reasoning about resources is to find the correct partition of resources when applying the (linear) conjunction right rule. In certain fragments of separation logic such as symbolic heap, this problem is simplified to AC-rewriting and can be solved relatively easily with existing techniques. However, in a logic with richer syntax, theorem provers often struggle to find the right partition of resources; this can be observed from the experiments of theorem provers for BBI and abstract separation logics [21,23,33]. To capture arbitrary interaction between additive connectives $(\wedge, \rightarrow)$ and multiplicative connectives $(*, -\!*)$, LS_{PASL} uses ternary relational atoms as the underlying data structure, which complicates the reasoning. This subsection proposes two-stage tactics for such situations, and gives two solutions for the second stage. Our techniques can also be adopted in other logic systems with ternary relations. Consider the following example:

Example 1. $(h_1, h_2 \triangleright h_3) \implies (h_4, h_5 \triangleright h_1) \implies (h_6, h_7 \triangleright h_2) \implies (h_8, h_9 \triangleright h_6) \implies (h_{10}, h_{11} \triangleright h_8) \implies \lfloor A \rfloor_{h_4} \implies \lfloor B \rfloor_{h_5} \implies \lfloor C \rfloor_{h_{10}} \implies \lfloor D \rfloor_{h_{11}} \implies \lfloor E \rfloor_{h_9} \implies \lfloor F \rfloor_{h_7} \implies \cdots \implies \neg \lfloor (((B*E)*(A*D))*C)*F \rfloor_{h_3} \implies \bot$

Recall Lemma lspasl-starr-der in Sect. 3. To apply it, we need to find an atom $(h_1, h_2 \triangleright h_0)$ which matches the labelled formula $\neg \lfloor A*B \rfloor_{h_0}$. The ternary relation represents a partition of the resource h_0, and only the correct partition will lead to a derivation. In separata, "non-inv-struct" applies identity, commutativity, and associativity without any direction. It may generate many atoms that are not needed and increase the search space. Thus the first problem is how to generate the exact set of ternary relational atoms for lspasl-starr applications. Let us take a closer look at the subgoal by viewing each ternary relational atom $(h, h' \triangleright h'')$ as a binary tree where h'' is the root and h, h' are leaf nodes. We then obtain the binary tree in Fig. 2 (left).

The first stage of the tactics is to analyse the structure of the $*$ formula and try to locate each piece of resource in the subgoal. In Example 1, it is easy to observe that A is true at world h_4 etc. Combined with the structure of the formula, we obtain the binary tree in Fig. 2 (right), which contains a few question marks that represent the worlds which are currently unknown. For instance, we do not know what is the combination of h_5 and h_9 in the subgoal;

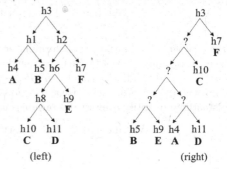

Fig. 2. Graphical representation of Example 1.

thus we should create a new ternary relational atom $\exists h.(h_5, h_9 \triangleright h)$ with a fresh symbol h, and try to find an instance of h later.

In a more general case, we first give an algorithm *findworld* (Algorithm 1) to find the world where a formula is true at and store all the new ternary relational atoms we create. We use "@" for concatenation of lists and "[]" for an empty list. The next step is to collect all the ternary relational atoms we have created to obtain Fig. 2 (right), as done in the algorithm *starstruct* (Algorithm 2).

Now that we know exactly the set of required ternary relational atoms, the second stage of the tactics is to derive Fig. 2 (right) from (left). We propose two solutions to the second stage. The first solution works for the separation algebra defined in Dockins et al.'s work [16], which is more general than the one used in

Data: subgoal, and a formula F
Result: a pair of a world and a list of ternary relational atoms
if $\lfloor F \rfloor_h$ is in subgoal for some h **then**
    return $(h, [])$;
else if $F \equiv A \wedge B$ *or* $F \equiv A \vee B$ *or* $F \equiv A \rightarrow B$ **then**
    $(h, l) \leftarrow$ findworld(subgoal, A);
    if $h \equiv NULL$ **then**
        return findworld(subgoal, B);
    else
        return (h, l);
else if $F \equiv \neg A$ *or* $F \equiv \exists x.A$ *or* $F = \forall x.A$ **then**
    return findworld(subgoal, A);
else if $F \equiv emp$ **then**
    return $(0, [])$;
else if $F = A * B$ **then**
    $(ha, la) \leftarrow$ findworld(subgoal, A); $(hb, lb) \leftarrow$ findworld(subgoal, B);
    if $(ha, hb \triangleright h)$ occurs in subgoal for some h **then**
        return $(h, la@lb)$;
    else
        Create a fresh variable h'; **return** $(h', la@lb@[(ha, hb \triangleright h')])$;
else
    return $(NULL, [])$;

Algorithm 1. The algorithm findworld.

Data: subgoal, and negated star formula $\neg \lfloor A*B \rfloor_h$
Result: a conjunction of ternary relational atoms
$(ha, la) \leftarrow$ findworld(subgoal, A); $(hb, lb) \leftarrow$ findworld(subgoal, B);
Make a conjunction of each ternary relational atom in $la@lb@[(ha, hb \triangleright h)]$ and
existentially quantify over all the variables created in findworld;

Algorithm 2. The algorithm starstruct.

this paper. A common property shared by Fig. 2 (left) and (right) is that the two binary trees have the same root and the same multiset of leaf nodes. It is easy to observe that a binary tree naturally corresponds to a list of ternary relational atoms. We can use the following lemma to derive Fig. 2 (right) where we say a node is *internal* if it is not the root nor a leaf node:

Lemma. *Given two binary trees t_1 and t_2 with the same root and the same multiset of leaf nodes. Suppose every internal node in t_2 is existentially quantified. There exists a sequence of lspasl-e-der and lspasl-a-der applications to derive t_2 from t_1.*

The intuition is that Fig. 2 (left) and (right) can be seen as parse trees of $*$ connected terms with different ways of bracketing. The two lemmas lspasl-e-der and lspasl-a-der correspond respectively to applications of commutativity and associativity of $*$. Those applications grant us a transformation from one bracketing to the other bracketing.

In the case that certain internal nodes in t_2 are not existentially quantified, by the construction in Algorithms 1 and 2, they must be existing worlds in the subgoal. Suppose the subgoal contains a binary tree in below (left) and also contains $(h_2, h_4 \triangleright h_5)$, and Algorithms 1 and 2 suggest to derive below (right). We can still use the above lemma to derive below (middle) from (left), where the question mark is an existentially quantified variable. We then instantiate the quantified variable to a fresh variable, e.g., h_6, and use lspasl-p-der (partial-functionality of $+$ and $\triangleright$) to unify h_6 and h_5, then derive below (right). This solution may be easy to implement in an external theorem prover, but we faced difficulties when implementing it in Isabelle/HOL. Specifically, whenever we use Algorithms 1 and 2 to obtain the atoms we need to derive, we have to prove that those atoms correspond to a binary tree. Since ternary relation is a definition in Isabelle/HOL, the proof of the tree representation is non-trivial and slow for large instances.

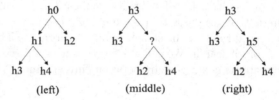

$$\text{(left)} \qquad \text{(middle)} \qquad \text{(right)}$$

The second solution is inspired by "forward reasoning" and "inverse method" [13]. This solution does not depend on the tree representation, but we shall describe it in terms of trees for simplicity. Instead of deriving Fig. 2 (right) from (left), we build (right) up from scratch using the information in (left). This can be seen as forward reasoning on ternary relational atoms. We start by choosing the bottom-most "sub-tree" $(h, h' \triangleright h'')$ in the tree to be derived. If we can prove that $h \# h'$, then there must be a world that represents the combination of h and h'. If h'' is not existentially quantified, we can use partial-functionality to derive that the combination of h and h' must be h''. Proving $h \# h'$ is the hard part. The intuition is that if h and h' are two leaf nodes in a (fragment of the) tree formed from the subgoal, then they must be "separated". For instance, from Fig. 2 (left), we should be able to derive $h_4 \# h_{10}$ since they are both leaf nodes. We should also be able to derive $h_4 \# h_6$ because they are leaf nodes of a fragment of the tree. We need to prove the following lemmas to reason about "separateness" of worlds:

disj-distri-tern: if $w \# z$ and $(x, y \triangleright z)$ then $w \# x$.
disj-distri-tern-rev: if $x \# y$ and $x \# z$ and $(y, z \triangleright w)$ then $x \# w$.
disj-comb: if $(x, y \triangleright z)$ and $x \# w$ and $y \# w$ then $z \# w$.
exist-comb: $x \# y$ implies $\exists z.(x, y \triangleright z)$.

Now we construct Fig. 2 (right) bottom-up using the algorithm *provetree* (Algorithm 3). The first step is to derive $h_5 \# h_9$ using lemmas disj-comb, disj-distri-tern, and disj-distri-tern-rev, and lspasl-e-der, and unfolding the definition of

Data: subgoal, and a tree t representing conjunctions of ternary relational atoms
Result: A proof that subgoal $\Longrightarrow t$
repeat
    Choose a lowest level ternary relational atom $(h, h' \rhd h'')$ in t;
    Prove $h \,\#\, h'$ using lemmas disj-comb, disj-distri-tern and disj-distri-tern-rev;
    if h'' is existentially quantified **then**
        Derive $\exists h''.(h, h' \rhd h'')$ using Lemma exist-comb;
    else
        Derive $\exists h'''.(h, h' \rhd h''')$ using Lemma exist-comb; Prove that $h''' = h''$;
    until All ternary relational atoms in t are covered;

Algorithm 3. The algorithm provetree.

the ternary relation. Next we obtain $\exists h.(h_5, h_9 \rhd h)$ using Lemma exist-comb. The part in Algorithm 3 where we show $h''' - h''$ can be done by applying lemmas lspasl-a-der and lspasl-e-der and unfolding the definition of the ternary relation. We repeat this process until we have derived the entire tree in Fig. 2 (right). Now we can use it in lspasl-starr-der applications to solve Example 1. The whole picture is that we guess the shape of the binary tree, and guide the application of structural rules by the structure of the $*$ formula. Thus we achieve a "formula-driven" proof search. We call the above tactics "starr-smart" and we extend separata by applying starr-smart between "invert" and "lsfasl-boxl-der".

4.2 Tactics for Magic Wand

Although $-\!*$ in general is very difficult to handle and it is often deemed as a source of non-recursive-enumerability in the heap model [6], we observe that many applications of $-\!*$ in [30] are of the form $(A * (B -\!* C)) \to (A' * C)$ where A is a $*$ connected formula which can be transformed to $A'*B$. Consider the following example:

$$(D * A * ((D * C) -\!* B) * C) \to (A * B)$$

Instead of deriving the correct way to split the resource according to the formula on the right hand side, we can use associativity and commutativity of $*$ and transform the left hand side into $A * (D * C) * ((D * C) -\!* B)$, which suffices to deduce the right hand side by the following lemma:

Lemma (magic-mp). $\lfloor C * (C -\!* B) \rfloor_h$ *implies* $\lfloor B \rfloor_h$.

Hence the key in this tactic is to transform a formula into the form $A*C*(C -\!* B)$ then simplify the formula. There are many ways to implement this, for simplicity we can analyse each $*$ connected formula $F \equiv F_1 * \cdots * F_n$, and for each $F_i \equiv (C -\!* B)$ where $C \equiv C_1 * \cdots * C_m$, we try to match F_j, $j \neq i$ with C_k. If each C_k can be successfully matched with a (different) F_j, we can then obtain a remainder R such that $F \equiv R * C * (C -\!* B)$. We then apply Lemma magic-mp to remove this occurrence of $-\!*$. We integrate this tactic into the sub-procedure "invert". The extension of separata with the tactics in Sects. 4.1 and 4.2 is called *starforce*.

Compared to separata, starforce applies structural rules and the rules for $*$ and $-*$ using specialised strategies, which often lead to better performance. However, in rare cases, starforce may be too aggressive and its intermediate tactics may get stuck. Therefore we leave both options to the user.

4.3 Tactics for the Heap Model

The separation algebra in this paper can be easily extended to capture pointers and potentially other data structures in the heap model. We shall focus on pointers here. For generality, we can define a points-to predicate as $'a \Rightarrow' v \Rightarrow' h \Rightarrow bool$ where $'a$ and $'v$ can be instantiated to *address* and *value* respectively in a concrete model, and $'h$ is the type of worlds in the separation algebra. We shall write this predicate as $\lfloor (a \mapsto v) \rfloor_h$, the intended meaning is that address a in heap h has value v. We can then give this relation some properties à la Hóu et al.'s work [24] to mimic pointers in the heap model:

Injection: if $\lfloor (a \mapsto v) \rfloor_{h_1}$ and $\lfloor (a \mapsto v) \rfloor_{h_2}$ then $h_1 = h_2$.
Non-emptiness: $\neg \lfloor (a \mapsto v) \rfloor_0$.
Not-larger-than-one: if $\lfloor (a \mapsto v) \rfloor_h$ and $(h_1, h_2 \triangleright h)$ then $h_1 = 0$ or $h_2 = 0$.
Address-disjointness: $\neg \lfloor (a_1 \mapsto v_1) \rfloor_{h_1}$ or $\neg \lfloor (a_1 \mapsto v_2) \rfloor_{h_2}$ or $\neg (h_1, h_2 \triangleright h)$.
Uniqueness: if $\lfloor (a_1 \mapsto v_1) \rfloor_h$ and $\lfloor (a_2 \mapsto v_2) \rfloor_h$ then $a_1 = a_2$ and $v_1 = v_2$.
Extensibility: for any h, v, there exist a, h_1, h_2 such that $(h_1, h \triangleright h_2)$ and $\lfloor (a \mapsto v) \rfloor_{h_1}$.

It is straightforward to prove corresponding lemmas for these properties and integrate the application of the lemmas into starforce, resulting in a new method *sepointer*.

5 Examples

This section demonstrates our implementation of the above tactics in Isabelle/HOL. Our proof methods separata, starforce, and sepointer can prove formulae *automatically* without human interaction. For space reasons we only show some examples. The source code and an extensive list of tested formulae can be accessed at [2].

Benchmark Examples. We show three BBI formulae from the previous benchmarks [21,33], these formulae are also valid in separation logic. The first one is very hard for existing BBI theorem provers, but it can be solved easily *by separata*, which combines the strength of the Isabelle engine and LS_{PASL} proof system.

$$(emp \to (p0 -* (((p0*(p0 -* p1))*(\neg p1)) -* (p0*(p0*((p0 -* p1)*(\neg p1))))))) \to$$
$$((((emp*p0)*(p0*((p0 -* p1)*(\neg p1)))) \to (((p0*p0)*(p0 -* p1))*(\neg p1))*emp)$$

Without separata, one could rely on Isabelle's *sledgehammer*, which will spend a few seconds to find a proof. There are also examples that sledgehammer fails to find proofs in 300s, but separata can solve them instantly:

$$\neg(((A*(C\mathbin{-\!\!*}(\neg((\neg(A\mathbin{-\!\!*}B))*C)))) \wedge (\neg B))*C)$$

In general, our Isabelle tactics can prove many complicated formulae which otherwise may be time consuming to find proofs in Isabelle.

We have tested our tactics on other benchmark formulae for BBI and PASL provers [21,33], both separata and starforce can prove those formulae automatically.

Example 1 in Sect. 4 is an instance that separata struggles but starforce can solve it easily. Similarly, starforce can easily prove the following formula, which is inspired by an example in seL4, using the tactic in Sect. 4.2:

$$(E*F*G*((C*Q*R)\mathbin{-\!\!*}B)*C*((G*H)\mathbin{-\!\!*}I)*H*((F*E)\mathbin{-\!\!*}Q)*A*R) \to (A*B*I)$$

The tactics for the heap model allow us to demonstrate more concrete examples. For instance, the following formula from the benchmark in [24] says that if the current heap can be split into two parts, one is $(e_1 \mapsto e_2)$ and the other part does not contain $(e_3 \mapsto e_4)$, and the current heap contains $(e_3 \mapsto e_4)$, then we deduce that $(e_3 \mapsto e_4)$ and $(e_1 \mapsto e_2)$ must be the same heap, therefore $e_1 = e_3$. This kind of reasoning about overlaid data structures requires applications of cross-split, and it is usually non-trivial to find proofs manually (Sledgehammer fails to find a proof in 300s).

$$(((e_1 \mapsto e_2) * \neg((e_3 \mapsto e_4) * \top)) \wedge ((e_3 \mapsto e_4) * \top)) \to (e_1 = e_3)$$

We can also prove some properties about "septraction" in separation logic with rely/guarantee where $A\mathbin{-\!\!\circledast}B \equiv \neg(A\mathbin{-\!\!*}\neg B)$, such as the formula below [38]:

$$((x \mapsto y) \mathbin{-\!\!\circledast} (z \mapsto w)) \to ((x = z) \wedge (y = w) \wedge emp)$$

Examples in seL4 Poofs. Klein et al. implemented separation algebra in Isabelle/HOL as a part of the renowned seL4 project. Many lemmas in their proofs related to separation algebra can now be proved with a single application of separata or starforce. The method *sep-solve* developed in the seL4 project fails to prove most of the examples we have tested for our tactics. Compared to the tactics developed in this paper, sep-solve is more ad-hoc. That is, our tactics are based on a more systematic proof theory (the labelled sequent calculus), whereas sep-solve focuses on special cases that are useful in practice. As a result, although sep-solve also have similar tactics to handle $\mathbin{-\!\!*}$, its treatment is different because it does not consider ternary relational atoms as its underlying data structure. Below is a lemma in the development of sep-solve:

lemma $(\bigwedge s.\lfloor Q \rfloor_s \Longrightarrow \lfloor Q' \rfloor_s) \Longrightarrow \lfloor R\mathbin{-\!\!*}R' \rfloor_s \Longrightarrow \lfloor (Q * R)\mathbin{-\!\!*}(Q' * R') \rfloor_s$
(1) **apply** *(erule sep-curry[rotated])*   *(2)* **apply** *(sep-select-asm 1 3)*
(3) **apply** *(sep-drule (direct) sep-mp-frame)*   *(4)* **by** *(sep-erule-full refl-imp, clarsimp)*

This proof can now be simplified to just "**by** separata". In cases like above, separata/starforce can be used as substitutes in the correctness proof of seL4.

6   Case Study

There are two ways to use our tactics: the user can prove that a logic is an instantiation of separation algebra, and directly apply our proof methods, as demonstrated in the semantics of actions; or the user can implement our tactics in another framework, as shown in our extension of the SL framework of Lammich and Meis [28].

Semantics of Ations. Our ongoing project involves integrating the semantics of actions in Feng's local rely/guarantee [17] in the CSimpl framework [35]. The assertion language of local rely/guarantee extends separation logic assertions with an additional semantic level to specify predicates over pairs (σ, σ') of states, called *actions*, which are represented by the states before and after the action. The semantics of actions redefines the separation logic operations in terms of Cartesian product of states. Additionally, the assertion language at the state level defines the separation logic operators for a state composed of three elements (l, s, i) to represent local, shared, and logical variables respectively. We represent both actions and states as products, and the core of the local rely guarantee's assertion language can be defined in CSimpl by showing that the Cartesian product of two heap-sep-algebras is a heap-sep-algebra. The instantiation of the product as a heap-sep-algebra involves proving the following properties:

zero-prod-def: $0 \equiv (0, 0)$.
plus-prod-def: $p_1 + p_2 \equiv ((fst\ p_1) + (fst\ p_2), (snd\ p_1) + (snd\ p_2))$.
sep-disj-prod-def: $p_1 \mathbin{\#} p_2 \equiv ((fst\ p_1) \mathbin{\#} (fst\ p_2)) \wedge ((snd\ p_1) \mathbin{\#} (snd\ p_2))$.

We then prove that the properties in Sects. 2.1 and 2.2 hold for pairs of actions. For an application, we use our tactics to prove the following lemma where

$$\lceil\ a\ \rceil \equiv (\lambda(\sigma,\sigma').\ (\sigma{=}\sigma') \wedge (a\ \sigma)) \text{ and } tran\text{-}Id \equiv \lceil\ \lambda s.\ True\ \rceil:$$

lemma assumes *a1:* $\lfloor A\ *\ tran\text{-}Id \rfloor_{(\sigma1,\sigma2)}$ **and** *a2:* $(\sigma1,\sigma' \rhd \sigma1') \wedge (\sigma2, \sigma' \rhd \sigma2')$
shows $\lfloor A\ *\ tran\text{-}Id \rfloor_{(\sigma1',\sigma2')}$
proof -   from *a2* **have** $((\sigma1,\sigma2), (\sigma',\sigma') \rhd (\sigma1',\sigma2'))$ **by** $(metis(full\text{-}types)$
$tern\text{-}dist1)$
  then show *?thesis* **using** *a1 id-pair-comb*   **apply** $(simp\ add\ :\ tran\text{-}Id$
$\text{-}def\ satis\text{-}def)$
  by *separata* **qed**

Here $\lceil a \rceil$ represents the action with equal initial and end states that satisfies a, so *tran_id* represents the identity relation. Before we use separata in semantics of actions we have to provide some domain knowledge that separata does not know, such as the first step which composes two ternary relational atoms into a ternary relational atom of pairs. We then need to unfold the definitions in the semantics of actions, and separata can solve the resultant subgoal quickly.

Lammich and Meis's SL Famework. In case the proof of instantiation of separation algebra is complex or Isabelle/HOL is not accessible, the user can implement our tactics in another framework (or even proof assistant). To demonstrate this we port separata to Lammich and Meis's SL framework [28] (source code at [2]). This process involves proving that the inference rules in LS_{PASL} are sound and adopting the applications of the rules in the new framework. Developing advanced tactics is feasible but time-consuming.

7   Related Work

Separation algebra was first defined as a cancellative, partial commutative monoid [11], and later formalized by Klein et al. in Isabelle/HOL [25]. Their definition includes a "separateness" relation # which is not used in other works such as [8,9,16]. We did, however, find this relation essential in developing tactics for automated reasoning (cf. Sect. 4). Later developments by Brotherston et al. [8] and Dockins et al. [16] added a few more properties in separation algebra, such as single unit, indivisible unit, disjointness, splittability, cross-split. We extend Klein et al.'s formalisation with all these properties except splittability because it does not hold in our applications. The proof theory for logics of separation algebra dates back to the Hilbert system and sequent calculus for Boolean bunched implications (BBI) [32]. The semantics of BBI is a generalised separation algebra: a non-deterministic commutative monoid [18]. The undecidability of BBI and other separation algebra induced logics [8,29] did not stop the development of semi-decision procedures, including display calculus [7], nested sequent calculus [33], and labelled sequent calculus [23]. Among these proof systems, nested sequent calculus and labelled sequent calculus are more suitable for automated reasoning. Hóu et al. developed labelled sequent calculi for propositional abstract separation logics [21] and corresponding theorem provers. Brotherson and Villard gave an axiomatisation for separation algebras using hybrid logic [9]. As far as we know, except Klein et al.'s work [25], none of these proof systems nor their proof search procedures have been formalised in a proof assistant.

Historically, the term "separation logic" refers to both the framework for reasoning about programs and the assertion logic in the framework. There have been numerous mechanisations of separation logic frameworks, but most of them focus on the reasoning of programs (e.g., [36]), whereas this paper focuses on the reasoning of assertions, so they are not directly comparable to this work. Moreover, most mechanisations of separation logic framework, e.g., Smallfoot [4], Holfoot [37], Myreen's rewriting tactics for SL [31], Ynot [15], Bedrock [14], and Charge! [3], only use a small subset of the assertion language, typically variants of symbolic heaps. Although some of those assertion logics are also induced from separation algebra, having a simpler syntax means that the reasoning task may be easier, and more efficient tactics, such as bi-abduction [10], can be developed for those logics. Consequently, the reasoning in those assertion logics is also not comparable to our work, which considers the full first-order assertion language. The few examples that use the full (or even higher-order) assertion language include Lammich and Meis's Isabelle/HOL formalisation of SL [28], Varming and Birkedal's formalisation of Higher-order SL (HOSL) [39], and the Iris project [26].

Lammich and Meis's SL framework includes a proof method *solve_entails* for assertions. A close inspection on the source code shows that it is mostly used to prove rather simple formulae such as $(A * emp) \rightarrow (A * \top)$ and $A * B \rightarrow B * A$ (although it can reason about some properties of lists). These formulae can be easily proved by our tactics. On the other hand, none of the examples shown in Sect. 5 can be proved by solve_entails. The interactive proof mode in Iris 3.0 [27] can solve many formulae in a restricted format, which is sufficient in their application. However, their tactics are not fully automatic. The formalisation of HOSL [39] also lacks automated proof methods.

This paper fills the gap of automated tactics for assertions in formalisations of SL. It is straightforward to adopt our tactics in other Isabelle/HOL formalisations; implementation in Coq should also be feasible since one can translate our tactics to Gallina and OCaml embedded code in Coq. Thus our work can be used to greatly improve the automation in SL mechanisations that involve the full language of assertions.

References

1. Facebook Infer. http://fbinfer.com/
2. Isabelle/HOL tactics for separation algebra. http://securify.scse.ntu.edu.sg/SoftVer/Separata
3. Bengtson, J., Jensen, J.B., Birkedal, L.: Charge! a framework for higher-order separation logic in Coq. In: Beringer, L., Felty, A. (eds.) ITP 2012. LNCS, vol. 7406, pp. 315–331. Springer, Heidelberg (2012). doi:10.1007/978-3-642-32347-8_21
4. Berdine, J., Calcagno, C., O'Hearn, P.W.: Smallfoot: modular automatic assertion checking with separation logic. In: Boer, F.S., Bonsangue, M.M., Graf, S., Roever, W.-P. (eds.) FMCO 2005. LNCS, vol. 4111, pp. 115–137. Springer, Heidelberg (2006). doi:10.1007/11804192_6
5. Berdine, J., Calcagno, C., O'Hearn, P.W.: Symbolic execution with separation logic. In: Yi, K. (ed.) APLAS 2005. LNCS, vol. 3780, pp. 52–68. Springer, Heidelberg (2005). doi:10.1007/11575467_5
6. Brochenin, R., Demri, S., Lozes, E.: On the almighty wand. In: Kaminski, M., Martini, S. (eds.) CSL 2008. LNCS, vol. 5213, pp. 323–338. Springer, Heidelberg (2008). doi:10.1007/978-3-540-87531-4_24
7. Brotherston, J.: A unified display proof theory for bunched logic. ENTCS **265**, 197–211 (2010)
8. Brotherston, J., Kanovich, M.: Undecidability of propositional separation logic and its neighbours. J. ACM **61**, 14:1–14:43 (2014)
9. Brotherston, J., Villard, J.: Parametric completeness for separation theories. In: POPL 2014, pp. 453–464 (2014)
10. Calcagno, C., Distefano, D., O'Hearn, P.W., Yang, H.: Compositional shape analysis by means of bi-abduction. J. ACM **58**(6), 1–66 (2011)
11. Calcagno, C., O'Hearn, P.W., Yang, H.: Local action and abstract separation logic. In: LICS 2007, pp. 366–378. IEEE (2007)
12. Cervesato, I., Hodas, J.S., Pfenning, F.: Efficient resource management for linear logic proof search. In: Dyckhoff, R., Herre, H., Schroeder-Heister, P. (eds.) ELP 1996. LNCS, vol. 1050, pp. 67–81. Springer, Heidelberg (1996). doi:10.1007/3-540-60983-0_5

13. Chaudhuri, K., Pfenning, F.: Focusing the inverse method for linear logic. In: Ong, L. (ed.) CSL 2005. LNCS, vol. 3634, pp. 200–215. Springer, Heidelberg (2005). doi:10.1007/11538363_15

14. Chlipala, A.: Mostly-automated verification of low-level programs in computational separation logic. In: PLDI 2011, pp. 234–245 (2011)

15. Chlipala, A., Malecha, G., Morrisett, G., Shinnar, A., Wisnesky, R.: Effective interactive proofs for higher-order imperative programs. In: ICFP 2009 (2009)

16. Dockins, R., Hobor, A., Appel, A.W.: A fresh look at separation algebras and share accounting. In: Hu, Z. (ed.) APLAS 2009. LNCS, vol. 5904, pp. 161–177. Springer, Heidelberg (2009). doi:10.1007/978-3-642-10672-9_13

17. Feng, X.: Local rely-guarantee reasoning. In POPL 2009, pp. 315–327. ACM (2009)

18. Galmiche, D., Larchey-Wendling, D.: Expressivity properties of Boolean BI through relational models. In: Arun-Kumar, S., Garg, N. (eds.) FSTTCS 2006. LNCS, vol. 4337, pp. 357–368. Springer, Heidelberg (2006). doi:10.1007/11944836_33

19. Ronghui, G., Shao, Z., Chen, H., Xiongnan, W., Kim, J., Sjöberg, V., Costanzo, D.: Certikos: an extensible architecture for building certified concurrent OS kernels. In OSDI 2016, pp. 653–669 (2016)

20. Hodas, J.S., López, P., Polakow, J., Stoilova, L., Pimentel, E.: A tag-frame system of resource management for proof search in linear-logic programming. In: Bradfield, J. (ed.) CSL 2002. LNCS, vol. 2471, pp. 167–182. Springer, Heidelberg (2002). doi:10.1007/3-540-45793-3_12

21. Hóu, Z., Clouston, R., Goré, R., Tiu, A.: Proof search for propositional abstract separation logics via labelled sequents. In: POPL 2014 (2014)

22. Hóu, Z., Goré, R., Tiu, A.: Automated theorem proving for assertions in separation logic with all connectives. In: Felty, A.P., Middeldorp, A. (eds.) CADE 2015. LNCS (LNAI), vol. 9195, pp. 501–516. Springer, Cham (2015). doi:10.1007/978-3-319-21401-6_34

23. Hóu, Z., Tiu, A., Goré, R.: A labelled sequent calculus for BBI: proof theory and proof search. In: Galmiche, D., Larchey-Wendling, D. (eds.) TABLEAUX 2013. LNCS, vol. 8123, pp. 172–187. Springer, Heidelberg (2013). doi:10.1007/978-3-642-40537-2_16

24. Hóu, Z., Tiu, A.: Completeness for a first-order abstract separation logic. In: Igarashi, A. (ed.) APLAS 2016. LNCS, vol. 10017, pp. 444–463. Springer, Cham (2016). doi:10.1007/978-3-319-47958-3_23

25. Klein, G., Kolanski, R., Boyton, A.: Mechanised separation algebra. In: Beringer, L., Felty, A. (eds.) ITP 2012. LNCS, vol. 7406, pp. 332–337. Springer, Heidelberg (2012). doi:10.1007/978-3-642-32347-8_22

26. Krebbers, R., Jung, R., Bizjak, A., Jourdan, J.-H., Dreyer, D., Birkedal, L.: The essence of higher-order concurrent separation logic. In: Yang, H. (ed.) ESOP 2017. LNCS, vol. 10201, pp. 696–723. Springer, Heidelberg (2017). doi:10.1007/978-3-662-54434-1_26

27. Krebbers, R., Timany, A., Birkedal, L.: Interactive proofs in higher-order concurrent separation logic. In: POPL 2017, pp. 205–217 (2017)

28. Lammich, P., Meis, R.: A separation logic framework for imperative HOL. In: AFP 2012 (2012)

29. Larchey-Wendling, D., Galmiche, D.: Non-deterministic phase semantics and the undecidability of Boolean BI. ACM TOCL 14(1), 6:1–6:41 (2013)

30. Murray, T., Matichuk, D., Brassil, M., Gammie, P., Bourke, T., Seefried, S., Lewis, C., Gao, X., Klein, G.: seL4: from general purpose to a proof of information flow enforcement. In: SP 2013, pp. 415–429, May 2013

31. Myreen, M.O.: Separation logic adapted for proofs by rewriting. In: Kaufmann, M., Paulson, L.C. (eds.) ITP 2010. LNCS, vol. 6172, pp. 485–489. Springer, Heidelberg (2010). doi:10.1007/978-3-642-14052-5_34

32. O'Hearn, P.W., Pym, D.J.: The logic of bunched implications. BSL **5**, 215–244 (1999)

33. Park, J., Seo, J., Park, S.: A theorem prover for Boolean BI. In: POPL 2013, pp. 219–232 (2013)

34. Reynolds, J.C.: Separation logic: a logic for shared mutable data structures. In LICS 2002, pp. 55–74 (2002)

35. Sanán, D., Zhao, Y., Hou, Z., Zhang, F., Tiu, A., Liu, Y.: CSimpl: a rely-guarantee-based framework for verifying concurrent programs. In: Legay, A., Margaria, T. (eds.) TACAS 2017. LNCS, vol. 10205, pp. 481–498. Springer, Heidelberg (2017). doi:10.1007/978-3-662-54577-5_28

36. Sergey, I., Nanevski, A., Banerjee, A.: Mechanized verification of fine-grained concurrent programs. In PLDI 2015, pp. 77–87 (2015)

37. Tuerk, T.: A formalisation of smallfoot in HOL. In: Berghofer, S., Nipkow, T., Urban, C., Wenzel, M. (eds.) TPHOLs 2009. LNCS, vol. 5674, pp. 469–484. Springer, Heidelberg (2009). doi:10.1007/978-3-642-03359-9_32

38. Vafeiadis, V., Parkinson, M.: A marriage of rely/guarantee and separation logic. In: Cambridge Technical report, vol. 687 (2007)

39. Varming, C., Birkedal, L.: Higher-order separation logic in Isabelle/HOLCF. ENTCS **218**, 371–389 (2008)

Categoricity Results for Second-Order ZF in Dependent Type Theory

Dominik Kirst[(✉)] and Gert Smolka[(✉)]

Saarland University, Saarbrücken, Germany
{kirst,smolka}@ps.uni-saarland.de

Abstract. We formalise the axiomatic set theory second-order ZF in the constructive type theory of Coq assuming excluded middle. In this setting we prove Zermelo's embedding theorem for models, categoricity in all cardinalities, and the correspondence of inner models and Grothendieck universes. Our results are based on an inductive definition of the cumulative hierarchy eliminating the need for ordinals and transfinite recursion.

1 Introduction

Second-order ZF is different from first-order ZF in that the replacement axiom quantifies over all relations at the class level. This is faithful to Zermelo's [22] informal view of axiomatic set theory and in sharp contrast to the standard first-order axiomatisation of ZF (cf. [6,8]). The difference between the two theories shows in the possibility of artificial and counterintuitive models of first-order ZF that are excluded by the more determined second-order ZF [17].

Zermelo [22] shows in an informal higher-order setting a little noticed embedding theorem saying that given two models of second-order ZF one embeds isomorphically into the other. From Zermelo's paper it is clear that different models of second-order ZF differ only in the height of their cumulative hierarchy and that higher models admit more Grothendieck universes [20] (i.e., sets closed under all set constructions).

The present paper studies second-order ZF in the constructive type theory of Coq [16] assuming excluded middle. We sharpen Zermelo's result by showing that second-order ZF is categorical in every cardinality, which means that equipotent models are always isomorphic. Using the fact that the height of a model is determined by its universes, we show that second-order ZF extended with an axiom fixing the number of universes to a finite n is categorical (i.e., all models are isomorphic).

For our results we employ the cumulative hierarchy, which is a well-ordered hierarchy of sets called stages such that every set appears in a stage and every universe appears as a stage. The usual way the cumulative hierarchy is established is through the ordinal hierarchy and transfinite recursion. We replace this long-winded first-order approach with a direct definition of the cumulative hierarchy as an inductive predicate, which leads to an elegant and compact development. While an inductive definition of the cumulative hierarchy has not been

© Springer International Publishing AG 2017
M. Ayala-Rincón and C.A. Muñoz (Eds.): ITP 2017, LNCS 10499, pp. 304–318, 2017.
DOI: 10.1007/978-3-319-66107-0_20

proposed before, inductive definitions of this form are known as tower constructions [12, 14]. Tower constructions go back to Zermelo [21] and Bourbaki [4], and are used by Smullyan and Fitting [14] to obtain the ordinal hierarchy.

The development of this paper is formalised and verified with the Coq proof assistant. Coq proves as an ideal tool for our research since types and thus models are first-class, inductive predicates and inductive proofs are well supported, and unnecessary assumptions (e.g. choice functions) are not built in. We assume excluded middle throughout the development and do not miss further built-in support for classical reasoning. The Coq development accompanying this paper has less than 1500 lines of code (about 500 for specifications and 1000 for proofs) and can be found at https://www.ps.uni-saarland.de/extras/itp17-sets. The theorems and definitions of the PDF version of this paper are hyperlinked with the Coq development.

The paper is organised as follows. We first discuss our formalisation of ZF and pay attention to the notion of inner models. Then, we study the cumulative hierarchy and prove that Grothendieck universes appear as stages. Next we prove the embedding theorem and show that ZF is categorical in every cardinality. Then we discuss categorical extensions of ZF. We end with remarks comparing our type-theoretic approach to ZF with the standard first-order approach.

2   Axiom System and Inner Models

We work in the type theory of Coq augmented by excluded middle for classical reasoning. Our model-theoretical approach is to study types that provide interpretations for the relations and constructors of set theory as follows:

Definition 1. *A set structure is a type M together with constants*

$$\_ \in \_ : M \to M \to \mathsf{Prop} \qquad \bigcup : M \to M$$
$$\emptyset : M \qquad\qquad \mathcal{P} : M \to M$$
$$\_@\_ : (M \to M \to \mathsf{Prop}) \to M \to M$$

for membership, empty set, union, power, and replacement.

Most of the following definitions rely on some fixed set structure M. We call the members $x, y, z, a, b, c : M$ **sets** and the members $p, q : M \to \mathsf{Prop}$ **classes**. Further, we use set-theoretical notation where convenient, for instance we write $x \in p$ if px and $x \subseteq p$ if $y \in p$ for all $y \in x$. We say that p and x **agree** if $p \subseteq x$ and $x \subseteq p$ and we call p **small** if there is some agreeing x. We take the freedom to identify a set x with the agreeing class $(\lambda y.\, y \in x)$.

ZF-like set theories assert every set to be free of infinitely descending $\in$-chains, in particular to be free of any $\in$-loops. This can be guaranteed by demanding all sets to contain a $\in$-least element, the so-called regularity axiom. From this assertion one can deduce the absence of infinitely descending $\in$-chains and hence an induction principle that implies $x \in p$ for all x if one can show that

$y \in p$ whenever $y \subseteq p$. Given a type theory that provides inductive predicates, $\in$-induction can be obtained with an inductive predicate defining well-ordered sets.

Definition 2. *We define the class of **well-founded sets** inductively by:*

$$\frac{x \subseteq WF}{x \in WF}$$

Then the induction principle of WF is exactly $\in$-induction and the wished axiom can be formulated as $x \in WF$ for all x. This and the other usual axioms of ZF yield the notion of a model:

Definition 3. *A set structure M is a **model (of ZF)** if*

> Ext : $\forall x, y. \, x \subseteq y \rightarrow y \subseteq x \rightarrow x = y$
> Eset : $\forall x. \, x \notin \emptyset$
> Union : $\forall x, z. \, z \in \bigcup x \leftrightarrow \exists y. \, z \in y \wedge y \in x$
> Power : $\forall x, y. \, y \in \mathcal{P}x \leftrightarrow y \subseteq x$
> Rep : $\forall R \in \mathcal{F}(M). \forall x, z. \, z \in R@x \leftrightarrow \exists y \in x. \, Ryz$
> WF : $\forall x. \, x \in WF$

*where $R \in \mathcal{F}(M)$ denotes that $R : M \rightarrow M \rightarrow$ Prop is a functional relation. We denote the predicate on structures expressing this collection of axioms by **ZF** and write $M \models$ **ZF** for **ZF** M.*

Apart from the inductive formulation of the foundation axiom, there are further ways in which our axiomatisation **ZF** differs from standard textbook presentations. Most importantly, we employ the second-order version of relational replacement which is strictly more expressive than any first-order scheme and results in a more determined model theory. Moreover, we do not assume the axiom of infinity because guaranteeing infinite sets is an unnecessary restriction for our investigation of models. Finally, we reconstruct the redundant notions of pairing, separation, and description instead of assuming them axiomaticly in order to study some introductory example constructions. The following definition of unordered pairs can be found in [15]:

Definition 4. *We define the **unordered pair** of x and y by:*

$$\{x, y\} := (\lambda ab. \, (a = \emptyset \wedge b = x) \vee (a = \mathcal{P}\emptyset \wedge b = y))@\mathcal{P}(\mathcal{P}\emptyset)$$

*As usual we abbreviate $\{x, x\}$ by $\{x\}$ and call such sets **singletons**.*

Lemma 5. $z \in \{x, y\}$ *if and only if $z = x$ or $z = y$.*

Proof. The given defining relation is obviously functional. So by applying Rep we know that $z \in \{x, y\}$ if and only if there is $z' \in \mathcal{P}(\mathcal{P}\emptyset)$ such that $z' = \emptyset$ and $z = x$ or $z' = \mathcal{P}\emptyset$ and $z = y$. This is equivalent to the statement $z = x$ or $z = y$ since we can simply pick z' to be the respective element of $\mathcal{P}(\mathcal{P}\emptyset)$. $\qquad\square$

The next notion we recover is separation, allowing for defining subsets of the form $x \cap p = \{\, y \in x \mid y \in p \,\}$ for a set x and a class p. By the strong replacement axiom we can show the separation axiom again in higher-order formulation.

Definition 6. *We define **separation** by* $x \cap p := (\lambda ab.\, a \in p \wedge a = b)@x$.

Lemma 7. $y \in x \cap p$ *if and only if* $y \in x$ *and* $y \in p$.

Proof. The defining relation is again functional by construction. So Rep states that $y \in x \cap p$ if and only if there is $z \in x$ such that $z \in p$ and $z = y$. This is equivalent to $y \in x$ and $y \in p$. $\qquad\square$

Finally, relational replacement implies the description principle in the form that we can construct a function that yields the witness of uniquely inhabited classes. The construction we employ can be found in [9]:

Definition 8. *We define a **description operator*** $\delta p := \bigcup((\lambda ab.\, b \in p)@\{\emptyset\})$.

Lemma 9. *If* p *is uniquely inhabited, then* $\delta p \in p$.

Proof. Let x be the unique inhabitant of p. By uniqueness we know that the relation $(\lambda ab.\, b \in p)$ is functional, so Rep implies that $(\lambda ab.\, b \in p)@\{\emptyset\} = \{x\}$ and Union implies that $\delta p = \bigcup\{x\} = x \in p$. $\qquad\square$

We note that functional replacement, i.e. the existence of a set $f@x$ for a function $f : M \to M$ and a set x is logically weaker than the relational replacement we work with. First, it is clear that such functions can be turned into functional relations by $(\lambda xy.\, fx = y)$. So relational replacement implies functional replacement and we will in fact use the latter where possible. Conversely, functional relations can only be turned into actual functions in the presence of a description operator. Hence description, which can be seen as a weak form of choice, must be assumed separately when opting for functional replacement.

At this point we can start discussing the model-theory of **ZF**. A first result is in direct contrast to the existence of countable models of first-order ZF guaranteed by the Löwenheim-Skolem Theorem. To this end, we employ the inductive data type $\mathbb{N}$ of natural numbers n for a compact formulation of the infinity axiom: we assume an injection $\overline{n}$ that maps numbers to sets together with a set ω that exactly contains the sets $\overline{n}$.

Lemma 10. *If* M *is a model of* **ZF** *with infinity, then* M *is uncountable.*

Proof. Suppose $f : \mathbb{N} \to M$ were a surjection from the inductive data type of natural numbers onto M. Then consider the set $X := \{\, \overline{n} \in \omega \mid \overline{n} \notin fn \,\}$. Since f is assumed surjective, there is $m : \mathbb{N}$ with $fm = X$. But this implies the contradiction $\overline{m} \in X \leftrightarrow \overline{m} \notin fm = X$. $\qquad\square$

When studying the cumulative hierarchy in the next section we will frequently encounter classes or, more specifically, sets that are closed under the set constructors. Such classes resemble actual models of **ZF** and we use the remainder of this section to make this correspondence formal.

Definition 11. *A class $p : M \to$ Prop is called **inner model** if the substructure of M consisting of the subtype induced by p and the correspondingly restricted set constructors is a model in the sense of Definition 3. We then write $p \models$ **ZF**.*

Definition 12. *A class p is **transitive** whenever $x \in y \in p$ implies $x \in p$ and **swelled** (following the wording in [14]) whenever $x \subseteq y \in p$ implies $x \in p$. Transitive and swelled sets are defined analogously.*

Definition 13. *A transitive class p with $\emptyset \in p$ is **ZF-closed** if for all $x \in p$:*

*(1) $\bigcup x \in p$ (**closure under union**),*
*(2) $\mathcal{P}x \in p$ (**closure under power**),*
*(3) $R@x \in p$ if $R \in \mathcal{F}(M)$ and $R@x \subseteq p$ (**closure under replacement**).*

*If p is small, then we call the agreeing set a (**Grothendieck) universe**.*

Lemma 14. *If p is ZF-closed, then $p \models$ **ZF**.*

Proof. Most axioms follow mechanically from the closure properties and transitivity. To establish WF we show that the well-foundedness of sets $x \in p$ passes on to the corresponding sets in the subtype by $\in$-induction.   □

3   Cumulative Hierarchy

It is a main concern of ZF-like set theories that the domain of sets can be stratified by a class of $\subseteq$-well-ordered stages. The resulting hierarchy yields a complexity measure for every set via the first stage including it, the so-called rank. One objective of our work is to illustrate that studying the cumulative hierarchy becomes very accessible in a dependent type theory with inductive predicates.

Definition 15. *We define the inductive class $\mathcal{S}$ of **stages** by the rules:*

$$\frac{x \in \mathcal{S}}{\mathcal{P}x \in \mathcal{S}} \qquad \frac{x \subseteq \mathcal{S}}{\bigcup x \in \mathcal{S}}$$

Fact 16. *The following hold:*

(1) $\emptyset$ is a stage.
(2) All stages are transitive.
(3) All stages are swelled.

Proof. We prove the respective statements in order.

(1) is by the second definitional rule as $\emptyset \subseteq \mathcal{S}$.
(2) is by stage induction using that power and union preserve transitivity.
(3) is again by stage induction.   □

The next fact expresses that union and separation maintain the complexity of a set while power and pairing constitute an actual rise.

Fact 17. *Let x be a stage, p a class and $a, b \in x$ then:*

(1) $\bigcup a \in x$
(2) $\mathcal{P}a \in \mathcal{P}x$
(3) $\{a, b\} \in \mathcal{P}x$
(4) $a \cap p \in x$

Proof. Again we show all statements independently.

(1) is by stage induction with transitivity used in the first case.
(2) is also by stage induction.
(3) is direct from Lemma 5.
(4) follows since x is swelled and $a \cap p \subseteq a$. □

We now show that the class $\mathcal{S}$ is well-ordered by $\subseteq$. Since $\subseteq$ is a partial order we just have to prove linearity and the existence of least elements, which bot An economical proof of linearity employs the following **double-induction principle** [14]:

Fact 18. *For a binary relation R on stages it holds that Rxy for all $x, y \in \mathcal{S}$ if*

(1) $R(\mathcal{P}x)y$ whenever Rxy and Ryx and
(2) $R(\bigcup x)y$ whenever Rzy for all $z \in x$.

Proof. By nested stage induction. □

Lemma 19. *If $x, y \in \mathcal{S}$, then either $x \subseteq y$ or $\mathcal{P}y \subseteq x$.*

Proof. By double-induction we just have to establish (1) and (2) for R instantiated by the statement that either $x \subseteq y$ or $\mathcal{P}y \subseteq x$. Then 1 is directly by case analysis on the assumptions Rxy and Ryx and using that $x \subseteq \mathcal{P}x$ for stages x. The second follows from a case distinction whether or not y is an upper bound for x in the sense that $z \subseteq y$ for all $z \in x$. If so, we know $(\bigcup x) \subseteq y$. If not, there is some $z \in x$ with $z \not\subseteq y$. So by the assumption Rzy only $\mathcal{P}y \subseteq z$ can be the case which implies $\mathcal{P}y \subseteq \bigcup x$. □

Fact 20. *The following alternative formulations of the linearity of stages hold:*

(1) $\subseteq$-linearity: $x \subseteq y$ or $y \subseteq x$
(2) $\in$-linearity: $x \subseteq y$ or $y \in x$
(3) trichotomy: $x \in y$ or $x = y$ or $y \in x$

Proof. (1) and (2) are by case distinction on Lemma 19. Then (3) is by (2). □

Lemma 21. *If p is an inhabited class of stages, then there exists a least stage in p. This means that there is $x \in p$ such that $x \subseteq y$ for all $y \in p$.*

Proof. Let $x \in p$. By $\in$-induction we can assume that every $y \in x$ with $y \in p$ admits a least stage in p. So if there is such a y there is nothing left to show. Conversely, suppose there is no $y \in x$ with $y \in p$. In this case we can show that x is already the least stage in p by $\in$-linearity. □

The second standard result about the cumulative hierarchy is that it exhausts the whole domain of sets and hence admits a total rank function.

Definition 22. *We call a $\in S$ the **rank** of a set x if $x \subseteq a$ but $x \notin a$. Since the rank is unique by trichotomy we can refer to it via a function ρ.*

Lemma 23. $\rho x = \bigcup \mathcal{P}@(\rho@x)$ *for every x. Thus every set has a rank.*

Proof. For a set x we can assume that every $y \in x$ has rank ρy by $\in$-induction. Then consider the stage $z := \bigcup \mathcal{P}@(\rho@x)$. Since for every $y \in x$ we know $y \in \mathcal{P}(\rho y)$, we deduce $x \subseteq z$. Moreover, suppose it were $x \in z$, so $x \in \mathcal{P}(\rho y)$ for some $y \in x$. Then this would imply the contradiction $y \in \rho(y)$, so we know $x \notin z$. Thus z is the rank of x. As a consequence, for every set x we know that $x \in \mathcal{P}(\rho x)$. Hence every set occurs in a stage.   □

Fact 24. *The hierarchy of stages exhausts all sets.*

Proof. Holds since every set x is an element of the stage $\mathcal{P}(\rho x)$.   □

We now turn to study classes of stages that are closed under some or all set constructors. The two introductory rules for stages already hint at the usual distinction of successor and limit stages. However, since we do not require x to contain an infinitely increasing chain in the second rule this distinction will not exactly mirror the non-exclusive rule pattern.

Definition 25. *We call $x \in S$ a **limit** if $x = \bigcup x$ and a **successor** if $x = \mathcal{P}y$ for some $y \in S$. Note that this means $\emptyset$ is a limit.*

Fact 26. *If $x \subseteq S$, then either $\bigcup x \in x$ or $x \subseteq \bigcup x$.*

Proof. Suppose it were $x \nsubseteq \bigcup x$ so there were $y \in x$ with $y \notin \bigcup x$. Then to establish $\bigcup x \in x$ it suffices to show that $y = \bigcup x$. Since $\bigcup x$ is the unique $\subseteq$-greatest element of x, it is enough to show that y is a $\subseteq$-greatest element, i.e. that $z \subseteq y$ for all $z \in x$. So let $z \in x$, then by linearity of stages it must be either $z \subseteq y$ or $y \in z$. The latter case implies $y \in \bigcup x$ contradicting the assumption.   □

Lemma 27. *Every stage is either a limit or a successor.*

Proof. Let x be a stage and apply stage induction. In the first case we know that x is a successor. In the second case we know that x is a set of stages that are either successors or limits and want to derive a decision for $\bigcup x$. Now we distinguish the two cases of Fact 26. If $\bigcup x \in x$, the inductive hypothesis yields the decision. If $x \subseteq \bigcup x$, it follows that $\bigcup x$ is a limit.   □

Lemma 28. *If x is an inhabited limit, then x is transitive, contains $\emptyset$, and is closed under union, power, pairing, and separation.*

Proof. Transitivity and closure under union and separation hold for arbitrary stages by Facts 16 and 17. Further, x must contain $\emptyset$ since it can be constructed from the set witnessing inhabitance by separation. The closure under power follows from the fact that every set $y \in x$ occurs in a stage $a \in x$. Then finally, the closure under pairing follows from Fact 17.   □

Hence, inhabited limits almost satisfy all conditions that constitute universes, only the closure under replacement is not necessarily given. So in order to study actual inner models one can examine the subclass of inhabited limits closed under replacement. In fact, this subclass turns out to be exactly the universes.

Lemma 29. *If $a \in u$ for a universe u, then $\rho a \in u$.*

Proof. By ϵ-induction we may assume that $\rho b \in u$ for all $b \in a$, so we know $\rho @ a \in u$ by the closure of u under replacement. Also, we know $\rho a = \bigcup \mathcal{P} @ (\rho @ a)$ by Lemma 23. Thus $\rho a \in u$ follows from the closure properties of u. $\square$

Lemma 30. *Universes are exactly inhabited limits closed under replacement.*

Proof. The direction from right to left is simple given that limits are already closed under all set constructors but replacement. Conversely, a universe is closed under replacement by definition and it is also easy to verify $u = \bigcup u$ given that for $x \in u$ we know $x \in \mathcal{P}(\rho x) \in u$ by the last lemma. So we just need to justify that u is a stage. We do this by showing that $u = \bigcup (u \cap S)$. The inclusion $u \supseteq \bigcup (u \cap S)$ is by transitivity. For the converse suppose $x \in u$. Then $x \subseteq \bigcup (u \cap S)$ again by knowing $x \in \mathcal{P}(\rho x) \in u$. $\square$

We remark that inhabited limits are models of the set theory Z which is usually defined to be ZF with pairing and separation instead of replacement. Also note that in our concrete axiomatisation **ZF** without infinity it is undecided whether there exists a universe, whereas assuming the existence of an infinite set allows for constructing the universe of all hereditarily finite sets.

4   Embedding Theorem

In this section we prove Zermelo's embedding theorem for models of second-order ZF given in [22]. Given two models M and N of **ZF**, we define a structure-preserving embedding $\approx$, called $\in$-bisimilarity, and prove it either total, surjective or both. We call this property of $\approx$ **maximality**. By convention, we let x, y, z range over the sets in M and a, b, c range over the sets in N in the remainder of this document.

Definition 31. *We define an inductive predicate $\approx: M \to N \to \mathsf{Prop}$ by*

$$\frac{\forall y \in x.\, \exists b \in a.\, y \approx b \qquad \forall b \in a.\, \exists y \in x.\, y \approx b}{x \approx a}$$

We call the first condition **(bounded) totality** *on x and a and write $x \triangleright a$. The second condition is called* **(bounded) surjectivity** *on x and a, written $x \triangleleft a$. We call $\approx$ $\in$-bisimilarity and if $x \approx a$ we call x and a **bisimilar**.*

The following lemma captures the symmetry present in the definition.

Lemma 32. *$x \approx a$ iff $a \approx x$ and $x \triangleright a$ iff $a \triangleleft x$.*

Proof. We first show that $a \approx x$ whenever $x \approx a$, the converse is symmetric. By $\in$-induction on x we may assume that $b \approx y$ whenever $y \approx b$ for some $y \in x$. Now assuming $x \approx a$ we show $a \triangleright x$. So for $b \in a$ we have to find $y \in x$ with $b \approx y$. By $x \triangleleft a$ we already know there is $y \in x$ with $y \approx b$. Then the inductive hypothesis implies $b \approx y$ as wished. That $x \triangleright a$ follows analogously and the second statement is a consequence of the first. $\square$

It turns out that $\approx$ is a partial $\in$-isomorphism between the models:

Lemma 33. *The relation $\approx$ is functional, injective, and respects membership.*

Proof. We show that $\approx$ is functional. By induction on $x \in WF$ we establish $a = a'$ whenever $x \approx a$ and $x \approx a'$. We show the inclusion $a \subseteq a'$, so first suppose $b \in a$. Since $x \triangleleft a$ there must be $y \in x$ with $y \approx b$. Moreover, since $x \triangleright a'$ there must be $b' \in a'$ with $y \approx b'$. By induction we know that $b = b'$ and hence $b \in a'$. The other inclusion is analogous and injectivity is by symmetry.

It remains to show that $\approx$ respects membership. Hence let $x \approx a$ and $x' \approx a'$ and suppose $x \in x'$. Then by $x' \triangleright a'$ there is $b \in a'$ with $x \approx b$. Hence $a = b$ by functionality of $\approx$ and thus $a \in a'$. $\square$

Since the other set constructors are uniquely determined by their members, it follows that they are also respected by the $\in$-bisimilarity:

Fact 34. $\emptyset \approx \emptyset$

Proof. Both $\emptyset \triangleright \emptyset$ and $\emptyset \triangleleft \emptyset$ hold vacuously. $\square$

Lemma 35. *If $x \approx a$, then $\bigcup x \approx \bigcup a$*

Proof. By symmetry (Lemma 32) we just have to prove $\bigcup x \triangleright \bigcup a$. So suppose $y \in \bigcup x$, so $y \in z \in x$. By $x \triangleright a$ we have $c \in a$ with $z \approx c$ and applying $z \triangleright c$ we have $b \in c$ with $y \approx b$. So $c \in b \in a$ and thus $b \in \bigcup a$. $\square$

Lemma 36. *If $x \approx a$, then $\mathcal{P}x \approx \mathcal{P}a$*

Proof. Again, we just show $\mathcal{P}x \triangleright \mathcal{P}a$. Hence let $y \in \mathcal{P}x$, so $y \subseteq x$. Then we can construct the image of y under $\approx$ by $b := \{c \in a \mid \exists z \in y.z \approx c\}$. Clearly $b \subseteq a$ so $b \in \mathcal{P}a$ and by $x \approx a$ it is easy to establish $y \approx b$. $\square$

Before we can state the corresponding lemma for replacement we first have to make precise how binary relations in one model are expressed in the other.

Definition 37. *For $R : M \to M \to \mathsf{Prop}$ we define $\overline{R} : N \to N \to \mathsf{Prop}$ by*

$$\overline{R}ab := \exists xy.\, x \approx a \wedge y \approx b \wedge Rxy$$

In particular, if $R \in \mathcal{F}(M)$ is functional then it follows that $\overline{R} \in \mathcal{F}(N)$.

Lemma 38. *If $x \approx a$, $R \in \mathcal{F}(M)$, and $R@x \subseteq \mathsf{dom}(\approx)$, then $R@x \approx \overline{R}@a$.*

Proof. We first show that $R@x \triangleright \overline{R}@a$, so let $y \in R@x$. Then by $R@x \subseteq \mathrm{dom}(\approx)$ there is b with $y \approx b$. It suffices to show $b \in \overline{R}@a$ which amounts to finding $c \in a$ with $\overline{R}cb$. Now by $y \in R@x$ there is $z \in x$ with Rzy. Hence there is $c \in a$ with $z \approx c$ since $x \triangleright a$. This implies $\overline{R}cb$.

We now show $R@x \triangleleft \overline{R}@a$, so let $b \in \overline{R}@a$. Then there is $c \in a$ with $\overline{R}cb$. By definition this already yields z and y with $z \approx c$, $y \approx b$, and Rzy. Since $\approx$ respects membership we know $z \in x$ and hence $y \in R@x$.   □

Note that these properties immediately imply the following:

Lemma 39. *If* $\mathrm{dom}(\approx)$ *is small, then it agrees with a universe.*

Proof. First, $\emptyset \in \mathrm{dom}(\approx)$ since $\emptyset \approx \emptyset$. Further, $\mathrm{dom}(\approx)$ is transitive by the totality part of $x \approx a$ for every $x \in \mathrm{dom}(\approx)$. The necessary closure properties of universes were established in the last lemmas.   □

The dual statement for $\mathrm{ran}(\approx)$ holds as well by symmetry. Now given that $\approx$ preserves all structure of the models, every internally specified property holds simultaneously for similar sets. In particular, $\approx$ preserves the notion of stages and universes:

Lemma 40. *If* $x \approx a$ *and* x *is a stage, then* a *is a stage.*

Proof. We show that all a with $x \approx a$ must be stages by stage induction on x. So suppose x is a stage and we have $\mathcal{P}x \approx b$. Since $x \in \mathcal{P}x$, by $\mathcal{P}x \triangleright b$ there is $a \in b$ with $x \approx a$. Then by induction a is a stage. Moreover, Lemma 36 implies that $\mathcal{P}x \approx \mathcal{P}a$. Then by functionality we know that b equals the stage $\mathcal{P}a$.

Now suppose x is a set of stages and we have $\bigcup x \approx b$. Since $\mathcal{P}(\mathcal{P}(\bigcup x)) \approx \mathcal{P}(\mathcal{P}b)$ by Lemma 36 and $x \in \mathcal{P}(\mathcal{P}(\bigcup x))$ there is some $a \in \mathcal{P}(\mathcal{P}b)$ with $x \approx a$. But then we know that $\bigcup x \approx \bigcup a$ by Lemma 35 and $b = \bigcup a$ by functionality, so it remains to show that a is a set of stages. Indeed, if we let $c \in a$ then $x \triangleleft a$ yields $y \in x$ with $y \approx c$ and since x is a set of stages we can apply the inductive hypothesis for y to establish that c is a stage.   □

Lemma 41. *If* $x \approx a$ *and* x *is a universe, then* a *is a universe.*

Proof. We first show that a is transitive, so let $c \in b \in a$. By bounded surjectivity there are $z \in y \in x$ with $z \approx c$ and $y \approx b$. Then $z \in x$ since x is transitive, which implies $c \in a$ since $\approx$ preserves membership.

The proofs that a is closed under the set constructors are all similar. Consider some $b \in a$, then for instance we show $\bigcup b$ in a. The assumption $x \approx a$ yields $y \in x$ with $y \approx b$. Since x is closed under union it follows $\bigcup y \in x$ and since $\bigcup y \approx \bigcup b$ by Lemma 35 it follows that $\bigcup b \in a$. The proof for power is completely analogous and for replacement one first mechanically verifies that $\overline{R}@y \subseteq x$ for every functional relation $R \in \mathcal{F}(N)$ with $R@b \subseteq a$.   □

In order to establish the maximality of $\approx$ we first prove it maximal on stages:

Lemma 42. *Either* $\mathcal{S}_M \subseteq \mathrm{dom}(\approx)$ *or* $\mathcal{S}_N \subseteq \mathrm{ran}(\approx)$.

Proof. Suppose there were $x \notin \mathsf{dom}(\approx)$ and $a \notin \mathsf{ran}(\approx)$, then we can in particular assume x and a to be the least such stages by Lemma 21. We will derive the contradiction $x \approx a$. By symmetry, we just have to show $x \triangleright a$ which we do by stage induction for x. The case $\mathcal{P}(x)$ for some stage x is impossible given that, by leastness of $\mathcal{P}x \notin \mathsf{dom}(\approx)$, necessarily $x \in \mathsf{dom}(\approx)$ holds which would, however, imply $\mathcal{P}x \in \mathsf{dom}(\approx)$ by Lemma 36.

In the case $\bigcup x$ for a set of stages x we may assume that $x \subseteq \bigcup x$ by Fact 26. Now suppose $y \in z \in x$, then we want to find $b \in W$ with $y \approx b$. We do case analysis whether or not $z \in \mathsf{dom}(\approx)$. If so, then there is c with $z \approx c$. Since $z \in x$ we know that z is a stage and so must be c by Lemma 40. Then by linearity it must be $c \in W$ and $z \triangleright c$ yields the wished $b \in W$ with $y \approx b$. If z were not in $\mathsf{dom}(\approx)$, we have $\bigcup x \subseteq z$ since $\bigcup x$ is the least stage not in the domain. But since $z \in x$ and $x \subseteq \bigcup x$ this yields $z \in z$ contradicting well-foundedness.   □

Theorem 43. *The relation $\approx$ is maximal, that is $M \subseteq \mathsf{dom}(\approx)$ or $N \subseteq \mathsf{ran}(\approx)$.*

Proof. Suppose $\approx$ were neither total nor surjective, so there were $x \notin \mathsf{dom}(\approx)$ and $a \notin \mathsf{ran}(\approx)$. By Fact 24 we know that $x \in \mathcal{P}(\rho x)$ and $a \in \mathcal{P}(\rho a)$. Then by Lemma 42 it is either $\mathcal{P}(\rho x) \in \mathsf{dom}(\approx)$ or $\mathcal{P}(\rho a) \in \mathsf{ran}(\approx)$. But then it follows either $x \in \mathsf{dom}(\approx)$ or $a \in \mathsf{ran}(\approx)$ contradicting the assumption.   □

From this theorem we can conclude that embeddebility is a linear pre-order on models of ZF. We can further strengthen the result by proving one side of $\approx$ small if $\approx$ is not already **full**, meaning both total and surjective.

Lemma 44. *If x is a stage with $x \notin \mathsf{dom}(\approx)$, then $\mathsf{dom}(\approx) \subseteq x$.*

Proof. Since $x \notin \mathsf{dom}(\approx)$ we know that $\approx$ is surjective by Theorem 43. So let $y \approx a$, then we want to show that $y \in a$. By exhaustiveness a occurs in some stage b and since $\approx$ is surjective there is z with $z \approx b$. Lemma 40 justifies that z is a stage. By linearity we have either $z \subseteq x$ or $x \in z$. In the former case we are done since $y \in z$ given that $\approx$ respects the membership $a \in b$. The other case is a contradiction since it implies $x \in \mathsf{dom}(\approx)$.   □

The dual holds for the stages of N and $\mathsf{ran}(\approx)$, hence we summarise:

Theorem 45. *Exactly one of the following statements holds:*

(1) $\approx$ is full, so $M \subseteq \mathsf{dom}(\approx)$ and $N \subseteq \mathsf{ran}(\approx)$.
(2) $\approx$ is surjective and $\mathsf{dom}(\approx)$ is small and a universe of M.
(3) $\approx$ is total and $\mathsf{ran}(\approx)$ is small and a universe of N.

Proof. Suppose $\approx$ were not full, then it is still maximal by Theorem 43. So for instance let $\approx$ be surjective but not total, then we show (2). Being not total, $\approx$ admits a stage x with $x \notin \mathsf{dom}(\approx)$. Then by Lemma 44 we know $\mathsf{dom}(\approx) \subseteq x$, so the domain is realised by $x \cap \mathsf{dom}(\approx)$. This set is a universe by Lemma 39. □

5   Categoricity Results

In the remainder of this work, we examine to what extent the model theory of **ZF** is determined and study categorical extensions. If $\approx$ is full for models M and N, we call M and N **isomorphic**. An axiomatisation is called **categorical** if all of its models are isomorphic. Without assuming any further axioms, we can prove **ZF** categorical in every cardinality:

Theorem 46. *Equipotent models of* **ZF** *are isomorphic.*

Proof. If models M and N are equipotent, we have a function $F : M \to N$ with inverse $G : N \to M$. Then from either of the cases (2) and (3) of Theorem 45 we can derive a contradiction. So for instance suppose $\approx$ is surjective and $X = \mathsf{dom}(\approx)$ is a universe of M. We use a variant of Cantor's argument where G simulates the surjection of X onto the power set of X. Hence define $Y := \{\, x \in X \mid x \notin G(ix) \,\}$ where i is the function obtained from $\approx$ by description. Then Y has preimage $y := i^{-1}(FY)$ and we know that $y \in X$ by surjectivity. Hence, by definition of Y we have $y \in Y$ iff $y \notin G(iy) = G(i(i^{-1}(FY))) = G(F(Y)) = Y$, contradiction. Thus case (1) holds and so $\approx$ is indeed full. $\square$

An internal way to determine the cardinality of models and hence to obtain full categoricity is to control the number of universes guaranteed by the axioms. For instance, one can add an axiom excluding the existence of any universe.

Definition 47. ZF$_0$ *is* **ZF** *plus the assertion that there exists no universe.*

Note that **ZF$_0$** axiomatises exactly the structure of hereditarily finite sets [1,13] and this is of course incompatible with an infinity axiom. That **ZF$_0$** is consistent relative to **ZF** is guaranteed:

Lemma 48. *Every model of* **ZF** *has an inner model without universes.*

Proof. Let M be a model of **ZF**. If M contains no universe, then the full class $(\lambda x. \top)$ is an inner model of **ZF$_0$**. Otherwise, if M contains a universe u, then we can assume u to be the least such universe since universes are stages by Lemma 30 and stages are well-ordered by Lemma 21. Then it follows that u constitutes an inner model of **ZF$_0$**. First, u is an inner model of **ZF** by Lemma 14. Secondly, if there were a universe u' in the sub-structure induced by u, then u' would be a universe that is smaller than u, contradiction. $\square$

Lemma 49. ZF$_0$ *is categorical.*

Proof. Again from either of the cases (2) and (3) of Theorem 45 we can derive a contradiction. So for instance suppose $\approx$ is surjective and $X = \mathsf{dom}(\approx)$ is a universe of M. This directly contradicts the minimality assumption of M. $\square$

The categoricity result for **ZF$_0$** can be generalised to axiomatisations that guarantee exactly n universes. Note that stating axioms of such a form presupposes an external notion of natural numbers, for instance given by the inductive type $\mathbb{N}$. We avoid employing further external structure such as lists to express finite cardinalities and instead make use of the linearity of universes as follows:

Definition 50. *We define* $\mathbf{ZF}_{n+1}$ *to be* $\mathbf{ZF}$ *plus the following assertions:*

(1) there exists a universe that contains at least n universes and
(2) there exists no universe that contains at least $n + 1$ universes.

The notion that a universe u contains at least n universes is defined recursively with trivial base case and where u is said to contain $n + 1$ universes if there is a universe $u' \in u$ that contains at least n universes.

Since it is undecided whether or not a given model contains a universe, we cannot construct inner models that satisfy $\mathbf{ZF}_{n+1}$ for any n. Due to the connection of universes and inaccessible cardinals (cf. [20]), $\mathbf{ZF}_{n+1}$ constitutes a rise in proof-theoretic strength over $\mathbf{ZF}_n$. Independent of the consistency question, we can still prove all models of $\mathbf{ZF}_n$ isomorphic for every n:

Lemma 51. $\mathbf{ZF}_n$ *is categorical for all n.*

Proof. We have already proven $\mathbf{ZF}_0$ categorical in Lemma 49 so we just have to consider $\mathbf{ZF}_{n+1}$. As in the two proofs above we suppose that $\approx$ is surjective as well as that $X = \mathrm{dom}(\approx)$ is a universe of M and derive a contradiction. In fact, we show that X contains at least $n + 1$ universes and hence violates (2) of the above definition for M. By (1) for N we know there is a universe $u \in N$ that contains at least n universes. Hence by surjectivity we know that $i^{-1}u \in X$, where i is again the function obtained from $\approx$. Then Lemma 41 implies that $i^{-1}u$ is a universe of M. Moreover, since $\approx$ preserves all structure, it follows that $i^{-1}u$ contains at least n universes as u did. But then X contains a universe that contains at least n universes, so it must contain at least $n + 1$ universes.   □

We remark that this process can be extended to transfinite ordinalities. For instance, one could consider axiomatisations $\mathbf{ZF}_W$ relative to a well-ordered type W with the axiom that W and the class of universes are order-isomorphic. Then it follows that $\mathbf{ZF}_W$ is categorical, subsuming our discussed examples.

6   Discussion

The formalisation of ZF in a type theory with inductive predicates as examined in this work differs from common textbook presentations (cf. [6, 8, 14]) in several ways, most importantly in the use of second-order replacement and the inductive definition of the cumulative hierarchy. Now we briefly outline some of the consequences.

Concerning the second-order version of the replacement axiom, it has been known since Zermelo [22] that second-order ZF admits the embedding theorem for models. It implies that models only vary in their external cardinality, i.e. the notion of cardinality defined by bijections on type level or, equivalently, in height of their cumulative hierarchy. Thus controlling these parameters induces categorical axiomatisations.

As a consequence of categoricity, all internal properties (including first-order undecided statements such as the axiom of choice or the continuum hypothesis) become semantically determined in that there exist no two models such that a property holds in the first but fails in the second (cf. [7,18]). This is strikingly different from the extremely undetermined situation in first-order ZF, where models can be arbitrarily incomparable and linearity of embeddability is only achieved in extremely controlled situations (cf. [5]). This is related to the fact that the inner models admitted by second-order ZF are necessarily universes whereas those of first-order ZF can be subsets of strictly less structure.

The main insight is that the second-order replacement axiom asserts the existence of all subsets of a given set contrarily to only the definable ones guaranteed by a first-order scheme. This fully determines the extent of the power set operation whereas it remains underspecified in first-order ZF. Concretely, first-order ZF admits counterexamples to Lemma 36. Furthermore, the notions of external cardinality induced by type bijections and internal cardinality induced by type bijections that can be encoded as sets coincide in second-order ZF since every bijection witnessing external equipotence of sets can be represented by a replacement set. That the two notions of cardinality differ for first-order set theory has been pointed out by Skolem [11]. The Löwenheim-Skolem Theorem implies the existence of a countable model of first-order ZF (that still contains internally uncountable sets) whereas models of second-order ZF with infinity are provably uncountable.

Inductive predicates make a set-theoretic notion of ordinals in their role as a carrier for transfinitely recursive definitions superfluous. Consider that commonly the cumulative stages are defined by $V_\alpha := \mathcal{P}^\alpha \emptyset$ using transfinite recursion on ordinals α. However, this presupposes at least a basic ordinal theory including the recursion theorem, making the cumulative hierarchy not immediately accessible. That this constitutes an unsatisfying situation has been addressed by Scott [10] where an axiomatisation of ZF is developed from the notion of rank as starting point.

In the textbook approach, the well-ordering of the stages V_α is inherited directly from the ordinals by showing $V_\alpha \subseteq V_\beta$ iff $\alpha \subseteq \beta$. Without presupposing ordinals, we have to prove linearity of $\subseteq$ and the existence of least $\subseteq$-elements directly. As it was illustrated in this work these direct proofs are not substantially harder than establishing the corresponding properties for ordinals.

We end with a remark on our future directions. We plan to first make the axiomatisations $\mathbf{ZF}_W$ precise and formalise the categoricity proof. Subsequently, we will turn to the consistency question and construct actual models following Aczel et al. [2], Werner [19], and Barras [3]. Note that all these implement a flavour of (constructive) second-order ZF whereas Paulson [9] develops classical first-order ZF using the proof assistant Isabelle. We conjecture that the type theory of Coq with excluded middle and a weak form of choice allows for constructing models of $\mathbf{ZF}_n$ for every n. Moreover, it would be interesting to formalise first-order ZF in type theory by making the additional syntax for predicates and relations explicit. Then the classical relative consistency results concerning choice and the continuum hypothesis can be examined.

References

1. Ackermann, W.: Die widerspruchsfreiheit der allgemeinen mengenlehre. Math. Ann. **114**, 305–315 (1937)
2. Aczel, P., Macintyre, A., Pacholski, L., Paris, J.: The type theoretic interpretation of constructive set theory. J. Symb. Log. **49**(1), 313–314 (1984)
3. Barras, B.: Sets in Coq, Coq in sets. J. Formaliz. Reason. **3**(1), 29–48 (2010)
4. Bourbaki, N.: Sur le théorème de Zorn. Arch. Math. **2**(6), 434–437 (1949)
5. Hamkins, J.D.: Every countable model of set theory embeds into its own constructible universe. J. Math. Log. **13**(02) (2013). http://www.worldscientific.com/doi/abs/10.1142/S0219061313500062
6. Hrbacek, K., Jech, T.: Introduction to Set Theory, Third Edition, Revised and Expanded. CRC Press, Boca Raton (1999)
7. Kreisel, G.: Two notes on the foundations of set-theory. Dialectica **23**(2), 93–114 (1969)
8. Kunen, K.: Set Theory an Introduction to Independence Proofs. Elsevier, Amsterdam (2014)
9. Paulson, L.C.: Set theory for verification: I. from foundations to functions. J. Autom. Reason. **11**(3), 353–389 (1993)
10. Scott, D.: Axiomatizing set theory. Proc. Symp. Pure Math. **13**, 207–214 (1974)
11. Skolem, T.: Some remarks on axiomatized set theory. In: van Heijenoort, J. (ed.) From Frege to Gödel: A Sourcebook in Mathematical Logic, pp. 290–301. toExcel, Lincoln, NE, USA (1922)
12. Smolka, G., Schäfer, S., Doczkal, C.: Transfinite constructions in classical type theory. In: Urban, C., Zhang, X. (eds.) ITP 2015. LNCS, vol. 9236, pp. 391–404. Springer, Cham (2015). doi:10.1007/978-3-319-22102-1_26
13. Smolka, G., Stark, K.: Hereditarily finite sets in constructive type theory. In: Blanchette, J.C., Merz, S. (eds.) ITP 2016. LNCS, vol. 9807, pp. 374–390. Springer, Cham (2016). doi:10.1007/978-3-319-43144-4_23
14. Smullyan, R., Fitting, M.: Set Theory and the Continuum Problem. Dover Books on Mathematics. Dover Publications, Mineola (2010)
15. Suppes, P.: Axiomatic Set Theory. Dover Books on Mathematics Series. Dover Publications, Mineola (1960)
16. The Coq Proof Assistant. http://coq.inria.fr
17. Uzquiano, G.: Models of second-order Zermelo set theory. Bull. Symb. Log. **5**(3), 289–302 (1999)
18. Väänänen, J.: Second-order logic or set theory? Bull. Symb. Log. **18**(1), 91–121 (2012)
19. Werner, B.: Sets in types, types in sets. In: Abadi, M., Ito, T. (eds.) TACS 1997. LNCS, vol. 1281, pp. 530–546. Springer, Heidelberg (1997). doi:10.1007/BFb0014566
20. Williams, N.H.: On Grothendieck universes. Compos. Math. **21**(1), 1–3 (1969)
21. Zermelo, E.: Neuer beweis für die möglichkeit einer wohlordnung. Math. Ann. **65**, 107–128 (1908)
22. Zermelo, E.: Über Grenzzahlen und Mengenbereiche: Neue Untersuchungen Über die Grundlagen der Mengenlehre. Fund. Math. **16**, 29–47 (1930)

Making PVS Accessible to Generic Services by Interpretation in a Universal Format

Michael Kohlhase[1], Dennis Müller[1(✉)], Sam Owre[2], and Florian Rabe[3]

[1] Computer Science, FAU Erlangen-Nürnberg, Erlangen, Germany
[2] SRI Palo Alto, Menlo Park, USA
d.mueller@kwarc.info
[3] Computer Science, Jacobs University Bremen, Bremen, Germany

Abstract. PVS is one of the most powerful proof assistant systems and its libraries of formalized mathematics are among the most comprehensive albeit under-appreciated ones. A characteristic feature of PVS is the use of a very rich mathematical and logical foundation, including e.g., record types, undecidable subtyping, and a deep integration of decision procedures. That makes it particularly difficult to develop integrations of PVS with other systems such as other reasoning tools or library management periphery.

This paper presents a translation of PVS and its libraries to the OMDoc/MMT framework that preserves the logical semantics and notations but makes further processing easy for third-party tools. OMDoc/MMT is a framework for formal knowledge that abstracts from logical foundations and concrete syntax to provide a universal representation format for formal libraries and interface layer for machine support. Our translation allows instantiating generic OMDoc/MMT-level tool support for the PVS library and enables future translations to libraries of other systems.

1 Introduction

Motivation. One of the most critical bottlenecks in the field of interactive theorem proving is the lack of interoperability between proof assistants and related tools. This leads to a duplication of efforts: both formalizations and auxiliary tool support (e.g., for automated proving, library management user interfaces) cannot be easily shared between systems.

In both areas, previous work has shown significant potential for knowledge sharing. Regarding formalizations, library translations such as [KW10, OS06, KS10] have been used to transport theorems across systems, and alignments have been used to match corresponding declarations in different libraries [GK14]. Regarding tool support, Isabelle's sledgehammer component [MP08] provides a generic way to integrate different automation tools, and Dedukti [BCH12] has been used as an independent proof checker for various proof assistant libraries. A great example is premise selection, e.g., based on machine-learning [KU15]: a single tool can be used for every proof assistant—provided the language and

© Springer International Publishing AG 2017
M. Ayala-Rincón and C.A. Muñoz (Eds.): ITP 2017, LNCS 10499, pp. 319–335, 2017.
DOI: 10.1007/978-3-319-66107-0_21

library are available in a universal format that can be plugged into the generic selection algorithm.

Unfortunately, the latter point—the universal format—is often prohibitively expensive for many interesting applications. Firstly, it is extremely difficult to design a universal format that strikes a good trade-off between simplicity and universality. And secondly, even in the presence of such a format, it is difficult to implement the export of a library into the universal format. Here it is important to realize that any export attempt is doomed that uses a custom parser or type checker for the library—only the internal data structures maintained by the proof assistant are informative enough for most use cases. Consequently, only expert developers can perform this step. Of these, each proof assistant community only has very few.

In previous work, the authors have developed such a universal format [Koh06, RK13, KR16] for formal knowledge: OMDoc is an XML language geared towards making formula structure and context dependencies explicit while remaining independent of the underlying logical formalism. We also built a strong implementation—the MMT system—and a number of generic services, for example [Rab14a, KS06]. We have already successfully applied our approach to Mizar in [Ian+13] and HOL Light in [KR14]. In both cases, we systematically (i) manually defined the logic of the proof assistant in a logical framework, and (ii) instrumented the proof assistant to export its libraries. The OMDoc/MMT language provides the semantics that ties together the three involved levels (logical framework, logic, and library) and the implementation provides a uniform high-level API for further processing. Critically, the exports systematically avoid any (deep) encoding of logical features. That is important so that further processing can work with the exact same structure apparent to a user of the proof assistant.

Contribution. We apply our approach to PVS [ORS92]: we present a definition of the PVS logic in OMDoc/MMT and an export feature for PVS libraries. We exemplify the latter by exporting the Nasa Library [Lan16b], the largest and most important library of PVS. The translated libraries are available at [PVS].

Finally, we present several applications that instantiate MMT-level services for PVS libraries. Notably, even though the export itself is our main contribution, these applications immediately yield added-value for PVS users. Firstly, we instantiate generic library management facilities for browsing both the content and theory graphs of PVS libraries. Secondly, our most advanced application instantiates MathWebSearch [KS06], a substitution tree–based search engine, for PVS libraries. Here users enter search queries and see search results inside PVS, and MathWebSearch performs the actual search; neither tool is aware of the respective other, and MMT provides the high-level interface that allows semantics-aware mediation between these tools.

Related Work. The Logosphere project [Pfe+03] already aimed at a similar export from PVS. Both the definition of the PVS logic in LF and the export of the library turned out to be too difficult at the time: the definition had to omit, e.g., record types and the module system, thus making any export impossible.

Independent of our work, Frederic Gilbert is pursuing a very similar export of PVS into Dedukti [BCH12] that appears to be unpublished as of this writing. Its primary interest is the independent verification of PVS libraries. An interesting difference to our approach is that Dedukti is a fixed, simple logical framework that requires a non-trivial (deep) encoding of some advanced PVS features (e.g., predicate subtyping); as we discuss in Sect. 3, our approach uses a more complex, adaptive logical framework that allows for translations of the PVS library without such encodings.

Hypatheon [Lan16a] uses SQL for indexing PVS theories and making them searchable via a GUI client. It renders proof-side assistance by finding suitable lemmas within PVS libraries, retrieving other declarations, and viewing the full theories that contain them. However, it has no access to the fully type-checked and disambiguated libraries.

Overview. We briefly recap PVS in Sect. 2. Then we describe the definition of the PVS logic in our framework in Sect. 3 and of the PVS library in Sect. 4. Building on this, we present our applications in Sect. 5 and conclude in Sect. 6.

2 Preliminaries

PVS [ORS92] is a verification system, combining language expressiveness with automated tools. The language is based on higher-order logic, and is strongly typed. The language includes types and terms such as: numbers, records, tuples, functions, quantifiers, and recursive definitions. Full predicate subtypes are supported, which makes type checking undecidable; PVS generates type obligations (TCCs) as artefacts of type checking. For example, division is defined such that the second argument is nonzero, where nonzero is defined:

nonzero_real: TYPE = {r: real | r /= 0}

Note that functions in PVS are total; partiality is only supported via subtyping.

Beyond this, the PVS language has structural subtypes (i.e., a record that adds new fields to a given record), dependent types for record, tuple, and functions, recursive and co-recursive datatypes, inductive and co-inductive definitions, theory interpretations, and theories as parameters, conversions, and judgements that provide control over the generation of proof obligations. Specifications are given as collections of parameterized theories, which consist of declarations and formulas, and are organized by means of imports.

The PVS prover is interactive, but with a large amount of automation built in. It is closely integrated with the type checker, and features a combination of decision procedures, BDDs, automatic simplification, rewriting, and induction. There are also rules for ground evaluation, random test case generation, model checking, and predicate abstraction. The prover may be extended with user-defined proof strategies.

PVS has been used as a platform for integration. It has a rich API, making it relatively easy to add new proof rules and integrate with other systems. Examples of this include the model checker, Duration Calculus, MONA, Maple, Ag, and Yices. The system is normally used through a customized Emacs interface, though it is possible to run it standalone (PVSio does this), and PVS features an XML-RPC server (developed independently of the work presented here) that will allow for more flexible interactions. PVS is open source, and is available at http://pvs.csl.sri.com.

As a running example, Fig. 1 gives a part of the PVS theory defining *equivalence closures* on a type T in its original syntax. PVS uses upper case for keywords and logical primitives; square brackets are used for type and round brackets for term arguments. The most important declarations in theories are (i) includes of other theories, e.g., the binary subset predicate subset? and the type equivalence of equivalence relations on T are included from the theories sets and relations (These includes are redundant in the PVS prelude and added here for clarity.), (ii) typed identifiers, possibly with definitions such as EquivClos, and (iii) named theorems (here with omitted proof) such as EquivClosSuperset.

VAR declarations are one of several non-logical declarations: they only declare variable types, which can then be omitted later on; here PRED[[T,T]] abbreviates the type of binary relations on T.

```
EquivalenceClosure[T : TYPE] : THEORY
BEGIN
   IMPORTING sets, relations
   R: VAR PRED[[T, T]]
   x, y : VAR T
   EquivClos(R) : equivalence[T] =
      { (x, y) | FORALL(S : equivalence[T]) : subset?(R, S) IMPLIES S(x, y) }
   EquivClosSuperset : LEMMA
      subset?(R, EquivClos(R))
   ...
END EquivalenceClosure
```

Fig. 1. The PVS prelude in the MathHub browser

3   Defining the PVS Logic in a Logical Framework

Defining the PVS logic in a logical framework is a significant challenge. Therefore, we start by giving an overview of the difficulties before describing our approach.

Difficulties. A logical framework like LF [HHO93], Dedukti [BCH12], or λ-Prolog [MN86] tends to admit very elegant definitions for a certain class of logics, but definitions can get awkward quickly if logics fall outside that fragment.

This often boils down to the question of shallow vs. deep encodings. The former represents a logic feature (e.g., subtyping) in terms of a corresponding framework feature, whereas the latter applies a logic encoding to remove the feature (e.g., encode subtyping in a logical framework without subtyping by using a subtyping predicate and coercion functions). Deep encodings have two disadvantages: (i) They destroy structure of the original formalization, often in a way that is not easily invertible and blows up the complexity of library translations. (ii) They require the library translation to apply non-trivial and error-prone steps that become part of the trusted code base. In fact, multiple logical frameworks (including Dedukti) were specifically designed to have a richer logical framework that allows for more logics to be defined elegantly.

Even if we ignore the proof theory (and thus the use of decision procedures) entirely, PVS is particularly challenging in this regard. The sequel describes the most important challenges.

The PVS typing relation is undecidable due to predicate subtyping: selecting a sub-type of α by giving a predicate p as in $\{x \in \alpha \mid p(x)\}$. Thus, a shallow encoding is impossible in any framework with a decidable typing relation. The most elegant solution is to design a new framework that allows for undecidable typing and then use a shallow encoding.

PVS uses anonymous record types (like in SML) as a primitive feature. This includes record subtyping and a complex variant of record/product/function updates. A deep encoding of anonymous record types is extremely awkward: the simplest encoding would be to introduce a new named product type for every occurrence of a record type in the library. Even then it is virtually impossible to formalize an axiom like "two records are equal if they agree up to reordering of fields" elegantly in a declarative logical framework. Therefore, the most feasible option again is to design a new framework that has anonymous record types as a primitive.

PVS uses several types of built-in literals, namely arbitrary-precision integers, rational numbers, and strings. Not every logical framework provides exactly the same set of built-in types and operations on them.

PVS allows for (co)inductive types. While these are relatively well-understood by now, most logical frameworks do not support them. And even if they do, they are unlikely to match the idiosyncrasies of PVS such as declaring a predicate subtype for every constructor. Again it is ultimately more promising to mimic PVS's idiosyncrasies in the logical framework so that we can use a shallow encoding.

PVS uses a module system that, while not terribly complex, does not align perfectly with the modularity primitives of existing logical frameworks. Concretely, theories are parametric, and a theory may import the same theory multiple times with different parameters, in which case any occurrence of an imported symbol is ambiguous. Simple deep encodings can duplicate the multiply-imported symbols or treat them as functions that are applied to the parameters. Both options seem feasible at first but ultimately do not scale well – already the PVS Prelude (the small library of PVS built-ins) causes difficulties. This led us (contrary to our original plans) to mimic PVS-style parametric imports in the logical framework as well to allow for a shallow encoding.

A Flexible Logical Framework. We have extensively investigated definitions of PVS in logical frameworks, going back more than 10 years when a first (unpublished) attempt to define PVS in LF was made by Schürmann as part of an ongoing collaboration. In the end, all of the above-mentioned difficulties pointed us in the same direction: the logical framework must adapt to the complexity of PVS – any attempt to adapt PVS to an existing logical framework (by designing an appropriate deep encoding) is likely to be doomed. This negative result is in itself a notable contribution of this paper. It is likely to apply also to similarly complex object logics such as Coq.

If done naively, developing a new framework that permits a shallow encoding would scale badly: it would lack mature implementation support and would not make future integrations of PVS with other provers any easier. Therefore, we have spent several years developing the MMT framework. It is born out of the tradition of logical frameworks but systematically allows future extensions of the framework. Its main strength is that such extensions, e.g., the features needed for PVS, can be added at comparatively low cost: whereas most logical frameworks would require reimplementing most parts starting from the kernel, MMT allows plugging in language features as a routine part of daily development. Importantly, all MMT level automation (including parsing, type reconstruction, and IDE, which are crucial for writing logic definitions in practice) is generic and thus remains applicable even when new language features are added. Moreover, MMT supports modular composition of language features so that all developments we made for PVS can be reused when working with other provers.

It is beyond the scope of this paper to present the architecture of MMT, and we only sketch the extension pathways most critical for PVS.

Firstly, MMT expressions are generic syntax trees including variable binding and labeling [Rab14b]. Besides constants (global identifiers) and bound variables (local identifiers), the leaves of the syntax tree may also be arbitrary literals [Rab15]. An MMT theory defining the language T declares one constant for each expression constructor of T. For example, the MMT theory for LF declares 4 symbols for type, λ, Π, and application.

Secondly, the key algorithms of the MMT kernel – including parsing, type reconstruction, and computation – are rule-based [Rab17]. Each rule is an object in the underlying programming language that can be generated from a declarative formulation or (in the general case) implemented directly. In either case, the current context determines which rules are used. For example, the MMT theory for LF declares three parsing rules, ten typing/equality rules, and one computation rule. Together, these are sufficient to recover type reconstruction for LF.

Thirdly, MMT allows for *derived* declarations [Ian17]. Each derived declaration indicates the language *feature* that defines its semantics. And the individual features can be easily implemented by MMT plugins, usually by elaborating derived declarations to more primitive ones. For example, we can declare the feature of inductive types, as a derived declaration containing the constructors in its body. Notably, while elaboration defines the meaning of the derived declaration, many MMT algorithms can work with the unelaborated version, e.g., by supplying appropriate induction rules to the type reconstruction algorithm.

Defining PVS. To define the language of PVS in MMT, we carried out two steps.

Firstly, we designed a logical framework that extends LF with three features: anonymous record types, predicate subtypes, and imports of multiple instances of the same parametric theory. We use MMT to build this framework modularly. LF (which already existed in MMT) and each of the three new features are defined in a separate MMT theory, each including a few constants and rules for them. Finally, we import all of them to obtain the new logical framework LFX. Then we use LFX to define the MMT theory for PVS. The sequel lists the constants and rules in this theory.

```
tp : type
expr : tp  type # 1 prec −1
tpjudg : {A} expr A  tp  type # 2 : 3

pvspi : {A} (expr A  tp)  tp # 2
fun_type : tp  tp  tp = [A,B] [x: expr A] B # 1 2
pvsapply : {A,f : expr A  tp} expr ( f)  {a:expr A} expr (f a) # 3 ( 4 ) prec −1000015
lambda : {A,f : expr A  tp} ({a:expr A}expr (f a))  expr ( f) # 3
```

Fig. 2. Some basic typing related symbols for PVS

We begin with a definition of PVS's higher-order logic using only LF features. This includes dependent product and function types[1], classical booleans, and the usual formula constructors (see Fig. 2). This is novel in how exactly it mirrors the syntax of PVS (e.g., PVS allows multiple aliases for primitive constants) but requires no special MMT features.

We declare three constants for the three types of built-in literals together with MMT rules for parsing and typing them. Using the new framework features, we give a shallow encoding of predicate subtyping (see Fig. 3 for the new typing rule), a shallow definition of anonymous record types, as well as new declarations for PVS-style inductive and co-inductive types.

4   Translating the PVS Library

The PVS library export requires three separate developments:

Firstly, PVS has been extended with an XML export. This is similar to the LaTeX extension in PVS, which is built on the Common Lisp Pretty Printing

---

[1] Contrary to typical dependently-typed languages, PVS does not allow declaring dependent *base* types, but predicate subtyping can be used to introduce types that depend on terms. Interestingly, this is neither weaker nor stronger than the dependent types in typical $\lambda\Pi$ calculi.

```
setsub : {A} (expr (A  bool))  tp #  1 | 2
rule rules?SetsubRule
```

```
object SetsubRule extends ComputationRule(PVSTheory.expr.path) {
  def apply(check: CheckingCallback)(tm: Term, covered: Boolean)
   (implicit stack: Stack, history: History): Option[Term]
  = tm match {
   case expr(PVSTheory.setsub(tp,prop)) =>
     Some(LFX.Subtyping.predsubtp(expr(tp),proof("internal_judgment",
       Lambda(doName,expr(tp),pvsapply(prop,OMV(doName),expr(tp),
                    bool.term)._1))))
   case _ => None
  }
}
```

Fig. 3. PVS-style predicate subtyping in MMT and the corresponding rule

facility. The XML export was developed in parallel with a Relax NG specification for the PVS XML files. Because PVS allows overloading of names, infers theory parameters, and automatically adds conversions, the XML generation is driven from the internal type-checked abstract syntax, rather than the parse tree. Thus the generated XML contains the fully type-checked form of a PVS specification with all overloading disambiguated. Future work on this will include the generation of XML forms for the proof trees.

Secondly, we documented the XML schema used by PVS as a set of inductive types in Scala (the programming language underlying MMT). We wrote a generic XML parser in Scala that generates a schema-specific parser from such a set of inductive types (see Fig. 5 for part of the specification). That way any change to the inductive types automatically changes the parser. While seemingly a minor implementation detail, this was critical for feasibility because the XML schema changed frequently along the way.

Thirdly, we wrote an MMT plugin that parses the XML files generated by PVS and turns them into MMT content. This includes creating various generic indexes that can be used later for searching the content.

All processing steps preserve source references, i.e., URLs that point to a location (file and line/column) in a source file (the quatruples of numbers at place= in Fig. 4 and <link rel="...?sourceRef" in Fig. 6).

The table in Fig. 7 gives an overview of the sizes of the involved libraries and the run times[2] of the conversion steps. We note that the XML encoding considerably increases the size of representations. This is due to two effects: the internal, disambiguated form contains significantly more information than the user syntax (e.g. theory parameter instances and reconstructed types), and XML as a machine-oriented format is naturally more verbose. Furthermore, OMDoc uses OpenMath for term structures, which again increases file size. In practice

---

[2] All numbers measured on standard laptops.

```
<theory place="6049_0_6075_22">
 <id>EquivalenceClosure</id>
 <const−decl place="6057_2_6058_75">
   <id>EquivClos</id>
   <arg−formals>
     <binding place="6057_12_6057_13">
       <id>R</id>
       <type−name>
         <id>PRED</id>
         <actuals>
           <tuple−type>
             <type−name><id>T</id></type−name>
             <type−name><id>T</id></type−name>
           </tuple−type>
         </actuals>
       </type−name>
     </binding>
   </arg−formals>
   <type−name place="6057_17_6057_31">
     <id>equivalence</id>
     <actuals>
       ...
```

Fig. 4. A part of the function EquivalenceClosure/EquivClos in XML

```
case class const_decl(
    named: ChainedDecl,
    arg_formals: List[bindings],
    tp: DeclaredType,
    _def: Option[Expr]
) extends Decl
```

Fig. 5. The scala-specification of PVS constant declarations for XML parsing

```
<omdoc>
 <theory name="EquivalenceClosure"
  base="http://pvs.csl.sri.com/prelude"
  meta="http://pvs.csl.sri.com/?PVS">
 <constant name="EquivClos">
  <type>
   <om:OMOBJ>
    <om:OMA>
     <om:OMS base="http://cds.omdoc.org/urtheories" module="LambdaPi" name="apply"/>
     <om:OMS base="http://pvs.csl.sri.com/" module="PVS" name="expr"/>
     <om:OMA>
      <om:OMS base="http://cds.omdoc.org/urtheories" module="LambdaPi" name="apply"/>
      <om:OMS base="http://pvs.csl.sri.com/" module="PVS" name="pvspi"/>
      ...
      <metadata>
       <link rel="http://cds.omdoc.org/mmt?metadata?sourceRef"
         resource="prelude/pvsxml/EquivalenceClosure.xml#−1.6057.17:−1.6057.31"/>
      </metadata>
     </om:OMA>
```

Fig. 6. A part of the function EquivalenceClosure/EquivClos in OMDoc

the file sizes are no problem for the MMT tools presented here, so we consider file sizes as a (small) price to be paid for interoperability and universal tool support.

	PVS source		PVS → XML		XML → OMDoc	
	size/gz	check time	result size/gz	run time	result size/gz	run time
Prelude	189.7/46.6kB	33s	23.5/.67MB	11s	83.3/1.6MB	3m41s
NASA Lib	1.9/.426MB	23m25s	387.2/8.9MB	3m11s	2.5/.04GB	58m56s

Fig. 7. File sizes of the PVS import at various stages

5   Applications

With the OMDoc/MMT translation of the PVS libraries, PVS gains access to library management facilities implemented at the OMDoc/MMT level. There are two ways to exploit this: publishing the converted PVS libraries on a dedicated server, like our MathHub system, or running the OMDoc/MMT toolstack locally alongside PVS. Both options offer similar functionality, the main difference is the intended audience: the first option is for outside users who want to access the PVS libraries, and the latter is for PVS users who develop new content or refactor the library.

MathHub [Ian+14,MH] bundles a GitLab-based repository manager with MMT and various periphery systems into a common, web-based user interface. We commit the exported PVS libraries as OMDoc/MMT files into the repository [GMP] as separate libraries — currently Prelude and NASA. MathHub has been configured to make these available via the (i) MathHub user interface, (ii) MMT presentation web server, (iii) MMT web services, and (iv) the MathWeb-Search daemon. All of these components give the user different ways of interacting with the system and PVS content. Below we explore three that are directly useful for PVS users.

The local workflow installs OMDoc/MMT tools on the same machine as PVS. In that case, users are able to browse the current version of the available PVS libraries including all experimental or private theories that are part of the current development. This also enables PVS to use OMDoc/MMT services as background tools that remain transparent to the PVS user.

In both workflows, OMDoc/MMT-based periphery systems become available to the PVS user that are either not provided by the PVS tools or in a much restricted way. We will go over the three most important ones in detail.

5.1   Browsing and Interaction

The transformed PVS content can be browsed interactively in the document-oriented MathHub presentation pages (theories as active documents) and in the

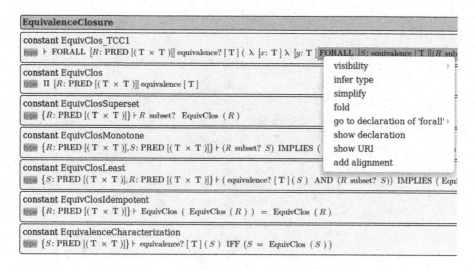

Fig. 8. The PVS prelude in the MMT browser

MMT web browser (see Fig. 8). Both allow interaction with the PVS content via a generic Javascript-based interface. This provides buttons to toggle the visibility of parts computed by PVS – e.g. omitted types and definitions – at the declaration level. The right-click menu shown in Fig. 8 is specific to the selected sub-formula (highlighted in gray); here we have eight applicable interactions which range from inferring the subformula type via definition lookup to management actions such as registering an alignment to concepts in other libraries. New interactions can be added as they become available in the MMT system.

The MMT instance in the local workflow provides the additional feature of inter-process communication between PVS and MMT as a new menu item: the action *navigate to this declaration in connected systems*. We implemented a listener for this action that forwards the command to PVS via an XML-RPC call at the default PVS port. Correspondingly, we implemented a case in the PVS server that opens the corresponding file in the PVS emacs system and navigates to the relevant line.

5.2   Graph Viewer

MathHub includes a theory graph viewer that allows interactive, web-based exploration of the OMDoc/MMT theory graphs. It builds on the `visjs` JavaScript visualization library [VJS], which uses the HTML5 canvas to layout and interact with graphs client-side in the browser.

PVS libraries make heavy use of theories as a structuring mechanism, which makes a graph viewer for PVS particularly attractive. Figure 9 shows the full graph in a central-gravity layout induced by the PVS prelude, where we have (manually) clustered the subgraphs for bit vectors and finite sets (the orange barrel-shaped nodes). The lower right corner shows a zoomed-in fragment.

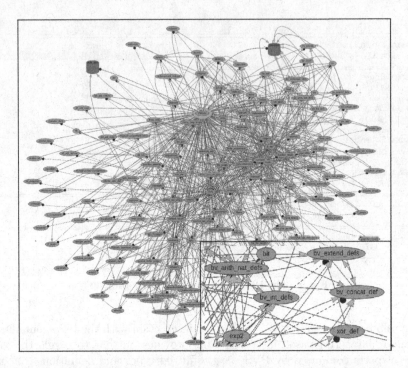

Fig. 9. The basic PVS libraries in the MathHub theory graph viewer (Color figure online)

The theory graph allows dragging nodes around to fine-tune the layout. Hovering over a node or edge triggers a preview of the theory. All nodes support the same context menu actions in the graph viewer as the corresponding theories do in the browser above. Thus, it is possible to select a theory in the graph viewer and then navigate to it in the browser or (if run locally) in the PVS system.

5.3 Search

MathWebSearch [KS06] is an OMDoc/MMT-level formula search engine that uses query variables for subterms and first-order unification as the query language. It is developed independently, but MMT includes a plugin for generating Math-WebSearch index files using its content MathML interface. Thus, any library available to MMT can be indexed and searched via MathWebSearch. Moreover, MMT includes a frontend for MathWebSearch so that search queries can be supplied in any format that MMT can understand, e.g., the XML format produced by PVS.

MMT exposes the search frontend both in its GUI for humans and as an HTTP service for other systems. Here we use the latter: We have added a feature to the PVS emacs interface that allows users to enter a search query in PVS syntax. PVS parses the query, type-checks it, and converts it to XML. The XML

```
<mws:query limitmin="0" answsize="1000" totalreq="yes"
  output="xml" xmlns:m="http://www.w3.org/1998/Math/MathML"
  xmlns:mws="http://www.mathweb.org/mws/ns">
 <mws:expr>
 <apply>
  <csymbol>http://cds.omdoc.org/urtheories?LambdaPi?apply</csymbol>
  <csymbol>http://pvs.csl.sri.com/?PVS?pvsapply</csymbol>
  <mws:qvar xmlns:mws="http://www.mathweb.org/mws/ns">l1</mws:qvar>
  <mws:qvar xmlns:mws="http://www.mathweb.org/mws/ns">l2</mws:qvar>
  <csymbol>http://pvs.csl.sri.com/prelude?EquivalenceClosure?EquivClos</csymbol>
  <mws:qvar xmlns:mws="http://www.mathweb.org/mws/ns">A</mws:qvar>
 </apply>
 </mws:expr>
</mws:query>
```

Fig. 10. A query for applications of EquivClos

is sent to Mmt, which acts as the mediator between the proof assistant—here PVS—and library management periphery—here MathWebSearch—and returns the search results to PVS.

```
[ {"lib_name" : "",
   "theory_name" : "EquivalenceClosure",
   "name" : "EquivClosMonotone",
   "Position" : "3_2_5_5_5_5"},
  {"lib_name" : "",
   "theory_name" : "EquivalenceClosure",
   "name" : "EquivalenceCharacterization",
   "Position" : "2_2_5_5_5_5"},
   ...
]
```

Fig. 11. A query result for Fig. 10

The PVS user enters the PVS query EquivClos(?A), where we have extended the PVS syntax with query variables like ?A. After OMDoc/Mmt translation, this becomes the MathWebSearch query in Fig. 10—note the additional symbols from LF introduced by the representation in the logical framework. The representation also introduces unknown meta-variables for the domain and range of the EquivClos function, which become the additional query variables l1 and l1. MathWebSearch returns a JSON record with all results, and we show the first two in Fig. 11: two occurrences of (instances of) EquivClos(?A) in two declarations in the theory EquivalenceClosure Fig. 1. The attribute lib_name is the name of the library; by PVS convention, it is empty for the Prelude. The attributes theory_name and name give the declaration that contains the match, and Position gives the path to its subterm that matched the query.

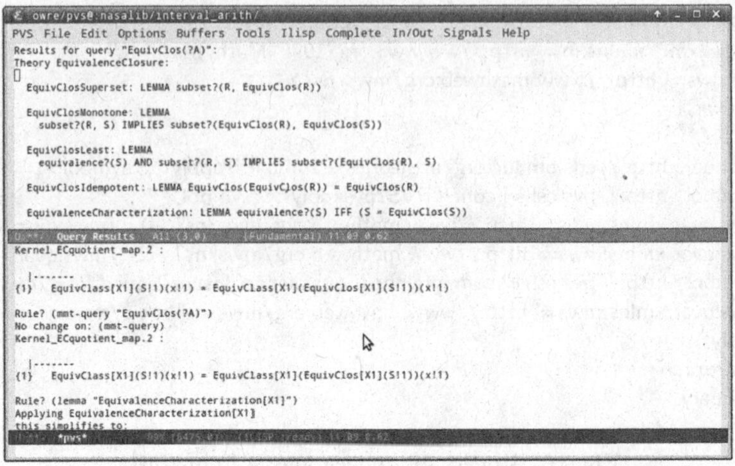

Fig. 12. Example for displaying the query result in PVS

Figure 12 shows what the query will look like while doing a PVS proof. The current implementation is just a proof-of-concept—for the mature version the part of PVS that sends the query to the MMT server and displays the results still has to be implemented thoroughly. But the remaining steps are straightforward.

Future work will exploit this functionality to search specifically for existing theorems that may be helpful in a specific part of an ongoing PVS proof.

6   Conclusion

The work reported in this paper contributes to avoiding duplication of efforts in the development of theorem proving systems, their libraries, and supporting periphery systems. Specifically, we have developed a representation of the PVS logic in the OMDoc/MMT representation format, as well as an automated translation of the PVS libraries into OMDoc/MMT; the result is available at [PVS].

In contrast to earlier representation and translation projects undertaken by us—e.g. Mizar and HOL Light—the PVS logic is much more challenging due to its highly expressive language features, which defy formalization in current logical frameworks like LF. Therefore we make use of the extensibility of the OMDoc/MMT system and implement several extensions of LF (LFX). In essence, we use the MMT system as a prototyping system for logical frameworks. Our experience with encoding the PVS logic is that critical features such as undecidable subtyping, record types, (co)inductive types and literals can naturally be expressed at this level. While LFX is less well-understood than established logical frameworks, it already proved very useful as a development tool. Most importantly, we use it to give shallow and therefore structure-preserving encodings of PVS features without having to forgo the advantages of logical frameworks.

This information architecture is essential for system interoperability. In our case we have shown that we can use the generic—i.e. language-independent— MMT tool chain for PVS. Concretely we have instantiated three generic periphery systems for PVS: a library browser, a theory graph viewer, and a search engine. Given the OMDoc/MMT representation of the PVS libraries, these directly work for PVS libraries and can be easily plugged together with the PVS system. This supplements and improves on the existing functionality that was designed specifically for PVS such as the Hypatheon system [Lan16a].

Our work immediately enables three kinds of future work. Firstly, it makes the PVS libraries available for existing generic services developed by other researchers. For example, it becomes much easier to apply machine learning–based premise selection as in [KU15] to PVS. Secondly, it applies also to all the other theorem proving libraries that have been translated to OMDoc/MMT: Besides HOL Light and Mizar, we also have experimental translations of TPS, TPTP, TPS, Focalize, Specware, IMPS, and Metamath, as well as several informal mathematical libraries including the OEIS and the SMGloM terminology base. Using flexible alignments [Kal+16] between the libraries, we can guide library developers to corresponding parts of other formalizations, approximately translate the content across libraries, or reuse notations (e.g. to show HOL Light content in a form that looks familiar to PVS users). Finally, while PVS does not store full proof terms, it stores enough information to export good proof sketches. Besides being an important sanity-check for the correctness of the translation, this would help transporting PVS proofs to other provers. We plan to revisit this issue after designing a good representation language for proof sketches.

Acknowledgements. This work has been partially funded by DFG under Grants KO 2428/13-1 and RA-18723-1. The authors gratefully acknowledge the contribution of Marcel Rupprecht, who has extended the graph viewer for this paper.

References

[BCH12] Boespflug, M., Carbonneaux, Q., Hermant, O.: The $\lambda\Pi$-calculus modulo as a universal proof language. In: Pichardie, D., Weber, T. (eds.) Proceedings of PxTP2012: Proof Exchange for Theorem Proving, pp. 28–43 (2012)

[GK14] Gauthier, T., Kaliszyk, C.: Matching concepts across HOL libraries. In: Watt, S.M., Davenport, J.H., Sexton, A.P., Sojka, P., Urban, J. (eds.) CICM 2014. LNCS, vol. 8543, pp. 267–281. Springer, Cham (2014). doi:10.1007/978-3-319-08434-3_20

[GMP] MathHub PVS Git Repository. http://gl.mathhub.info/PVS. Accessed 11 Apr 2017

[HHO93] Harper, R., Honsell, F., Plotkin, G.: A framework for defining logics. J. Assoc. Comput. Mach. 40(1), 143–184 (1993)

[Ian+13] Iancu, M., et al.: The Mizar mathematical library in OMDoc: translation and applications. J. Automated Reason. 50(2), 191–202 (2013). doi:10.1007/s10817-012-9271-4

[Ian+14] Iancu, M., Jucovschi, C., Kohlhase, M., Wiesing, T.: System description: MathHub.info. In: Watt, S.M., Davenport, J.H., Sexton, A.P., Sojka, P., Urban, J. (eds.) CICM 2014. LNCS, vol. 8543, pp. 431–434. Springer, Cham (2014). doi:10.1007/978-3-319-08434-3_33. http://kwarc.info/kohlhase/papers/cicm14-mathhub.pdf. ISBN 978-3-319-08433-6

[Ian17] Iancu, M.: Towards flexiformal mathematics. Ph.D. thesis. Jacobs University, Bremen (2017)

[Kal+16] Kaliszyk, C., et al.: A standard for aligning mathematical concepts. In: Kohlhase, M. et al. (eds.) Intelligent Computer Mathematics – Work in Progress Papers (2016). http://kwarc.info/kohlhase/papers/cicmwip16-alignments.pdf

[Koh06] Kohlhase, M.: OMDoc: An Open Markup Format for Mathematical Documents (Version 1.2). Lecture Notes in Artificial Intelligence, vol. 4180. Springer, Heidelberg (2006)

[KR14] Kaliszyk, C., Rabe, F.: Towards knowledge management for HOL light In: Watt, S.M., Davenport, J.H., Sexton, A.P., Sojka, P., Urban, J. (eds.) CICM 2014. LNCS, vol. 8543, pp. 357–372. Springer, Cham (2014). doi:10.1007/978-3-319-08434-3_26. http://kwarc.info/frabe/Research/KR_hollight_14.pdf. ISBN 978-3-319-08433-6

[KR16] Kohlhase, M., Rabe, F.: QED reloaded: towards a pluralistic formal library of mathematical knowledge. J. Formalized Reason. 9(1), 201–234 (2016)

[KS10] Krauss, A., Schropp, A.: A mechanized translation from higher-order logic to set theory. In: Kaufmann, M., Paulson, L.C. (eds.) ITP 2010. LNCS, vol. 6172, pp. 323–338. Springer, Heidelberg (2010). doi:10.1007/978-3-642-14052-5_23

[KU15] Kaliszyk, C., Urban, J.: HOL(y)Hammer: online ATP service for HOL light. Math. Comput. Sci. 9(1), 5–22 (2015)

[KW10] Keller, C., Werner, B.: Importing HOL light into Coq. In: Kaufmann, M., Paulson, L.C. (eds.) ITP 2010. LNCS, vol. 6172, pp. 307–322. Springer, Heidelberg (2010). doi:10.1007/978-3-642-14052-5_22

[KS06] Kohlhase, M., Sucan, I.: A search engine for mathematical formulae. In: Calmet, J., Ida, T., Wang, D. (eds.) AISC 2006. LNCS, vol. 4120, pp. 241–253. Springer, Heidelberg (2006). doi:10.1007/11856290_21

[Lan16a] NASA Langley. Hypatheon: A Database Capability for PVS Theories (2016). https://shemesh.larc.nasa.gov/people/bld/hypatheon.html

[Lan16b] NASA Langley. NASA PVS Library (2016). http://shemesh.larc.nasa.gov/fm/ftp/larc/PVS-library/pvslib.html

[MH] MathHub.info: Active Mathematics. http://mathhub.info. Accessed 28 Jan 2014

[MN86] Miller, D.A., Nadathur, G.: Higher-order logic programming. In: Shapiro, E. (ed.) ICLP 1986. LNCS, vol. 225, pp. 448–462. Springer, Heidelberg (1986). doi:10.1007/3-540-16492-8_94

[MP08] Meng, J., Paulson, L.: Translating higher-order clauses to first-order clauses. J. Automated Reason. 40(1), 35–60 (2008)

[ORS92] Owre, S., Rushby, J.M., Shankar, N.: PVS: a prototype verification system. In: Kapur, D. (ed.) CADE 1992. LNCS, vol. 607, pp. 748–752. Springer, Heidelberg (1992). doi:10.1007/3-540-55602-8_217

[OS06] Obua, S., Skalberg, S.: Importing HOL into Isabelle/HOL. In: Furbach, U., Shankar, N. (eds.) IJCAR 2006. LNCS, vol. 4130, pp. 298–302. Springer, Heidelberg (2006). doi:10.1007/11814771_27

[Pfe+03] Pfenning, F., et al.: The Logosphere Project (2003). http://www.logosphere.
          org/

[PVS]   The PVS libraries in OMDoc/MMT format. https://gl.mathhub.info/PVS.
          Accessed 29 May 2017

[Rab14a] Rabe, F.: A logic-independent IDE. In: Benzmüller, C., Woltzenlogel Paleo,
          B. (eds.) Workshop on User Interfaces for Theorem Provers, pp. 48–60
          (2014). Elsevier

[Rab14b] Rabe, F.: How to identify, translate, and combine logics? J. Logic Comput.
          (2014). doi:10.1093/logcom/exu079

[Rab15]   Rabe, F.: Generic literals. In: Kerber, M., Carette, J., Kaliszyk, C., Rabe,
          F., Sorge, V. (eds.) CICM 2015. LNCS, vol. 9150, pp. 102–117. Springer,
          Cham (2015). doi:10.1007/978-3-319-20615-8_7

[Rab17]   Rabe, F.: A Modular Type Reconstruction Algorithm (2017). http://kwarc.
          info/frabe/Research/rabe_recon_17.pdf

[RK13]   Rabe, F., Kohlhase, M.: A scalable module system. Inf. Comput. **230**(1),
          1–54 (2013)

[VJS]   vis.js - A dynamic, browser based visualization library. http://visjs.org.
          Accessed 04 May 2017

[Wat+14] Watt, S.M., et al. (eds.) Intelligent Computer Mathematics. LNCS, vol.
          8543. Springer, Heidelberg (2014). doi:10.1007/978-3-319-08434-3. ISBN
          978-3-319-08433-6

Formally Verified Safe Vertical Maneuvers for Non-deterministic, Accelerating Aircraft Dynamics

Yanni Kouskoulas[1]([⊠]), Daniel Genin[1], Aurora Schmidt[1],
and Jean-Baptiste Jeannin[2]

[1] The Johns Hopkins University Applied Physics Laboratory, Laurel, USA
yanni.Kouskoulas@jhuapl.edu
[2] Samsung Research America, Cambridge, USA

Abstract. We present the formally verified predicate and strategy used to independently evaluate the safety of the final version (Run 15) of the FAAs next-generation air-traffic collision avoidance system, ACAS X. This approach is a general one that can analyze simultaneous vertical and horizontal maneuvers issued by aircraft collision avoidance systems. The predicate is specialized to analyze sequences of vertical maneuvers, and in the horizontal dimension is modular, allowing it to be safely composed with separately analyzed horizontal dynamics. Unlike previous efforts, this approach enables analysis of aircraft that are turning, and accelerating non-deterministically. It can also analyze the safety of coordinated advisories, and encounters with more than two aircraft. We provide results on the safety evaluation of ACAS X coordinated collision avoidance on a subset of the system state space. This approach can also be used to establish the safety of vertical collision avoidance maneuvers for other systems with complex dynamics.

1 Introduction

As air travel increases and the airspace grows more crowded, existing air traffic management mechanisms such as altitude separation and manned air-traffic control are expected to experience significant stress. For decades, the Traffic Collision Avoidance System (TCAS) [3], first put into operation in the 1970s, has been the system of last resort, making mid-air collisions rare events. To address limitations that have been identified in TCAS, and to safely handle additional congestion and new participants expected in the future, the US Federal Aviation Administration (FAA), along with international partners, is developing a drop-in replacement, the next-generation Collision Avoidance System called ACAS X [9]. Like TCAS, ACAS X is intended to provide a final measure of safety, giving

This work was supported by the Federal Aviation Administration (FAA) Traffic-Alert & Collision Avoidance System (TCAS) Program Office (PO) AJM-233: Volpe National Transportation Systems Center Contract No. DTRT5715D30011.

© Springer International Publishing AG 2017
M. Ayala-Rincón and C.A. Muñoz (Eds.): ITP 2017, LNCS 10499, pp. 336–353, 2017.
DOI: 10.1007/978-3-319-66107-0_22

advice that helps prevent mid-air collisions when all other preventive measures have failed.

In 2013, our group was designated as the independent verification and validation (V&V) team for ACAS X. We began developing an independent approach to formally verify the safety of the overall ACAS X system, either to establish guaranteed safety under certain operating conditions, or to identify different categories of problems and bring them to the attention of ACAS X developers and the FAA. This proved to be challenging for a number of reasons, including that ACAS X has very complicated behavior, and does not have a precisely stated set of requirements – informally, its goal is to provide an improvement over TCAS, both in terms of safety and alerting behavior. In addition, the system has an enormous state space – over 28×10^{12} state points – and complex logic based on the massive lookup table and complementary run-time components. This analysis, detailed in [6,7] has been so successful that we were able to find hundreds of millions of straight-line flight (i.e., the simplest possible) unsafe conditions in early versions of the system that were not identified by the standard simulation and testing approaches.

Our previous efforts were fundamentally limited to analyzing intruders that flew in a straight line, without any acceleration or maneuvering. The analysis also could not address the safety of an own-ship aircraft (our term for the aircraft in which the observer travels) that turns or makes any sort of horizontal maneuvers; previous analysis limited own-ship non-determinism to vertical motion.

The present work describes a new approach to vertical safety analysis that allows us to analyze the safety of encounters where both the intruder and own-ship are independently accelerating non-deterministically in the vertical and horizontal directions. To do this, we create a vertical safety predicate that relaxes the assumption of constant, relative horizontal velocity that in our previous work restricted us from analyzing horizontal acceleration in maneuvers such as turns. Our predicate has parameters that describe horizontal safety, but is not limited to any particular horizontal dynamics; it can be composed with any horizontal motion that has been correctly analyzed. With the development of appropriate analysis for different horizontal dynamics, this approach could also assess the safety of non-deterministic, accelerating horizontal and vertical dog-fight-like maneuvers.

The main contribution of this paper is in providing a predicate to analyze the safety of vertical advisories during turns and in the presence of non-deterministically accelerating intruders. All the theorems in this paper and safety predicates for vertical motion are formally verified, meaning that their correctness is ensured via a machine-checked mathematical proof.[1]

The rest of the paper is organized as follows: Sects. 2 and 3 provide an overview of how to use the predicate by analyzing safety for an example encounter, describing the parameterization of pilot behavior and horizontal dynamics; Sect. 4 provides a detailed description of the development of vertical

---

[1] Proofs can be viewed and downloaded at https://bitbucket.org/ykouskoulas/vert_safety_proofs/src/.

safety predicates; Sect. 5 discusses issues related to formalizing our guarantees; Sects. 6 and 7 describes how we extend our safety proofs to a real system, and our results; and Sects. 8 and 9 describe related work, and conclude.

2   Overview

This section presents an overview of the logic of our approach, starting with its basic properties and walking through an illustrative example of how it would be used in practice.

Safety Property. The logic of this approach comes from the definition of *safety* used in this analysis; it allows us to decompose the safety analysis into two steps that we can treat seperately: a horizontal problem, and a vertical problem.

For this work, safety between two aircraft means that one aircraft doesn't come within a certain vertically oriented cylinder with radius r_p and half-height h_p centered on the other aircraft. This definition includes exact collision as well as any dangerously close approach between two aircraft, and is referred to by the aviation community as a Near Mid-Air Collision (NMAC). We call this volume the NMAC puck due to the resemblance with a hockey puck. Aircraft trajectories have uncertainty associated with them, and the puck represents the volume in which the other aircraft location might be found. Entering it represents, in the worst case, an actual collision.

We define *horizontal conflict* as the condition where the horizontal projections of the two aircraft come within the horizontal bounds of a puck centered on one of them; *vertical conflict* is when their vertical projections come within the vertical bounds of a puck, also centered on one of them. The two aircraft have an NMAC only if they are in horizontal and vertical conflict simultaneously.

To formalize our safety property, we define $J(t) = J_x(t)\hat{x} + J_y(t)\hat{y} + J_z(t)\hat{z}$ to be the trajectory of the ownship, and $K(t) = K_x(t)\hat{x} + K_y(t)\hat{y} + K_z(t)\hat{z}$ to be the trajectory of the intruder, both in a Cartesian coordinate system with x, y and z axes aligned to east, north and up, respectively. We have horizontal conflict whenever

$$C_h(t) \equiv |((J_x(t) - K_x(t))\,\hat{x} + (J_y(t) - K_y(t))\,\hat{y}| \le r_p \tag{1}$$

is true. We have vertical conflict when

$$C_v(t) \equiv |J_z(t) - K_z(t)| \le h_p \tag{2}$$

is true. An NMAC occurs at time t only when:

$$C_h(t) \wedge C_v(t) \tag{3}$$

We will first analyze the horizontal dynamics to determine the timing of the encounter, i.e. when the aircraft come together. We call this timing a parameterization of horizontal safety, because it establishes safety within a series of time intervals. Subsequently, the safe-by-design logic can be used to establish safety for a sequence of independent, non-deterministic, vertical maneuvers made by the pilot of each aircraft outside of these intervals. For each safety evaluation, we must choose a sequence and timing of vertical maneuvers for each aircraft, and it is under these assumptions that we can establish safety or the possibility of collision. The following paragraphs go through these steps to apply safety analysis for a specific example.

Parameterizing Horizontal Safety. To parameterize horizontal safety, we must analyze the horizontal motion of the two aircraft and identify time intervals in which the probability of the aircrafts' horizontal projections coming into proximity (i.e., horizontal conflict as defined in Eq. 1) is non-zero. Through this horizontal parameterization, we establish safety outside these intervals, because when the aircraft are far away from each other $C_h(t)$ is false, and Eq. 3 cannot be satisfied – there is no possibility of immediate collision.

We index each time interval of possible horizontal conflict using index i, and define t_{ei} and t_{xi} as times of earliest entry into and latest exit from conditions where horizontal conflict is possible, for interval i. This defines a set of time intervals, $V_i = [t_{ei}, t_{xi}]$, and their union $V = \bigcup_{i \in \{1...n\}} V_i$, during which safety must be established through the absence of vertical conflict.

Consider the example of two aircraft whose horizontal trajectories follow deterministic circular paths, as shown in Fig. 1, where the speed of the own-ship is chosen by the pilot. To simplify our example, we assume that the speed of the

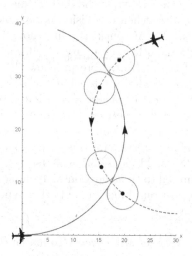

Fig. 1. Example horizontal turning trajectories, projected onto horizontal cartesian coordinate system, viewed from above. The own-ship trajectory is represented by a solid line, and the intruder is represented by a dashed line. Circles and green color indicates the extent of trajectory segments where collision is possible. (Color figure online)

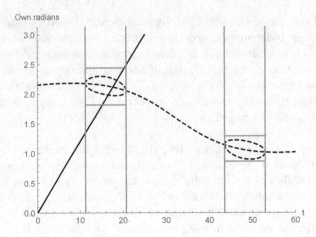

Fig. 2. Analysis of encounter timing for one possible combination of ground speeds. Positions are given as a radian measure on the own-ship's trajectory circle, and ground speeds are assumed constant for this particular scenario. The dashed line is the intruder's center projected on the own-ship's trajectory, and when it intersects the trajectory, the extent of that intersection is plotted above and below the center. Vertical lines correspond to the beginning and end of time intervals when a collision is possible, and to the disks in Fig. 1.

intruder is known, although this is not required for the analysis in general. The solid line represents the own-ship while the dashed line represents the intruder aircraft. One way to visualize when a collision is possible is to imagine a disk representing the top of the NMAC puck traveling along one of the trajectories. When that disk intersects the other trajectory, a collision is possible, depending on the relative speeds of the aircraft. Here we show the disk on the intruder's trajectory at the four points where it touches the own-ship's trajectory, and highlight the parts of its trajectory where a collision is possible. Figure 2 illustrates the timing analysis that is necessary to compute the horizontal conflict interval. Assuming the intruder's ground speed is known and consistent with Fig. 2, the horizontal conflict intervals for this geometry can be read off the plot to determine that $V = [11.2\,\mathrm{s},\ 20.7\,\mathrm{s}] \cup [43.5\,\mathrm{s},\ 53.0\,\mathrm{s}]$.

Our analysis is not limited to these horizontal dynamics; we can also establish safety for more complex horizontal motion and other types of non-determinism, as long as we can compute V.

Sequence and Timing of Vertical Maneuvers. To match common flight patterns and the ACAS X advisory system, the vertical dynamics of each aircraft is modeled by a sequence of non-deterministic maneuvers, specified by allowed acceleration and velocity ranges. By combining maneuvers it is possible to represent a variety of behaviors, including straight line flight, choice of one of a series of actions (where the decision is unknown at the time of safety analysis), unrestricted vertical flight, compliance with an ACAS X vertical advisory, delayed

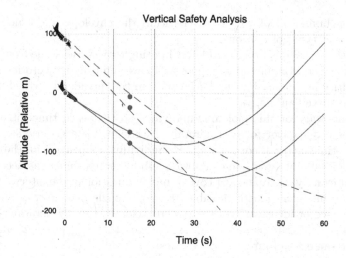

Fig. 3. Bounding envelopes for vertical motion of ownship and intruder (dashed). Horizontal conflict intervals are indicated by vertical lines. Safety is guaranteed despite any maneuvers the pilots may make that cause variations in vertical acceleration and velocity, or variations in horizontal ground speed of the ownship, within assumed dynamics.

compliance with an ACAS X advisory, a reversal of vertical motion direction to ensure safety, or straight line flight followed by a level-off maneuver. Thus, the proposed dynamics captures many, if not most, operationally relevant aircraft encounter scenarios.

For our example from Fig. 1, assume the intruder starts above the own-ship, the aircraft are descending, with the intruder diving towards the ground. The own-ship engages in a vertical chase for the first 15 s of the encounter, diving at a less extreme rate, and then follows advice to sharply accelerate upwards, eventually crossing altitudes with the intruder.

Vertical Safety Predicate. Once we have analyzed horizontal dynamics, and chosen a pilot model (i.e. a sequence of vertical pilot timing and actions) we can apply the vertical safety predicate to establish whether we can definitively avoid collision under our assumptions. Figure 3 illustrates the extent of vertical motion in our example scenario from Fig. 1 by plotting the most extreme vertical trajectories of the own-ship and intruder. These boundaries describe a reachable envelope of altitudes for each moment in time. Our predicate Ψ, described in Sect. 4, guarantees safety for this geometry under our assumptions, and the figure illustrates the intuition behind the predicate's logic, confirming that the aircraft are safely separated vertically during both horizontal conflict intervals.

This envelope introduces non-determinism in our model, the representation of uncertain vertical motion in the future. Even though the limiting trajectories of our envelope are piecewise polynomial and our dynamics are simple, our dynamics are not limited to piecewise polynomial trajectories. This model allows us to represent a continuous family of irregular trajectories within our acceleration limits, all of which travel within the envelope but which include many different types of motion.

While the ownship's upper and lower limiting trajectories issue from a single point at time zero, the intruder's upper and lower limiting trajectories bound a range of altitudes, indicating the uncertainty in the intruder's vertical position, e.g., due to surveillance error.

Time intervals for this plot are subdivided so that each time interval contains a single maneuver for each aircraft. Although, for a generic sequence of maneuvers, time intervals corresponding to individual maneuvers for the ownship and intruder will not agree, we can always subdivide maneuvers as necessary to ensure that exactly one maneuver covers the full duration of the interval for both ownship and intruder. This is possible because a single maneuver of duration d and a sequence of identical (with regard to velocity and acceleration bounds) maneuvers with durations $d_0, d_1, d_2, \ldots, d_n$, such that $\sum_{i=1}^{n} d_i = d$ encompass exactly the same set of aircraft trajectories.

3   Modeling and Assumptions

Modeling Non-deterministic Vertical Maneuvers. Each vertical maneuver is defined by a duration of time d the maneuver is in effect, and a range of vertical velocities, $[v_{\min}, v_{\max}]$. During the maneuver, the pilot accelerates the aircraft with the intention of bringing vertical velocity into the specified range. Acceleration is non-deterministic, and each maneuver has a set of four limiting vertical accelerations $a_{\min} \leq a_a < 0 < a_b \leq a_{\max}$. The subscripts a and b indicate the maximum and minimum acceleration allowed when the aircraft is above and below the target range of vertical velocities, respectively. During a maneuver, the pilot can choose to follow any acceleration $a(t)$, that is continuous, integrable, and satisfies

$$
\begin{aligned}
\forall t, (v(t) > v_{\max} &\rightarrow a_{\min} \leq a(t) \leq a_a) \wedge \\
(v(t) = v_{\max} &\rightarrow a_{\min} \leq a(t) \leq 0) \wedge \\
(v_{\min} < v(t) < v_{\max} &\rightarrow a_{\min} \leq a(t) \leq a_{\max}) \wedge \\
(v(t) = v_{\min} &\rightarrow 0 \leq a(t) \leq a_{\max}) \wedge \\
(v(t) < v_{\min} &\rightarrow a_b \leq a(t) \leq a_{\max})
\end{aligned}
\tag{4}
$$

where $v(t) = \int_0^t a(t)\mathrm{d}t + v(0)$ is the velocity of the aircraft.

In the Coq formalization, we prove the following properties about pilot behavior:

Theorem 1 (Pilot-model vertical compliance). *The constraints on $a(t)$ given in Eq. 4 ensure that when the aircraft is below (above) the target range of vertical velocities, it will accelerate towards it with acceleration a_b (a_a) until it is within its bounds.*

Theorem 2 (Pilot-model maintains vertical compliance). *The constraints on $a(t)$ given in Eq. 4 ensure that once the aircraft has entered the allowed range of vertical velocities, the aircraft will stay within that range.*

There are sequences of maneuvers and certain geometries where it is impossible for a pilot to follow Eq. 4 while maintaining continuous acceleration. For example, compliance with one maneuver may require the pilot to increase vertical velocity by maintaining a positive acceleration, which may abruptly change to a requirement to decrease vertical velocity by maintaining negative acceleration, at the beginning of the next maneuver. In this case there will be trajectories with $a(t)$ satisfying the requirements of the first maneuver that will have no continuous extension to the second maneuver.

In order to ensure that any sequence of maneuvers individually satisfying Eq. 4 can be followed while maintaining acceleration $a(t)$ that is continuous (i.e. has a derivative) everywhere, we introduce the concept of an *auxiliary maneuver* for every pair of consecutive maneuvers. The auxiliary maneuver provides a finite time window to allow acceleration to transition continuously from one maneuver to the next, thus avoiding potential discontinuous changes in acceleration at the boundary between maneuvers. This simple device dramatically simplifies analysis by removing the need for additional restrictions that would otherwise be necessary to enforce the global continuity of $a(t)$.

Given a pair of maneuvers with target vertical velocity intervals $[v_{\min}, v_{\max}]$ and $[w_{\min}, w_{\max}]$, and acceleration bounds $a_{\min} \leq a_a < 0 < a_b \leq a_{\max}$ and $b_{\min} \leq b_a < 0 < b_b \leq b_{\max}$, respectively, the matching auxiliary maneuver is given by a target velocity interval $[\min(v_{\min}, w_{\min}), \max(v_{\max}, w_{\max})]$ and the acceleration bounds are $\min(a_{\min}, b_{\min}) \leq \max(a_a, b_a) < 0 < \min(a_b, b_b) \leq \max(a_{\max}, b_{\max})$. The minimal duration of an auxiliary maneuver is bounded below only by the limits on the derivative of the aircraft's acceleration, sometimes also referred to as jerk.

To simplify the formal safety proofs, we have chosen to assume that $a(t)$ is continuous – a natural assumption from the point of view of physics – and treat auxiliary maneuvers as undistinguished from other maneuvers. The alternative would be to have done the safety proofs for $a(t)$ that would be allowed to become discontinuous at the beginning of each maneuver. However, we did not pursue this approach since it is simultaneously less realistic and more difficult to implement in Coq.

4   Vertical Safety Predicates

In this section, we develop formally-verified, quantifier-free predicates establishing pairwise safety between two aircraft. We do this for arbitrary sequences of vertical maneuvers, where both pilots are accelerating non-deterministically. The predicates are also constructed in a modular fashion so they can be composed with a separate analysis of horizontal motion to ensure overall safety of an encounter.

Vertical Safety Predicates. To guarantee vertical separation between two aircraft, we establish a bounding envelope that contains all altitudes reachable by each aircraft for each sequential maneuver as a function of time. We then construct a predicate that computes a bounding envelope for each aircraft separately according to the initial position of each, and ensures that the envelopes don't overlap during V, the vertical conflict intervals. We establish the safety of this predicate via formal proofs in Coq.

The bounding envelope for a single aircraft executing a single maneuver (Eq. 4) depends on the initial range of vertical positions and velocities of the aircraft at the start of the maneuver as well as the maneuver velocity and acceleration bounds. In the time-altitude domain, edges of the bounding envelope are given by the upper and lower limiting trajectories. These trajectories originate from the extremes of the initial velocity and position ranges, and follow the extreme values of acceleration and velocity allowed by the maneuver. Specifically, limiting trajectories have the following form

$$J_z(t) = \begin{cases} \left(\frac{a}{2}t^2 + v_0 t + z_0\right) \hat{z} & \text{if } 0 \le t < t_r \\ \left(v_\ell t - t_r \frac{(v_t - v_0)}{2} + z_0\right) \hat{z} \text{ if } t_r \le t \end{cases} \tag{5}$$

where v_0 and z_0 are the initial vertical velocity and position of the aircraft, v_t is the matching extreme of the velocity range of the maneuver, and $t_r = \frac{v_t - v_0}{a}$ is the time when the limiting trajectory reaches the maneuver velocity range. So we have

$$(v_t, a) = \begin{cases} (v_{\max}, a_a) & \text{if } v_0 > v_{\max} \\ (v_{\max}, a_{\max}) & \text{if } v_0 \le v_{\max} \end{cases} \tag{6}$$

for the upper limiting trajectory and

$$(v_t, a) = \begin{cases} (v_{\min}, a_{\min}) & \text{if } v_0 > v_{\min} \\ (v_{\min}, a_b) & \text{if } v_0 \le v_{\min} \end{cases} \tag{7}$$

for the lower limiting trajectory. In the Coq formalization, we prove

Theorem 3. *An aircraft following an arbitrary trajectory satisfying the constraints of Eq. 4 remains within the altitude envelope bounded above and below by limiting trajectories determined by Eqs. 5, 6 and 7.*

Once upper and lower limiting trajectories are constructed we have an envelope of altitudes over time reachable by a non-deterministically maneuvering aircraft, with boundaries that are described piecewise by polynomials of at most degree two.

So far, we have been describing the dynamics for one aircraft, but we can use this model for each aircraft in the encounter, plotting reachable envelopes vs. time, and allowing us to visualize the uncertainty in position and relationship between aircrafts at each moment. Figure 3 provides a visual example of upper and lower limiting trajectories for ownship (solid lines) and intruder (dashed lines) aircraft.

To develop quantifier-free predicates that indicate the absence of vertical conflict for a pair of aircraft, we take the difference of their opposite limiting trajectories (lower-upper and upper-lower), and then compute whether the resulting polynomial is positive. Physically, this means the aircraft are safely separated. We first define the predicate

$$\Gamma((A, B, C), t_b, t_e) \equiv t_b \leq t_e \rightarrow$$
$$(A > 0 \wedge ((0 \leq D \wedge (R_1 > t_e \vee R_2 < t_b)) \vee D < 0) \vee$$
$$A < 0 \wedge (0 < D \wedge R_2 < t_b \wedge R_1 > t_e) \vee \qquad (8)$$
$$A = 0 \wedge (B > 0 \wedge -C/B < t_b \vee B < 0 \wedge -C/B > t_e \vee$$
$$B = 0 \wedge C > 0))$$

to compute whether an arbitrary polynomial $At^2 + Bt + C$ represented by the vector of its coefficients (A, B, C) is positive over the interval $[t_b, t_e]$, where the subscripts b and e represent the beginning and ending times of the interval. In this predicate, we define $D \equiv B^2 - 4AC$, $R_1 \equiv \frac{(-B-\sqrt{D})}{2A}$, and $R_2 \equiv \frac{(-B+\sqrt{D})}{2A}$ – the expressions for the discriminant and roots of a quadratic. The predicate is made of a disjunction of three clauses, which analyze the polynomial when second order coefficient A is positive, zero, or negative. If A is non-zero there are two cases corresponding to an upward, $A > 0$, or downward, $A < 0$, extending parabola with at most two roots. If $A = 0$ the polynomial is linear with at most one root. The rest of the logic compares the location of the roots with the end points of the time interval $[t_b, t_e]$ and determines whether the curve is positive in that interval. We formalize and prove the following theorem in Coq:

Theorem 4 (Safely separated second-order polynomial interval). *The predicate $\Gamma((A, B, C), t_b, t_e)$ computes whether a polynomial $At^2 + Bt + C$ is positive over the interval $[t_b, t_e]$.*

Each limiting trajectory within each maneuver is a piecewise function composed of at most two pieces: a quadratic piece, corresponding to the aircraft accelerating toward the maneuver's target velocity range, and a linear piece, corresponding to the aircraft maintaining one of the extremal velocities in the maneuver's target velocity range. Either of these pieces could be missing depending on the state of the aircraft at the beginning of the maneuver and the maneuver's duration. We next define a predicate

$$\Phi(Q_1, L_1, t_{t1}, Q_2, L_2, t_{t2}, t_b, t_e) \equiv$$
$$\Gamma(Q_1 - Q_2 - P, \max(t_b, 0), \min(t_e, t_{t1}, t_{t2})) \wedge$$
$$\Gamma(L_1 - L_2 - P, \max(t_b, t_{t1}, t_{t2}), t_e) \wedge$$
$$(t_{t1} > t_{t2} \rightarrow \qquad (9)$$
$$\Gamma(Q_1 - L_2 - P, \max(t_b, \min(t_{t1}, t_{t2})), \min(t_e, \max(t_{t1}, t_{t2})))) \wedge$$
$$(t_{t1} < t_{t2} \rightarrow$$
$$\Gamma(L_1 - Q_2 - P, \max(t_b, \min(t_{t1}, t_{t2})), \min(t_e, \max(t_{t1}, t_{t2}))))$$

to compute whether two limiting trajectories described by Q_1, L_1, and Q_2, L_2 are safely separated in interval $[t_b, t_e]$. In this predicate, $P = (0, 0, h_p)$ and h_p is the half-height of the NMAC puck. Each Q_i and L_i is a 3-vector containing the coefficients of the polynomials corresponding to the quadratic and linear pieces of trajectory i, respectively. The times t_{t1} and t_{t2} are the times when each respective trajectory transitions from one piece to the next. The predicate Φ computes the separation and determines whether it is adequate, (i.e. $> h_p$) for all points in the interval of interest, ensuring that the correct polynomial is used for each trajectory at each point. Given that each limiting trajectory is composed of at most two pieces, there are four possible combinations of polynomials that appear in the analysis: (Q_1, Q_2), (Q_1, L_2), (L_1, Q_2), (L_1, L_2). Each of these possibilities corresponds to one term of the conjunction in the definition of Φ. The predicate Φ has four instances of Γ, since it establishes safety for the different pieces (linear and quadratic) of a trajectory for an entire maneuver.

We formalize, and prove the following theorem in Coq:

Theorem 5 (Safely separated trajectory interval, above). *The predicate*

$$\Phi((\alpha_1, \beta_1, \gamma_1), (\delta_1, \epsilon_1, \zeta_1), t_{t1}, (\alpha_2, \beta_2, \gamma_2), (\delta_2, \epsilon_2, \zeta_2), t_{t2}, t_e, t_x) \tag{10}$$

computes whether a trajectory

$$T_1(t) = \begin{cases} (\alpha_1 t^2 + \beta_1 t + \gamma_1) & 0 \le t < t_{t1} \\ (\delta_1 t^2 + \epsilon_1 t + \zeta_1) & t_{t1} \le t \end{cases} \tag{11}$$

is safely separated and above trajectory

$$T_2(t) = \begin{cases} (\alpha_2 t^2 + \beta_2 t + \gamma_2) & 0 \le t < t_{t2} \\ (\delta_2 t^2 + \epsilon_2 t + \zeta_2) & t_{t2} \le t \end{cases} \tag{12}$$

by a distance of h_p over the interval $[t_b, t_e]$.

Consider an aircraft executing a sequence of m maneuvers, defined by minimum and maximum velocity bounds ($[v_{min1}, v_{max1}], [v_{min2}, v_{max2}], \ldots,$ $[v_{minm}, v_{maxm}]$), for durations ($d_1, d_2, \ldots, d_m$), each maneuver having an envelope of possible trajectories bounded by Eq. 5. We define $\{t_{mi}\}$ as the set of times that identify the start of each maneuver. We also assume the aircraft have horizontal dynamics for which there are n time intervals ($[t_{e1}, t_{x1}], [t_{e2}, t_{x2}], \ldots, [t_{en}, t_{xn}]$) when the probability of horizontal conflict is non-zero. For convenience, we compute a set of times (τ_{mn}, υ_{mn}) that are the entry and exit times for conflict interval n, intersecting maneuver m, relative to the starting time of the maneuver:

$$(\tau_{mn}, \upsilon_{mn}) = \begin{cases} (\max(0, t_{en}), \min(d_1, t_{xn})) & \text{for } m = 1 \\ (\max(0, t_{en}) - \sum_{i=1}^{m-1} d_i, \min(d_m, t_{xn} - \sum_{i=1}^{m-1} d_i)) & \\ & \text{for } m > 1 \end{cases} \tag{13}$$

For each aircraft there is an upper and lower bounding trajectory; each of these bounding trajectories has a quadratic and a linear piece for each maneuver. We define Q and L to be 3-dimensional vectors representing the quadratic and linear parts of the bounding trajectory for a single maneuver and a single aircraft, and the time t_r to indicate when each limiting trajectory transitions between the quadratic and linear pieces. Each of these quantities uses a superscript with a tag to represent which aircraft (own or intruder), an up or down arrow indicating whether the bound is a trajectory that bounds the aircraft from above or below, respectively. Each variable also has a subscript index i that identifies the maneuver it describes.

So collectively, $Q_i^{\mathrm{Own}\uparrow}$, $L_i^{\mathrm{Own}\uparrow}$, and $t_{ri}^{\mathrm{Own}\uparrow}$ represent the upper limiting trajectory for the ownship for maneuver i, and $Q_i^{\mathrm{Own}\downarrow}$, $L_i^{\mathrm{Own}\downarrow}$, and $t_{ri}^{\mathrm{Own}\downarrow}$ to describe the lower limiting trajectory for the ownship in the same way. These vectors contain the second, first, and zeroth order coefficients from Eq. 5. So

$$
\begin{aligned}
Q_i^{\mathrm{Own}\uparrow} &\equiv \left(\tfrac{a}{2}, v_0, z_0\right) & Q_i^{\mathrm{Own}\downarrow} &\equiv \left(\tfrac{a}{2}, v_0, z_0\right) \\
L_i^{\mathrm{Own}\uparrow} &\equiv \left(0, v_{\mathrm{maxi}}, z_0 - \tfrac{(v_{\mathrm{maxi}}-v_0)^2}{2a}\right) & L_i^{\mathrm{Own}\downarrow} &\equiv \left(0, v_{\mathrm{mini}}, z_0 - \tfrac{(v_{\mathrm{mini}}-v_0)^2}{2a}\right) & (14) \\
t_{ri}^{\mathrm{Own}\uparrow} &\equiv \tfrac{v_{\mathrm{maxi}}-v_0}{a} & t_{ri}^{\mathrm{Own}\downarrow} &\equiv \tfrac{v_{\mathrm{mini}}-v_0}{a}
\end{aligned}
$$

represents upper and lower bounding trajectories for the own-ship. The initial conditions v_0 and z_0 are set so that velocity and position are continuous at the boundary between the different maneuvers, and a is set according to Eqs. 6 and 7.

Similarly, we define $Q_i^{\mathrm{Int}\uparrow}$, $L_i^{\mathrm{Int}\uparrow}$, $t_i^{\mathrm{Int}\uparrow}$, $Q_i^{\mathrm{Int}\downarrow}$, $L_i^{\mathrm{Int}\downarrow}$, and $t_{ri}^{\mathrm{Int}\downarrow}$ to describe the upper and lower limiting trajectories of the intruder aircraft, replacing parameters with the ones appropriate for that aircraft.

Finally, we define the predicate

$$
\Psi = \bigwedge_{j \in \{1,\dots,n\}} \left(\left(\bigwedge_{i \in \{1,\dots,m\}} \Phi(Q_i^{\mathrm{Own}\downarrow}, L_i^{\mathrm{Own}\downarrow}, t_{ri}^{\mathrm{Own}\downarrow}, Q_i^{\mathrm{Int}\uparrow}, L_i^{\mathrm{Int}\uparrow}, t_{ri}^{\mathrm{Int}\uparrow}, \tau_{ij}, v_{ij}) \right) \vee \left(\bigwedge_{i \in \{1,\dots,m\}} \Phi(Q_i^{\mathrm{Int}\downarrow}, L_i^{\mathrm{Int}\downarrow}, t_{ri}^{\mathrm{Int}\downarrow}, Q_i^{\mathrm{Own}\uparrow}, L_i^{\mathrm{Own}\uparrow}, t_{ri}^{\mathrm{Own}\uparrow}, \tau_{ij}, v_{ij}) \right) \right) \tag{15}
$$

that helps establish safety between aircraft during a series of horizontal conflict intervals, as they follow a series of maneuvers. Its construction mirrors the following logic. An encounter is safe if each of its horizontal conflict intervals is safe; the outer conjunction over j ensures safety for each interval. Each conflict interval is safe if either the own-ship is always safely above the intruder, or vice versa; the left side and right side of the disjunction account for these two possibilities. One aircraft is safely above the other if they are safely separated during each of the maneuvers in the conflict interval; the inner conjunction over i accounts for each maneuver. We formalize and prove the following theorem in Coq:

Theorem 6 (Safely separated vertical trajectories). *The predicate Ψ computes whether a particular encounter is safe (i.e. collision-free) according to Eq. 15, for n time intervals $([t_{e1}, t_{x1}], [t_{e2}, t_{x2}], \dots, [t_{en}, t_{xn}])$ during a sequence of m maneuvers $([v_{min1}, v_{max1}], [v_{min2}, v_{max2}], \dots, [v_{minm}, v_{maxm}])$, with respective durations given by $(d_1, d_2, \dots, d_m)$.*

5   Formalizing Guarantees

We used Coq to formalize our proofs for this work, and this had both advantages and disadvantages compared with KeYmaera, which we had used previously. (A version of KeYmaera with scripting capabilities was unavailable for use since the system was between versions at the time of this work.) The immediate disadvantages of this change were that we could not concisely express our system using the specialized terms used for hybrid programs, and we did not have access to the reasoning strategies made available in differential-dynamic logic (dL), since presently there is no mechanization of dL in the Coq environment. Consequently, we expressed our model in terms of the more general framework of inductive constructions using higher order logic and Coq's expressive system of dependent types, and had to develop a set of lemmas about non-deterministic vertical motion from scratch, using Coq's Real library. The immediate advantage of this change was access to the well-developed scripting and automation capabilities of the relatively mature Coq environment, and the potential for integrating our present work with proofs that reason about trajectories involving trigonometric functions, as might be required for some types of non-deterministic horizontal turning behavior.

6   Extending Safety Guarantees to ACAS X

Our initial objective was to use this predicate formally verify that whenever possible, the system provided sequences of advice to the pilot that guaranteed safety and absence of collision under our acceleration assumptions.

ACAS X's complicated behavior is contained in a data structure that when uncompressed more than five hundred megabytes in size. The table is an optimal policy that minimizes costs associated with a Markov decision process representing the aircraft encounter. Reasoning about the table is challenging. There is discretization in the MDP, undersampling in the state space, and the logic of the table is related to optimizing a set of weights, whose relationship with actual safety in the real world is not straightforward.

The approach we took to formal verification treats the logic as opaque. Instead of creating a model of ACAS X that faithfully reproduces its details and quirks and trying to load it into a prover, we instead focused on evaluating its behavior throughout the state space. We developed the model described in Sect. 4, an independent logic for a collision-avoidance system that is safe-by-design. We prove it to be safe everywhere, and extend proofs about its safety to proofs about the safety of the real system. This extension is done via exploration of the system's state space, and comparison of the behavior at state points in the table to the allowable range of geometrically safe behaviors identified by our logic. The states in the table definitively determine the system's behavior in the continuous state space – the score function at off-table states are interpolated from the table's values in a local neighborhood. To evaluate each state, our predicate evaluates the future possibilities, taking into account pilot non-determinism, sensor noise, and delay in the system, using the envelopes we previously described, acceleration limets, and the parameters of the NMAC puck. This approach makes it possible to do formal verification and draw conclusions about ACAS X over the entirety of its state space, but also makes the logic reusable for other collision avoidance systems.

To formally verify the system in its entirety with this approach, we would need to do two things: first, we would run the logic over all of the table's states, and then we would have to develop guarantees about off-table points in the state space. Proofs and reasoning would have to be developed to fill in the rest of the state space after the table's states were evaluated.

We ran an comprehensive evaluation of all the table's states in an earlier version of the system for straight-line trajectories. Our first comprehensive run took nearly a month to set up and run on our local cluster, returned so many examples of unsafe behavior that we had difficulty characterizing them. The initial results were that we quickly proved the system was not safe, and identified where. We almost immediately found areas where it gave unsafe advice, but where advice was possible that would guarantee safety.

Since we had counterexamples that will not be resolved, we could not prove safety comprehensively. At this point, we switched our focus from making comprehensive guarantees about the system's behavior to making local guarantees of safety or dangerous conditions, and characterizing the safety tradeoff made during its design.

7   Application to ACAS X Coordination Logic

This section describes how the vertical safety predicate was used to evaluate safety of ACAS X, for encounters where both aircraft are equipped with ACAS X and are executing coordinated vertical safety maneuvers simultaneously. This analysis was not possible earlier, because the previous framework we used [6, 7] was fundamentally limited to analyzing a non-accelerating intruder; even vertical maneuvers for the intruder were not analyzable.

Using our new framework, we analyzed the advice generated by a prototype of ACAS X on a subset of the system's behavior table cut-points. We first collected the advisories that ACAS X issues on the chosen state space samples by querying ACAS X for both the ownship and intruder aircraft advisories.

The pilots of each aircraft are assumed to begin responding to an advisory 5 s after the first advisory is issued, and 3 s after each subsequent advisory. The safety predicate Ψ is evaluated at each selected state point with the harvested advisories assigned to the ownship and intruder accordingly. The horizontal motion model chosen here is the deterministic straight line model.

We called the state points where Ψ fails with the ACAS X advisories but succeeds with another set of ownship and intruder advisories *counterexamples*. A counterexample is a point in the state space where ACAS X issues advisories that are not guaranteed to be safe according to Ψ but there are other advisories that would guarantee safety. In the terminology of [6,7], Ψ is a *safeable* predicate.

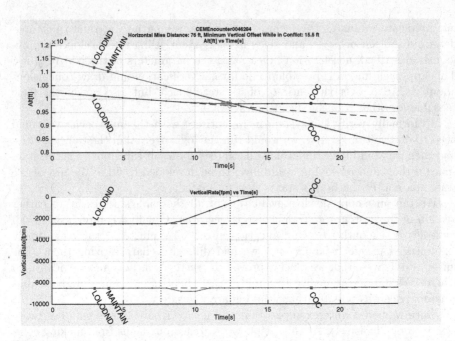

Fig. 4. An example NMAC found where the set of ACAS X advisories does not prevent a close approach in altitude during the period where the aircraft are within 500 ft horizontally, denoted by vertical dashed lines. Ownship and intruder trajectories are shown in blue and red, respectively. (Color figure online)

Recall, that a state point advisory combination is safeable if it is safe or can be made safe in the future by issuing additional advisories after a limited delay.

Of the 589,560 state points examined, 29,295 were identified as safeable counterexamples. To identify the most dangerous state space configurations the safeable counterexample set was further run through full ACAS X simulations with nominal trajectory accelerations set to zero. The result was a set of 3,301 state points where the system issued advice that created NMACs.

Examining the above set of dangerous aircraft configurations in terms of their state space coordinates, we observed a striking pattern—all of them had a low or moderate horizontal closing speed of between 10 and 200 ft/s. In practice, this means that the aircraft will remain in horizontal proximity for an extended period of time. For example, at the horizontal closing rate of 100 ft/s it can take the aircraft up to 10 s to clear the horizontal projection of the NMAC region.

Figure 4 shows conditions found by our analysis where ACAS X advice that does not guarantee safety. The two aircraft follow nearly parallel horizontal paths that cross at a very small angle (not shown). The intruder aircraft (red track) descends rapidly at −8500 ft/min, while the ownship (blue track) descends at a more moderate rate of −2500 ft/min. The dotted vertical lines indicate the time interval during which the aircraft are within 500 ft of each other and, hence, must maintain vertical separation of at least 100 ft to avoid NMAC.

The resolution advisories issued by ACAS X—DO NOT DESCEND (DND) and MAINTAIN VERTICAL SPEED (MAINTAIN), for the ownship and intruder aircraft respectively – result in an NMAC at time 12 s. The dotted blue line indicates the straight line continuation of the ownship trajectory that would have occurred with no advisory. To guarantee safety, ACAS X could continue to advise DO NOT DESCEND to the intruder, while advising the ownship to MAINTAIN vertical velocity.

These results pointed to an important flaw in the system assumptions about the possible range of durations of horizontal proximity. The problems stemming from slow horizontal closing configurations are actively being addressed in the final ACAS X system.

8   Related Work

Many efforts have explored developing correct and comprehensive guarantees about collision avoidance decisions over a system's state space. This paper improves on these because it develops guaranteed geometric safety under more realistic dynamics. The ACAS X system logic [8] is based on a policy that results from optimizing a Markov Decision Process (MDP) using value iteration to minimize a set of costs; [2,10] analyze the state space of a similar MDP using probabilistic model checking and an adaptive Monte Carlo tree search respectively, to identify undesirable behavior. Collision avoidance algorithms are developed for both horizontal and vertical motion in 3D in [13,14] for polynomial trajectories with a finite time horizon, and formally verified with PVS. TCAS, the predecessor for ACAS X. Its resolution advisories have been formalized in PVS. In [12], the logic for TCAS is formalized in PVS and used to identify straight-line encounter geometries that generate advisories in a noiseless environment.

There are a number of simulation approaches [1,5] that allow for more precise description of dynamics than the present work. However are limited to evaluating safety for a finite number of trajectories.

Prior efforts that match our dynamics as well as providing a formal proof of safety can be found in [4,11,15,16]. All these use a hybrid system model to develop safe horizontal maneuvers, unlike the present work which develops vertical maneuvers, and is applied to a practical system.

The most closely related work is [6,7]. We retain the overall approach to verification, very similar non-deterministic dynamics, and the idea of computing reachable envelopes to make guarantees about a range of future possibilities. The present work differs because it can analyze the safety of encounters with each aircraft making independent sequences of non-deterministic maneuvers, including acceleration, turns, and pilot delay. The proofs here are formalized in Coq.

9   Conclusion

This framework and the detailed vertical predicates offer a flexible approach to a formally verified analysis of the safety of a collision avoidance system. It relaxes restrictive assumptions about acceleration and horizontal motion and allows us to ensure the safety of a wider variety of pilot behavior and ACAS X system conditions than before. This analysis can ensure the safety of intruders that accelerate vertically, aircraft that make horizontal turns, coordinated ACAS X advisories, and multi-threat encounters. Its flexibility extends, further, to ensuring safe vertical motion in the presence of mixed horizontal and vertical advisories.

Acknowledgments. We gratefully acknowledge Neal Suchy and Josh Silbermann for their leadership and support. We thank André Platzer, Ryan Gardner and Christopher Rouff for their comments and technical discussion.

References

1. Chludzinski, B.J.: Evaluation of TCAS II version 7.1 using the FAA fast-time encounter generator model. Technical report ATC-346, MIT Lincoln Laboratory (2009)
2. Essen, C., Giannakopoulou, D.: Analyzing the next generation airborne collision avoidance system. In: Ábrahám, E., Havelund, K. (eds.) TACAS 2014. LNCS, vol. 8413, pp. 620–635. Springer, Heidelberg (2014). doi:10.1007/978-3-642-54862-8_54
3. Federal Aviation Administration: Introduction to TCAS II, Version 7.1 (2011)
4. Ghorbal, K., Jeannin, J.B., Zawadzki, E., Platzer, A., Gordon, G.J., Capell, P.: Hybrid theorem proving of aerospace systems: applications and challenges. J. Aerosp. Inf. Syst. **11**, 202–713 (2014)
5. Holland, J.E., Kochenderfer, M.J., Olson, W.A.: Optimizing the next generation collision avoidance system for safe, suitable, and acceptable operational performance. Air Traffic Control Q. **21**, 275–297 (2014)
6. Jeannin, J., Ghorbal, K., Kouskoulas, Y., Gardner, R., Schmidt, A., Zawadzki, E., Platzer, A.: Formal verification of ACAS X, an industrial airborne collision avoidance system. In: Girault, A., Guan, N. (eds.) 2015 International Conference on Embedded Software, EMSOFT 2015, Amsterdam, The Netherlands, 4–9 October 2015. ACM (2015)
7. Jeannin, J.-B., Ghorbal, K., Kouskoulas, Y., Gardner, R., Schmidt, A., Zawadzki, E., Platzer, A.: A formally verified hybrid system for the next-generation airborne collision avoidance system. In: Baier, C., Tinelli, C. (eds.) TACAS 2015. LNCS, vol. 9035, pp. 21–36. Springer, Heidelberg (2015). doi:10.1007/978-3-662-46681-0_2
8. Kochenderfer, M.J., Chryssanthacopoulos, J.P.: Robust airborne collision avoidance through dynamic programming. Technical report ATC-371, MIT Lincoln Laboratory (2010)
9. Kochenderfer, M.J., Holland, J.E., Chryssanthacopoulos, J.P.: Next generation airborne collision avoidance system. Lincoln Lab. J. **19**(1), 17–33 (2012)
10. Lee, R., Kochenderfer, M.J., Mengshoel, O.J., Brat, G.P., Owen, M.P.: Adaptive stress testing of airborne collision avoidance systems. In: 2015 IEEE/AIAA 34th Digital Avionics Systems Conference (DASC), p. 6C2-1. IEEE (2015)
11. Loos, S.M., Renshaw, D.W., Platzer, A.: Formal verification of distributed aircraft controllers. In: HSCC, pp. 125–130. ACM (2013). doi:10.1145/2461328.2461350
12. Muñoz, C., Narkawicz, A., Chamberlain, J.: A TCAS-II resolution advisory detection algorithm. In: Proceedings of the AIAA Guidance Navigation, and Control Conference and Exhibit 2013, AIAA-2013-4622, Boston, Massachusetts (2013)

13. Narkawicz, A., Muñoz, C.: Formal verification of conflict detection algorithms for arbitrary trajectories. Reliab. Comput. **17**, 209–237 (2012)
14. Narkawicz, A., Muñoz, C.: A formally verified conflict detection algorithm for polynomial trajectories. In: Proceedings of the 2015 AIAA Infotech@ Aerospace Conference, Kissimmee, Florida (2015)
15. Platzer, A., Clarke, E.M.: Formal verification of curved flight collision avoidance maneuvers: a case study. In: Cavalcanti, A., Dams, D.R. (eds.) FM 2009. LNCS, vol. 5850, pp. 547–562. Springer, Heidelberg (2009). doi:10.1007/978-3-642-05089-3_35
16. Tomlin, C., Pappas, G.J., Sastry, S.: Conflict resolution for air traffic management: a study in multiagent hybrid systems. IEEE Trans. Autom. Control **43**(4), 509–521 (1998)

Using Abstract Stobjs in ACL2 to Compute Matrix Normal Forms

Laureano Lambán[1], Francisco J. Martín-Mateos[2]([⊠]), Julio Rubio[1],
and José-Luis Ruiz-Reina[2]

[1] Department of Mathematics and Computation,
University of La Rioja, Logroño, Spain
{lalamban,julio.rubio}@unirioja.es

[2] Department of Computer Science and Artificial Intelligence,
University of Sevilla, Seville, Spain
{fjesus,jruiz}@us.es

Abstract. We present here an application of abstract single threaded objects (*abstract stobjs*) in the ACL2 theorem prover, to define a formally verified algorithm that given a matrix with elements in the ring of integers, computes an equivalent matrix in column echelon form. Abstract stobjs allow us to define a sound logical interface between matrices defined as lists of lists, convenient for reasoning but inefficient, and matrices represented as unidimensional stobjs arrays, which implement accesses and (destructive) updates in constant time. Also, by means of the abstract stobjs mechanism, we use a more convenient logical representation of the transformation matrix, as a sequence of elemental transformations. Although we describe here a particular normalization algorithm, we think this approach could be useful to obtain formally verified and efficient executable implementations of a number of matrix normal form algorithms.

Keywords: Matrices · ACL2 · Abstract stobjs · Matrix normal forms

1 Introduction

Computing normal forms of matrices is a wide subject which presents many applications in different areas of Mathematics. For instance, one of the fundamental processes in Linear Algebra is the resolution of systems of linear equations, and the constructive methods to carry that task out are based on the computation of triangular forms of a given matrix. In the same way, Smith normal form, a particular kind of equivalent diagonal matrix, plays an essential role in the theory of finitely generated modules over a ring and, in particular, it is a key result to determine the structure of a finitely generated abelian group. Smith form also provides a well-known method for finding integer solutions of

Supported by Ministerio de Ciencia e Innovación, projects TIN2013-41086-P and MTM2014-54151-P.

M. Ayala-Rincón and C.A. Muñoz (Eds.): ITP 2017, LNCS 10499, pp. 354–370, 2017.
DOI: 10.1007/978-3-319-66107-0_23

systems of linear Diophantine equations [11]. The key point of all these procedures is to ensure that the output matrix (a reduced form) preserves some of the fundamental invariants of the input matrix such as the row (column) space, the rank, the determinant, the elementary divisors and so on.

There exists a huge range of algorithmic methods for computing normal forms of matrices [12], which are based on well established mathematical results. Nevertheless, it is advisable to have verified programs available in order to avoid the possible inaccuracies which can occur during the path from algorithms to programs. The aim is the paper is to propose a data structure and a logical infrastructure to implement formally verified matrix normal forms algorithms, in the ACL2 theorem prover, with special emphasis on how to efficiently execute the verified algorithms.

The ACL2 system [1] is at the same time a programming language, a logic for reasoning about models implemented in the language, and a theorem prover for that logic. The programming language is an extension of an applicative subset of Common Lisp, and thus the verified algorithms can be executed, under certain conditions, in the underlying Common Lisp. ACL2 has several features mainly devoted to get an efficient execution of the algorithms, in a sound way with respect to the logic. Abstract single-threaded objects [1,7] is one of those features, providing a sound logical connection between efficient concrete data structures and more abstract data structures, convenient for reasoning. We propose here to use this feature to implement and formally verify matrix algorithms for computing normal forms.

In particular, we describe in this paper a formally verified implementation of an algorithm to compute a column echelon form of a matrix with elements in the ring of integers. This formalization is done as an initial step for developing computational homological algebra in the ACL2 system and in particular to calculating (persistent) homology [10]. But although we describe here the formalization of a specific normalization algorithm, we think this approach could be generalized to other normalization algorithms as well.

The organization of the paper is as follows. The next section is devoted to describe a formalization of matrices in ACL2, represented as lists of lists, and also a representation for matrix normalization problems. This representation is natural for reasoning, but has inefficiencies due to the applicative nature of Lisp lists. Section 3 describes how we can compute using more efficient data structures, and still have the more natural representation for reasoning, by means of ACL2 abstract single-threaded objects. In Sect. 4, we illustrate this infrastructure describing how we formally verified an algorithm for computing a column echelon form for integer matrices. The paper ends with some discussion about related work and conclusions. Due to the lack of space, we will omit some ACL2 definitions and skip some technical details (for example, all the functions declarations). The complete source files containing the ACL2 formalization are accessible at: http://www.glc.us.es/fmartin/acl2/mast-cef.

2   A Data Structure for Reasoning in the Logic

In this section, we describe a data structure that can be used to define a matrix normal form algorithm. This data representation is suitable for reasoning, but inefficient for execution, as we will see. We will refer to this as the *abstract representation*.

2.1   Matrices as Lists of Lists

A very natural way to represent a 2-dimensional matrix in ACL2 is as a list whose elements are lists of the same length, each one representing a row of the matrix. For example, the list '((1 0 0 0) (0 1 0 0) (0 0 1 0) (0 0 0 1)) represents the identity matrix of dimension 4. The following function matp is the recognizer for well-formed matrices represented as lists of lists:

```
(defun matp-aux (A ncols)
  (cond ((atom A) (equal A nil))
        (t (and (true-listp (first A))
                (equal (len (first A)) ncols)
                (matp-aux (rest A) ncols)))))

(defun matp (A)
  (if (atom A)
      (equal A nil)
      (and (consp (first A))
           (matp-aux A (len (first A))))))
```

Note that if (matp A), then the number of rows of A is given by its length, and the number of columns by the length of (for instance) its first element. In our formalization, these are defined by the functions nrows-m and ncols-m, respectively. We have also defined the function (matp-dim A m n) checking that A is a matrix of a given size $m \times n$. As we have said in the introduction, the algorithm we have formalized is restricted to matrices with elements in the ring of integers; the function integer-matp (and integer-matp-dim) recognizes the ACL2 object that are matp and with all its elements being integers.

Accessing and updating matrix elements is done via nth and update-nth, respectively, as defined by the following functions aref-m and update-aref-m:

```
(defun aref-m (A i j)
  (nth j (nth i A)))

(defun update-aref-m (i j val A)
  (update-nth i (update-nth j val (nth i A)) A))
```

Using this representation, these operations are not done in constant time, and updating is not destructive, since it follows the usual "update by copy" semantics of applicative lists. This is a drawback if we want efficient algorithms on matrices. In the next section we will show how to address this issue.

A typical definition scheme for matrix operations or matrix properties is by means of two nested loops, the outer iterating on its rows indices, and the inner on its column indices for a fixed row. In our formalization, this is done using two recursive functions. The following definition of the product of two matrices illustrates this recursion scheme:

```
(defun matrix-product-row-col (A B P i j cA cP)
  (cond ((or (not (natp j)) (not (natp cP))) P)
        ((>= j cP) P)
        (t (let ((P1 (update-aref-m i j
                                    (mp-res-i-j A B i j 0 cA) P)))
             (matrix-product-row-col A B P1 i (1+ j) cA cP)))))

(defun matrix-product-row (A B P i rP cA cP)
  (cond ((or (not (natp i)) (not (natp rP))) P)
        ((>= i rP) P)
        (t (let ((P1 (matrix-product-row-col A B P i 0 cA cP)))
             (matrix-product-row A B P1 (1+ i) rP cA cP)))))
```

Here mp-res-i-j implements the sum $\sum_k a_{ik}b_{kj}$, and P is a matrix with the same number of rows as A and the same number columns as B, where we store the resulting matrix product. Thus, matrix product is defined by the following function:

```
(defun matrix-product (A B)
  (let* ((rA (nrows-m A))
         (cA (ncols-m A))
         (cB (ncols-m B))
         (P (initialize-mat rA cB nil)))
    (matrix-product-row A B P 0 rA cA cB)))
```

Using this representation for matrices, we proved a number of well-known algebraic properties of matrix operations. For example, the following are the statements for product associativity and right identity (where matrix-id defines the identity matrix of a given dimension):

```
(defthm matrix-product-associative
  (implies (and (matp A) (matp B)
                (equal (nrows-m B) (ncols-m A))
                (equal (nrows-m C) (ncols-m B)))
           (equal (matrix-product (matrix-product A B) C)
                  (matrix-product A (matrix-product B C)))))

(defthm matrix-product-right-identity
  (implies (integer-matp-dim A (len A) n)
           (equal (matrix-product A (matrix-id n)) A)))
```

A general technique we used to prove most of these algebraic properties is based on the property that (equal P Q) if P and Q are matrices of the same dimension $m \times n$ such that $p_{ij} = q_{ij}$ for $0 <= i < m, 0 <= j < n$. We proved this property in a general way using the ACL2 encapsulation mechanism, and then we use it by functional instantiation, after proving the corresponding algebraic property for the individual entries of both sides of the equality. See the book `matrices-lists-of-lists.lisp` in the supporting materials, for details.

2.2 An Abstract Representation for Matrix Normal Form Computation

Algorithms that compute matrix normal forms, often compute also transformation matrices that relate the original matrix with its normal form. For example, in the algorithm we describe in Sect. 4, the goal is to obtain, for a given matrix A, a matrix H in a desired normal form and an invertible transformation matrix[1] T such that $A \cdot T = H$. A general description of a matrix normal form algorithm could be the following: we operate on two matrices, initially the original matrix and the identity matrix; at every step, an elementary transformation (or *operator*) is applied to the first matrix and the same transformation is applied to the second matrix; when the algorithm stops, we have H and T with the desired properties.

We now explain a possible data structure for such algorithms, which turns out to be natural for reasoning. First, we will represent the matrix A being transformed, using the list of lists representation described in the previous subsection. For the transformation matrix T we adopt a different approach: although the executable algorithm will deal with the whole matrix, in the logic it will be more convenient to see that transformation matrix as a list of operators, describing the sequence of elementary transformations carried out; and each operator will be a short description of the transformation. The reason is that it is easier to prove the properties of the transformation matrix, if we explicitly have the sequence of elementary transformations that this matrix represents.

For our concrete normal form algorithm described in Sect. 4, it turns out that only one type of elementary transformation is needed[2]: given two distinct column indices c1 and c2 and four integers x1, x2, y1 and y2, this transformation replaces column c1 by the linear combination of column c1 times x1 plus column c2 times x2, and also replaces column c2 by the linear combination of column c1 times y1 plus column c2 times y2. We will call this operator a *linear combination of columns (lcc)*, and in the logic it will be represented as the list (c1 c2 x1 x2 y1 y2). In our formalization, the function (lcc-op l n) checks if l is such

---

[1] Some algorithms for computing matrix normal forms, like the Smith normal form, need to compute two transformation matrices, but similar ideas would apply in that case.

[2] Of course, other normal forms algorithms needs different elementary transformations, and possible more than one. But again, the same ideas described here could be applied in such cases.

operator, where c1 and c2 are less than n. And (lcc-op-seq seq n) checks if
seq is a list of lcc operators.

The above considerations lead us to the following predicate mast$ap, recognizing the data representation we have just described (the prefix $a is for
"abstract"):

```
(defun mast$ap (x)
  (and (true-listp x)
       (equal (len x) 2)
       (let ((A (first x))
             (seq (second x)))
         (cond ((atom A) (and (equal A nil) (equal seq nil)))
               (t (and (integer-matp-dim A (nrows-m A) (ncols-m A))
                       (lcc-op-seq seq (ncols-m A))))))))
```

We have defined a number of functions that operate on this data structure.
The main operation is linear combination of columns. For that, we first need to
define the function lin-comb-cols-1st, which effectively carries out the linear
combination of columns on a given matrix. Note that here we have an extra
parameter max-r, which indicates a row index. This allows us to perform the
linear combination of columns only until that row, but not below (the reason is
that during the transformation process, we will be sure that there will only be
zeros below a given row):

```
(defun lin-comb-cols-1st-rows (A c1 c2 r max-r x1 x2 y1 y2)
  (cond ((or (not (natp max-r)) (not (natp r))) A)
        ((> r max-r) A)
        (t (let* ((Arc1 (aref-m A r c1))
                  (Arc2 (aref-m A r c2))
                  (nArc1 (+ (* x1 Arc1) (* x2 Arc2)))
                  (nArc2 (+ (* y1 Arc1) (* y2 Arc2)))
                  (nA (update-aref-m r c2 nArc2
                                     (update-aref-m r c1 nArc1 A))))
             (lin-comb-cols-1st-rows nA c1 c2 (1+ r) max-r
                                     x1 x2 y1 y2)))))

(defun lin-comb-cols-1st (A c1 c2 max-r x1 x2 y1 y2)
  (lin-comb-cols-1st-rows A c1 c2 0 max-r x1 x2 y1 y2))
```

Now, the following function implements the lcc transformation on our
abstract representation. Note that the transformation is only effectively carried
out on the first matrix:

```
(defun lin-comb-cols$a (mast$a c1 c2 max-r x1 x2 y1 y2)
  (list (lin-comb-cols-1st (first mast$a) c1 c2 max-r x1 x2 y1 y2)
        (cons (list c1 c2 x1 x2 y1 y2) (second mast$a))))
```

We would like to define our matrix normal form algorithm using this and other functions defined on the abstract representation, but as we have said we can improve execution if we do not use applicative lists. And also, probably, if we were not interested in formal verification, we wouldn't have dealt with lcc operators, but with the whole transformation matrix instead.

3   Using Abstract Stobjs to Represent Matrices

So let us now define an executable and efficient data structure representation, and see how we can relate it to the abstract representation described above. Efficient execution is achieved in the ACL2 system mainly by means of two features: guards and single threaded objects. The *guard* of a function is a specification of its intended domain. Although functions in the ACL2 logic are total, guards provide a way to specify and verify the inputs for which the function can be safely executed directly in the underlying raw Common Lisp. A *guard-verified* function respects the guards of all the functions that it calls (including itself in case of a recursive function). All the functions involved in the algorithm of Sect. 4 have been guard-verified.

The second feature related to efficient execution is provided by single threaded objects (*stobjs*). These are data structures that allow accessing and updating in constant time, and destructive updates on them. When an object is declared to be single-threaded, ACL2 enforces certain syntactic restrictions on its use, ensuring that in every moment, only one copy of the object is needed (for example, one of these restrictions requires that if a function updates a stobj, then it has to return the stobj). With these restrictions, the destructive updates are consistent with the applicative functional semantics of ACL2.

Therefore, it would be good if we can execute our matrix algorithms using stobjs. Nevertheless, although we can use arrays as fields of a stobj, those arrays have to be 1-dimensional and accessing and updating the array is only allowed via elementary operations, so reasoning directly using this representation could be difficult. Fortunately, another ACL2 feature, abstract stobjs, will allow us to define an alternative logical interface for the stobj.

3.1   A Stobj for Computing Matrix Normal Forms

Before describing the abstract stobj we have used, let us show the corresponding stobj, where the execution will take place (we will call this the *concrete representation*). In ACL2, a stobj is defined, using `defstobj`, as a structure with a number of fields, where each field can be either of array type or of non-array type. In our case, we will define a stobj with two 1-dimensional array fields, each one storing the elements of a 2-dimensional matrix, in linearized form. The idea is that one of the 1-dimensional arrays stores the matrix being transformed, and the other stores the transformation matrix. We also need two non-array fields, to store the number of rows and the number of columns of the first matrix. The following defines this stobj (the `$c` suffix is for *concrete*):

```
(defstobj mast$c
  (nrows$c  :type (integer 0 *) :initially 0)
  (ncols$c  :type (integer 0 *) :initially 0)
  (matrix$c :type (array integer (0)) :initially 0 :resizable t)
  (trans$c  :type (array integer (0)) :initially 0 :resizable t))
```

Array fields in stobjs are defined in the logic as ordinary lists, but for execution in the underlying Lisp, raw Lisp arrays are used. The effect of this ACL2 form is to introduce the stobj mast$c and its associated recognizers, creator, accessors, updaters, and length and resize functions for its fields. For example, given an index i, (matrix$ci i mast$c) and (update-matrix$ci i v mast$c) respectively access and update (with value v) the i-th cell of the matrix$c array. Similar functions are defined for the trans$c array. These operations are executed in constant time and the update is destructive (at the price of syntactic restrictions on the use of the stobj). Logically speaking, they are defined in terms of nth and update-nth.

We have defined a number of functions operating on this concrete representation. Let us show, for example, how we implement the linear combination of columns. First, the following function performs that operation on the first matrix (we omit some technical details):

```
(defexec lin-comb-cols-matrix$c-rows
                        (mast$c i j s r max-r x1 x2 y1 y2)
  ...
  (cond ((> r max-r) mast$c)
    (t (let* ((mat-i (mat$ci i mast$c))
              (mat-j (mat$ci j mast$c))
              (new-mat-i (+ (* x1 mat-i) (* x2 mat-j)))
              (new-mat-j (+ (* y1 mat-i) (* y2 mat-j))))
          (seq mast$c
               (update-mat$ci i new-mat-i mast$c)
               (update-mat$ci j new-mat-j mast$c)
               (lin-comb-cols-matrix$c-rows mast$c (+ i s) (+ j s)
                          s (1+ r) max-r x1 x2 y1 y2)))))))

(defun lin-comb-cols-matrix$c (mast$c c1 c2 max-r x1 x2 y1 y2)
  (lin-comb-cols-matrix$c-rows
     mast$c c1 c2 (ncols$c mast$c) 0 max-r x1 x2 y1 y2))
```

Here i and j are indices of positions in the 1-dimensional array (initially, c1 and c2, respectively), and r is the current row of the corresponding 2-dimensional array (initially 0). Note that to move to the next row in both columns, we add s (the number of columns) to both indices.

In a very similar way, we define a function lin-comb-cols-trans$c that performs the same operation on the trans$c 1-dimensional array. And finally, we sequentially apply both transformations (note that the operation on the transformation matrix is performed until the last row):

```
(defun lin-comb-cols$c (mast$c c1 c2 max-r x1 x2 y1 y2)
  (seq mast$c
       (lin-comb-cols-matrix$c mast$c c1 c2 max-r x1 x2 y1 y2)
       (lin-comb-cols-trans$c mast$c c1 c2
                              (1- (ncols$c mast$c)) x1 x2 y1 y2)))
```

3.2   The Abstract Stobj

Until now, we have defined an abstract representation (convenient for reasoning), and also a concrete representation (suitable for execution). In both representations, we have defined functions that perform the main operations needed for our matrix normal form algorithm. Now we can combine the best of both representations, thanks to abstract stobjs.

But before we have to introduce a (non-executable) *correspondence predicate*, describing in what sense the concrete and the abstract representations are related. Basically: the concrete representation stores the size of the matrix in the abstract representation; the first 1-dimensional array of the concrete representation is a linearized version of the matrix of the abstract one; and the second 1-dimensional matrix of the concrete representation is a linearized version of the result of applying the sequence of lcc operators of the abstract representation, to the identity matrix:

```
(defun-nx mast$corr (mast$c mast$a)
  (let ((nrows (len (first mast$a)))
        (ncols (len (first (first mast$a)))))
    (and (equal nrows (nth 0 mast$c))
         (equal ncols (nth 1 mast$c))
         (equal (append-lst (first mast$a)) (nth 2 mast$c))
         (equal (append-lst (apply-lcc-op-seq (second mast$a)
                                              (matrix-id ncols)))
                (nth 3 mast$c)))))
```

Here (append-lst ls) is a function that given a list of lists ls, concatenates all of them into one single list. And apply-lcc-op-seq is a function that applies a sequence of lcc operators to a given matrix. Here it is its definition:

```
(defun apply-lcc-op (op A)
  (let ((c1 (nth 0 op)) (c2 (nth 1 op)) (x1 (nth 2 op))
        (x2 (nth 3 op)) (y1 (nth 4 op)) (y2 (nth 5 op)))
    (lin-comb-cols-lst A c1 c2 (1- (nrows-m A)) x1 x2 y1 y2)))

(defun apply-lcc-op-seq (seq A)
  (cond ((endp seq) A)
        (t (apply-lcc-op (first seq)
                         (apply-lcc-op-seq (rest seq) A)))))
```

Note that this function is only for specification. In particular, we apply
lin-comb-cols-1st to all the rows of the matrix and not only until a given
row, since that optimization will only make sense in the particular implementation of a normalization algorithm.

Now we can define the abstract stobj that provides a sound logical connection between both representations. In ACL2, a `defabsstobj` event defines an
abstract single-threaded object that is proven to satisfy a given invariant property, and that can only be accessed or updated by some given functions called
exports. These functions have an abstract definition that ACL2 uses for reasoning and a different concrete implementation that ACL2 uses for execution on
a corresponding concrete stobj. In our case, this is the abstract stobj we have
defined:

```
(defabsstobj mast
  :exports (initialize-mast
            nrows ncols
            aref-mat
            lin-comb-cols
            get-mat
            get-trans))
```

Here `initialize-mast` is a function that given an initial matrix A (as a list
of lists), stores it in the abstract stobj. The abstract definition for this export
is straightforward (simply returns (list A nil)), but the concrete executable
definition is far more difficult, since it has to store each element of A and each
element of the identity matrix in the corresponding 1-dimensional arrays of the
stobj. As for lin-comb-cols, we have already presented its abstract and concrete definitions. These two exports update the abstract stobj, and the rest of
the exports are only accessors: nrows and ncols give the number of rows and
columns of the first matrix, aref-mat access to an element of the first matrix
by its row and column indices; and get-mat and get-trans return, respectively,
the first and the second matrices, as list of lists. Note that again this is easy
for the abstract representation (especially get-mat) but it is not trivial for the
concrete definitions.

Unless specified, the names for the corresponding concrete stobj, the correspondence predicate, and for the abstract and concrete functions associated
with each export, are obtained appending the suffixes $a (abstract) or $c (concrete) to the names given in the defabsstobj. To accept a defabsstobj event,
all these corresponding abstract and concrete functions have to be previously
defined, their guards verified, and also a number of proof obligations automatically generated by the event must be already proved. These proof obligations
guarantee that the correspondence between the abstract and the concrete representation, the recognizer property, and the guards of the exports are preserved
after updating the stobj, and also that the abstract and the concrete corresponding accessors return the same values. That is, the proof obligations essentially
guarantee that reasoning with the abstract representation and executing with

the concrete representation is logically sound. See `matrices-abstobj.lisp` for the statements of all these proof obligations and a proof of them.

Once this abstract stobj is defined, we can use it as the data structure for a matrix normal form algorithm, provided that the single-threadedness syntactic restrictions are met. The only primitive functions we can use to access or update the abstract stobj are the exports. We emphasize that when proving theorems about the algorithm, ACL2 uses the abstract definitions of the exports (that is, the ones with the `$a` suffix); but for execution, it uses the concrete data structure and definitions (that is, the ones with the `$c` suffix).

4   An Algorithm to Compute a Column Echelon Form

We illustrate how we can use the described absstobj framework, by means of a verified implementation of an algorithm that given a matrix of integers A, computes an equivalent integer matrix C that it is in column echelon form, together with a unimodular transformation integer matrix T such that $A \cdot T = C$. We say that a matrix C is in *column echelon form* if zero columns of C precede nonzero columns and, for each nonzero column of C, its leading entry (the last nonzero element of the column) is above the leading entries of the following columns. This notion of column echelon form is not exactly the same as other classical echelon forms usually defined in the literature, such as Hermite or Howell forms. Nevertheless, as we have said in the introduction, this has to be considered in the context of developing ACL2 programs to compute homology groups of chain complexes, and it turns out that this simple echelon transformation is suitable for this task. And anyway, our main purpose here is to illustrate with this example how we can apply the absstobj infrastructure just described.

Although the algorithm is implemented for integer matrices, it could be generalized to matrices in a more general class of rings, namely, the class of Bézout domains. Roughly speaking, a Bézout domain is an integral domain where every finite ideal is principal. This property is equivalent to the existence of an explicit Greatest Common Divisor (*gcd*) operation providing the Bézout identity of every pair of elements: if d is the gcd of two elements a and b, there exist two elements x and y such that $d = ax + by$. Note that in a ring we do not have in general the inverse of an element, so we cannot apply here usual techniques employed when the entries are in a field (like Gaussian elimination).

4.1   Definition of the Algorithm

Let us now present the ACL2 implementation of the column echelon form algorithm. First, a key ingredient is the extended Euclides algorithm which, besides the greatest common divisor of two integers, computes the coefficients of the Bézout identity. In particular, we have defined a function (`bezout a b`) such that given two integers a and b, returns a tuple of integers (g s_1 t_1 s_2 t_2) such that $g = gcd(a, b), s_1 a + t_1 b = d$ and $s_2 a + t_2 b = 0$. Note that these properties can be expressed in matrix form:

$$(a \ b) \cdot \begin{pmatrix} s_2 \ s_1 \\ t_2 \ t_1 \end{pmatrix} = (0 \ \ gcd(a,b))$$

This 2×2 matrix has the property that it is unimodular (determinant 1 or -1) and thus invertible *in the ring of integers*. It is an elementary transformation matrix that can be also easily generalized to size $n \times n$, in such a way that right multiplication by this elementary matrix is just like applying a lcc operator. Essentially, the algorithm iteratively applies this transformation with the aim of obtaining the zero entries needed in the echelon form. This is done from the last row to the first one, and in every row, from a given column to the first one.

The following functions implement the algorithm operating on the abstract stobj mast. This means that the only elementary operations we can apply to mast are the exports specified in its defabsstobj. The first function is cef-bezout-row-col below, which given a row index (- i 1) and column indices (- c 1) and (- j 1), apply the lcc transformation on those columns, and thus obtaining a zero in the position of row (- i 1) and column (- c 1), using as pivot the entry of the same row and column (- j 1). This is done when we already know that the entries of the given columns that are below the given row are already zero, so it is justified to do the linear combination only until that row:

```
(defun cef-bezout-row-col (mast c i j)
  (mv-let (g s1 t1 s2 t2)
      (bezout (aref-mat mast (- i 1) (- c 1))
              (aref-mat mast (- i 1) (- j 1)))
    (lin-comb-cols mast (- c 1) (- j 1) (- i 1) s2 t2 s1 t1)))
```

Given the position of a pivot, this lcc transformation is applied for all the columns to the left, obtaining zeros in the row of the pivot, until the column of the pivot. This recursive process is carried out by the function cef-reduct-row-col and initiated by the function cef-reduct-row, from a given pair of row and column indices:

```
(defun cef-reduct-row-col (mast c i j)
  (cond ((zp c) mast)
        (t (seq mast
                (cef-bezout-row-col mast c i j)
                (cef-reduct-row-col mast (- c 1) i j)))))

(defun cef-reduct-row (mast i j)
  (cond ((zp j) mast)
        (t (cef-reduct-row-col mast (- j 1) i j))))
```

To get the echelon form, we iteratively apply this process from the last row to the first one. We also have to take into account that the column of the pivot is changing from one row to the next, depending on the result obtained after reducing that row. If we have a zero in the position of the pivot, the column of the pivot is unchanged. Otherwise is decremented by one:

```
(defun cef-row-col (mast i j)
  (cond ((or (zp i) (zp j)) mast)
        (t (let ((mast (cef-reduct-row mast i j)))
             (if (= (aref-mat mast (- i 1) (- j 1)) 0)
                 (cef-row-col mast (- i 1) j)
                 (cef-row-col mast (- i 1) (- j 1)))))))
```

Given an input matrix A (represented as lists of lists). The algorithm is initiated calling the export `initialize-mast`, and then the function `cef-row-col`, starting in the last row and columns:

```
(defun cef (A mast)
  (seq mast
       (initialize-mast mast A)
       (cef-row-col mast (nrows mast) (ncols mast))))
```

Note that the above function `cef` receives as input the `mast` abstract stobj and thus, due to the single-threadedness requirements, it has to return also the abstract stobj. Nevertheless, we can define a function `cef-matrix` in which the input and output are not explicitly connected to the stobj. This can be done using `mast` locally (by means of `with-local-stobj`), and finally returning the computed matrices represented as lists of lists (using the exports `get-mat` and `get-trans`):

```
(defun cef-matrix (A)
  (with-local-stobj mast
    (mv-let (mast mat trans)
            (seq mast
                 (cef A mast)
                 (mv mast (get-mat mast) (get-trans mast)))
            (mv mat trans))))
```

4.2   Main Theorems Proved

We proved in ACL2 the following theorems, stating that given an integer matrix A, the algorithm `cef-matrix` computes an equivalent integer matrix that is in column echelon normal form:

```
(defthm cef-cef-matrix
  (implies (integer-matp A)
           (let ((H (first (cef-matrix A))))
             (and (integer-matp-dim H (nrows-m A) (ncols-m A))
                  (cef-p H)))))

(defthm matrix-product-cef-matrix
  (implies (integer-matp A)
           (let ((H (first (cef-matrix A)))
```

```
                 (TR (second (cef-matrix A))))
             (and (integer-matp-dim TR (ncols-m A) (ncols-m A))
                  (equal (matrix-product A TR) H)))))

(defthm inverse-matrix-cef-matrix
   (implies (integer-matp A)
            (let ((TR (second (cef-matrix A)))
                  (TR-INV (cef-matrix-transinv A)))
               (and (equal (matrix-product TR TR-INV)
                           (matrix-id (ncols-m A)))
                    (equal (matrix-product TR-INV TR)
                           (matrix-id (ncols-m A)))))))
```

In the first of three above theorems, the function cef-p is a predicate checking that a matrix is in column echelon form. The result is proved by defining a more general invariant about the form of the matrix during the transformation process; the stopping condition of the algorithm and this invariant implies the theorem.

The second theorem establishes that the second matrix computed by the algorithm is indeed the transformation matrix. This is also an invariant of the process, and note that we have to deal also with the fact that we do the linear combination only until a given row, since from that row on, we have zeros. Additionally, we need to prove the relation between the linear combination of columns carried out by lin-comb-cols and the matrix product by the elementary transformation matrix that can be obtained from a lcc operator.

Finally the third theorem establishes that the transformation matrix is invertible, where (cef-matrix-transinv A) is a function that obtains the inverse of the transformation matrix computed by the algorithm. We emphasize that the abstract representation is specially convenient, among other reasons, for defining this function and proving the theorem. This is its definition:

```
(defun-nx cef-matrix-transinv (A)
  (let ((res (cef A '(nil nil))))
    (apply-lcc-op-seq
     (inv-lcc-op-seq (second res)) (matrix-id (ncols-m A)))))
```

Given a lcc operator whose coefficients have been obtained as the result of an application of the extended Euclides algorithm, then we can prove that there exists a corresponding lcc operator describing the inverse linear combination (that is, the operator is invertible). Given a sequence of lcc invertible operators, the function inv-lcc-op-seq, obtains the reversed sequence of the inverses of each operator. We apply this function to the sequence of operators stored in the second element of the final abstract stobj computed by the algorithm, and then we apply this inverse sequence to the identity matrix. Note that we are taking advantage from the fact that our abstract representation contains the lcc operators explicitly (although our executable concrete representation deals only with the transformation matrix, not with the abstract operators).

For details about the ACL2 proof of these theorems, we urge the interested reader to consult the supporting materials, books `matrices-abstobj-properties.lisp` and `cef-mast.lisp`.

4.3   Experimental Results

To check how this formally verified abstract stobj implementation influences the execution performance of the algorithm, we tested it on several random matrices of different sizes. We compared it to two other implementations of the same algorithm: an analogous unverified ACL2[3] implementation, that uses matrices represented as applicative lists of lists, instead of the abstract stobj; and also an iterative version of the same algorithm in Python 3, using (mutable) lists, which have accesses and updates in constant time. For each size, we generated a number of matrices, and averaged the execution time obtained.

Table 1. Execution times for random matrices

Size	10	20	30	40
List	0.00	0.01	0.54	153.83
Mast	0.00	0.01	0.53	151.96
Python	0.00	0.01	0.59	55.90

Table 2. Execution times for random first column based matrices

Size	160	170	180	190	200
List	32.82	42.07	53.36	65.79	82.62
Mast	0.19	0.23	0.27	0.33	0.38
Python	2.62	3.10	3.60	4.71	5.81

In the Table 1, we show the execution time for random matrices until size 40×40. We see that for sizes below 30×30, the execution times are good for the three implementations. Nevertheless, for sizes 40×40 and bigger, the execution times become unacceptable for both ACL2 implementations, and even for the Python implementation. Nevertheless, we conjecture that the data structures used are not responsible of this slow down: this algorithm and other dealing with integers matrices, usually generate very big numbers. A naive treatment of the arithmetic operations is not enough for dealing with this complexity (and the techniques usually applied [6,12] are out of the scope of this paper).

To concentrate on how the data structures used really influence the execution times, we generate matrices of sizes until 200×200, in which only the first column is random, and the rest of the columns are multiples of the first one. In this way, the arithmetic operations are very straightforward, and the execution times essentially come from accessing and updating the arrays. These execution times are shown in the Table 2. We can see that the applicative ACL2 version is also very slow for that sizes, but the ACL2 abstract stobjs implementation is fast, and even better than the Python implementation.

---

[3] We used ACL2 Version 7.2 compiled with SBCL 1.2.16.

5   Related Work and Conclusions

We have presented in this paper an approach to formally verify matrix normal form algorithms, while still having efficient data structures for execution. For that, we use the ACL2 system and in particular abstract single-threaded objects, which allow both a convenient logical representation of data and a more efficient concrete representation for execution. We have illustrated this approach showing an ACL2 formal verification of an algorithm to compute echelon forms of integer matrices.

Several formalizations in which matrix algebra plays an important role have been presented in most of theorem provers. For example, using the Coq system [4,8,9] or in Isabelle [2,3]. In all these works, the emphasis is mainly put in the formalization, and in particular they formalize more general results with respect to the algebraic structures involved. In [2] it is also described how to speed-up execution times of the formalized algorithm, first by data type refinements and then by generating code to be executed in a functional programming language. In our case, the approach is different: since ACL2 is built on top of Common Lisp and the logic formalizes an applicative subset of it, we reason directly on the final implementation and execution and reasoning is carried out on the same system.

In addition to stobjs, ACL2 provides 2-dimensional arrays, which under reasonable assumptions provide access in constant time to the entries of the array. This data structures is used in [5] to formalize some common operations and properties of matrices in ACL2. However, the stobj approach is generally more efficient when there are updates [1].

We think abstract stobjs provide a suitable framework for dealing with matrices in ACL2. They provide a clean separation between the data structures used for execution, and the properties of the algorithms that operate on them. In particular, we think the approach shown here for a concrete matrix normalization algorithm can be applied in general to other algorithms that compute normal forms of matrices.

It is worth noting that previous to the introduction of abstract stobjs in ACL2, it was also possible to have a similar formalization strategy: we could have defined two different versions of the algorithm (abstract and concrete, stobj based), prove the main properties of the abstract algorithm and then prove that both versions compute the same results. Now, abstract stobjs provide sound and enhanced support from the system, to carry out this proof strategy: first, we can specify in advance the elementary operations (exports) that will be allowed to operate on the data structures; and second, once introduced, we can concentrate on the abstract definitions, to reason about the properties of the algorithms that use it. A significant downside of the older approach was that one had to prove the correspondence between every newly introduced concrete and abstract function, whereas all such work is done once and for all when using abstract stobjs, thereby easing the maintenance of a formally verified ACL2 implementation.

References

1. ACL2 version 7.4. http://www.cs.utexas.edu/users/moore/acl2/
2. Aransay, J., Divasón, J.: Formalisation in higher-order logic and code generation to functional languages of the Gauss-Jordan algorithm. J. Funct. Program. **25**(9), 1–21 (2015)
3. Aransay, J., Divasón, J.: Formalization of the computation of the echelon form of a matrix in Isabelle/HOL. Form. Asp. Comput. **28**, 1005–1026 (2016)
4. Cano, G., Cohen, C., Dénès, M., Mörtberg, A., Siles, V.: Formalized linear algebra over elementary divisor rings in Coq logical methods in computer. Science **12**(2), 1–29 (2016)
5. Cowles, J., Gamboa, R., Van Baalen, J.: Using ACL2 arrays to formalize matrix algebra. In: Proceedings of ACL2 2003 (2003)
6. Domich, P.D., Kannan, R., Trotter Jr., L.E.: Hermite normal form computation using modulo determinant arithmetic. Math. Oper. Res. **12**, 50–69 (1987)
7. Goel, S., Hunt Jr., W.A., Kaufmann, M.: Abstract stobjs and their application to ISA modeling. In: Proceedings of ACL2 2013, pp. 54–69 (2013)
8. Gonthier, G.: Point-free, set-free concrete linear algebra. In: van Eekelen, M., Geuvers, H., Schmaltz, J., Wiedijk, F. (eds.) ITP 2011. LNCS, vol. 6898, pp. 103–118. Springer, Heidelberg (2011). doi:10.1007/978-3-642-22863-6_10
9. Heras, J., Coquand, T., Mörtberg, A., Siles, V.: Computing persistent homology within Coq/SSReflect. ACM Trans. Comput. Log. **14**(4), 1–26 (2013)
10. Lambán, L., Martín-Mateos, F.-J., Rubio, J., Ruiz-Reina, J.-L.: Towards a verifiable topology of data. In: Proceedings of EACA-2016, pp. 113–116 (2016)
11. Newman, M.: The Smith normal form. Linear Algebra Appl. **254**, 367–381 (1997)
12. Storjohann, A.: Algorithms for matrix canonical forms. Ph.D. thesis, Swiss Federal Institute of Technology, Zurich (2013)

Typing Total Recursive Functions in Coq

Dominique Larchey-Wendling[✉]

LORIA – CNRS, Nancy, France
dominique.larchey-wendling@loria.fr

Abstract. We present a (relatively) short mechanized proof that Coq types any recursive function which is provably total in Coq. The well-founded (and terminating) induction scheme, which is the foundation of Coq recursion, is maximal. We implement an unbounded minimization scheme for decidable predicates. It can also be used to reify a whole category of undecidable predicates. This development is purely constructive and requires no axiom. Hence it can be integrated into any project that might assume additional axioms.

1 Introduction

This paper contains a mechanization in Coq of the result that any total recursive function can be represented by a Coq term. A short slogan could be *Coq types any total recursive function,* but that would be a bit misleading because the term *total* might also refer to the *meta-theoretical level* (see Sect. 7).

The theory of partial recursive (or μ-recursive) functions describes the class of recursive functions by an inductive scheme: it is the least set of partial functions $\mathbb{N}^k \longrightarrow \mathbb{N}$ containing constant functions, zero, successor and closed under composition, recursion and *unbounded minimization* [9]. Forbidding minimization (implemented by the μ operator) leads to the sub-class of primitive recursive functions which are total functions $\mathbb{N}^k \longrightarrow \mathbb{N}$. Coq has all the recursive schemes except unbounded minimization so it is relatively straightforward to show that any primitive recursive function $f : \mathbb{N}^k \longrightarrow \mathbb{N}$ can be represented by a Coq term $t_f : \mathcal{N}^k \to \mathcal{N}$ where $\mathcal{N}$ is a short notation for the Coq type nat of Peano natural numbers. To represent all partial recursive functions $\mathbb{N}^k \longrightarrow \mathbb{N}$ by Coq terms, we would first need to deal with partiality and change the type into $\mathcal{N}^k \to$ option $\mathcal{N}$ (for instance) because (axiom-free) Coq only contains total functions; so here the term None : option $\mathcal{N}$ represents the undefined value. Unfortunately, this does not work because Coq (axiom-free) meta-level normalization would transform such an encoding into a solution of the *Halting problem*.

Then, from a theoretical standpoint one question remains: which are the functions that Coq can represent in the type $\mathcal{N}^k \to \mathcal{N}$. In this paper, we give a mechanized proof that formally answers of half of the question:

The type $\mathcal{N}^k \to \mathcal{N}$ contains every recursive function of arity k which can be proved total in Coq.

Work partially supported by the TICAMORE project (ANR grant 16-CE91-0002).

M. Ayala-Rincón and C.A. Muñoz (Eds.): ITP 2017, LNCS 10499, pp. 371–388, 2017.
DOI: 10.1007/978-3-319-66107-0_24

Such a result was hinted in [2] but we believe that mechanizing the suggested approach implies a lot of work (see Sect. 2). This property of totality of Coq can compared to the characterization of System F definable functions as those which are provably total in AF_2 [5]. Besides the fact that AF_2 and Coq are different logical frameworks, the main difference here is that we mechanize the result inside of Coq itself whereas the AF_2 characterization is proved at the meta-theoretical level.

Before the detailed description of our contributions, we want to insist on different meanings of the *notion of function* that should not be confused:

- The μ-recursive schemes are the constructors of an inductive type of *algorithms* which are the "source code" and can be interpreted as partial function $\mathbb{N}^k \longrightarrow \mathbb{N}$ in Set theory or as predicates $\mathcal{N}^k \to \mathcal{N} \to \mathtt{Prop}$ in Coq;
- The *Set-theoretic notion of partial function* is a *graph/relation* between elements and their images. μ-recursive functions should not be understood independently of the algorithm that implements theses relations: it is impossible to recover an algorithm from the data of the graph alone;
- Then Coq has *function types* $A{\to}B$ which is a related but nevertheless entirely different notion of function and we rather call them predicates here.

Now let us give a more detailed description of the result we have obtained. We define a dependent family of types $\mathcal{A}_k$ representing recursive algorithms of arity $k : \mathcal{N}$. An algorithm $f : \mathcal{A}_k$ defines a (partial) recursive function denoted $[\![f]\!]$ and which is represented in Coq as a predicate $[\![f]\!] : \mathcal{N}^k \to \mathcal{N} \to \mathtt{Prop}$:

The proposition $[\![f]\!]\ \boldsymbol{v}\ x$ reads as: the computation of the algorithm f from the input k-tuple $\boldsymbol{v}$ terminates and results in x.

The implementation of the relation $[\![f]\!]$ is a simple exercise. It is more difficult to show that whenever the relation $(\boldsymbol{v}, x) \mapsto [\![f]\!]\ \boldsymbol{v}\ x$ between the input $\boldsymbol{v}$ and the result x is *total*, then there is a term $t_f : \mathcal{N}^k \to \mathcal{N}$ (effectively computable from f) such that the result of the computation of f on the input $\boldsymbol{v}$ is $(t_f\ \boldsymbol{v})$ for any $\boldsymbol{v} : \mathcal{N}^k$. This is precisely what we show in the following formal statement:

$$\forall (k : \mathcal{N})(f : \mathcal{A}_k),\ (\forall \boldsymbol{v}, \exists x, [\![f]\!]\ \boldsymbol{v}\ x) \to \{t_f : \mathcal{N}^k \to \mathcal{N} \mid \forall \boldsymbol{v}, [\![f]\!]\ \boldsymbol{v}\ (t_f\ \boldsymbol{v})\} \quad \text{(CiT)}$$

The statement means that if $[\![f]\!]$ represents a *total function* $(\forall \boldsymbol{v}, \exists x, [\![f]\!]\ \boldsymbol{v}\ x)$, then it can be *effectively transformed* into a Coq term $t_f : \mathcal{N}^k \to \mathcal{N}$ such that $(t_f\ \boldsymbol{v})$ is the value computed by the recursive function $[\![f]\!]$ on the input $\boldsymbol{v}$.

As we already pointed out, "Coq is Total" (CiT) is only one half of the characterization of the predicates that are definable in the type $\mathcal{N}^k \to \mathcal{N}$. The other half of the characterization, i.e. any predicate of type $\mathcal{N}^k \to \mathcal{N}$ corresponds to a μ-recursive function, while meta-theoretically provable for axiom-free Coq, cannot not be proved within Coq itself; see Sect. 7.

We will call *reification* the process of transforming a non-informative predicate like $P : \forall \boldsymbol{v}, \exists x, [\![f]\!]\ \boldsymbol{v}\ x$ into an informative predicate $Q : \forall \boldsymbol{v}, \{x \mid [\![f]\!]\ \boldsymbol{v}\ x\}$.[1]

---

[1] From which the term $t_f := \boldsymbol{v} \mapsto \mathtt{proj1\_sig}(Q\ \boldsymbol{v})$ of (CiT) is trivially derived.

In its general form, reification is a map from inhabited X : Prop to X : Type; it transforms a non-informative proof of existence of a witness into an effective witness. In a proof system like HOL for instance, reification is built-in by Hilbert's epsilon operator. On the contrary, because of its constructive design, Coq does not allow unrestricted reification. If needed in its full generality, it requires the addition of specific axioms as discussed in Sects. 3.1 and 8.

One of the originalities of this work is that the proof we develop is purely constructive (axiom free) and *avoids the detour through small-step operational semantics*, that is the use of a model of computation on an encoded representation of recursive functions. For instance, programs are represented by numbers (Gödel coding) in the proof of the *S-m-n* theorem [13]. It is also possible to use other models of computations such as register machines (or Turing machines) or even λ-calculus as in [8] or in our own dependently typed implementation [7] of Krivine's reference textbook [6]; see Sect. 7.

In Sect. 2, we present an overview of the consequences of the use of small-step operational semantics and how we avoid it. In Sect. 3 we describe how to implement unbounded minimization of inhabited decidable predicates in Coq. Section 4 presents the inductive types we need for our development, most notably the dependent type $\mathcal{A}_k$ of recursive algorithms of arity k and Sect. 5 defines three different but equivalent semantics for $\mathcal{A}_k$, in particular a decidable *cost aware big-step semantics* which is the critical ingredient to avoid small-step semantics. Section 6 concludes with the formal statement of (CiT) and its proof. In Sect. 7, we discuss related work and/or alternative approaches. In Sect. 8, we describe how to reify undecidable predicates (under some assumptions of course), in particular, *provability* predicates, *normalizability* predicates and even arbitrary *recursively enumerable predicates*. Section 9 lists some details of the implementation and how it is split into different libraries.

To shorten notations, we recall that we denote by $\mathcal{N}$ the inductively defined Coq type nat of natural numbers. The μ-recursive scheme of composition requires the use of k-tuples which we implement as vectors. Vectors are typeset in a bold font such as in $\boldsymbol{v} : \mathcal{N}^k$ and they correspond to a polymorphic dependent type described in Sect. 4. Π-types are denoted with a $\forall$ symbol. We denote Σ-types with their usual Coq notations, which are $(\exists x, P\ x)$: Prop for non-informative existential quantification, $\{x \mid P\ x\}$: Set for informative existential quantification, or even $\{x : X \,\&\, P\ x\}$: Type when $P : X \to$ Type carries information as well. These Σ-types are inductively defined in modules Logic and Specif of the standard library. The interpretation of the different existential quantifiers of Coq is discussed in Sect. 3.1.

2   Avoiding Small-Step Operational Semantics

In this section we give a high level view of our strategy to obtain a mechanized proof of the typability of total recursive functions in Coq. Let us first discuss the approach which is outlined in [2] (Sect. 4.4, p. 685).

1. By *Kleene's normal form theorem* [9], every recursive function can be obtained by the minimization of a primitive (hence total) recursive function;
2. Every primitive recursive function can directly be typed in Coq. The primitive recursion scheme is precisely the recursor `nat_rec` corresponding to the inductive type `nat` (denoted $\mathcal{N}$ in this paper);
3. The outermost minimization could be implemented by a "specific minimization function" defined by mutual structural recursion.

Items 1 and 2 are results which should not come as a surprise to anyone knowledgeable of μ-recursion theory and basic Coq programming. These observations were already made in [2]. Their approach to minimization (i.e. Item 3) seems[2] however distinct from what we propose as Item 3' here:

3'. Minimizations of inhabited and decidable predicates of type $\mathcal{N} \to$ Prop can be implemented in (axiom free) Coq.

Item 3' could be considered as a bit surprising. Indeed, inductive type theory and hence Coq prohibits unbounded minimization. Hence we did not suspect that Coq could have such a property. When it first came to our attention, we realized that it provided a direct path towards a proof that Coq "had" any total recursive function. Critical for our approach, Item 3' is described in Sect. 3.

Despite its apparent straightforwardness, this three steps approach (with either Item 3 or Item 3') is difficult to implement because of Item 1. Indeed, let us describe more precisely what it implies. Kleene's normal form theorem involves the T primitive recursive predicate which decides whether a given (encoding of a) computation corresponds to a given (encoding of a) program code or not. For this, you need a *small-step operational semantics* (a model of computation), say for instance Minsky (or counter) machines, and a compiler from recursive functions code to Minsky machines. You need of course a correctness proof for that compiler. Since the T predicate operates on natural numbers $\mathcal{N}$, all these data-structures should be encoded in $\mathcal{N}$ which complicates proofs further. Then the T predicate should answer the following question: does this given encoding of a sequence of states correspond to the execution of that given encoding of a Minsky machine. Most importantly, the T predicate should be proved primitive recursive and correct w.r.t. this specification. Programming using primitive recursive schemes is really cumbersome and virtually nobody does this.

Compared to the above three steps approach, the trick which is used in this paper is to merge Items 1 and 2. Instead of showing that recursive functions are minimizations of primitive recursive functions, it is sufficient to show that *recursive functions are minimizations of Coq definable predicates*. From this point of view, it is possible to completely avoid the encoding/decoding phases from/to $\mathcal{N}$ but more importantly, we do not need a small-step semantics any more; we can replace it with a *decidable big-step semantics*: this avoids the implementation of a model of computation and thus, the proof of correctness of a compiler.

---

[2] It is difficult to use a word more accurate than "seems" because the relevant discussion in [2] is just a short outline of an approach, not a proof or an actual implementation.

Our mechanization proceeds in the following steps. We define an inductive predicate denoted $[f; v] \dashv\!\langle\alpha\rangle\!\rangle x$ and called *cost aware big-step semantics*. It reads as: the recursive algorithm f terminates on input v and outputs x at cost α. This relation is functional/deterministic in both α and x. We show the equivalence $[\![f]\!] \, v \, x \iff \exists\alpha, [f; v] \dashv\!\langle\alpha\rangle\!\rangle x$. We establish the central result of decidability of cost aware big-step semantics *when α is fixed:* for any f, v and α, either x together with a proof of $[f; v] \dashv\!\langle\alpha\rangle\!\rangle x$ can be computed (i.e. $\{x \mid [f; v] \dashv\!\langle\alpha\rangle\!\rangle x\}$), or (an informative "or") a proof that no such x exists can be computed (i.e. $\neg\exists x, [f; v] \dashv\!\langle\alpha\rangle\!\rangle x$). These results are combined in the following way: from a proof of definedness $(\exists x, [\![f]\!] \, v \, x)$, we deduce $\exists x \exists\alpha, [f; v] \dashv\!\langle\alpha\rangle\!\rangle x$. Equivalently we get $\exists\alpha, \texttt{inhabited} \ \{x \mid [f; v] \dashv\!\langle\alpha\rangle\!\rangle x\}$. By unbounded minimization of inhabited decidable predicates (see Sect. 3), we reify the proposition $\exists\alpha, \texttt{inhabited} \ \{x \mid [f; v] \dashv\!\langle\alpha\rangle\!\rangle x\}$ into the predicate $\{\alpha \ \& \ \{x \mid [f; v] \dashv\!\langle\alpha\rangle\!\rangle x\}\}$. Then we extract α, x and a proof that $[f; v] \dashv\!\langle\alpha\rangle\!\rangle x$, hence $[\![f]\!] \, v \, x$, showing that the computed value x is the output value of f on input v.

3   Reifying $\exists P$ into ΣP for $P : \mathcal{N} \to \{\texttt{Prop}, \texttt{Type}\}$

In this section, we describe a way to reify non-informative inhabited decidable predicates of type $P : \mathcal{N} \to \texttt{Prop}$. So we show how to constructively build a value $n : \mathcal{N}$ and a proof term $t : P \, n$. We use an unbounded (but still well-founded) minimization algorithm whose termination is guaranteed by a proof of inhabitation $\exists n, P \, n$. The mechanization occurs in the file nat_minimizer.v which is nearly self-contained. In a way, this shows that Coq has unbounded minimization of inhabited and decidable predicates, whereas the theory of recursive functions has unbounded minimization of partial recursive functions. In Sect. 3.3, we also reify informative decidable predicates $P : \mathcal{N} \to \texttt{Type}$ that are inhabited, i.e. verifying $\exists n, \texttt{inhabited} \, (P \, n)$.

3.1   Existential Quantification in Coq

Let us recall the usual interpretation of the existential quantifiers that are available in Coq. In Type Theory, they are called Σ-types over a index type X:

- for $P : X \to \texttt{Prop}$, the expression $\exists x : X, P \, x$ (or $\texttt{ex} \, P$) is of type $\texttt{Prop}$ and a term of that type is only a proof that there exists $x : X$ which satisfies $P \, x$. The witness x need not be effective. It can be obtained by non-constructive means. For instance, the proof may use axioms in $\texttt{Prop}$ such as the excluded middle (typically). We say that the predicate $\exists x : X, P \, x$ is *non-informative*;
- for $P : X \to \texttt{Prop}$, the expression $\{x : X \mid P \, x\}$ (or $\texttt{sig} \, P$) is of type $\texttt{Set}/\texttt{Type}$ and a proof term for it is an (effective) term x together with a proof of $P \, x$ (x must be described by purely constructive methods). We say that the predicate $\{x : X \mid P \, x\}$ is *informative*;
- for $P : X \to \texttt{Type}$, the expression $\{x : X \ \& \ P \, x\}$ (or $\texttt{sigT} \, P$) is of type $\texttt{Type}$. It carries both an effective witness x such that $P \, x$ is inhabited and an effective inhabitant of $P \, x$. The predicate $\{x : X \ \& \ P \, x\}$ is *fully informative*.

When the computational content of terms is extracted, the sub-terms of type Prop are pruned and their code does not impact the extracted terms: this property is called *proof irrelevance*. It implies that adding axioms in Prop will only allow to show more (termination) properties but it will not change the behaviour of terms. However, proof irrelevance is not preserved by adding axioms in Type.

The *Constructive Indefinite Description axiom* as stated in Coq standard library module ChoiceFacts can reify any non-informative predicate $\exists P$:

$$\forall(X:\text{Type})(P:X\rightarrow\text{Prop}),(\exists x:X, P\ x)\rightarrow\{x:X\mid P\ x\} \qquad \text{(CID)}$$

It provides an (axiomatic) transformation of $\exists P$ (i.e. $\exists x, P\ x$ in Coq) into ΣP (i.e. $\{x\mid P\ x\}$ in Coq). The type $\forall X:\text{Type}, \text{inhabited}\ X\rightarrow X$ provides an equivalent definition of (CID) where $\text{inhabited}:\text{Type}\rightarrow\text{Prop}$ is the "hidding predicate" of the Logic module; see file cid.v and Sect. 3.3.

Assuming the axiom (CID) creates an "artificial" bridge between two separate worlds.[3] Some would even claim that such an axiom is at odds with the design philosophy of Coq: the default bridges that exist between the non-informative sort Prop and the informative sorts Set/Type were carefully introduced by Coq designers to be "constructively" safe; in particular, to ensure that extraction is proof irrelevant. Assuming (CID) would not be inconsistent with extraction but it would leave a hole in the extracted terms that make use of it. Moreover, assuming (CID), one can easily derive a proof of $\forall A\ B:\text{Prop}, A\vee B\rightarrow\{A\}+\{B\}$ and thus, a statement like $\forall x, \{P\ x\}+\{\neg P\ x\}$ cannot be interpreted as "P is decidable" anymore. This is well explained in [3] together with the relations between (CID) and Hilbert's epsilon operator. You will also find a summary of the incompatibilities between (CID) and other features or axioms in Coq.

3.2   The Case of Predicates of Type $\mathcal{N}\rightarrow$ Prop

We describe a way to implement an instance of (CID) constructively but of course, that proof requires additional assumptions: we require that P is a decidable predicate that ranges over $\mathcal{N}$ instead of an arbitrary type X. We do not extract the missing information x but instead, we generate it using a well-founded algorithm that first transforms the non-informative inhabitation predicate $\exists x:\mathcal{N}, P\ x$ into a *termination certificate* for a well-founded minimization algorithm that sequentially enumerates natural numbers in *ascending order*.

Recall the definition of the non-informative *accessibility* predicate from the Wf module of the Coq standard library:

```
Inductive Acc {X : Type} (R : X → X → Prop) (x : X) :=
  | Acc_intro : (∀y : X, R y x → Acc R y) → Acc R x
```

We write Acc R instead of Acc X R because the parameter X is declared implicit.

---

[3] Of course this statement is of philosophical nature. We do not claim that assuming additional axiom is evil, but carelessly adding axioms is a recipe for inconsistencies.

We assume a predicate $P : \mathcal{N} \to \texttt{Prop}$ and we suppose that P is decidable (in Coq) with a decision term H_P. We define a binary relation $R : \mathcal{N} \to \mathcal{N} \to \texttt{Prop}$ and we show the following results:

```
Variables  (P : N → Prop) (H_P : ∀n : N, {P n} + {¬P n})
Let  R (n m : N)  :=  (n = 1 + m) ∧ ¬P m
Let  P_Acc_R       :  ∀n : N, P n → Acc R n
Let  Acc_R_dec     :  ∀n : N, Acc R (1 + n) → Acc R n
Let  Acc_R_zero    :  ∀n : N, Acc R n → Acc R 0
Let  Acc_P         :  ∀n : N, Acc R n → {x : N | P x}
```

which all have straightforward proofs except for Acc_P. That last one is done by induction on the accessibility predicate Acc R n. The proof term Acc_P uses the decision term H_P to choose between stopping and moving on to the successor: it stops when H_P n returns "true," i.e. left T with $T : P$ n; it loops on $1 + n$ when H_P n returns "false," i.e. right F with $F : \neg P$ n. We analyse the term:

```
Let  Acc_inv (n : N) (H_n : Acc R n) : ∀m, R m n → Acc R m :=
     match H_n with Acc_intro _ H'_n ↦ H'_n end

Fixpoint Acc_P (n : N) (H_n : Acc R n) : {x : N | P x} :=
     match H_P n with
     | left T  ↦ exist _ n T
     | right F ↦ Acc_P (1 + n) (Acc_inv _ H_n _ (conj eq_refl F))
     end.
```

The recursion cannot be based on the argument n because it would not be structurally well-founded in that case and the Coq type-checker would reject it. The recursion is based on the Acc R n predicate. The definition is split in two parts to make it more readable; Acc_inv is from the module Wf of the standard library. The term Acc_P is a typical example of fixpoint by induction over an ad hoc predicate (see [2] or *the Coq'Art* [1] p. 428). The Fix_F fixpoint operator of the Wf module of the Coq standard library is defined this way as well. The *cover-induction principle* as defined in [4] uses a similar idea.

As a consequence, we can reify decidable and inhabited predicates over $\mathcal{N}$:

Theorem nat_reify $(P : \mathcal{N} \to \texttt{Prop})$:
$$\left(\forall n : \mathcal{N}, \{P\, n\} + \{\neg P\, n\}\right) \to \left(\exists n : \mathcal{N}, P\, n\right) \to \{n : \mathcal{N} \mid P\, n\}$$

The proof is now simple: using P_Acc_R and Acc_R_zero, from $\exists n, P\,n$ we deduce Acc R 0, and thus $\{x : \mathcal{N} \mid P\, x\}$ using Acc_P.

Considering this somewhat unexpected result, maybe some further clarifications about the proof of **nat_reify** are mandatory at this stage. The witness n which is contained in the hypothesis $\exists n, P\, n$ of sort **Prop** is not informative and thus cannot be extracted to build a term of sort **Type**. As this remarks seems contradictory with what we show, we insist on the fact that we do not extract the witness n contained in the hypothesis by inspection of its term. Instead, we compute the minimum value m which satisfies $P\,m$ by testing all cases in sequence: $P\,0$?, $P\,1$?, ... until we reach the first value m which satisfies $P\,m$ (the decidability of P is required for that). To ensure that such a computation is well-founded, we use the non-informative witness n contained in $\exists n, P\,n$ as a *bound on the search space*; but a bound in sort Prop: we encode n into the accessibility predicate $A_n : \mathtt{Acc}\ R\ 0$ which is then used as a certificate for the well-foundedness of the computation of $\mathtt{Acc\_P}\ 0\ A_n$.

3.3   Reification of Predicates of Type $\mathcal{N} \to$ Type

We now generalize the previous result nat_reify to predicates in $\mathcal{N} \to$ Type instead of just $\mathcal{N} \to$ Prop. But we first need to introduce two predicates:

$$\mathtt{Inductive}\quad \mathtt{inhabited}\ (P : \mathtt{Type}) : \mathtt{Prop} := \mathtt{inhabits} : P \to \mathtt{inhabited}\ P$$
$$\mathtt{Definition}\quad \mathtt{decidable\_t}\ (P : \mathtt{Type}) : \mathtt{Type} := P + P \to \mathtt{False}$$

where inhabited is from the standard library (module Logic) and decidable_t is an informative version of the decidable predicate of the Decidable module of the standard library. Their intuitive meaning is the following:

- inhabited P hides the information of the witness of P. Whereas a term of type P is a witness that P is inhabited, a term of type inhabited P hides the witness by the use of the non-informative constructor inhabits;
- decidable_t P means that either a term of type P is given or a proof that P is void is given. The predicate is informative and contains a Boolean choice (represented by the $+$) which tells whether P is inhabited or not. But it may also contain an effective witness that P is inhabited.

We can now lift the theorem nat_reify that operates on $\mathcal{N} \to$ Prop to informative predicates of type $\mathcal{N} \to$ Type in the following way:

Theorem nat_reify_t $(P : \mathcal{N} \to \mathtt{Type})$:
$$(\forall n, \mathtt{decidable\_t}\,(P\ n)) \to (\exists n, \mathtt{inhabited}\,(P\ n)) \to \{n : \mathcal{N}\,\&\,P\ n\}$$

The proof is only a slight variation from the $\mathcal{N} \to$ Prop case. Notice that the result type $\{n : \mathcal{N}\,\&\,P\ n\}$ contains the reified value n for which $P\,n$ is inhabited, but it also contains the effective witness that $P\,n$ is not void. On the contrary, in the hypothesis $\exists n, \mathtt{inhabited}\,(P\,n)$ neither n nor the witness that $P\,n$ is inhabited have to be provided by effective means.

4   Dependent Types for Recursive Algorithms

So far, we have only encountered datatypes which originate in the Coq standard library, and that are imported by default when loading Coq, most notably $\mathcal{N}$ which is a least solution of the fixpoint equation $\mathcal{N} \equiv \{0\} + \{\mathtt{S}\ n \mid n : \mathcal{N}\}$. We will need the type of vectors VectorDef.t and the type of positions Fin.t that also belong to the standard library module Vector. However, the standard library only contains a small fraction of the results that we use for these datatypes. Moreover, *the implementation of some functions of the* Vector *module is incompatible with how we intend to use them.* Typically, the definition of VectorDef.nth which selects a component of a vector by its position does not type-check in our succinct definition of the upcoming recalg_rect recursor: the definition of VectorDef.nth makes Coq unable to certify the structural decrease of recursive sub-calls which is mandatory for Fixpoint definitions. As a consequence, we use our own vectors and positions libraries. This represents little overhead compared to extending the standard libraries in the Vector module.

We define three types that depend on a parameter $k : \mathcal{N}$ representing an arity. First the type of *positions*

$$\mathsf{pos}\,0 \equiv \emptyset \qquad \mathsf{pos}(1 + k) \equiv \{\mathsf{fst}\} + \{\mathsf{nxt}\,p \mid p : \mathsf{pos}\,k\}$$

which is isomorphic to $\mathsf{pos}\,k \equiv \{i : \mathcal{N} \mid i < k\}$ but avoids carrying the proof term $i < k$. The library pos.v contains the inductive definitions of the type $\mathsf{pos}\,k$ and the tools to manipulate positions smoothly: an inversion tactic pos_inv, maps $\mathsf{pos2nat} : \mathsf{pos}\,k \to \mathcal{N}$ and $\mathsf{nat2pos} : \forall i, i < k \to \mathsf{pos}\,k$, etc. To shorten the notations in this paper, $\overline{p}$ denotes $\mathsf{pos2nat}\,p$, the natural number below k which corresponds to p.

Positions of $\mathsf{pos}\,k$ mainly serve as coordinates for accessing the components of *vectors of arity* k

$$X^0 \equiv \{\mathsf{vec\_nil}\} \qquad X^{1+k} \equiv X \times X^k$$

where X^k is a compact notation for $\mathsf{vec}\,X\,k$. The type is polymorphic in X and dependent on $k : \mathcal{N}$. We will write terms of type X^k in a bold font like with $\boldsymbol{v}$ or $\boldsymbol{w}$ to remind the reader that these are vectors. Given a position $p : \mathsf{pos}\,k$ and a vector $\boldsymbol{v} : X^k$, we write $\boldsymbol{v}_p : X$ for the p-th component of $\boldsymbol{v}$, a short-cut for $\mathsf{vec\_pos}\,\boldsymbol{v}\,p$. $\mathsf{vec\_pos}$ is obtained from the "correspondence" $X^k \equiv \mathsf{pos}\,k \to X$. Notice however that the type X^k enjoys an extensional equality (i.e. $\boldsymbol{v} = \boldsymbol{w}$ whenever $\boldsymbol{v}_p = \boldsymbol{w}_p$ holds for any $p : \mathsf{pos}\,k$) whereas the function type $\mathsf{pos}\,k \to X$ does not. The file vec.v contains the inductive definition of the type of vectors together with the tools to smoothly manipulate vectors and their components where coordinates can either be positions of type $\mathsf{pos}\,k$ or natural number $i : \mathcal{N}$ satisfying $i < k$. The constructors are $\mathsf{vec\_nil} : X^0$ and $\mathsf{vec\_cons} : X \to X^k \to X^{1+k}$ and $\mathsf{vec\_cons}\,x\,\boldsymbol{v}$ is denoted $x\#\boldsymbol{v}$ here. The converse operations are $\mathsf{vec\_head} : X^{1+k} \to X$ and $\mathsf{vec\_tail} : X^{1+k} \to X^k$.

With positions and (polymorphic) vectors, we can now introduce the inductive type of *recursive algorithms of arity* k denoted by $\mathcal{A}_k$ which is defined by the rules of Fig. 1 and implemented in the file recalg.v. The notation $\mathcal{A}_k$ is a short-cut for **recalg** k. Notice that $\mathcal{A}_k$ is a dependent type (of sort Set). It is the least type which contains *constants* of arity 0, *zero* and *succ* of arity 1, *projections* at every arity k for each possible coordinate, and which is closed under the *composition, primitive recursion* and *unbounded minimization* schemes. $\mathcal{A}_k$ itself does not carry the semantics of those recursive algorithms: it corresponds to the source code. We will give a meaning/semantics to those recursive algorithms in Sect. 5 so that they correspond to the usual notion of recursive functions.

$$\frac{n : \mathcal{N}}{\mathsf{cst}_n : \mathcal{A}_0} \qquad \frac{}{\mathsf{zero} : \mathcal{A}_1} \qquad \frac{}{\mathsf{succ} : \mathcal{A}_1} \qquad \frac{p : \mathsf{pos}\, k}{\mathsf{proj}_p : \mathcal{A}_k}$$

$$\frac{f : \mathcal{A}_k \quad g : \mathcal{A}_i^k}{\mathsf{comp}\, f\, g : \mathcal{A}_i} \qquad \frac{f : \mathcal{A}_k \quad g : \mathcal{A}_{2+k}}{\mathsf{rec}\, f\, g : \mathcal{A}_{1+k}} \qquad \frac{f : \mathcal{A}_{1+k}}{\mathsf{min}\, f : \mathcal{A}_k}$$

Fig. 1. The type $\mathcal{A}_k$ of recursive algorithms of arity k.

To be able to compute with or prove properties of terms of type $\mathcal{A}_k$, we implement a general fully dependent recursion scheme `recalg_rect` described in the file recalg.v. This principle is not automatically generated by Coq because of the nested induction between the types $\mathcal{A}_k$ and vec _ k which occurs in the constructor comp $f\, g$. The definition of `recalg_rect` looks simple but it only works well because vec_pos was carefully designed to allow the Coq type-checker to detect nested recursive calls as structurally simpler: using the "equivalent" VectorDef.nth instead of vec_pos prohibits successful type-checking. We also show the injectivity of the constructors of the type $\mathcal{A}_k$. Some require the use of the Eqdep_dec module of the standard library because of the dependently typed context. For example, the statement of the injectivity of the constructor comp $f\, g$ involves type castings eq_rect (or heterogenous equality):

Fact `ra_comp_inj` $k\, k'\, i\, (f : \mathcal{A}_k)\, (f' : \mathcal{A}_{k'})\, (g : \mathcal{A}_i^k)\, (g' : \mathcal{A}_i^{k'})$:

$$\mathsf{comp}\, f\, g = \mathsf{comp}\, f'\, g' \to \exists e : k = k', \wedge \begin{cases} \mathsf{eq\_rect}\, \_\, \_\, f\, \_\, e = f' \\ \mathsf{eq\_rect}\, \_\, \_\, g\, \_\, e = g' \end{cases}$$

5   A Decidable Semantics for Recursive Algorithms

In this section, we define three equivalent semantics for recursive algorithms. First the standard *relational semantics* defined by recursion on $f : \mathcal{A}_k$, then an equivalent *big-step semantics* defined by a set of inductive rules. Those two semantics cannot be decided. Then we define a refinement of big-step semantics by annotating it with a cost. By constraining the cost, we obtain a decidable semantics for recursive algorithms $\mathcal{A}_k$.

5.1   Relational and Big-Step Semantics

We define relational semantics $[\![f : \mathcal{A}_k]\!] : \mathcal{N}^k \to \mathcal{N} \to \mathsf{Prop}$ of recursive algorithms by structural recursion on $f : \mathcal{A}_k$ so as to satisfy the fixpoint equations of Fig. 2 where $[\![f]\!]$ is a notation for ra_rel f; the fixpoint equations ra_rel_fix_* are proved in the file ra_rel.v. Without preparation, such a definition could be quite technical because of the nested recursion between the type $\mathcal{A}_k$ and the type vec $\mathcal{A}_i$ k of the parameter g in the constructor comp $f\, g$. Using our general recursion principle recalg_rect, the code is straightforward; but see the remark

$$\llbracket \text{cst}_n \rrbracket\_ x \iff n = x \qquad\qquad \llbracket \text{zero} \rrbracket\_ x \iff 0 = x$$

$$\llbracket \text{succ} \rrbracket v\, x \iff 1 + v_{\text{fst}} = x \qquad \llbracket \text{proj}_p \rrbracket v\, x \iff v_p = x$$

$$\llbracket \text{comp}\, f\, g \rrbracket v\, x \iff \exists w, \llbracket f \rrbracket w\, x \wedge \forall p, \llbracket g_p \rrbracket v\, w_p$$

$$\llbracket \text{rec}\, f\, g \rrbracket (0 \# v)\, x \iff \llbracket f \rrbracket v\, x$$

$$\llbracket \text{rec}\, f\, g \rrbracket (1 + n \# v)\, x \iff \exists y, \llbracket \text{rec}\, f\, g \rrbracket (n \# v)\, y \wedge \llbracket g \rrbracket (n \# y \# v)\, x$$

$$\llbracket \text{min}\, f \rrbracket v\, x \iff \exists w, \llbracket f \rrbracket (x \# v)\, 0 \wedge \forall p : \text{pos}\, x, \llbracket f \rrbracket (\overline{p} \# v)\, (1 + w_p)$$

Fig. 2. Relational semantics `ra_rel` for recursive algorithms of $\mathcal{A}_k$.

about `recalg_rect` in Sect. 4. We explicitly mention the type $p : \text{pos}\, x$ in the definition of $\llbracket \text{min}\, f \rrbracket$ because it is the only type which does not depend on the type of f: the dependent parameter x is the result of the computation.

The big-step semantics for recursive algorithms in $\mathcal{A}_k$ is an inductive relation of type $\text{ra\_bs} : \forall k, \mathcal{A}_k \to \mathcal{N}^k \to \mathcal{N} \to \text{Prop}$ and we denote $[f; v] \rightsquigarrow x$ for $(\text{ra\_bs}\ k\ f\ v\ x)$; the parameter k is implicit in the notation. $[f; v] \rightsquigarrow x$ intuitively means that the computation of f starting from input v yields the result x. We define big-step semantics in file ra_bs.v by the inductive rules of Fig. 3. We point out that the rule corresponding to $[\text{min}\, f; v] \rightsquigarrow x$ is of unbounded arity but still finitary because $\text{pos}\, x$ is a finite type. These rules are similar to those used to define the semantics of Partial Recursive Functions in [13] except that thanks to our dependent typing, we do not need to specify *well-formedness conditions*. In ra_sem_eq.v, we show that big-step semantics is equivalent to relational semantics:

Theorem `ra_bs_correct` $k\ (f : \mathcal{A}_k)\ (v : \mathcal{N}^k)\ x : \llbracket f \rrbracket v\, x \iff [f; v] \rightsquigarrow x$

However big-step semantics has the advantage of being defined by a set of inductive rules instead of being defined by recursion on $f : \mathcal{A}_k$.

Relational and big-step semantics are not recursive/computable relations: this is an instance of the *Halting problem*. As such, these relations cannot be implemented by a Coq evaluation function $\text{ra\_rel\_eval} : \mathcal{A}_k \to \mathcal{N}^k \to \text{option}\, \mathcal{N}$ satisfying $\text{ra\_rel\_eval}\ f\ v = \text{Some}\ x \iff \llbracket f \rrbracket v\, x$ for any f, v and x. Indeed, when it is axiom free, Coq has normalisation which implies that the functions that can be defined in it are total recursive at the meta-theoretical level. Nevertheless big-step semantics as presented in Fig. 3 is an intermediate step towards a decidable semantics for $\mathcal{A}_k$.

5.2 Cost Aware Big-Step Semantics

The cost aware big-step semantics for recursive algorithms in $\mathcal{A}_k$ is defined as an inductive predicate of type $\text{ra\_ca} : \forall k, \mathcal{A}_k \to \mathcal{N}^k \to \mathcal{N} \to \mathcal{N} \to \text{Prop}$. We denote $(\text{ra\_ca}\ k\ f\ v\ \alpha\ x)$ by $[f; v] \dashv \alpha \rangle\!\rangle\, x$ where the argument k is implicit in the notation. $[f; v] \dashv \alpha \rangle\!\rangle\, x$ intuitively means that the computation of f on input v yields the result x and costs α. We define the predicate ra_ca in file ra_ca.v by the rules of Fig. 4. It is interesting to compare these rules with those of conventional big-step semantics ra_bs of Fig. 3. The very simple but nonetheless powerful idea

$$\overline{[\text{cst}_n;v] \rightsquigarrow n} \qquad \overline{[\text{zero};v] \rightsquigarrow 0} \qquad \overline{[\text{succ};v] \rightsquigarrow 1 + v_{\text{fst}}} \qquad \overline{[\text{proj}_p;v] \rightsquigarrow v_p}$$

$$\frac{[f;v] \rightsquigarrow x}{[\text{rec } f\, g; 0\#v] \rightsquigarrow x} \qquad \frac{[\text{rec } f\, g; n\#v] \rightsquigarrow y \quad [g; n\#y\#v] \rightsquigarrow x}{[\text{rec } f\, g; 1 + n\#v] \rightsquigarrow x}$$

$$\frac{[f;w] \rightsquigarrow x \quad \forall p, [g_p;v] \rightsquigarrow w_p}{[\text{comp } f\, g; v] \rightsquigarrow x} \qquad \frac{[f;x\#v] \rightsquigarrow 0 \quad \forall p : \text{pos } x, [f; \overline{p}\#v] \rightsquigarrow 1 + w_p}{[\min f; v] \rightsquigarrow x}$$

Fig. 3. Big-step semantics ra_bs for recursive algorithms of $\mathcal{A}_k$.

$$\overline{[\text{cst}_n;v] -\!(1\!)\!\!\rangle n} \qquad \overline{[\text{zero};v] -\!(1\!)\!\!\rangle 0} \qquad \overline{[\text{succ};v] -\!(1\!)\!\!\rangle 1 + v_{\text{fst}}} \qquad \overline{[\text{proj}_p;v] -\!(1\!)\!\!\rangle v_p}$$

$$\frac{[f;v] -\!(\alpha\!)\!\!\rangle x}{[\text{rec } f\, g; 0\#v] -\!(1 + \alpha\!)\!\!\rangle x} \qquad \frac{[\text{rec } f\, g; n\#v] -\!(\alpha\!)\!\!\rangle y \quad [g; n\#y\#v] -\!(\beta\!)\!\!\rangle x}{[\text{rec } f\, g; 1 + n\#v] -\!(1 + \alpha + \beta\!)\!\!\rangle x}$$

$$\frac{[f;w] -\!(\alpha\!)\!\!\rangle x \quad \forall p, [g_p;v] -\!(\beta_p\!)\!\!\rangle w_p}{[\text{comp } f\, g; v] -\!(1 + \alpha + \Sigma\beta\!)\!\!\rangle x} \qquad \frac{[f;x\#v] -\!(\alpha\!)\!\!\rangle 0 \quad \forall p : \text{pos } x, [f; \overline{p}\#v] -\!(\beta_p\!)\!\!\rangle 1 + w_p}{[\min f; v] -\!(1 + \alpha + \Sigma\beta\!)\!\!\rangle x}$$

Fig. 4. Cost aware big-step semantic ra_ca for recursive algorithms of $\mathcal{A}_k$.

to get decidability is to decorate big-step semantics with a cost and to constrain computations by a cost that must be exactly matched. This is how we realize the principle of our proof that Coq contains total recursive functions: we avoid a small-step semantics (Kleene's T predicate) and replace it with a big-step semantics for recursive algorithm that is nevertheless decidable.

We show the equivalence of relational and cost aware big-step semantics

Theorem ra_ca_correct $(k:\mathcal{N})\ (f:\mathcal{A}_k)\ (v:\mathcal{N}^k)\ (x:\mathcal{N})$:
$$[\![f]\!]\ v\ x \iff \exists \alpha : \mathcal{N}, [f;v] -\!(\alpha\!)\!\!\rangle x$$

in file ra_sem_eq.v. The proof is circular in style: ra_ca implies ra_bs implies ra_rel implies $\exists$ra_ca and all these three implications are proved by induction on the obvious inductive parameter. Do not feel puzzled by a statement of equivalence between a decidable and an undecidable semantics, because it is the quantifier $\exists \alpha$ in ra_ca_correct which brings undecidability.

Inversion lemmas named ra_ca_*_inv are essential tools to prove the high-level properties of Sect. 5.3. They allow case analysis on the last step of an inductive term depending on the shape of the conclusion. Here is the inversion lemma of one rule:

Lemma ra_ca_rec_S_inv $(k:\mathcal{N})\ (f:\mathcal{A}_k)\ (g:\mathcal{A}_{2+k})\ (v:\mathcal{N}^k)\ (n\ \gamma\ x:\mathcal{N})$:
$$[\text{rec } f\, g; 1 + n\#v] -\!(\gamma\!)\!\!\rangle x \rightarrow \exists y\ \alpha\ \beta, \wedge \begin{cases} \gamma = 1 + \alpha + \beta \\ [\text{rec } f\, g; n\#v] -\!(\alpha\!)\!\!\rangle y \\ [g; n\#y\#v] -\!(\beta\!)\!\!\rangle x \end{cases}$$

Such results could be difficult to establish if improperly prepared. In our opinion, the easiest way to prove it is to implement a global inversion lemma that encompasses the whole set of rules of Fig. 4. Then a lemma like `ra_ca_rec_S_inv` can be proved by applying the global inversion lemma and discriminate between incompatible constructors of type $\mathcal{A}_k$ (in most cases) or use injectivity of thoses constructors (in one case). The global inversion lemma is quite complicated to write because of dependent types. It would fill nearly half of this page (see lemma `ra_ca_inv` in file `ra_ca.v`). However it is actually trivial to prove, a "reversed" situation which is rare enough to be noticed.

5.3   Properties of Cost Aware Big-Step Semantics

The annotation of cost in the rules of Fig. 4 satisfies the following paradigm: the cost of a compound computation is greater than the sum of the costs of its sub-computations. Hence, we can derive that no computation is free of charge:

Theorem `ra_ca_cost` k $(f : \mathcal{A}_k)$ $(v : \mathcal{N}^k)$ $(\alpha\, x : \mathcal{N})$: $[f; v] \dashv\!\!\!\langle\!\langle \alpha \rangle\!\rangle\, x \to 0 < \alpha$

The proof is by immediate case analysis on $[f; v] \dashv\!\!\!\langle\!\langle \alpha \rangle\!\rangle\, x$. The cost and results given by cost aware big-step semantics are unique (provided they exist)

Theorem `ra_ca_fun` $(k : \mathcal{N})$ $(f : \mathcal{A}_k)$ $(v : \mathcal{N}^k)$ $(\alpha\, \beta\, x\, y : \mathcal{N})$:
$$[f; v] \dashv\!\!\!\langle\!\langle \alpha \rangle\!\rangle\, x \to [f; v] \dashv\!\!\!\langle\!\langle \beta \rangle\!\rangle\, y \to \alpha = \beta \wedge x = y$$

The proof is by induction on $[f; v] \dashv\!\!\!\langle\!\langle \alpha \rangle\!\rangle\, x$ together with inversion lemmas `ra_ca_*_inv` to decompose $[f; v] \dashv\!\!\!\langle\!\langle \beta \rangle\!\rangle\, y$. Inversion lemmas are the central ingredient of this proof.

Now the key result: cost aware big-step semantics is decidable (in sort **Type**, see Sect. 3.3) *when the cost is fixed*

Theorem `ra_ca_decidable_t` $(k : \mathcal{N})$ $(f : \mathcal{A}_k)$ $(v : \mathcal{N}^k)$ $(\alpha : \mathcal{N})$:
$$\texttt{decidable\_t}\, \big\{ x \mid [f; v] \dashv\!\!\!\langle\!\langle \alpha \rangle\!\rangle\, x \big\}$$

Its proof is the most complicated of our whole development. It proceeds by induction on $f : \mathcal{A}_k$ and uses inversion lemmas `ra_ca_*_inv`, functionality `ra_ca_fun` as well as a small *decidability library* to lift decidability arguments over (finitely) quantified statements. The central constituents of that library are:

Lemma `decidable_t_bounded` $(P : \mathcal{N} \to \textbf{Type})$:
    $(\forall n : \mathcal{N},\ \texttt{decidable\_t}\,(P\ n))$
    $\to \forall n : \mathcal{N},\ \texttt{decidable\_t}\, \{ i : \mathcal{N}\, \&\, i < n \times P\ i \}$
Lemma `vec_sum_decide_t` $(n : \mathcal{N})$ $(\boldsymbol{P} : (\mathcal{N} \to \textbf{Type})^n)$:
    $(\forall (p : \texttt{pos}\ n)\ (i : \mathcal{N}),\ \texttt{decidable\_t}\,(\boldsymbol{P}_p\ i))$
    $\to \forall m : \mathcal{N},\ \texttt{decidable\_t}\, \{ \boldsymbol{v} : \mathcal{N}^n\, \&\, \varSigma \boldsymbol{v} = m \times \forall p,\ \boldsymbol{P}_p\ \boldsymbol{v}_p \}$

Lemma vec_sum_unbounded_decide_t $(P : \mathcal{N} \to \mathcal{N} \to \text{Type})$:
$\quad (\forall n\, i : \mathcal{N},\ \text{decidable\_t}\,(P\, n\, i))$
$\quad \to (\forall n : \mathcal{N},\ P\, n\, 0 \to \text{False})$
$\quad \to \forall m : \mathcal{N},\ \text{decidable\_t}\,\{n : \mathcal{N}\, \&\, \{q : \mathcal{N}^n\, \&\, \Sigma q = m \times \forall p, P\, \overline{p}\, q_p\}\}$

Some comments about the intuitive meaning of such results could be useful. Recall that decidability has to be understood over Type (as opposed to Prop):

- decidable_t_bounded states that whenever $P\, n$ is decidable for any n, then given a bound m, it is decidable whether there exists $i < m$ such that $P\, i$ holds. Hence bounded existential quantification inherits decidability;
- vec_sum_decide_t states that if P is a pos $n \times \mathcal{N}$ indexed family of decidable predicates, then it is decidable whether there exists vector $v : \mathcal{N}^n$ (of length n) which satisfies $P_p\, v_p$ for each of its components (indexed by $p : \text{pos}\, n$), and such that the sum of the components of v is a fixed value m. This express the decidability of some kind of universal quantification bounded by the length of a vector;
- vec_sum_unbounded_decide_t states that if P is a $\mathcal{N} \times \mathcal{N}$ indexed family of decidable predicates such that $P\, \_0$ is never satisfied, then it is decidable whether there exists a vector q of arbitrary length which satisfies P at every component and such that the sum of those components is a fixed value m. This is a variant of vec_sum_decide_t but for unbounded vector length, only the sum of the components acts as a bound.

Once ra_ca_decidable_t is established, we combine it with ra_ca_fun to easily define a bounded computation function for recursive algorithms, as is done for instance at the end of file ra_ca_props.v:

Definition   ra_ca_eval $(k : \mathcal{N})\ (f : \mathcal{A}_k)\ (v : \mathcal{N}^k)\ (\alpha : \mathcal{N}) : \text{option}\, \mathcal{N}$

Proposition   ra_ca_eval_prop $(k : \mathcal{N})\ (f : \mathcal{A}_k)\ (v : \mathcal{N}^k)\ (\alpha\, x : \mathcal{N})$:
$\quad\quad [f ; v] \,\text{--}(\!\alpha\rangle\!\rangle\, x \iff \text{ra\_ca\_eval}\, f\, v\, \alpha = \text{Some}\, x$

Notice that the function ra_ca_eval could be proved primitive recursive with proper encoding of $\mathcal{A}_k$ into $\mathcal{N}$ but the whole point of this work is to avoid having to program with primitive recursive schemes.

6   The Totality of Coq

In this section, we conclude our proof that Coq contains all the recursive functions for which totality can be established in Coq. We assume an arity $k : \mathcal{N}$ and a recursive algorithm $f : \mathcal{A}_k$ which is supposed to be total:

Variables $(k : \mathcal{N})\ (f : \mathcal{A}_k)\ (H_f : \forall v : \mathcal{N}^k,\ \exists x : \mathcal{N},\ [\![f]\!]\, v\, x)$

Mimicking Coq sectioning mechanism, these assumptions hold for the rest of the current section. We first show that given an input vector $v : \mathcal{N}^k$, both a cost $\alpha : \mathcal{N}$ and a result $x : \mathcal{N}$ can be computed constructively:

Let coq_f $(v : \mathcal{N}^k) : \{\alpha : \mathcal{N}\, \&\, \{x : \mathcal{N} \mid [f ; v] \,\text{--}(\!\alpha\rangle\!\rangle\, x\}\}$

The proof uses unbounded minimization as implemented in `nat_reify_t` to find a cost α such that $\{x:\mathcal{N} \mid [f;v] \dashv\!\langle\!\langle\alpha\rangle\!\rangle\, x\}$ is an inhabited type. This can be decided for each possible cost thanks to `ra_ca_decidable_t`. Recall that `nat_reify_t` tries 0, then 1, then 2, etc. until it finds the one which is guaranteed to exist. The warranty is provided by a combination of H_f and `ra_ca_correct`.

To obtain the predicate $t:\mathcal{N}^k \to \mathcal{N}$ that realizes $[\![f]\!]$, we simply permute x and α in `coq_f` v. We define $t := v \mapsto \mathtt{proj1\_sig}(\mathtt{projT2}(\mathtt{coq\_f}\ v))$. Using `projT1(coq_f v)`, `proj2_sig(projT2(coq_f v))` and `ra_ca_correct`, it is trivial to show that $t\ v$ satisfies $[\![f]\!]\ v\ (t\ v)$. Hence, closing the section and discharging the local assumptions, we deduce the totality theorem.

Theorem `Coq_is_total` $(k:\mathcal{N})$ $(f:\mathcal{A}_k)$:
$$\left(\forall v:\mathcal{N}^k,\ \exists x:\mathcal{N},\ [\![f]\!]\ v\ x\right) \to \left\{t:\mathcal{N}^k \to \mathcal{N} \mid \forall v:\mathcal{N}^k,\ [\![f]\!]\ v\ (t\ v)\right\}$$

7   Discussion: Other Approaches, Church Thesis

Comparing our method with the approach based on Kleene's normal form theorem (Sect. 2), we remark that the introduction of small-step semantics would only be used to measure the length (or cost) of computations. Since there is at most one computation from a given input in deterministic models of computation, any computation can be recovered from its number of steps *by primitive recursive means*. Hence the idea of short-cutting small-step semantics by a cost.

It is not surprising that the Kleene's normal form approach was only suggested in [2]. Mechanizing a Turing complete model of computation is bound to be a lengthy development. Mainly because translating between elementary models of computation resembles writing programs in assembly language that you moreover have to specify and prove correct. And unsurprisingly, such developments are relatively rare and recent, with the notable exception of [13] which formalizes computability notions in Coq. μ-recursive functions are not dependently typed in [13] (so there is a well-formedness predicate) and they are not compiled into a model of execution. In [12] however, the same author presents a compiler from μ-recursive functions to Unlimited Register Machines, proved correct in HOL. Turing machines, Abacus machines and μ-recursive functions are implemented in [11] with the aim of been able to characterize decidability in HOL. The development in [8] approaches computability in HOL4 through λ-calculus also with the aim at the mechanization of computability arguments. We recently published online a constructive implementation in (axiom-free) Coq [7] of an significant portion of Krivine's textbook [6] on λ-calculus, including a translation from μ-recursive functions to λ-terms with dependent types in Coq. Actually, this gave us a first mechanized proof that Coq contained any total recursive function by using leftmost β-reduction strategy to compute normal forms. But it requires the introduction of intersection type systems, a development of more than 25 000 lines of code.

Now, what about a characterization of the functions of type $\mathcal{N} \to \mathcal{N}$ definable in Coq? Or else, is such a converse statement of (CiT)

$$\forall(k:\mathcal{N})\ (g:\mathcal{N}^k \to \mathcal{N}), \exists f:\mathcal{A}_k, \forall \boldsymbol{v}:\mathcal{N}^k,\ [\![f]\!]\ \boldsymbol{v}\ (g\ \boldsymbol{v}) \qquad \text{(ChT)}$$

provable in Coq? It is not too difficult to see that (ChT) does not hold in a model of Coq where function types contain the full set of set theoretic functions like in [10], because it contains non-computable functions. However, it is for us an open question whether a statement like (ChT) could be satisfied in a model of Coq, for instance in an effective model.

In such a case, the statement (ChT) would be independent of (axiom free) Coq: (ChT) would be both unprovable and unrefutable in Coq. We think (ChT) very much expresses an internal form of *Church thesis* in Coq: the functions which are typable in Coq are exactly the total recursive functions. The problem which such a statement is that the notion of totality is not independent from the logical framework in which such a totality is expressed and some frameworks are more expressive that others, e.g., Set theory defines more total recursive functions that Peano arithmetic. It is not clear how (ChT) could be used to simplify undecidability proofs in Coq.

8   Reifying Undecidable Predicates

In Sect. 3, we did explain how to reify the non-informative predicate $(\exists n, P\ n)$ into the informative predicate $\{n \mid P\ n\}$, for P of type $\mathcal{N} \to \texttt{Prop}$. This occurred under an important restriction: P is assumed Coq-decidable there. The Coq term nat_reify that implements this transformation is nevertheless used in Sect. 6 to reify the undecidable "computes into" predicate ra_bs. This predicate is first represented as an existential quantification of the decidable precidate ra_ca, which is basically a bounded version of ra_bs. Then nat_reify is used to compute the bound by minimization. Without entering in the full details, we introduce some of the developments that can be found in the file applications.v.

We describe how to reify other kinds of undecidable predicates. For instance, we can reify undecidable predicates that can be bounded in some broad sense. Consider a predicate $P : X \to \texttt{Prop}$ for which we assume the following: P is equivalent to $\bigcup_n (Q\ n)$ for some $Q : \mathcal{N} \to X \to \texttt{Prop}$ such that $Q\ n$ is (informatively) finite for any $n : \mathcal{N}$. Then, the predicate $\exists P$ can be reified into ΣP:

Variables	$(X : \texttt{Type})\ (P : X \to \texttt{Prop})\ (Q : \mathcal{N} \to X \to \texttt{Prop})$
	$(H_P : \forall x,\ P\ x \iff \exists n,\ Q\ n\ x)$
	$(H_Q : \forall n,\ \{l : \texttt{list}\ X \mid \forall x,\ \texttt{In}\ x\ l \iff Q\ n\ x\})$
Theorem	weighted_reif : $(\exists x : X, P\ x) \to \{x : X \mid P\ x\}$

The idea of the proof is simply that the first parameter of Q is a weight of type $\mathcal{N}$ and that for a given weight n, there are only finitely many elements x that satisfy $Q\ n\ x$ (hence $P\ x$). The weight n such that $\exists x, Q\ n\ x$ is reified using nat_reify, then the value x is computed as the first element of the list given by H_Q n. The hypothesis $\exists x, P x$ ensures that the list given by H_Q n is not empty.

Among its direct applications, such a weighted reification scheme can be used to reify *provability* predicates for arbitrary logics, at least those where formulæ and proofs can be encoded as natural numbers. This very low restriction allows to cover a very wide range of logics, with the notable exception of infinitary logics (where either formulæ are infinite or some rules have an infinite number of premisses). Hence, one can compute a proof of a statement provided such a proof exists. Another application is the reification of the *normalizable* predicate for any reduction (i.e. binary) relation which is finitary (i.e. with finite direct images). This applies in particular to β-reduction in λ-calculus.

To conclude, we implement a judicious remark of one of the reviewers. He points out that we can derive a proof of *Markov's principle for recursively enumerable predicates* over $\mathcal{N}^k$ (instead of just decidable ones). These are predicates of the form $v \mapsto [\![f]\!]\ v\ 0$ for some μ-recursive f function of arity k.

Theorem re_reify k $(f : \mathcal{A}_k)$: $\left(\exists v : \mathcal{N}^k,\ [\![f]\!]\ v\ 0\right) \to \left\{v : \mathcal{N}^k \mid [\![f]\!]\ v\ 0\right\}$

Hence if a recursively enumerable predicate can be proved inhabited, possibly using 1-consistent axioms in sort **Prop** such as e.g. excluded middle, then a witness of that inhabitation can be computed.

9   The Structure of the Coq Source Code

The implementation involves around 4 500 lines of Coq code. It has been tested and should compile under Coq 8.5pl3 and Coq 8.6. It is available under a Free Software license at https://github.com/DmxLarchey/Coq-is-total.

More than half of the code belongs to the utils.v utilities library, mostly in files pos.v, vec.v and tree.v. These could be shrunk further because they contain some code which is not necessary to fulfil the central goal of the paper. The files directly relevant to this development are:

utils.v The library of utilities that regroups notations.v, tac_utils.v, list_utils.v, pos.v, nat_utils.v, vec.v, finite.v and tree.v;

nat_minimizer.v The reification of $\exists P$ to ΣP by unbounded minimization of decidable predicates of types $\mathcal{N} \to$ Prop and $\mathcal{N} \to$ Type, see Sect. 3;

recalg.v The dependently typed definition of recursive algorithms with a general recursion principle and the injectivity of type constructors, see Sect. 4;

a_{rel,bs,ca}.v The definitions of relational, big-step and cost aware big-step semantics, with inversion lemmas, see Sects. 5.1 and 5.2;

ra_sem_eq.v The proof of equivalence between the three previous semantics, see Sects. 5.1 and 5.2;

ra_ca_props.v High-level results about cost aware big-step semantics, mainly its functionality and its decidability, see Sect. 5.3;

decidable_t.v The decidability library to lift decision arguments to finitely quantified statements, see Sect. 5.3;

coq_is_total.v The file that implements Sect. 6, which shows that any provably total recursive function can be represented by a Coq term;

applications.v The file that implements Sect. 8, reification of (undecidable) weighted predicates, provability predicates, normalizability predicates and recursively enumerable predicates.

References

1. Bertot, Y., Castéran, P.: Interactive Theorem Proving and Program Development - Coq'Art: The Calculus of Inductive Constructions. Texts in Theoretical Computer Science. An EATCS Series. Springer, Heidelberg (2004)
2. Bove, A., Capretta, V.: Modelling general recursion in type theory. Math. Struct. Comput. Sci. **15**(4), 671–708 (2005)
3. Castéran, P.: Utilisation en Coq de l'opérateur de description (2007). http://jfla. inria.fr/2007/actes/PDF/03_casteran.pdf
4. Coen, C.S., Valentini, S.: General recursion and formal topology. In: Partiality and Recursion in Interactive Theorem Provers, PAR@ITP 2010, EPiC Series, Edinburgh, UK, 15 July 2010, vol. 5, pp. 71–82. EasyChair (2010)
5. Girard, J.Y., Taylor, P., Lafont, Y.: Proofs and Types. Cambridge University Press, New York (1989)
6. Krivine, J.: Lambda-Calculus, Types and Models. Ellis Horwood Series in Computers and Their Applications. Ellis Horwood, Masson (1993)
7. Larchey-Wendling, D.: A constructive mechanization of Lambda Calculus in Coq (2017). http://www.loria.fr/~larchey/Lambda_Calculus
8. Norrish, M.: Mechanised computability theory. In: Eekelen, M., Geuvers, H., Schmaltz, J., Wiedijk, F. (eds.) ITP 2011. LNCS, vol. 6898, pp. 297–311. Springer, Heidelberg (2011). doi:10.1007/978-3-642-22863-6_22
9. Soare, R.I.: Recursively Enumerable Sets and Degrees. Springer-Verlag New York Inc., New York (1987)
10. Werner, B.: Sets in types, types in sets. In: Abadi, M., Ito, T. (eds.) TACS 1997. LNCS, vol. 1281, pp. 530–546. Springer, Heidelberg (1997). doi:10.1007/BFb0014566
11. Xu, J., Zhang, X., Urban, C.: Mechanising turing machines and computability theory in Isabelle/HOL. In: Blazy, S., Paulin-Mohring, C., Pichardie, D. (eds.) ITP 2013. LNCS, vol. 7998, pp. 147–162. Springer, Heidelberg (2013). doi:10.1007/978-3-642-39634-2_13
12. Zammit, V.: A mechanisation of computability theory in HOL. In: Goos, G., Hartmanis, J., Leeuwen, J., Wright, J., Grundy, J., Harrison, J. (eds.) TPHOLs 1996. LNCS, vol. 1125, pp. 431–446. Springer, Heidelberg (1996). doi:10.1007/BFb0105420
13. Zammit, V.: A proof of the S-m-n theorem in Coq. Technical report, The Computing Laboratory, The University of Kent, Canterbury, Kent, UK, March 1997. http://kar.kent.ac.uk/21524/

Effect Polymorphism in Higher-Order Logic (Proof Pearl)

Andreas Lochbihler[(✉)]

Institute of Information Security, Department of
Computer Science, ETH Zurich, Zurich, Switzerland
andreas.lochbihler@inf.ethz.ch

Abstract. The notion of a *monad* cannot be expressed within higher-order logic (HOL) due to type system restrictions. We show that if a monad is used with values of only one type, this notion *can* be formalised in HOL. Based on this idea, we develop a library of effect specifications and implementations of monads and monad transformers. Hence, we can abstract over the concrete monad in HOL definitions and thus use the same definition for different (combinations of) effects. We illustrate the usefulness of effect polymorphism with a monadic interpreter.

1 Introduction

Monads have become a standard way to write effectful programs in pure functional languages [25]. In proof assistants, they provide a widely-used abstraction for modelling and reasoning about effects [3,4,14,17]. Abstractly, a monad consists of a type constructor τ and two polymorphic operations, return $:: \alpha \Rightarrow \alpha\ \tau$ for embedding values and bind $:: \alpha\ \tau \Rightarrow (\alpha \Rightarrow \beta\ \tau) \Rightarrow \beta\ \tau$ for sequencing (written $\ggg$ infix), satisfying three monad laws:

$$1.\ (m \ggg f) \ggg g = m \ggg (\lambda x.\ f\ x \ggg g)$$

$$2.\ \text{return } x \ggg f = f\ x \qquad 3.\ m \ggg \text{return} = m$$

Yet, the notion of a monad cannot be expressed as a formula in higher-order logic (HOL) [8] as there are no type constructor variables like τ in HOL and the sequencing operation bind occurs with three different type instances in the first law. Thus, only concrete monad instances have been used to model side effects of HOL functions. In fact, monad definitions for different effects abound in HOL, e.g., a state-error monad [3], non-determinism with errors and divergence [14], probabilistic choice [4], and probabilistic resumptions with errors [17]. Each of these formalisations fixes τ to a particular type (constructor) and develops its own reasoning infrastructure. This approach achieves *value polymorphism*, i.e., one monad can be used with varying types of values, but not *effect polymorphism* where one function can be used with different monads.

In this paper, we give up value polymorphism in favour of effect polymorphism. The idea is to fix the type of values to some type α_0. Then, the monad

© Springer International Publishing AG 2017
M. Ayala-Rincón and C.A. Muñoz (Eds.): ITP 2017, LNCS 10499, pp. 389–409, 2017.
DOI: 10.1007/978-3-319-66107-0_25

type constructor τ is applied only to α_0, which an ordinary HOL type variable μ can represent. So, the monad operations have the HOL types return :: $\alpha_0 \Rightarrow \mu$ and bind :: $\mu \Rightarrow (\alpha_0 \Rightarrow \mu) \Rightarrow \mu$. This notion of a monad can be formalised within HOL. In detail, we present an Isabelle/HOL library (available online [18]) for different monadic effects and their algebraic specification. All effects are also implemented as value-monomorphic monads and monad transformers. Using Isabelle's module system [1], function definitions can be made abstractly and later specialised to several concrete monads. As our running example, we formalise and reason about a monadic interpreter for a small language. The library has been used in a larger project to define and reason about parsers and serialisers for security protocols.

Contributions. We show the advantages of trading in value polymorphism for effect polymorphism. First, HOL functions with effects can be defined in an abstract monadic setting (Sect. 2) and reasoned about in the style of Gibbons and Hinze [6]. This preserves the level of abstraction that the monad notion provides. As the definitions need not commit to a concrete monad, we can use them in richer effect contexts, too—simply by combining our modular effect specifications. When a concrete monad instance is needed, it can be easily obtained by interpretation using Isabelle's module system.

Second, as HOL can express the notion of a value-monomorphic monad, we have also formalised several monad transformers [15,21] in HOL (Sect. 3). Thus, there is no need to define the monad and derive the reasoning principles for each combination of effects, as is current practice with value polymorphism. Instead, it suffices to formalise every effect only once as a transformer and combine them modularly.

Third, relations between different instances can be proven using the theory of representation independence (Sect. 4) as supported by the Transfer package [10]. This makes it possible to switch in the middle of a bigger proof from a complicated monad to a simpler one.

2   Abstract Value-Monomorphic Monads in HOL

In this section, we formalise value-monomorphic monads and monad transformers for several types of effects. A monadic interpreter for an arithmetic language will be used throughout as a running example. The language, adapted from Nipkow and Klein [22], consists of integer constants, variables, addition, and division.

```
datatype v exp = Const int | Var v | (v exp) ⊕ (v exp) | (v exp) ⊘ (v exp)
```

We formalise the concept of a monad using Isabelle's module system of locales [1]. The locale monad below fixes the two monad operations return and bind (written infix as $\gg=$) and assumes that the monad laws hold. It will collect definitions of functions, which use the monad operations, and theorems about them, whose proofs can use the monad laws. Every locale also defines a predicate of the same name that collects all the assumptions. When a user interprets the locale with more concrete operations and has discharged the assumptions for these operations, every definition and theorem inside the locale context is specialised to these operations. Although the type of values is a type variable α, α is fixed inside the locale. Instantiations may still replace α with any other HOL type. In other words, the locale monad formalises a *monomorphic* monad, but leaves the type of values unspecified. As usual, $m \gg m'$ abbreviates $m \ggg (\lambda_-.\ m')$.

locale monad = fixes return :: $\alpha \Rightarrow \mu$ and bind :: $\mu \Rightarrow (\alpha \Rightarrow \mu) \Rightarrow \mu$ (infixr $\ggg$)
   assumes BIND-ASSOC: $(m \ggg f) \ggg g = m \ggg (\lambda x.\ f\ x \ggg g)$
     and RETURN-BIND: return $x \ggg f = f\ x$
     and BIND-RETURN: $x \ggg$ return $= x$

Monads become useful only when effect-specific operations are available. In the remainder of this section, we formalise monadic operations for different types of effects and their properties. For each effect, we introduce a new locale in Isabelle that extends the locale monad, fixes the new operations, and specifies their properties. A locale extension inherits parameters and assumptions. This leads to a modular design: if several effects are needed, one merely combines the relevant locales in a multi-extension.

2.1 Failure and Exception

Failures are one of the simplest effects and widely used. A failure aborts the computation immediately. The locale monad-fail given below formalises the failure effect fail :: μ. It assumes that a failure propagates from the left hand side of bind. In contrast, there is no assumption about how fail behaves on the right hand side. Otherwise, if monad fail also assumed $m \ggg (\lambda_-.\ \text{fail}) = \text{fail}$, then fail would undo any effect of m. Although the standard implementation of failures using the option type satisfies this additional law, many other monad implementations do not, e.g., resumptions. Note that there is no need to delay the evaluation of fail in HOL because HOL has no execution semantics.

locale monad-fail = monad + fixes fail :: μ
   assumes FAIL-BIND: fail $\ggg f = $ fail

As a first example, we define the monadic interpreter eval :: $(\nu \Rightarrow \mu) \Rightarrow$ ν exp $\Rightarrow \mu$ for arithmetic expressions by primitive recursion using these abstract monad operations inside the locale monad-fail.[1] The first argument is an interpretation function $E :: \nu \Rightarrow \mu$ for the variables. The evaluation fails when a division by zero occurs.

---

[1] Type variables that appear in the signature of locale parameters are fixed for the whole locale. In particular, the value type α cannot be instantiated inside the locale monad or its extension monad-fail. The interpreter eval, however, returns ints. For this reason, eval is defined in an extension of monad-fail that merely specialises α to int. For readability, we usually omit this detail in this paper.

```
primrec (in monad-fail) eval :: (ν ⇒ μ) ⇒ ν exp ⇒ μ where
  eval E (Const i) = return i
| eval E (Var x)   = E x
| eval E (e₁ ⊕ e₂) = eval E e₁ ≫= (λi₁. eval E e₂ ≫= (λi₂. return (i₁ + i₂)))
| eval E (e₁ ⊘ e₂) =
    eval E e₁ ≫= (λi₁. eval E e₂ ≫= (λi₂. if i₂ = 0 then fail else return (i₁ div i₂)))
```

Note that evaluating a variable can have an effect μ, which is necessary to obtain a compositional interpreter. Let subst :: $(\nu \Rightarrow \nu'$ exp$) \Rightarrow \nu$ exp $\Rightarrow \nu'$ exp be the substitution function for exp. That is, subst σ e replaces every Var x in e with σ x. Then, the following compositionality statement holds (proven by induction on e and term rewriting with the definitions), where function composition $\circ$ is defined as $(f \circ g)(x) = f$ $(g$ $x)$.

lemma COMPOSITIONALITY: eval E (subst σ e) = eval (eval $E \circ \sigma$) e
  by *induction simp-all*

We refer to failures as exceptions whenever there is an operator catch :: $\mu \Rightarrow \mu \Rightarrow \mu$ to handle them. Following Gibbons and Hinze [6], the locale monad-catch assumes that catch and fail form a monoid and that returns are not handled. It inherits FAIL-BIND and the monad laws by extending the locale monad-fail. No properties about catch and bind are assumed because in general exception handling does not distribute over sequencing.

```
locale monad-catch = monad-fail + fixes catch :: μ ⇒ μ ⇒ μ
  assumes FAIL-CATCH:   catch fail m = m
     and CATCH-FAIL:    catch m fail = m
     and CATCH-CATCH:   catch (catch m₁ m₂) m₃ = catch m₁ (catch m₂ m₃)
     and RETURN-CATCH:  catch (return x) m = return x
```

2.2 State

Stateful computations use operations to read (get) and replace (put) the state of type σ. In a value-polymorphic setting, get :: σ τ and put :: $\sigma \Rightarrow$ unit τ are usually computations that return the state or () inhabiting the singleton type unit. Without value-polymorphism, these types cannot be formalised in the HOL setting because we cannot apply τ to different value types. Instead, our operations additionally take a continuation: get :: $(\sigma \Rightarrow \mu) \Rightarrow \mu$ and put :: $\sigma \Rightarrow \mu \Rightarrow \mu$. In a value-polymorphic setting, both signatures are equivalent. Passing the continuation return as in get return and λs. put s (return ()) yields the conventional operations. Conversely, our operations get f and put s m can be implemented as get $\gg= f$ and put $s \gg m$ using conventional get and put. The locale monad-state collects the properties get and put must satisfy:

locale monad-state = monad + **fixes** get :: $(\sigma \Rightarrow \mu) \Rightarrow \mu$ and put :: $\sigma \Rightarrow \mu \Rightarrow \mu$
  assumes PUT-GET: put s (get f) = put s (f s)
    and GET-GET: get $(\lambda s.\ \text{get}\ (f\ s))$ = get $(\lambda s.\ f\ s\ s)$
    and PUT-PUT: put s (put s' m) = put s' m
    and GET-PUT: get $(\lambda s.\ \text{put}\ s\ m)$ = m
    and GET-CONST: get $(\lambda_-.\ m)$ = m
    and BIND-GET: get $f \ggg g$ = get $(\lambda s.\ f\ s \ggg g)$
    and BIND-PUT: put s $m \ggg f$ = put s $(m \ggg f)$

The first four assumptions adapt Gibbons' and Hinze's axioms for the state operations [6] to the new signature. The fifth, GET-CONST, additionally specifies that get can be discarded if the state is not used. The last two assumptions, BIND-GET and BIND-PUT, demand that get and put distribute over bind. In the conventional value-polymorphic setting, where the continuations are applied using bind, these two are subsumed by the monad laws. In the remainder of this paper, get and put always take continuations.

    A state update function update can be implemented abstractly for all state monads. Like put, update takes a continuation m.

definition (**in** monad-state) update :: $(\sigma \Rightarrow \sigma) \Rightarrow \mu \Rightarrow \mu$ **where**
  update f m = get $(\lambda s.\ \text{put}\ (f\ s)\ m)$

The expected properties of update can be derived from monad-state's assumptions by term rewriting. For example,

lemma UPDATE-ID: update id m = m
  by (*simp add*: UPDATE-DEF GET-PUT)

lemma UPDATE-UPDATE: update f (update g m) = update $(g \circ f)$ m
  by (*simp add*: UPDATE-DEF PUT-GET PUT-PUT)

lemma UPDATE-BIND: update f $m \ggg g$ = update f $(m \ggg g)$
  by (*simp add*: UPDATE-DEF BIND-GET BIND-PUT)

    As an example, we implement a memoisation operator memo using the state operations. To that end, the state must be refined to a lookup table, which we model as a map of type $\beta \rightharpoonup \alpha = \beta \Rightarrow \alpha$ option. The definition uses the function $\lambda t.\ t(x \mapsto y)$ that takes a map t and updates it to associate x with y, leaving the other associations as they are; formally, $t(x \mapsto y) = (\lambda x'.\ \text{if}\ x = x'\ \text{then Some}\ y\ \text{else}\ t\ x')$.

definition (**in** monad-state) memo :: $(\beta \Rightarrow \mu) \Rightarrow \beta \Rightarrow \mu$ **where**
  memo f x = get $(\lambda table.$
    case $table$ x **of** Some $y \Rightarrow$ return y
         | None $\Rightarrow$ f $x \ggg (\lambda y.\ \text{update}\ (\lambda t.\ t(x \mapsto y))\ (\text{return}\ y)))$

A memoisation operator should satisfy three important properties. First, it should evaluate the memoised function at most on the given argument, not on others. This can be expressed as a congruence rule, which holds independently of the monad laws by definition:

lemma MEMO-CONG: $f\,x = g\,x \longrightarrow$ memo $f\,x =$ memo $g\,x$

Second, memoisation should be idempotent, i.e., if a function is already being memoised, then there is no point in memoising it once more.

lemma MEMO-IDEM: memo (memo f) $x =$ memo $f\,x$

The mechanised proof of MEMO-IDEM in Isabelle needs only two steps, which are justified by term rewriting with the properties of the monad operations and the case operator. Every assumption about get and put except GET-PUT is needed.

Third, the memoisation operator should indeed evaluate f on x at most once. As memo $f\,x$ memoises only the result of $f\,x$, but not the effect of evaluating $f\,x$, the next lemma captures this correctness property. Its proof is similar to MEMO-IDEM's.

lemma CORRECT: memo $f\,x \ggg (\lambda a.$ memo $f\,x \ggg g\,a) =$ memo $f\,x \ggg (\lambda a.g\,a\,a)$

2.3 Probabilistic Choice

Randomised computations are built from an operation ¢ for probabilistic choice. The probabilities are specified using probability mass functions (type π pmf) [7], i.e., discrete probability distributions. Binary probabilistic choice, which is often used in the literature [5, 6, 24], is less general as it leads to finite distributions. Continuous distributions would work, too, but they would clutter the theorems and proofs with measurability conditions.

Like the state operations, ¢ :: π pmf $\Rightarrow (\pi \Rightarrow \mu) \Rightarrow \mu$ takes a continuation to separate the type of probabilistic choices π from the type of values. The locale monad-prob assumes the following properties, where supp p denotes the support of p:

- sampling from the one-point distribution dirac x has no effect (SAMPLE-DIRAC),
- sequencing bind$_{\text{pmf}}$ in the probability monad yields sequencing (SAMPLE-BIND),
- sampling can be discarded if the result is unused (SAMPLE-CONST),
- sampling from independent distributions commutes (SAMPLE-COMM, independence is formalised by p and q not taking y and x as an argument, respectively),
- sampling calls the continuation only on values in p's support (SAMPLE-CONG), and
- sampling distributes over both sides of bind (BIND-SAMPLE$_1$, BIND-SAMPLE$_2$).

locale monad-prob = monad + **fixes** ¢ :: π pmf $\Rightarrow (\pi \Rightarrow \mu) \Rightarrow \mu$
  assumes SAMPLE-DIRAC: ¢ (dirac x) $f = f\,x$
    and SAMPLE-BIND: ¢ (bind$_{\text{pmf}}\,p\,f$) $g =$ ¢ $p\,(\lambda x.$ ¢ $(f\,x)\,g)$
    and SAMPLE-CONST: ¢ $p\,(\lambda \_.\,m) = m$
    and SAMPLE-COMM: ¢ $p\,(\lambda x.$ ¢ $q\,(f\,x)) =$ ¢ $q\,(\lambda y.$ ¢ $p\,(\lambda x.\,f\,x\,y))$
    and SAMPLE-CONG: $(\forall x \in$ supp $p.\,f\,x = g\,x) \longrightarrow$ ¢ $p\,f =$ ¢ $p\,g$
    and BIND-SAMPLE$_1$:   ¢ $p\,f \ggg g =$ ¢ $p\,(\lambda x.\,f\,x \ggg g)$
    and BIND-SAMPLE$_2$: $m \ggg (\lambda x.$ ¢ $p\,(f\,x)) =$ ¢ $p\,(\lambda y.\,m \ggg (\lambda x.\,f\,x\,y))$

2.4   Combining Abstract Monads

Formalising monads in this abstract way has the advantage that the different effects can be easily combined. In the running example, suppose that the variables represent independent random variables. Then, expressions are probabilistic computations and evaluation computes the joint probability distribution. For example, if x_1 and x_2 represent coin flips with 1 representing heads and 0 tails, then $\mathsf{Var}\ x_1 \oplus \mathsf{Var}\ x_2$ represents the probability distribution of the number of heads.

Here is a first attempt. Let $X :: \nu \Rightarrow \mathsf{int\ pmf}$ specify the distribution $X\ x$ for each random variable x. Combining the locales for failures and probabilistic choices, we let the variable environment do the sampling, where sample-var $X\ x = \math鈳}\ (X\ x)$ return:

locale monad-fail-prob = monad-fail + monad-prob

definition (**in** monad-fail-prob) wrong :: $(\nu \Rightarrow \mathsf{int\ pmf}) \Rightarrow \nu\ \mathsf{exp} \Rightarrow \mu$ **where**
  wrong $X\ e$ = eval (sample-var X) e

As the name suggests, wrong does not achieve what we intended. If a variable occurs multiple times in e, say $e = \mathsf{Var}\ x \oplus \mathsf{Var}\ x$, then wrong $X\ e$ samples x afresh for each occurrence. So, if $X\ x = $ uniform $\{0,1\}$, i.e., x is a coin flip, wrong $X\ e$ computes the probability distribution given by $0 \mapsto 1/4, 1 \mapsto 1/2, 2 \mapsto 1/4$ instead of $0 \mapsto 1/2, 2 \mapsto 1/2$. Clearly, we should sample every variable at most once. Memoising the variable evaluation achieves that. So, we additionally need state operations.

locale monad-fail-prob-state = monad-fail-prob + monad-state +
  assumes SAMPLE-GET: $\math鈳\ p\ (\lambda x.\ \mathsf{get}\ (f\ x)) = \mathsf{get}\ (\lambda s.\ \math鈳\ p\ (\lambda x.\ f\ x\ s))$

definition (**in** monad-fail-prob-state) lazy :: $(\nu \Rightarrow \mathsf{int\ pmf}) \Rightarrow \nu\ \mathsf{exp} \Rightarrow \mu$ **where**
  lazy $X\ e$ = eval (memo (sample-var X)) e

The interpreter lazy samples a variable only when needed. For example, in $e_0 = (\mathsf{Const}\ 1 \oslash \mathsf{Const}\ 0) \oplus \mathsf{Var}\ x_0$, the division by zero makes the evaluation fail before x_0 is sampled.

The locale monad-fail-prob-state adds an assumption that $\math鈳$ distributes over get. Such distributivity assumptions are typically needed because of the continuation parameters, which break the separation between effects and sequencing. Their format is as follows: If two operations f_1 and f_2 with continuations do not interact, then we assume $f_1\ (\lambda x.\ f_2\ (g\ x)) = f_2\ (\lambda y.\ f_1\ (\lambda x.\ g\ x\ y))$. Sometimes, such assumptions follow from existing assumptions. For example, SAMPLE-PUT follows from BIND-SAMPLE2 and put $s\ m = $ put s (return x) $\gg m$ for all x. A similar law holds for update.

lemma SAMPLE-PUT: $\math鈳\ p\ (\lambda x.\ \mathsf{put}\ s\ (f\ x)) = \mathsf{put}\ s\ (\math鈳\ p\ f)$

In contrast, SAMPLE-GET does not follow from the other assumptions due to the restriction to monomorphic values. The state of type σ, which get passes to its continuation, may carry more information than a value can hold. Indeed, in the case of lazy, the type int of values is countable, but the state type $v \rightharpoonup$ int is not if the type of variables is infinite. As put passes no information to its continuation, put's continuation can be pushed into bind as shown above. Still, put needs its continuation; otherwise, it would have to create a return value out of nothing, which would cause problems later (§4). Moreover, there is no need to explicitly specify how fail interacts with get and ¢ as get $(\lambda\_.\ \text{fail}) = \text{fail}$ and ¢ p $(\lambda\_.\ \text{fail}) = \text{fail}$ are special cases of GET-CONST and SAMPLE-CONST.

Instead of lazy sampling, we can also sample all variables eagerly. Let vars e return the (finite) set of variables in e. Then, the interpreter eager with eager sampling is defined as follows (all three definitions live in the locale monad-fail-prob-state):

definition sample-vars :: $(v \Rightarrow \text{int pmf}) \Rightarrow v$ set $\Rightarrow \mu \Rightarrow \mu$ where
  sample-vars X A m = fold $(\lambda x\ m.\ \text{memo (sample-var } X)\ x \gg m)\ m\ A$

definition lookup :: $v \Rightarrow \mu$ where
  lookup x = get $(\lambda s.\ \text{case } s\ x\ \text{of None} \Rightarrow \text{fail} \mid \text{Some } i \Rightarrow \text{return } i)$

definition eager :: $(v \Rightarrow \text{int pmf}) \Rightarrow v$ exp $\Rightarrow \mu$ where
  eager X e = sample-vars X (vars e) (eval lookup e)

where fold is the fold operator for finite sets [23]. The operator fold f requires that the folding function f is left-commutative, i.e., $f\ x\ (f\ y\ z) = f\ y\ (f\ x\ z)$ for all x, y, and z. In our case, $f = \lambda x\ m.\ \text{memo (sample-var } X)\ x \gg m$ is left-commutative by the following lemma about memo whose assumptions sample-var X satisfies by RETURN-BIND, BIND-SAMPLE$_1$, BIND-SAMPLE$_2$, and SAMPLE-GET. Moreover, by CORRECT, it is also idempotent, i.e., $f\ x \circ f\ x = f\ x$.

lemma MEMO-COMMUTE:
  $(\forall m\ x\ g.\ m \ggg (\lambda a.\ f\ x \ggg g\ a) = f\ x \ggg (\lambda b.\ m \ggg (\lambda a.\ g\ a\ b)))$
  $\longrightarrow$ $(\forall x\ g.\ \text{get } (\lambda s.\ f\ x \ggg g\ s) = f\ x \ggg (\lambda a.\ \text{get } (\lambda s.\ g\ s\ a)))$
  $\longrightarrow$ memo $f\ x \ggg (\lambda a.\ \text{memo } f\ y \ggg (\lambda b.\ g\ a\ b)) =$
  memo $f\ y \ggg (\lambda b.\ \text{memo } f\ x \ggg (\lambda a.\ g\ a\ b))$

This lemma and CORRECT illustrate the typical form of monadic statements. The assumptions and conclusions take a continuation g for the remainder of the program. This way, the statements are easier to apply because they are in normal form with respect to BIND-ASSOC. This observation also holds in a value-polymorphic setting.

Now, the question is whether eager and lazy sampling are equivalent. In general, the answer is no. For example, for e_0 from above, eager X e_0 samples and memoises the variable x_0, but lazy X e_0 does not. Thus, there are contexts that distinguish the two. If we extend monad-fail-prob-state with exception handling from monad-catch such that

CATCH-GET: catch (get f) m_2 = get $(\lambda s.\ \text{catch } (f\ s)\ m_2)$
CATCH-PUT: catch (put s m) m_2 = put s (catch m m_2)

then the two can be distinguished:

$$\mathsf{catch}\ (\mathsf{lazy}\ X\ e_0)\ (\mathsf{lookup}\ x_0) = \mathsf{fail}$$
$$\mathsf{catch}\ (\mathsf{eager}\ X\ e_0)\ (\mathsf{lookup}\ x_0) = \mathsf{memo}\ (\mathsf{sample\text{-}var}\ X)\ x_0$$

In contrast, if we assume that failures erase state updates, then the two *are* equivalent:

theorem LAZY-EAGER: $(\forall s.\ \mathsf{put}\ s\ \mathsf{fail} = \mathsf{fail}) \longrightarrow \mathsf{lazy}\ X\ e = \mathsf{eager}\ X\ e$

Proof. The proof consists of three steps proven by induction on e. First, by idempotence and left-commutativity, sample-vars $X\ V$ commutes with lazy $X\ e$ for any finite V:

$$\forall g.\ \mathsf{sample\text{-}vars}\ X\ V\ (\mathsf{lazy}\ X\ e \ggg g) = \mathsf{lazy}\ X\ e \ggg (\lambda i.\ \mathsf{sample\text{-}vars}\ X\ V\ (g\ i)) \quad (1)$$

Here, put s fail = fail ensures that all state updates are lost if a division by zero occurs. The next two steps will use (1) in the inductive cases for $\oplus$ and $\oslash$ to bring together the sampling of the variables and the evaluation of the subexpressions. Second,

$$\mathsf{lazy}\ X\ e \ggg g = \mathsf{sample\text{-}vars}\ X\ (\mathsf{vars}\ e)\ (\mathsf{lazy}\ X\ e \ggg g) \quad (2)$$

shows that the sampling can be done first, which holds by CORRECT. Finally,

$$\mathsf{sample\text{-}vars}\ X\ V\ (\mathsf{lazy}\ X\ e \ggg g) = \mathsf{sample\text{-}vars}\ X\ V\ (\mathsf{eval}\ \mathsf{lookup}\ e \ggg g) \quad (3)$$

holds for any finite set V with vars $e \subseteq V$. Here, Var x is the interesting case, which follows from $\forall g.$ memo $f\ x \ggg (\lambda i.\ \mathsf{lookup}\ x \ggg g\ i) = \mathsf{memo}\ f\ x \ggg$ $(\lambda i.\ g\ i\ i)$ and CORRECT. Taking $V =$ vars e and $g =$ return, (2) and (3) prove the lemma. □

In Sect. 3.5, we show that some monads satisfy LAZY-EAGER's assumption, but not all.

2.5   Further Abstract Monads

Apart from exceptions, state, and probabilistic choice, we have formalised effect specifications for non-deterministic choice alt :: $\mu \Rightarrow \mu \Rightarrow \mu$, the reader and writer monads with ask :: $(\rho \Rightarrow \mu) \Rightarrow \mu$ and tell :: $\omega \Rightarrow \mu \Rightarrow \mu$, and resumptions with pause :: $o \Rightarrow (\iota \Rightarrow \mu) \Rightarrow \mu$. We do not present them in detail as the examples in this paper do not require them.

Moreover, we formalise as locales the notions of a commutative monad, where bind satisfies $m_1 \ggg (\lambda x.\ m_2 \ggg f\ x) = m_2 \ggg (\lambda y.\ m_1 \ggg (\lambda x.\ f\ x\ y))$, and of a discardable monad, where the law $m \gg m' = m'$ makes it possible to drop a computation whose result is not used.

3   Implementations of Monads and Monad Transformers

In the previous section, we specified the properties of monadic operations abstractly. Now, we provide monad implementations that satisfy these specifications. Some effects are implemented as monad transformers [15,21], which allow us to compose implementations of different effects almost as modularly as the locales specifying them abstractly. In particular, we analyse whether the transformers preserve the specifications of the other effects. All our implementations are polymorphic in the values such that they can be used with any value type, although by the value-monomorphism restriction, each usage must individually commit to one value type.

3.1   The Identity Monad

The simplest monad implementation in our library is the identity monad ident, which models the absence of all effects. It is not really useful in itself, but will be an important building block when combining monads using transformers. The datatype α ident is a copy of α with constructor Ident and selector run-ident. To distinguish the abstract monad operations from their implementations, we subscript the latter with the implementation type. The lemma states that $return_{ident}$ and $bind_{ident}$ satisfy the assumption of the locale monad. Additionally, the identity monad is commutative and discardable.

datatype α ident = Ident (run-ident: α)
definition $return_{ident} :: \alpha \Rightarrow \alpha$ ident where $return_{ident}$ = Ident
definition $bind_{ident} :: \alpha$ ident $\Rightarrow (\alpha \Rightarrow \alpha$ ident) $\Rightarrow \alpha$ ident where
   $m \ggeq_{ident} f = f$ (run-ident m)

lemma monad $return_{ident}$ $bind_{ident}$

3.2   The Probability Monad

The probability monad α prob is another basic building block. We use discrete probability distributions [7] and Giry's probability monad operations dirac and $bind_{pmf}$, which we already used in the abstract specification in Sect. 2.3. Then, probabilistic choice $\mathcal{C}_{prob}$ is just monadic sequencing on α pmf. The probability monad is commutative and discardable.

type-synonym α prob = α pmf
definition $return_{prob} :: \alpha \Rightarrow \alpha$ prob where $return_{prob}$ = dirac
definition $bind_{prob} :: \alpha$ prob $\Rightarrow (\alpha \Rightarrow \alpha$ prob) $\Rightarrow \alpha$ prob where $bind_{prob}$ = $bind_{pmf}$
definition $\mathcal{C}_{prob} :: \pi$ pmf $\Rightarrow (\pi \Rightarrow \alpha$ prob) $\Rightarrow \alpha$ prob where $\mathcal{C}_{prob}$ = $bind_{pmf}$

lemma monad-prob $return_{prob}$ $bind_{prob}$ $\mathcal{C}_{prob}$

3.3   The Failure and Exception Monad Transformer

Failures and exception handling are implemented as a monad transformer. Thus, these effects can be added to any monad τ. In the value-polymorphic setting, the failure monad transformer takes a monad τ and defines a type constructor failT such that β failT is isomorphic to $(\beta$ option$)$ τ. That is, the transformer specialises the value type α of the inner monad to β option. In our value-monomorphic setting, the type variable μ represents the application of τ to the value type, i.e., β option. So, μ failT is just a copy of μ:

datatype μ failT = FailT (run-fail: μ)

As failT's operations depend on the inner monad, we fix abstract operations return and bind in an unnamed context and define failT's operations in terms of them. The line on the left indicates the scope of the context. At the end, which is marked by $\llcorner$, the fixed operations become additional arguments of the defined functions. Values in the inner monad now have type α option. The definitions themselves are standard [21].

context fixes return :: α option $\Rightarrow \mu$ **and** bind :: $\mu \Rightarrow (\alpha$ option $\Rightarrow \mu) \Rightarrow \mu$

> **definition** $\text{return}_{\text{failT}}$ :: $\alpha \Rightarrow \mu$ failT **where**
> $\text{return}_{\text{failT}}$ x = FailT (return (Some x))
>
> **definition** $\text{bind}_{\text{failT}}$ :: μ failT $\Rightarrow (\alpha \Rightarrow \mu$ failT$) \Rightarrow \mu$ failT **where**
> $m \ggg_{\text{failT}} f$ = FailT (run-fail $m \ggg$
>   (λx. case x of None $\Rightarrow$ return None | Some $y \Rightarrow$ run-fail (f y)))
>
> **definition** $\text{fail}_{\text{failT}}$ :: μ failT **where** $\text{fail}_{\text{failT}}$ = FailT (return None)
>
> **definition** $\text{catch}_{\text{failT}}$ :: μ failT $\Rightarrow \mu$ failT $\Rightarrow \mu$ failT **where**
> $\text{catch}_{\text{failT}}$ m_1 m_2 = FailT (run-fail $m_1 \ggg$
>   (λx. case x of None $\Rightarrow$ run-fail m_2 | Some _ $\Rightarrow$ return x))

If return and bind form a monad, so do $\text{return}_{\text{failT}}$ and $\text{bind}_{\text{failT}}$, and $\text{fail}_{\text{failT}}$ and $\text{catch}_{\text{failT}}$ satisfy the effect specification from Sect. 2.1, too. The next lemma expresses this.

> **lemma** monad-catch $\text{return}_{\text{failT}}$ $\text{bind}_{\text{failT}}$ $\text{fail}_{\text{failT}}$ $\text{catch}_{\text{failT}}$
>   **if** monad return bind

Clearly, we want to keep using the existing effects of the inner monad. So, we must lift their operations to failT and prove that their specifications are preserved. The lifting is not hard; the continuations of the operations are transformed in the same way as $\text{bind}_{\text{failT}}$ does. Here, we only show how to lift the state operations, where the locale monad-catch-state extends monad-catch and monad-state with CATCH-GET and CATCH-PUT. Moreover, failT also lifts ¢, alt, ask, tell, and pause, preserving their specifications. It is commutative if the inner monad is commutative and discardable.

> **context fixes** get :: $(\sigma \Rightarrow \mu) \Rightarrow \mu$ **and** put :: $\sigma \Rightarrow \mu \Rightarrow \mu$

```
definition get_failT :: (σ ⇒ μ failT) ⇒ μ failT where
  get_failT f = FailT (get (λs. run-fail (f s)))
definition put_failT :: σ ⇒ μ failT ⇒ μ failT where
  put_failT s m = FailT (put s (run-fail m))
lemma monad-catch-state return_failT bind_failT fail_failT catch_failT get_failT put_failT
  if monad-state return bind get put
```

From now on, as the context scope has ended, $\text{return}_{\text{failT}}$ and $\text{bind}_{\text{failT}}$ take the inner monad's operations return and bind as additional arguments. For example, we obtain a plain failure monad by applying failT to ident. Interpreting the locale monad-fail for $\text{return}_F = \text{return}_{\text{failT}}$ $\text{return}_{\text{ident}}$ and $\text{bind}_F = \text{bind}_{\text{failT}}$ $\text{return}_{\text{ident}}$ $\text{bind}_{\text{ident}}$ and $\text{fail}_F = \text{fail}_{\text{failT}}$ $\text{return}_{\text{ident}}$ yields an executable version of the interpreter eval from Sect. 2.1, which we refer to as eval_F. Then, Isabelle's code generator and term rewriter both evaluate

$$\text{eval}_F \ (\lambda x. \ \text{return}_F \ (((\lambda\_. \ 0)(x_0 := 5)) \ x)) \ (\text{Var} \ x_0 \oplus \text{Const} \ 7)$$

to FailT (Ident (Some 12)). Given some variable environment $Y :: \nu \Rightarrow \text{int},$[2] we obtain a textbook-style interpreter [22, Sect. 3.1.2] as run-ident (run-fail(eval_F (return[fail.] ∘ Y) e)).

3.4   The State Monad Transformer

The state monad transformer adds the effects of a state monad to some inner monad. The formalisation follows the same ideas as for failT, so we only mention the important points. The state monad transformer transforms a monad $\alpha \ \tau$ into the type $\sigma \Rightarrow (\alpha \times \sigma) \ \tau$ where σ is the type of states. So, in HOL, the type of values of the inner monad becomes $\alpha \times \sigma$ and μ represents $(\alpha \times \sigma) \ \tau$.

```
datatype (σ, μ) stateT = StateT (run-state: σ ⇒ μ)
```

Like for failT, the state monad operations $\text{return}_{\text{stateT}}$ and $\text{bind}_{\text{state}}$ depend on inner monad operations return and bind. With $\text{get}_{\text{stateT}}$ and $\text{put}_{\text{stateT}}$ defined in the obvious way, the transformer satisfies the specification monad-state for state monads.

```
context fixes return :: α × σ ⇒ μ and bind :: μ ⇒ (α × σ ⇒ μ) ⇒ μ
  definition return_stateT :: α ⇒ (σ, μ) stateT where
    return_stateT x = StateT (λs. return (x, s))
  definition bind_stateT :: (σ, μ) stateT ⇒ (α ⇒ (σ, μ) stateT) ⇒ (σ, μ) stateT where
    m ≫=_stateT f = StateT (λs. run-state f s ≫= (λ(x, s'). run-state (f x) s'))
  definition get_stateT :: (σ ⇒ (σ, μ) stateT) ⇒ (σ, μ) stateT where
    get_stateT f = StateT (λs. run-state (f s) s)
```

---

[2] Such environments can be nicely handled by applying a reader monad transformer on top (Sect. 4).

```
definition put_stateT :: σ ⇒ (σ, μ) stateT ⇒ (σ, μ) stateT where
  put_stateT s m = StateT (λ_. run-state m s)
lemma monad-state return_stateT bind_stateT get_stateT put_stateT
  if monad return bind
```

The state monad transformer lifts the other effect operations fail, ¢, ask, tell, alt, and pause according to their specifications. But catch cannot be lifted through stateT such that CATCH-GET and CATCH-PUT from Sect. 2.4 hold. As our exceptions carry no information, the inner monad cannot pass the state updates before the failure to the handler.

3.5 Composing Monads with Transformers

Composing the two monad transformers failT and stateT with the monad prob, we can now instantiate the probabilistic interpreter from Sect. 2.4. As is well known, the order of composition matters. If we first apply failT to prob and then stateT (SFP for short), the resulting interpreter $\mathsf{eval_{SFP}}\ E\ e :: (v \to \mathsf{int}, (\mathsf{int} \times (v \to \mathsf{int}))$ option prob failT) stateT nests the result state of type $v \to \mathsf{int}$ inside the option type for failures, i.e., failures do not return a new state. Thus, failures erase state updates, i.e., $\mathsf{put_{SFP}}\ s\ \mathsf{fail_{SFP}} = \mathsf{fail_{SFP}}$, and lazy and eager sampling are equivalent (LAZY-EAGER). Conversely, if we apply failT after stateT to prob (FSP for short), then $\mathsf{eval_{FSP}}\ E\ e :: (v \to \mathsf{int}, (\mathsf{int}\ \mathsf{option} \times (v \to \mathsf{int}))$ prob) stateT failT and failures do return a new state as only the result type int sits inside option. In particular, $\mathsf{put_{SFP}}\ s\ \mathsf{fail_{FSP}} \neq \mathsf{fail_{FSP}}$ in general, and lazy and eager sampling are not equivalent. We will consider the SFP case further in Sect. 4.

3.6 Further Monads and Monad Transformers

Apart from the monad implementations presented so far, our library provides implementations also for the other types of effects mentioned in Sect. 2.5. In particular, non-deterministic choice is implemented as a monad transformer based on finite multisets, which works only for commutative inner monads. Moreover, we define a reader (readT) and a writer (writerT) monad transformer. The reader monad transformer differs from stateT only in that no updates are possible. Thus, (ρ, μ) readT leaves the type of values of the inner monad unchanged, as no new state must be returned.

```
datatype (ρ, μ) readT = ReadT (run-read: ρ ⇒ μ)
context fixes return :: α ⇒ μ and bind :: μ ⇒ (α ⇒ μ) ⇒ μ
  definition return_readT :: α ⇒ (ρ, μ) readT where
    return_readT x = ReadT (λ_. return x)
  definition bind_readT :: (ρ, μ) readT ⇒ (α ⇒ (ρ, μ) readT) ⇒ (ρ, μ) readT where
    m ⪢_readT f = ReadT (λr. run-read m r ⪢ (λx. run-read (f x) r))
```

definition $\mathsf{ask_{readT}}$ :: $(\rho \Rightarrow (\rho, \mu)\ \mathsf{readT}) \Rightarrow (\rho, \mu)\ \mathsf{readT}$ where
  $\mathsf{ask_{readT}}\ f = \mathsf{ReadT}\ (\lambda r.\ \mathsf{run\text{-}read}\ (f\ r)\ r)$
definition $\mathsf{fail_{readT}}$ :: $(\mu \Rightarrow (\rho, \mu)\ \mathsf{readT})$ where $\mathsf{fail_{readT}}\ \mathsf{fail} = \mathsf{ReadT}\ (\lambda\_.\ \mathsf{fail})$

Resumptions are formalised as a plain monad using the codatatype

codatatype (o, ι, α) resumption $=$ Done α | Pause o $(\iota \Rightarrow (o, \iota, \alpha)$ resumption$)$

Unfortunately, we cannot define resumptions as a monad transformer in HOL despite the restriction to monomorphic values. The reason is that for a transformer with inner monad τ, the second argument of the constructor Pause would have to be of type $\iota \Rightarrow (o, \iota, \alpha)$ resumption τ, i.e., the codatatype would recurse through the unspecified type constructor τ. This is not supported by Isabelle's codatatype package [2] and, in fact, for some choices of τ, e.g., unbounded nondeterminism, the resumption transformer type does not exist in HOL at all. For the same reason, we cannot have other monad transformers that have similar recursive implementation types. Therefore, we fail to modulary construct all combinations of effects. For example, probabilistic resumptions with failures [17] are out of reach and must still be constructed from scratch.

3.7   Overloading the Monad Operations

When several monad transformers are composed, the monad operations quickly become large HOL terms as the transformer's operations take the inner monad's as explicit arguments. These large terms must be handled by the inference kernel, the type checker, the parser, and the pretty-printer, even if locale interpretations hide them from the user using abbreviations. To improve readability and the processing time of Isabelle, our library also defines the operations as single constants which are overloaded for the different monad implementations using recursion on types [26]. As overloading does not need these explicit arguments, it thus avoids the processing times for unification, type checking, and (un)folding of abbreviations. Yet, Isabelle's check against cyclic definitions [13] fails to see that the resulting dependencies must be acyclic (as the inner monad is always a type argument of the outer monad). So, we moved these overloaded definitions to a separate file and marked them as unchecked.[3] Overloading is just a syntactic convenience, on which the library and the examples in this paper do not rely. If users want to use it, they are responsible for not exploiting these unchecked dependencies.

---

[3] Isabelle's `adhoc-overloading` feature, which resolves overloading during type checking, cannot be used either as it does not support recursive resolutions. For example, resolving return :: $\alpha \Rightarrow \alpha$ option ident failT takes two steps: first to $\mathsf{return_{failT}}$ return and then to $\mathsf{return_{failT}}$ $\mathsf{return_{ident}}$. The second step fails due to the intricate interleaving of type checking and resolution. Even if this is just an implementation issue, resolving overloading during type checking prevents definitions that are generic in the monad, which general overloading supports.

4   Moving Between Monad Instances

Once all variables have been sampled eagerly, the evaluation of the expression itself is deterministic. Thus, the actual evaluation need not be done in a monad as complex as FSP or SFP. It suffices to work in a reader-failure monad with operations fail and ask, which we obtain by applying the monad transformers readT and failT to ident (RFI for short). Such simpler monads have the advantage that reasoning becomes easier as more laws hold. We now explain how the theory of representation independence [20] can be used to move between different monad instances by going from SFP to RFI. This ultimately yields a theorem that characterises eval_{SFP} in terms of eval_{RFI}. So, in general, this approach makes it possible to switch in the middle of a bigger proof from a complicated monad to a much simpler one.

Let us first deal with sampling. To go from α prob to β ident, we use a relation $\mathbb{IP}(A)$ between α ident and β prob since relations work better with higher-order functions than equations. Following Huffman and Kunčar [10], we call such relations correspondence relations. It is parametrised by a relation A between the values, which we will use later to express the differences in the values due to the monad transformers changing the value type of the inner monad. In detail, $\mathbb{IP}(A)$ relates a value Ident x to the one-point distribution dirac y iff A relates x to y. Then, the monad operations of ident and prob respect this relation. Respectfulness is formalised using the function relator $A \mapsto B$ defined by $(f, g) \in A \mapsto B$ iff $(x, y) \in A$ implies $(f(x), g(y)) \in B$ for all x and y. Then, the monad operations respecting $\mathbb{IP}(A)$ is expressed by the following two conditions:

- $(\text{return}_{\text{ident}}, \text{return}_{\text{prob}}) \in A \mapsto \mathbb{IP}(A)$ and
- $(\text{bind}_{\text{ident}}, \text{bind}_{\text{prob}}) \in \mathbb{IP}(A) \mapsto (A \mapsto \mathbb{IP}(A)) \mapsto \mathbb{IP}(A)$.

Note the similarity between the relations and the types of the monad operations, where A and $\mathbb{IP}$ take the roles of the type variables for values and of the monad type constructor, respectively. As the monad transformers failT and stateT are relationally parametric in the inner monad and eval is parametric in the monad, we prove the following relation between the evaluators automatically using Isabelle/HOL's Transfer prover [10]

$$(\text{eval}_{\text{SFP}} \ \text{lookup}_{\text{SFP}} \ e, \text{eval}_{\text{SFI}} \ \text{lookup}_{\text{SFI}} \ e) \in \text{rel}_{\text{stateT}} \ (\text{rel}_{\text{failT}} \ (\mathbb{IP}(=))) \qquad (4)$$

where SFI refers to the state-failure-identity composition of monads, $(=)$ is the identity relation, and $\text{rel}_{\text{stateT}}$ and $\text{rel}_{\text{failT}}$ are the relators for the datatypes stateT and failT [2]. Formally, the relators lift relations on the inner monad to relations on the transformed monad. For example, $(m_1, m_2) \in \text{rel}_{\text{stateT}} \ M$ iff (run-state $m_1 \ s$, run-state $m_2 \ s$) $\in M$ for all s, and $(m_1, m_2) \in \text{rel}_{\text{failT}} \ M$ iff (run-fail m_1, run-fail m_2) $\in M$. Intuitively, (4) states that in the monads SFP and SFI, eval behaves the same with respect to states updates and failure and the results are the same; in particular, the evaluation is deterministic.

In the following, we use the property of a relator rel that if M is the graph Gr f of a function f, then rel M is the graph of the function map f, where map is the canonical map function for the relator. For example, $\mathsf{map}_{\mathsf{failT}}\ f = \mathsf{FailT} \circ f \circ \mathsf{run\text{-}fail}$, so

$$\mathsf{rel}_{\mathsf{failT}}\ (\mathsf{Gr}\ f) = \mathsf{Gr}\ (\mathsf{map}_{\mathsf{failT}}\ f) \tag{5}$$

where $(x, y) \in \mathsf{Gr}\ f$ iff $f\ x = y$. Isabelle's datatype package automatically proves these relator-graph identities. The correspondence relation $\mathbb{IP}$ satisfies a similar law: $\mathbb{IP}(\mathsf{Gr}\ f) = \mathsf{Gr}\ (\mathsf{map}_{\mathbb{IP}}\ f)$ where $\mathsf{map}_{\mathbb{IP}}\ f = \mathsf{dirac} \circ f \circ \mathsf{run\text{-}ident}$.

Having eliminated probabilities, we next switch from the state monad transformer to the reader monad transformer. We again define a correspondence relation $\mathbb{RS}(s, M)$ between readT and stateT. It takes as parameters the environment s and the correspondence relation M between the inner monads. It relates the two monadic values m_1 and m_2 iff M relates the results of running m_1 and m_2 on s, i.e., $(\mathsf{run\text{-}read}\ m_1\ s, \mathsf{run\text{-}state}\ m_2\ s) \in M$. Again, we show that the monad operations respect $\mathbb{RS}(s, M)$ as formalised below. As readT and stateT are monad transformers, we assume that the operations of the inner monads respect M. These assumptions can be expressed using $\mapsto$ since the inner operations are arguments to readT's and stateT's operations. Here, $A \lessdot s$ adapts the relation A on values to stateT's change of the value type from α to $\alpha \times \sigma$; $(x, (y, s')) \in A \lessdot s$ iff $(x, y) \in A$ and $s' = s$, i.e., A relates the results and the state is not updated.

- $(\mathsf{return}_{\mathsf{readT}}, \mathsf{return}_{\mathsf{stateT}}) \in (A \lessdot s \mapsto M) \mapsto A \mapsto \mathbb{RS}(s, M)$,
- $(\mathsf{bind}_{\mathsf{readT}}, \mathsf{bind}_{\mathsf{stateT}}) \in$
  $(M \mapsto (A \lessdot s \mapsto M) \mapsto M) \mapsto \mathbb{RS}(s, M) \mapsto (A \mapsto \mathbb{RS}(s, M)) \mapsto \mathbb{RS}(s, M)$,
- $(\mathsf{ask}_{\mathsf{readT}}, \mathsf{get}_{\mathsf{stateT}}) \in (\{(s, s)\} \mapsto \mathbb{RS}(s, M)) \mapsto \mathbb{RS}(s, M)$, and
- $(\mathsf{fail}_{\mathsf{readT}}, \mathsf{fail}_{\mathsf{stateT}}) \in M \mapsto \mathbb{RS}(s, M)$,

Then, by representation independence, the Transfer package automatically proves the following relation between $\mathsf{eval}_{\mathrm{RFI}}$ and $\mathsf{eval}_{\mathrm{SFI}}$, where $\mathsf{lookup}_{\mathrm{RFI}}$ uses $\mathsf{ask}_{\mathsf{readT}}$ instead of $\mathsf{get}_{\mathsf{stateT}}$, and $\mathsf{rel}_{\mathsf{ident}}$ and $\mathsf{rel}_{\mathsf{option}}$ are the relators for the datatypes ident and option.

$$(\mathsf{eval}_{\mathrm{RFI}}\ \mathsf{lookup}_{\mathrm{RFI}}\ e, \mathsf{eval}_{\mathrm{SFI}}\ \mathsf{lookup}_{\mathrm{SFI}}\ e) \in \mathbb{RS}(s, \mathsf{rel}_{\mathsf{failT}}\ (\mathsf{rel}_{\mathsf{ident}}\ (\mathsf{rel}_{\mathsf{option}}\ (= \lessdot s))))$$

This says that running eval in RFI and SFI computes the same result, has the same behaviour with respect to state queries and failures, and does not update the state.

Actually, we can go from SFP directly to RFI, without the monad SFI as a stepping stone, thanks to $\mathbb{IP}$ taking a relation on the value types:

$$(\mathsf{eval}_{\mathrm{RFI}}\ \mathsf{lookup}_{\mathrm{RFI}}\ e, \mathsf{eval}_{\mathrm{SFP}}\ \mathsf{lookup}_{\mathrm{SFP}}\ e) \in \mathbb{RS}(s, \mathsf{rel}_{\mathsf{failT}}\ (\mathbb{IP}(\mathsf{rel}_{\mathsf{option}}\ (= \lessdot s)))) \tag{6}$$

As $= \lessdot s$ is the graph of $\lambda a.\ (a, s)$, using only the graph properties like (5) of $\mathbb{IP}$ and the relators, and using $\mathbb{RS}$'s definition, we derive the characterisation of $\mathsf{eval}_{\mathrm{SFP}}$ from (6):

run-state (eval$_{\text{SFP}}$ lookup$_{\text{SFP}}$ e) $s =$

  map$_{\text{failT}}$ (map$_{\mathbb{IP}}$ (map$_{\text{option}}$ ($\lambda a.\ (a, s)$))) (run-read (eval$_{\text{RFI}}$ lookup$_{\text{RFI}}$ e) s)

where map$_{\text{failT}}$ and map$_{\text{option}}$ are the canonical map functions for failT and option. Thus, instead of reasoning about eval$_{\text{SFP}}$ in SFP, we can conduct our proofs in the simpler monad RFI. For example, as RFI is commutative, subexpressions can be evaluated in any order. Thus, we get the following identity expressing the reversed evaluation order (and a similar one for $\oslash$).[4]

eval$_{\text{RFI}}$ E ($e_1 \oplus e_2$) = eval$_{\text{RFI}}$ $E\, e_2$ $\ggg_{\text{RFI}}$ ($\lambda j.$ eval$_{\text{RFI}}$ $E\, e_1$ $\ggg_{\text{RFI}}$ ($\lambda i.$ return$_{\text{RFI}}$ $(i + j)$))

In summary, we have demonstrated a generic approach to switch from a complicated monad to a much simpler one. Conceptually, the correspondence relations $\mathbb{IP}$ and $\mathbb{RS}$ just embed one monad or monad transformer (ident and readT) in a richer one (prob and stateT). It is precisely this embedding that ultimately yields the map functions in the characterisation. In this functional view, the respectfulness conditions express that the embedding is a monad homomorphism. Yet, we use relations for the embedding instead of functions because only relations work for higher-order operations in a compositional way.

The reader may wonder why we go through all the trouble of defining correspondence relations and showing respectfulness and parametricity. Indeed, in this example, it would probably have been easier to simply perform an induction over expressions and prove the equation directly. The advantage of our approach is that it does not rely on the concrete definition of eval. It suffices to know that eval is parametric in the monad, which Isabelle derives automatically from the definition. This automated approach therefore scales to arbitrarily complicated monadic functions whereas induction proofs do not. Moreover, note that the correspondence relations and respectfulness lemmas only depend on the monads. They can therefore be reused for other monadic functions.

5   Related Work

Huffman et al. [9,11] formalise the concept of value-polymorphic monads and several monad transformers in Isabelle/HOLCF, the domain theory library of Isabelle/HOL. They circumvent HOL's type system restrictions by projecting everything into HOLCF's universal domain of computable values. That is, they trade in HOL's set-theoretic model with its simple reasoning rules for a domain-theoretic model with ubiquitous $\bot$ values and strictness side conditions. This way, they can define a resumption monad transformer (for computable continuations). Being tied to domain theory, their library cannot be used to model effects of *plain* HOL functions, which is our goal, the strictness assumptions make their laws and proofs more complicated than ours, and functions defined with HOLCF

---

[4] Following the "as abstract as possible" spirit of this paper, we actually proved the identities in the locale of commutative monads and showed that readT is commutative if its inner monad is.

do not work with Isabelle's code generator. Still, their idea of projecting everything into a universal type could also be adapted to plain HOL, albeit only for a restricted class of monads; achieving a similar level of automation and modularity would require a lot more effort than our approach, which uses only existing features of Isabelle.

Gibbons and Hinze [6] axiomatize monads and effects using Haskell-style type constructor classes and use the algebraic specification to prove identities between Haskell programs, similar to our abstract locales in Sect. 2. Their specification of state effects omits GET-CONST, but they later assume that it holds [6, Sect. 10.2]. Being value-polymorphic, their operations do not need our continuations and the laws are therefore simpler. In particular, no new assumptions are typically needed when monad specifications are combined. In contrast, our continuations sometimes require interaction assumptions like SAMPLE-GET. Gibbons and Hinze only consider reasoning in the abstract setting and do not discuss the transition to concrete implementations and the relations between implementations. Also, they do not prove that monad implementations satisfy their specifications. Later, Jeuring et al. [12] showed that the implementations in Haskell do not satisfy them because of strictness issues similar to the ones in Huffman's work.

Lobo Vesga [16] formalised some of Gibbons' and Hinze's examples in Agda. She does not need assumptions for the continuations like we do as value-polymorphic monads can be directly expressed in Agda. Like Gibbons and Hinze, she does not study the connection between specifications and implementations. Thanks to the good proof automation in Isabelle, our mechanised proofs are much shorter than hers, which are as detailed as Gibbons' and Hinze's pen-and-paper proofs.

Lochbihler and Schneider [19] implemented support for equational reasoning about applicative functors, which are more general than monads. They focus on lifting identities on values to a concrete applicative functor. Reasoning with abstract applicative functors is not supported. Like monads, the concept of an applicative functor cannot be expressed as a predicate in HOL. Moreover, the applicative operations do not admit value monomorphisation like monads do, as the type of $\diamond$ contains applications of the functor type constructor τ to $\alpha \Rightarrow \beta$, α, and β. So, monads seem to be the right choice, even though we could have defined the interpreter eval applicatively (but not, e.g., memoisation).

6 · Conclusion

We have presented a library of abstract monadic effect specifications and their implementations as monads and monad transformers in Isabelle/HOL. We illustrated its usage and the elegance of reasoning using a monadic interpreter. The type system of HOL forced us to restrict the monads to monomorphic values. Monomorphic values work well when the reasoning involves only a few monadic functions like in our running example. In larger projects, this restriction can become a limiting factor. Nevertheless, in our project on formalising computational soundness results,[5] we successfully formalised and reasoned about several complicated serialisers and parsers for symbolic messages of security protocols. In that work, reasoning abstractly about effects and being able to move from one monad instance to another were crucial. More concretely, the serialiser converts symbolic protocol messages into bitstrings. The challenges were similar to those of our interpreter eval. Serialisation may fail when the symbolic message is not well-formed, similar to division by zero in the interpreter. When serialisation encounters a new nonce, it randomly samples a fresh bitstring, which must also be used for serialising further occurrences of the same nonce. We formalised this similar to the memoisation of variable evaluation in the interpreter. A further challenge not present in the interpreter was that the serialiser must also record the serialisation of all subexpressions such that the parser can map bitstrings generated by the serialiser back to symbolic messages without calling a decryption oracle or inverting a cryptographic hash function. The construction relied on the invariant that the recorded values were indeed generated by the serialiser, but such an invariant cannot be expressed easily for a probabilistic, stateful function. We therefore formalised also the switch from lazy to eager sampling for the serialiser (lazy sampling was needed to push the randomisation of encryptions into an encryption oracle) and the switch to a read-only version without recording of results using techniques similar to our example in Sect. 4.

Instead of specifying effects abstractly and composing them using monad transformers, we obviously could have formalised everything in a sufficiently rich monad that covers all the effects of interest, e.g., continuations. Then, there would be no need for abstract specifications as we could work directly with a concrete monad as usual, where our reasoning on the abstract level could be mimicked. But we would deprive ourselves of the option of going to a specific monad that covers precisely the effects needed. Such specialisation has two advantages: First, as shown in Sect. 4, simpler monads satisfy more laws, e.g., commutativity, which make the proofs easier. Second, concrete monads can have dedicated setups for reasoning and proof automation that are not available in the abstract setting. Our library achieves the best of both worlds. We can reason abstractly and thus achieve generality. When this gets too cumbersome or impossible, we can switch to a concrete monad, continuing to use the abstract properties already proven.

In the long run, we can imagine a definitional package for monads and monad transformers that composes concrete value-polymorphic monad transformers. Similar to how Isabelle's datatype package composes bounded natural functors [2], such a package must perform the construction and the derivation of all laws afresh for every concrete combination of monads, as value-polymorphic monads lie beyond HOL's expressiveness. When combined with a reinterpretation framework for theories, we could model effects and reason about them abstractly and concretely without the restriction to monomorphic values.

---

[5] http://www.infsec.ethz.ch/research/projects/FCSPI.html.

Acknowledgements. We thank Dmitriy Traytel and the anonymous reviewers for suggesting many improvements to the presentation. This work is supported by the Swiss National Science Foundation grant 153217 "Formalising Computational Soundness for Protocol Implementations".

References

1. Ballarin, C.: Locales: a module system for mathematical theories. J. Automat. Reason. **52**(2), 123–153 (2014)
2. Blanchette, J.C., Hölzl, J., Lochbihler, A., Panny, L., Popescu, A., Traytel, D.: Truly modular (co)datatypes for Isabelle/HOL. In: Klein, G., Gamboa, R. (eds.) ITP 2014. LNCS, vol. 8558, pp. 93–110. Springer, Cham (2014). doi:10.1007/978-3-319-08970-6_7
3. Bulwahn, L., Krauss, A., Haftmann, F., Erkök, L., Matthews, J.: Imperative functional programming with Isabelle/HOL. In: Ait Mohamed, O., Muñoz, C., Tahar, S. (eds.) TPHOLs 2008. LNCS, vol. 5170, pp. 134–149. Springer, Heidelberg (2008). doi:10.1007/978-3-540-71067-7_14
4. Eberl, M., Hölzl, J., Nipkow, T.: A verified compiler for probability density functions. In: Vitek, J. (ed.) ESOP 2015. LNCS, vol. 9032, pp. 80–104. Springer, Heidelberg (2015). doi:10.1007/978-3-662-46669-8_4
5. Erwig, M., Kollmansberger, S.: Functional pearls: probabilistic functional programming in Haskell. J. Funct. Program. **16**, 21–34 (2006)
6. Gibbons, J., Hinze, R.: Just do it: simple monadic equational reasoning. In: ICFP 2011, pp. 2–14. ACM (2011)
7. Hölzl, J., Lochbihler, A., Traytel, D.: A formalized hierarchy of probabilistic system types. In: Urban, C., Zhang, X. (eds.) ITP 2015. LNCS, vol. 9236, pp. 203–220. Springer, Cham (2015). doi:10.1007/978-3-319-22102-1_13
8. Homeier, P.V.: The HOL-Omega logic. In: Berghofer, S., Nipkow, T., Urban, C., Wenzel, M. (eds.) TPHOLs 2009. LNCS, vol. 5674, pp. 244–259. Springer, Heidelberg (2009). doi:10.1007/978-3-642-03359-9_18
9. Huffman, B.: Formal verification of monad transformers. In: ICFP 2012, pp. 15–16. ACM (2012)
10. Huffman, B., Kunčar, O.: Lifting and Transfer: a modular design for quotients in Isabelle/HOL. In: Gonthier, G., Norrish, M. (eds.) CPP 2013. LNCS, vol. 8307, pp. 131–146. Springer, Cham (2013). doi:10.1007/978-3-319-03545-1_9
11. Huffman, B., Matthews, J., White, P.: Axiomatic constructor classes in Isabelle/HOLCF. In: Hurd, J., Melham, T. (eds.) TPHOLs 2005. LNCS, vol. 3603, pp. 147–162. Springer, Heidelberg (2005). doi:10.1007/11541868_10
12. Jeuring, J., Jansson, P., Amaral, C.: Testing type class laws. In: Haskell 2012, pp. 49–60. ACM (2012)
13. Kunčar, O.: Correctness of Isabelle's cyclicity checker: implementability of overloading in proof assistants. In: CPP 2015, pp. 85–94. ACM (2015)
14. Lammich, P., Tuerk, T.: Applying data refinement for monadic programs to Hopcroft's algorithm. In: Beringer, L., Felty, A. (eds.) ITP 2012. LNCS, vol. 7406, pp. 166–182. Springer, Heidelberg (2012). doi:10.1007/978-3-642-32347-8_12
15. Liang, S., Hudak, P., Jones, M.: Monad transformers and modular interpreters. In: POPL 1995, pp. 333–343. ACM (1995)
16. Lobo Vesga, E.: Hacia la formalización del razonamiento ecuacional sobre mónadas. Technical report, Universidad EAFIT (2013). http://hdl.handle.net/10784/4554

17. Lochbihler, A.: Probabilistic functions and cryptographic oracles in higher order logic. In: Thiemann, P. (ed.) ESOP 2016. LNCS, vol. 9632, pp. 503–531. Springer, Heidelberg (2016). doi:10.1007/978-3-662-49498-1_20

18. Lochbihler, A.: Effect polymorphism in higher-order logic. Archive of Formal Proofs (2017). Formal proof development. http://isa-afp.org/entries/Monomorphic_Monad.shtml

19. Lochbihler, A., Schneider, J.: Equational reasoning with applicative functors. In: Blanchette, J.C., Merz, S. (eds.) ITP 2016. LNCS, vol. 9807, pp. 252–273. Springer, Cham (2016). doi:10.1007/978-3-319-43144-4_16

20. Mitchell, J.C.: Representation independence and data abstraction. In: POPL 1986, pp. 263–276. ACM (1986)

21. Moggi, E.: An abstract view of programming languages. Technical report ECS-LFCS-90-113, LFCS, School of Informatics, University of Edinburgh (1990)

22. Nipkow, T., Klein, G.: Concrete Semantics. Springer, Cham (2014). doi:10.1007/978-3-319-10542-0

23. Nipkow, T., Paulson, L.C.: Proof pearl: defining functions over finite sets. In: Hurd, J., Melham, T. (eds.) TPHOLs 2005. LNCS, vol. 3603, pp. 385–396. Springer, Heidelberg (2005). doi:10.1007/11541868_25

24. Ramsey, N., Pfeffer, A.: Stochastic lambda calculus and monads of probability distributions. In: POPL 2002, pp. 154–165. ACM (2002)

25. Wadler, P.: Monads for functional programming. In: Jeuring, J., Meijer, E. (eds.) AFP 1995. LNCS, vol. 925, pp. 24–52. Springer, Heidelberg (1995). doi:10.1007/3-540-59451-5_2

26. Wenzel, M.: Type classes and overloading in higher-order logic. In: Gunter, E.L., Felty, A. (eds.) TPHOLs 1997. LNCS, vol. 1275, pp. 307–322. Springer, Heidelberg (1997). doi:10.1007/BFb0028402

Schulze Voting as Evidence Carrying Computation

Dirk Pattinson[✉] and Mukesh Tiwari[✉]

The Australian National University, Canberra, Australia
{dirk.pattinson,u5935541}@anu.edu.au

Abstract. The correctness of vote counting in electronic election is one of the main pillars that engenders trust in electronic elections. However, the present state of the art in vote counting leaves much to be desired: while some jurisdictions publish the source code of vote counting code, others treat the code as commercial in confidence. None of the systems in use today applies any formal verification. In this paper, we formally specify the so-called Schulze method, a vote counting scheme that is gaining popularity on the open source community. The cornerstone of our formalisation is a (dependent, inductive) type that represents all correct executions of the vote counting scheme. Every inhabitant of this type not only gives a final result, but also all intermediate steps that lead to this result, and can so be externally verified. As a consequence, we do not even need to trust the execution of the (verified) algorithm: the correctness of a particular run of the vote counting code can be verified on the basis of the evidence for correctness that is produced along with determination of election winners.

1 Introduction

The Schulze Method [16] is a vote counting scheme that elects a single winner, based on preferential votes. While no preferential voting scheme can guarantee *all* desirable properties that one would like to impose due to Arrow's theorem [2], the Schulze method offers a good compromise, with a number of important properties already established in Schulze's original paper. A quantitative comparison of voting methods [15] also shows that Schulze voting is better (in a game theoretic sense) than others, more established, systems, and the Schulze Method is rapidly gaining popularity in the open software community. It is being used, for example in the Wikimedia Foundation's board elections with approximately 3,000 votes and 15 candidates [20], the Gentoo council and the OpenStack community (with both fewer votes and candidates).

The method itself rests on the relative *margins* between two candidates, i.e. the number of voters that prefer one candidate over another. The margin induces an ordering between candidates, where a candidate c is more preferred than d, if more voters prefer c over d than vice versa. One can construct simple examples (see e.g. [15]) where this order does not have a maximal element (a

© Springer International Publishing AG 2017
M. Ayala-Rincón and C.A. Muñoz (Eds.): ITP 2017, LNCS 10499, pp. 410–426, 2017.
DOI: 10.1007/978-3-319-66107-0_26

so-called *Condorcet Winner*). Schulze's observation is that this ordering *can* be made transitive by considering sequences of candidates (called *paths*). Given candidates c and d, a *path* between c and d is a sequence of candidates $p = (c, c_1, \ldots, c_n, d)$ that joins c and d, and the *strength* of a path is the minimal margin between adjacent nodes. This induces the *generalised margin* between candidates c and d as the strength of the strongest path that joins c and d. A candidate c then wins a Schulze count if the generalised margin between c and any other candidate d is at least as large as the generalised margin between d and c.

This paper presents a formal specification of the Schulze method, together with the proof that winners can always be determined which we extract to obtain a provably correct implementation of the Schulze method. The crucial aspect of our formalisation is that the vote counting protocol itself is represented as a dependent inductive type that represents all (correct) partial executions of the protocol. A complete execution can then be understood as a state of vote counting where election winners have been determined. Our main theorem then asserts that an inhabitant of this type exists, for all possible sets of incoming ballots. Crucially, every such inhabitant contains enough information to (independently) verify the correctness of the election result, and can be thought of as a *certificate* for the count.

From a computational perspective, we view tallying not merely as a function that delivers a result, but instead as a function that delivers a result, *together* with evidence that allows us to verify correctness. In other words, we augment verified correctness of an algorithm with the means to verify each particular *execution*.

From the perspective of electronic voting, this means that we no longer need to trust the hardware and software that was employed to obtain the election result, as the generated certificate can be verified independently. In the literature on electronic voting, this is known as *verifiability* and has been recognised as one of the cornerstones for building trust in election outcomes [7], and is the only answer to key questions such as the possibility of hardware malfunctions, or indeed running the very software that has been claimed to count votes correctly.

The certificate that is produced by each run of our extracted Schulze vote tallying algorithm consists of two parts. The first part details the individual steps of constructing the margin function, based on the set of all ballots cast. The second part presents evidence for the determination of winners, based on generalised margins. For the construction of the margin function, every ballot is processed in turn, with the margin between each pair of votes updated accordingly. The heart of our work lies in this second part of the certificate. To demonstrate that candidate c is an election winner, we have to demonstrate that the generalised margin between c and every other candidate d is at least as large as the generalised margin between d and c. Given that the generalised margin between two candidates c and d is determined in terms of paths $c, c_1, \ldots, c_n, d$ that join c and d, we need to exhibit

– evidence for the existence of a path p from c to d

– evidence for the fact that *no* path q from d to c is stronger than p

where the strength of a path $p = (c_0, \ldots, c_{n+1})$ is the minimum $\min\{m(c_i, c_{i+1}) \mid 0 \leq i \leq n\}$ of the margins between adjacent nodes. While evidently a path itself is evidence for its existence, the *non-existence* of paths with certain properties is more difficult to establish. Here, we use a coinductive approach. As existence of a path with a given strength between two candidates can be easily phrased as an inductive definition, the *complement* of this predicate arises as a greatest fixpoint, or equivalently as a coinductively defined predicate (see e.g. [10]). This allows us to witness the non-existence of paths by exhibiting co-closed sets.

Our formalisation takes place inside the Coq proof assistant [5] that we chose mainly because of its well-developed extraction mechanism and because it allows us to represent the Schulze voting scheme very concisely as a dependent inductive type. Interestingly, we make no use of Coq's mechanism of defining coinductive types [4]: as we are dealing with decidable predicates (formulated as boolean valued functions) only, it is simpler to directly introduce co-closed sets and establish their respective properties.

We take a propositions-as-types approach to synthesising a programme that computes election winners, together with accompanying evidence. That is, our main theorem states that winners (and certificates) exist for any set of initial ballots. As our proof is purely constructive, this amounts to an algorithm that computes witnesses for the existential quantifier. This allows us to use Coq's program extraction facility [12] to generate Haskell and OCaml code that we then compile into an executable, and use it to count votes according to the Schulze method. We report on experimental result and conclude with further work and a general reflection on our method.

Related Work. The idea of requiring that computations provide not only results, but also proofs attesting to the correctness of the computation is not new, and has been put forward in [1] for computations in general, and in [17] in the context of electronic voting. The general difficulty here is the precise nature of certificates, as different computations require a different type of evidence, and our conceptual contribution is to harness coinduction, more precisely co-closed sets as evidence for membership in the complement of inductively defined sets. Our approach is orthogonal to Necula's proof carrying code [13], where every executable (not every execution) is equipped with formal guarantees. Formal specification and verification of vote counting schemes has been done e.g. in [3,8] but none of the methods produce independently verifiable results. The idea of formalising a voting protocol as a type has been put forward in [14] where a variant of single transferable vote has been analysed. While the Schulze method has been analysed e.g. from the point of manipulation [9], this paper appears to be the first to present a formal specification (and a certificate-producing, verified implementation) of the Schulze method in a theorem prover.

Coq Sources. All Coq sources and the benchmarks used in the preparation of this paper are at http://users.cecs.anu.edu.au/~dpattinson/Sofware/.

2   Formal Specification of Schulze Voting

We begin with an informal description of Schulze voting. Schulze voting is *preferential* in the sense that every voter gets to express their preference about candidates in the form of a rank ordered list. Here, we allow voters to be indifferent about candidates but require voters to express preferences over *all* candidates. This requirement can be relaxed and we can consider e.g. unranked candidates as tied for the last position.

Given a set of ballots s and candidate set C, one constructs the margin function $m : C \times C \to \mathbb{Z}$. Given two candidates $c, d \in C$, the *margin* of c over d is the number of voters that prefer c over d, minus the number of voters that prefer d over c. In symbols

$$m(c,d) = \sharp\{b \in s \mid c >_b d\} - \sharp\{b \in s \mid d >_b c\}$$

where $\sharp$ denotes cardinality and $>_b$ is the strict (preference) ordering given by the ballot $b \in s$. A (directed) *path* from candidate c to candidate d is a sequence $p \equiv c_0, \ldots, c_{n+1}$ of candidates with $c_0 = c$ and $c_{n+1} = d$ ($n \geq 0$), and the *strength* $\mathsf{st}(p)$ of this path is the minimum margin of adjacent nodes, i.e.

$$\mathsf{st}(c_0, \ldots, c_{n+1}) = \min\{m(c_i, c_{i+1}) \mid 0 \leq i \leq n\}.$$

Note that the strength of a path may be negative. The Schulze method stipulates that a candidate $c \in C$ is a *winner* of the election with margin function m if, for all other candidates $d \in C$, there exists a number $k \in Z$ such that

- there is a path p from c to d with strength $\mathsf{st}(p) \geq k$
- all paths q from d to c have strength $\mathsf{st}(q) \leq k$.

Informally speaking, we can say that candidate c *beats* candidate d if there's a path p from c to d which is stronger than any path from d to c. Using this terminology, a candidate c is a winner if c cannot be beaten by any (other) candidate.

Remark 1. There are multiple formulations of the Schulze method in the literature. Schulze's original paper [16] only considers paths where adjacent nodes have to be distinct, and [9] only considers simple paths, i.e. paths without repeated nodes. Here, we consider *all* paths. It is easy to see that all three definitions are equivalent, i.e. they produce the same set of winners.

Our (Coq) formalisation takes a finite and non-empty type of candidates as given which we assume has decidable equality. For our purposes, the easiest way of stipulating that a type is finite is to require existence of a list containing all inhabitants of this type.

```
Parameter cand : Type.
Parameter cand_all : list cand.
Hypothesis cand_fin : forall c: cand, In c cand_all.
Hypothesis dec_cand : forall n m : cand, {n = m} + {n <> m}.
Hypothesis cand_inh : cand_all <> nil.
```

For the specification of winners of Schulze elections, we take the margin function as given for the moment (and later construct it from the incoming ballots). In Coq, this is conveniently expressed as a variable:

```
Variable marg : cand -> cand -> Z.
```

We formalise the notion of path and strength of a path by means of a single (but ternary) inductive proposition that asserts the existence of a path of strength $\geq k$ between two candidates, for $k \in \mathbb{Z}$.

```
Inductive Path (k: Z) : cand -> cand -> Prop :=
 | unit c d : marg c d >= k -> Path k c d
 | cons  c d e : marg c d >= k -> Path k d e -> Path k c e.
```

Using these definitions, we obtain the following notion of winning (and dually, losing) a Schulze election:

```
Definition wins_prop (c: cand) := forall d : cand, exists k : Z,
  Path k c d /\ (forall l, Path l d c -> l <= k).
```

```
Definition loses_prop (c : cand) := exists k: Z, exists  d: cand,
  Path k d c /\ (forall l, Path l c d -> l < k).
```

We reflect the fact that the above are *propositions* in the name of the definitions, in anticipation of type-level definitions of these notions later. The main reason for having equivalent type-level versions of the above is that purely propositional information is discarded during program extraction, unlike the type-level notions of winning and losing that represent evidence of the correctness of the determination of winners.

That is, our goal is to not only compute winners and losers according to the definition above, but also to provide independently verifiable evidence of the correctness of our computation. The propositional definitions of winning and losing above serve as a reference to calibrate their type level counterparts, and we demonstrate the equivalence between propositional and type-level conditions in the next section.

3   A Scrutiny Sheet for the Schulze Method

How can we *know* that, say, a candidate c in fact wins a Schulze election, and that, say, d is not a winner? One way would be to simply re-run an independent implementation of the method (usually hoping that results would be confirmed). But what happens if results diverge?

One major aspect of this paper is that we can answer this question by not only computing the set of winners, but in fact presenting *evidence* for the fact that a particular candidate does or does not win. In the context of electronic vote counting, this is known as a *scrutiny sheet*: a tabulation of all relevant data that allows us to verify the election outcome. Again drawing on an already computed margin function, to demonstrate that a candidate c wins, we need to exhibit an integer k for all competitors d, together with

- evidence for the existence of a path from c to d with strength $\geq k$
- evidence for the non-existence of a path from d to c that is stronger than k.

The first item is straight forward, as a path itself is evidence for the existence of a path, and the notion of path is inductively defined. For the second item, we need to produce evidence of membership in the *complement* of an inductively defined set.

Mathematically, given $k \in Z$ and a margin function $m : C \times C \to Z$, the pairs $(c, d) \in C \times C$ for which there exists a path of strength $\geq k$ that joins both are precisely the elements of the least fixpoint $\mathsf{LFP}(V_k)$ of the monotone operator $V_k : \mathsf{Pow}(C \times C) \to \mathsf{Pow}(C \times C)$, defined by

$$V_k(R) = \{(c, e) \in C^2 \mid m(c, e) \geq k \text{ or } (m(c, d) \geq k \text{ and } (d, e) \in R \text{ for some } d \in C)\}.$$

It is easy to see that this operator is indeed monotone, and that the least fixpoint exists, e.g. using Kleene's theorem [18]. To show that there is *no* path between d and c of strength $> k$, we therefore need to establish that $(d, c) \notin \mathsf{LFP}(V_{k+1})$.

By duality between least and greatest fixpoints, we have that

$$(c, d) \in C \times C \setminus \mathsf{LFP}(V_{k+1}) \iff (c, d) \in \mathsf{GFP}(W_{k+1})$$

where for arbitrary k, $W_k : \mathsf{Pow}(C \times C) \to \mathsf{Pow}(C \times C)$ is the operator dual to V_k, i.e.

$$W_k(R) = C \times C \setminus (V_k(C \times C \setminus R))$$

and $\mathsf{GFP}(W_k)$ is the greatest fixpoint of W_k. As a consequence, to demonstrate that there is *no* path of strength $\geq k$ between candidates d and c, we need to demonstrate that $(d, c) \in \mathsf{GFP}(W_{k+1})$. By the Knaster-Tarski fixpoint theorem [19], this greatest fixpoint is the supremum of all W_{k+1}-coclosed sets, that is, sets $R \subseteq C \times C$ for which $R \subseteq W_{k+1}(R)$. That is, to demonstrate that $(d, c) \in \mathsf{GFP}(W_{k+1})$, we need to exhibit a W_{k+1}-coclosed set R with $(d, c) \in R$. If we unfold the definitions, we have

$$W_k(R) = \{(c, e) \in C^2 \mid m(c, e) < k \text{ and } (m(c, d) < k \text{ or } (d, c) \in R \text{ for all } d \in C)\}$$

so that given *any* fixpoint R of W_k and $(c, e) \in C^2$, we know that (i) the margin between c and e is $< k$ so that there's no edge (or unit path) between c and e, and (ii) for any choice of midpoint d, either the margin between c and d is $< k$ (so that $c, d, \ldots$ cannot be the start of a path of strength $\geq k$) or we don't have a path between d and e of strength $\geq k$. We use the following terminology:

Definition 1. *Let $R \subseteq C \times C$ be a subset and $k \in \mathbb{Z}$. Then R is W_k-coclosed, or simply k-coclosed, if $R \subseteq W_k(R)$.*

Mathematically, the operator W_k acts on subsets of $C \times C$ that we think of as predicates. In Coq, we formalise these predicates as boolean valued functions and obtain the following definitions where we isolate the function `marg_lt` (that determines whether the margin between two candidates is less than a given integer) for clarity:

```
Definition marg_lt (k : Z) (p : (cand * cand)) :=
  Zlt_bool (marg (fst p) (snd p)) k.

Definition W (k : Z) (p: cand * cand -> bool) (x: cand * cand) :=
  andb
    (marg_lt k x)
    (forallb (fun m => orb (marg_lt k (fst x, m)) (p (m, snd x))) cand_all).
```

In order to formulate type-level definitions, we need to promote the notion of path from a Coq proposition to a proper type, and formulate the notion of k-coclosed predicate.

```
Definition coclosed (k : Z) (f : (cand * cand) -> bool) :=
  forall x, f x = true -> W k f x = true.

Inductive PathT (k: Z) : cand -> cand -> Type :=
| unitT : forall c d, marg c d >= k -> PathT k c d
| consT : forall c d e, marg c d >= k -> PathT k d e -> PathT k c e.
```

The only difference between type level paths (of type PathT) and (propositional) paths defined earlier is the fact that the former are proper types, not propositions, and are therefore not erased during extraction. Given the above, we have the following type-level definitions of winning (and dually, non-winning) for Schulze counting:

```
Definition wins_type c := forall d : cand, existsT (k : Z),
  ((PathT k c d) * (existsT (f : (cand * cand) -> bool),
    f (d, c) = true /\ coclosed (k + 1) f))%type.

Definition loses_type (c : cand) := existsT (k : Z) (d : cand),
  ((PathT k d c) * (existsT (f : (cand * cand) -> bool),
    f (c, d) = true /\ coclosed k f))%type.
```

The main result of this section is that type level and propositional evidence for winning (and dually, not winning) a Schulze election can be reconstructed from one another.

```
Lemma wins_type_prop : forall c, wins_type c -> wins_prop c.

Lemma wins_prop_type : forall c, wins_prop c -> wins_type c.
```

The different nature of the two propositions doesn't allow us to claim an equivalence between both notions, as biimplication is a propositional connective.

The proof of the first statement is completely straightforward, as the type carries all the information needed to establish the propositional winning condition. For the second statement above, we introduce an intermediate lemma based on the *iterated margin function* $M_k : C \times C \to \mathbb{Z}$. Intuitively, $M_k(c, d)$ is the strength of the strongest path between c and d of length $\leq k + 1$. Formally, $M_0(c, d) = m(c, d)$ and

$$M_{i+1}(c, d) = \max\{M_i(c, d), \max\{\min\{m(c, e), M_i(e, d) \mid e \in C\}\}\}$$

for $i \geq 0$. The iterated margin function, as defined above, allows for paths of arbitrary length. It is intuitively clear (and we establish this fact formally) that paths with repeated nodes do not contribute to the maximal strength of a path. Therefore, the iterated margin function stabilises at the n-th iteration (where n is the number of candidates). The formal proof loosely follows the evident pen-and-paper proof given for example in [6] that is based on cutting out segments of paths between repeated nodes and so establishes that a fixpoint is reached.

```
Lemma iterated_marg_fp: forall (c d : cand) (n : nat),
    M n c d <= M (length cand_all) c d.
```

That is, the *generalised margin*, i.e. the strength of the strongest (possibly infinite) path between two candidates is effectively computable.

This allows us to relate the propositional winning conditions to the iterated margin function and showing that a candidate c is winning implies that the generalised margin between this candidate and any other candidate d is at least as large as the generalised margin between d and c.

```
Lemma wins_prop_iterated_marg (c : cand) : wins_prop c ->
    forall d, M (length cand_all) d c <= M (length cand_all) c d.
```

This condition on iterated margins can in turn be used to establish the type-level winning condition, thus closing the loop to the type level winning condition.

```
Lemma iterated_marg_wins_type (c : cand) : (forall d,
    M (length cand_all) d c <= M (length cand_all) c d) ->
    wins_type c.
```

The crucial part of establishing the type-level winning conditions in the proof of the lemma above is the construction of a co-closed set. First note that M (length cand_all) is precisely the generalised margin function. Writing g for this function, we assume that $g(c, d) \geq g(d, c)$ for all candidates d, and given d, we need to construct a $k + 1$-coclosed set S where $k = g(c, d)$. One option is to put $S = \{(x, y) \mid g(x, y) < k + 1\}$. As every i-coclosed set is also j-coclosed for $i \leq j$, the set $S' = \{(x, y) \mid g(x, y) < g(d, c) + 1\}$ is also $k + 1$-coclosed and (in general) of smaller cardinality. We therefore witness the existence of a $k + 1$-coclosed set with S' as this leads to certificates that are smaller in size and therefore easier to check.

We note that the difference between the type-level and the propositional definition of winning is in fact more than a mere reformulation. As remarked before, one difference is that purely propositional evidence is erased during program extraction so that using just the propositional definitions, we would obtain a determination of election winners, but no additional information that substantiates this (and that can be verified independently). The second difference is conceptual: it is easy to verify that a set is indeed coclosed as this just involves a finite (and small) amount of data, whereas the fact that *all* paths between two candidates don't exceed a certain strength is impossible to ascertain, given that there are infinitely many paths.

In summary, determining that a particular candidate wins an election based on the wins_type notion of winning, the extracted program will *additionally* deliver, for all other candidates,

- an integer k and a path from the winning candidate to the other candidate
- a co-closed set that witnesses that no path reverse path of strength $> k$ exists.

It is precisely this additional data (on top of merely declaring a set of election winners) that allows for scrutiny of the process, as it provides an orthogonal approach to verifying the correctness of the computation: both checking that the given path has a certain strength, and that a set is indeed coclosed, is easy to verify. We reflect more on this in Sect. 7, and present an example of a full scrutiny sheet in the next section, when we join the type-level winning condition with the construction of the margin function from the given ballots.

4   Schulze Voting as Inductive Type

Up to now, we have described the specification of Schulze voting relative to a given margin function. We now describe the specification (and computation) of the margin function given a profile (set) of ballots. Our formalisation describes an individual *count* as a type with the interpretation that all inhabitants of this type are correct executions of the vote counting algorithm. In the original paper describing the Schulze method [16], a ballot is a linear preorder over the set of candidates.

In practice, ballots are implemented by asking voters to put numerical preferences against the names of candidates as illustrated by the image on the right. The most natural representation of a ballot is therefore a function $b : C \to \mathbb{N}$ that assigns a natural number (the preference) for each candidate, and we recover a strict linear preorder $<_b$ on candidates by setting $c <_b d$ if $b(c) > b(d)$.

Rank all candidates in order of preference

- 4 Lando Calrissian
- 3 Boba Fett
- 1 Mace Windu
- 2 Poe Dameron
- 2 Maz Kanata

As preferences are usually numbered beginning with 1, we interpret a preference of 0 as the voter failing to designate a preference for a candidate as this allows us to also accommodate invalid ballots. This is clearly a design decision, and we could have formalised ballots as functions $b : C \to 1 + \mathbb{N}$ (with 1 being the unit type) but it would add little to our analysis.

```
Definition ballot := cand -> nat.
```

The count of an individual election is then parameterised by the list of ballots cast, and is represented as a dependent inductive type. More precisely, we have a type State that represents either an intermediate stage of constructing the margin function or the determination of the final election result:

```
Inductive State: Type :=
| partial: (list ballot * list ballot)  -> (cand -> cand -> Z)
  -> State
| winners: (cand -> bool) ->  State.
```

The interpretation of this type is that a state either consists of two lists of ballots and a margin function, representing

- the set of ballots counted so far, and the set of invalid ballots seen so far
- the margin function constructed so far

or, to signify that winners have been determined, a boolean function that determines the set of winners.

The type that formalises correct counting of votes according to the Schulze method is parameterised by the profile of ballots cast (that we formalise as a list), and depends on the type State. That is to say, an inhabitant of the type Count n, for n of type State, represents a correct execution of the voting protocol up to state n. This state generally represents intermediate stages of the construction of the margin function, with the exception of the final step where the election winners are being determined. The inductive type takes the following shape:

```
Inductive Count (bs : list ballot) : State -> Type :=
| ax us m : us = bs -> (forall c d, m c d = 0) ->
    Count bs (partial (us, [])) m)        (* zero margin        *)
| cvalid u us m nm inbs : Count bs (partial (u :: us, inbs) m) ->
    (forall c, (u c > 0)%nat) ->          (* u is valid         *)
    (forall c d : cand,
      ((u c < u d) -> nm c d = m c d + 1) (* c preferred to d *) /\
      ((u c = u d) -> nm c d = m c d)      (* c, d rank equal    *) /\
      ((u c > u d) -> nm c d = m c d - 1))(* d preferred to c *) ->
    Count bs (partial (us, inbs) nm)
| cinvalid u us m inbs : Count bs (partial (u :: us, inbs) m) ->
    (exists c, (u c = 0)%nat)             (* u is invalid       *) ->
    Count bs (partial (us, u :: inbs) m)
| fin m inbs w (d : (forall c, (wins_type m c) + (loses_type m c))):
    Count bs (partial ([], inbs) m)       (* no ballots left    *) ->
    (forall c, w c = true <-> (exists x, d c = inl x)) ->
    (forall c, w c = false <-> (exists x, d c = inr x)) ->
    Count bs (winners w).
```

The intuition here is simple: the first constructor, ax, initiates the construction of the margin function, and we ensure that all ballots are uncounted, no ballots are invalid (yet), and the margin function is constantly zero. The second constructor, cvalid, updates the margin function according to a valid ballot (all candidates have preferences marked against their name), and removes the ballot from the list of uncounted ballots. The constructor `cinvalid` moves an invalid ballot (where one or more candidates aren't ranked) to the list of invalid ballots, and the last constructor `fin` applies only if the margin function is completely constructed (no more uncounted ballots). In its arguments, `w: cand -> bool` is the function that determines election winners, and `d` is a function that delivers, for every candidate, type-level evidence of winning or losing, consistent with `w`. Given this, we can conclude the count and declare `w` to be the set of winners (or more precisely, those candidates for which `w` evaluates to `true`).

Together with the equivalence of the propositional notions of winning or losing a Schulze count with their type-level counterparts, every inhabitant of the type `Count b (winners w)` then represents a correct count of ballots `b` leading to the boolean predicate `w: Cand -> bool` that determines the winners of the election with initial set `b` of ballots.

The crucial aspect of our formalisation of executions of Schulze counting is that the transcript of the count is represented by a type that is *not* a proposition. As a consequence, extraction delivers a program that produces the (set of) election winner(s), *together* with the evidence recorded in the type to enable independent verification.

Remark 2. In the previous section, we have given the definition of election winners relative to a given margin function. The inductive data type `Count` ties this in with the computation of the margin function. It is therefore reasonable to ask for a formal proof that – given a list `bs` of ballots – the existence of an inhabitant `Count bs (winners w)` is equivalent to the winning condition `wins_prop`, where the margin function is obtained from the ballots `bs`. This requires us to formulate a correctness predicate `is_marg: list ballot -> (cand -> cand -> Z) -> Prop` that links ballots and margin. The most natural way of achieving this is to formulate this predicate inductively, closely mirroring the clauses `cvalid` and `cinvalid` in the defintion of `Count`. This makes the formal proof that relates `Count` with the winning condition almost trivial.

5   All Schulze Election Have Winners

The main theorem, the proof which we describe in this section, is that all elections according to the Schulze method engender a boolean-valued function `w: Cand -> bool` that determines precisely which candidates are winners of the election, together with type-level evidence of this. Note that a Schulze election can have more than one winner, the simplest (but not the only) example being when no ballots at all have been cast. The theorem that we establish (and later extract as a program) simply states that for every incoming set of ballots, there is a boolean function that determines the election winners, together with an inhabitant of the type `Count` that witnesses the correctness of the execution of the count. In Coq, we use a type-level existential quantifier `existsT` where `existsT (x:A), P` stands for $\Sigma_{x:A}P$.

```
Theorem schulze_winners: forall (bs : list ballot),
   existsT (f : cand -> bool), Count bs (winners f).
```

The first step in the proof is elementary: We show that for any given list of ballots we can reach a state of the count where there are no more uncounted ballots, i.e. the margin function has been fully constructed.

The second step relies on the iterated margin function already discussed in Sect. 3. As $M_n(c, d)$ (for n being the number of candidates) is the strength of the strongest path between c and d, we construct a boolean function w such that $w(c) = \texttt{true}$ if and only if $M_n(c, d) \geq M_n(d, c)$ for all $d \in C$. We then construct the type-level evidence required in the constructor fin using the lemmma iterated_marg_wins_type described earlier.

Coq's extraction mechanism then allows us to turn this into a provably correct program. When extracting, all purely propositional information is erased and given a set of incoming ballots, the ensuing program produces an inhabitant of the (extracted) type Count that records the construction of the margin function, together with (type level) evidence of correctness of the determination of winners. That is, we see the individual steps of the construction of the margin function (one step per ballot) and once all ballots are exhausted, the determination of winners, together with paths and co-closed sets. The following is the transcript of a Schulze election where we have added wrappers to pretty-print the information content. This is the (full) scrutiny sheet promised in Sect. 3.

```
V: [A3 B1 C2 D4,..], I: [], M: [AB:0 AC:0 AD:0 BC:0 BD:0 CD:0]
-----------------------------------------------------------------
V: [A1 B0 C4 D3,..], I: [], M: [AB:-1 AC:-1 AD:1 BC:1 BD:1 CD:1]
-----------------------------------------------------------------
V: [A3 B1 C2 D4,..], I: [A1 B0 C4 D3], M: [AB:-1 AC:-1 AD:1 BC:1 BD:1 CD:1]
-----------------------------------------------------------------

                  . . .

-----------------------------------------------------------------
V: [A1 B3 C2 D4], I: [A1 B0 C4 D3], M: [AB:2 AC:2 AD:8 BC:5 BD:8 CD:8]
-----------------------------------------------------------------
V: [], I: [A1 B0 C4 D3], M: [AB:3 AC:3 AD:9 BC:4 BD:9 CD:9]
-----------------------------------------------------------------
winning: A
   for B: path A --> B of strength 3, 4-coclosed set:
      [(B,A),(C,A),(C,B),(D,A),(D,B),(D,C)]
   for C: path A --> C of strength 3, 4-coclosed set:
      [(B,A),(C,A),(C,B),(D,A),(D,B),(D,C)]
   for D: path A --> D of strength 9, 10-coclosed set:
      [(D,A),(D,B),(D,C)]
losing: B
  exists A: path A --> B of strength 3, 3-coclosed set:
     [(A,A),(B,A),(B,B),(C,A),(C,B),(C,C),(D,A),(D,B),(D,C),(D,D)]
losing: C
  exists A: path A --> C of strength 3, 3-coclosed set:
     [(A,A),(B,A),(B,B),(C,A),(C,B),(C,C),(D,A),(D,B),(D,C),(D,D)]
losing: D
  exists A: path A --> D of strength 9, 9-coclosed set:
     [(A,A),(A,B),(A,C),(B,A),(B,B),(B,C),(C,A),(C,B),(C,C),(D,A),(D,B),
       (D,C),(D,D)]
```

Here, we assume four candidates, A, B, C and D and a ballot of the form A3 B2 C4 D1 signifies that D is the most preferred candidate (the first preference), followed by B (second preference), A and C. In every line, we only display the first uncounted ballot (condensing the remainder of the ballots to an ellipsis), followed by votes that we have deemed to be invalid. We display the partially constructed margin function on the right. Note that the margin function satisfies $m(x, y) = -m(y, x)$ and $m(x, x) = 0$ so that the margins displayed allow us to reconstruct the entire margin function. In the construction of the margin function, we begin with the constant zero function, and going from one line to the next, the new margin function arises by updating according to the first ballot. This corresponds to the constructor cvalid and cinvalid being applied recursively: we see an invalid ballot being set aside in the step from the second to the third line, all other ballots are valid. Once the margin function is fully constructed (there are no more uncounted ballots), we display the evidence provided in the constructor fin: we present evidence of winning (losing) for all winning (losing) candidates. In order to actually verify the computed result, a third party observer would have to

1. Check the correctness of the individual steps of computing the margin
2. For winners, verify that the claimed paths exist with the claimed strength, and check that the claimed sets are indeed coclosed.

Contrary to re-running a different implementation on the same ballots, our scrutiny sheet provides an *orthogonal* perspective on the data and how it was used to determine the election result.

6   Experimental Results

Coq's built in extraction mechanism extracts into both Haskell and Ocaml, and allows to extract Coq types into built in (or user defined) types in the target programming language.

We have evaluated our approach by extracting the entire Coq development into both Haskell and OCaml, with all types defined by Coq extracted as is, i.e. in particular using Coq's unary representation of natural numbers. The results for the Haskell extraction are displayed in Fig. 1(a) using a logarithmic scale. Profiling the executable reveals that a large portion of time is being spent comparing natural numbers (that Coq represents in unary) for size. In Fig. 1(b), we have extracted Coq's natural number type to the (native) Haskell type Int of integers, and the comparison function to the Haskell native comparison operator (<=). The use of native integers has resulted in a nearly tenfold speedup as seen in the figure on the right. While extraction of Coq data types into their Haskell counterparts potentially jeopardises correctness of the code, the fact that we produce a transcript of the code (a scrutiny sheet) that can (and should!) be checked for correctness externally alleviates the risk of erroneous results that can be produced that way.

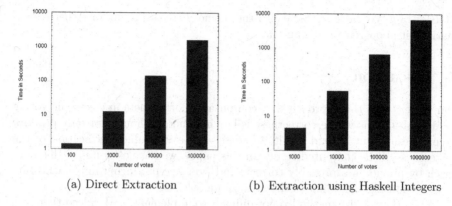

(a) Direct Extraction   (b) Extraction using Haskell Integers

Fig. 1. Experimental results: Haskell

To obtain better quantitative data on scalability, we have subsequently extracted into OCaml and fine-tuned some of the proofs so that the extracted code avoids re-computation. In particular, we have *not* extracted any Coq type to its Ocaml counterpart.

Other than a remarkable speedup compared to the Haskell extraction, we note that for a constant number of candidates, the computation time is almost independent of the number of ballots as witnessed by the graph on the left (Fig. 2(a)) where we are counting votes for 10 candidates. This is due to the fact that the Schulze method is linear in the number of ballots but cubic in the number of candidates. We currently suspect that the disproportionately long runtime displayed in the two rightmost bars is due to swapping when ballots are being read. In the right hand graph (Fig. 2(b)), we are counting 100,000 ballots and increase the number of candidates, witnessing this cubic behaviour.

All graphs have been produced assuming four candidates and (the same) randomly generated ballots on an Intel i7 2.6 GHz Linux desktop computer with

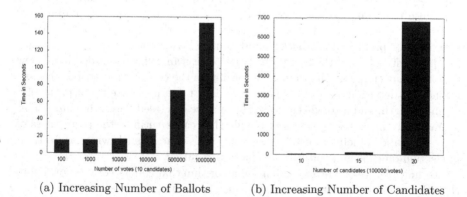

(a) Increasing Number of Ballots   (b) Increasing Number of Candidates

Fig. 2. Experimental results: OCaml

8 GB of ram. We have not analysed the memory consumption for either benchmark as it appeared to be minimal.

7   Discussion

Our paper takes the approach that computation of winners in electronic voting (and in situations where correctness is key in general) should not only produce an end result, but an end result, *together* with a verifiable justification of the correctness of the computed result. In this paper, we have exemplified this approach by providing a provably correct, and evidence-producing implementation of vote counting according to the Schulze method.

While the Schulze method is not difficult to implement, and indeed there are many freely available implementations, comparing the results between different implementations can give some level of assurance for correctness only in case the results agree. If there is a discrepancy, a certificate for the correctness of the count allows to adjudicate between different implementations, as the certificate can be checked with relatively little computational effort.

From the perspective of computational complexity, checking a transcript for correctness is of the same complexity as computing the set of winners, as our certificates are cubic in size, so that certificate checking is not less complex than the actual computation.

However, publishing an independently verifiable certificate that attests the individual steps of the computation helps to increase *trust* in the computed election outcome. Typically, the use of technology in elections increases the amount of trust that we need to place both in technological artefacts, and in people. It raises questions that range from fundamental aspects, such as proper testing and/or verification of the software, to very practical questions, e.g. whether the correct version of the software has been run. On the contrast, publishing a certificate of the count dramatically reduces the amount of trust that we need to place into both people and technology: the ability to publish a verifiable justification of the correctness of the count allows a large number of individuals to scrutinise the count. While only moderate programming skills are required to check the validity of a certificate (the transcript of the count), even individuals without any programming background can at least spot-check the transcript: for the construction of the margin function, everything that is needed is to show that the respective margins change according to the counted ballot. For the correctness of determination of winners, it is easy to verify existence of paths of a given strength, and also whether certain sets are co-closed – even by hand! This dramatically increases the class of people that can scrutinise the correctness of the count, and so helps to establish a trust basis that is much wider as no trust in election officials and software artefacts is required.

Technically, we do not *implement* an algorithm that counts votes according to the Schulze method. Instead, we give a specification of the Schulze winning conditions (`wins_prop` in Sect. 2) in terms of an already computed margin function that (we hope) can immediately be seen to be correct, and then show that those winning conditions are equivalent to the existence of inhabitants of types that carry verifiable evidence (wins_type). We then join the (type level) winning conditions with an inductive type that details the construction of the margin function in an inductive type. Via propositions-as-types, a provably correct vote counting function is then equivalent the proposition that there exists an inhabitant of Count for every set of ballots. Coq's extraction mechanism then allows us to extract a functional programs that produce election winners, together with verifiable certificates.

The approach taken in this paper, i.e. the formalisation of a voting protocol as a (dependent) inductive type, can be applied, and has been applied to other voting protocols, notably variants of single transferable vote [14]. As voting protocols differ substantially in detail, there is limited potential for code re-use.

8   Conclusion and Further Work

This paper has presented a formalisation of the Schulze method for counting preferential ballots. Our formalisation focuses on the correct execution of the method. One appealing aspect of the Schulze method is that it meets lots of desirable criteria of vote counting systems such as monotonicity, the Condorcet property or reversal symmetry. We leave the verification of these for future work.

In our formalisation of vote counting, there is a one-to-one correspondence between correct executions of the protocol, and inhabitants of a (dependent) inductive type. In our Coq development, we have used the propositions-as-types approach, and have constructed an existence proof, from which we have generated code. An alternative approach would be to implement a function that directly constructs inhabitants, and obtain a detailed performance comparison between both approaches. While we anticipate that a direct implementation brings performance benefits, our experimental evaluation shows that even with very little optimisation (Sect. 6), extracting vote counting program from an existence proof allows us to count a relatively large number of ballots already.

Finally, we remark that extracting Coq developments into a programming language itself is a non-verified process which could still introduce errors in our code. The most promising way to alleviate this is to independently implement (and verify) a certificate verifier, possibly in a language such as CakeML [11] that is guaranteed to be correct to the machine level.

References

1. Arkoudas, K., Rinard, M.C.: Deductive runtime certification. Electr. Notes Theoret. Comput. Sci. **113**, 45–63 (2005)
2. Arrow, K.J.: A difficulty in the concept of social welfare. J. Polit. Econ. **58**(4), 328–346 (1950)
3. Beckert, B., Goré, R., Schürmann, C., Bormer, T., Wang, J.: Verifying voting schemes. J. Inf. Secur. Appl. **19**(2), 115–129 (2014)
4. Bertot, Y.: Coinduction in Coq. CoRR, abs/cs/0603119 (2006)

5. Bertot, Y., Castéran, P., Huet, G., Paulin-Mohring, C.: Interactive Theorem Proving and Program Development: Coq'Art the Calculus of Inductive Constructions. Texts in Theoretical Computer Science. Springer, Berlin (2004). doi:10.1007/978-3-662-07964-5
6. Carré, B.A.: An algebra for network routing problems. IMA J. Appl. Math. **7**(3), 273 (1971)
7. Chaum, D.: Secret-ballot receipts: true voter-verifiable elections. IEEE Secur. Privacy **2**(1), 38–47 (2004)
8. Cochran, D., Kiniry, J.: Votail: a formally specified and verified ballot counting system for Irish PR-STV elections. In: Pre-proceedings of 1st International Conference on Formal Verification of Object-Oriented Software (FoVeOOS) (2010)
9. Hemaspaandra, L.A., Lavaee, R., Menton, C.: Schulze and ranked-pairs voting are fixed-parameter tractable to bribe, manipulate, and control. Ann. Math. Artif. Intell. **77**(3–4), 191–223 (2016)
10. Kozen, D., Silva, A.: Practical coinduction. Math. Struct. Comput. Sci. 1–21 (2016)
11. Kumar, R., Myreen, M.O., Norrish, M., Owens, S.: CakeML: a verified implementation of ML. In: Jagannathan, S., Sewell, P. (eds.) Proceedings of POPL 2014, pp. 179–192. ACM (2014)
12. Letouzey, P.: Extraction in Coq: an overview. In: Beckmann, A., Dimitracopoulos, C., Löwe, B. (eds.) CiE 2008. LNCS, vol. 5028, pp. 359–369. Springer, Heidelberg (2008). doi:10.1007/978-3-540-69407-6_39
13. Necula, G.C.: Proof-carrying code. In: Lee, P., Henglein, F., Jones, N.D. (eds.) Proceedings of POPL 1997, pp. 106–119. ACM Press (1997)
14. Pattinson, D., Schürmann, C.: Vote counting as mathematical proof. In: Pfahringer, B., Renz, J. (eds.) AI 2015. LNCS, vol. 9457, pp. 464–475. Springer, Cham (2015). doi:10.1007/978-3-319-26350-2_41
15. Rivest, R.L., Shen, E.: An optimal single-winner preferential voting system based on game theory. In: Conitzer, V., Rothe, J. (eds.) Proceedins of COMSOC 2010. Duesseldorf University Press (2010)
16. Schulze, M.: A new monotonic, clone-independent, reversal symmetric, and condorcet-consistent single-winner election method. Soc. Choice Welf. **36**(2), 267–303 (2011)
17. Schürmann, C.: Electronic elections: trust through engineering. In: Proceedings of RE-VOTE 2009, pp. 38–46. IEEE Computer Society (2009)
18. Stoltenberg-Hansen, V., Lindström, I., Griffor, E.: Mathematical Theory of Domains. Cambridge Tracts in Theoretical Computer Science, vol. 22. Cambridge University Press, Cambridge (1994)
19. Tarski, A.: A lattice-theoretical fixpoint theorem and its applications. Pac. J. Math. **5**(2), 285–309 (1955)
20. The Wikimedia Foundation. Wikimedia Foundation Board Election Results (2011). https://meta.wikimedia.org/wiki/Wikimedia_Foundation_elections/Board_elections/2011/Results/en. Accessed 30 May 2017

Verified Spilling and Translation Validation with Repair

Julian Rosemann$^{(\boxtimes)}$, Sigurd Schneider, and Sebastian Hack

Saarland Informatics Campus, Saarland University, Saarbrücken, Germany
rosemann@stud.uni-saarland.de

Abstract. Spilling is a mandatory translation phase in every compiler back-end. It decides whether and where a value is stored in a register or in memory and has therefore a significant impact on performance. In this paper, we study spilling in the setting of a verified compiler with a term-based intermediate representation that provides an alternative way to realize SSA. We devise a permissive correctness criterion to accommodate many SSA-based spilling algorithms and prove the criterion sound. As case study, we verify two basic spilling algorithms. Finally, we show that our criterion is decidable by deriving a translation validator that repairs spilling information if necessary. We show that the validator always produces a valid spilling, and that the validator does not alter valid spilling information. Our results are formalized in Coq as part of the LVC compiler project.

1 Introduction

Spilling is an important translation phase mandatory in every compiler back-end. It deals with the problem that there is an unbounded number of variables in the source program, but only finitely many registers in any processor. After successful spilling, the set of live variables at every program point is covered by the union of an unbounded set (the memory) and a set bounded by an integer k (the registers). We call k the *register bound*. Spilling must ensure the value of a variable resides in a register whenever an instruction uses it. For this purpose, spilling inserts store instructions (spills), which copy the values of variables from the registers to the memory, and load instructions (loads), which copy the values of variables from the memory to the registers. For performance, it is crucial that few load and spill instructions are executed, because register access is at least an order of magnitude faster than memory access. Introducing spills and loads also increases the code size, which is not desirable for performance.

As an example, consider the source program given in Listing 1. The program on the left needs at least three registers. The middle and right programs are different spilled forms of the left program, and each requires only two registers. Note that the decision whether x or y is spilled in the first line determines how many spills and loads are necessary in the continuation of the program.

Spilling determines whether a variable resides in a register at a program point, but does not determine the register. *Register assignment* assigns variables

© Springer International Publishing AG 2017
M. Ayala-Rincón and C.A. Muñoz (Eds.): ITP 2017, LNCS 10499, pp. 427–443, 2017.
DOI: 10.1007/978-3-319-66107-0_27

```
                        let X := x in          let Y := y in
let z := x + y in       let z := x + y in      let z := x + y in
                                               let X := x in
                                               let y := Y in

if z ≥ y                if z ≥ y               if z ≥ y
   then                    then                   then
                             let x := X in           let x := X in
     x + z                   x + z                   x + z
   else                    else                   else
     z                       z                      z
```

Listing 1: A program (left) and two spilled forms of the same program. Lowercase variables denote registers, uppercase variables denote spill slots.

to specific registers. Spilling and register assignment together form the register allocation phase. In the literature, register allocation is often treated as a single problem, without phase separation between spilling and register assignment. In this work, we leverage that the number of simultaneously live variables equals the register pressure to decouple spilling from register assignment. This is possible for static single assignment (SSA) programs [7], and programs in the intermediate language IL [11] used in the verified compiler LVC[1]. IL realizes SSA in a term-based setting by interpreting variable definition as binding with scope.

We develop a small framework for verification of spilling based on an inductively defined correctness criterion. The criterion is formulated relative to spilling information (i.e. which variable is spilled/loaded where) and liveness information. If spilling information satisfies the criterion, it can be used to obtain a program that meets the register bound, and in which variables are in the registers whenever they are used, and which is equivalent to the original program. To verify a spilling algorithm, it suffices to prove that every produced spilling information satisfies the criterion.

It is difficult to formally state what optimal spilling is. Minimizing loads and spills is not necessarily the most effective approach, because reducing the loads and spills at frequently passed program points is more important than anywhere else. Properties of different processor architectures further complicate the problem. Our correctness criterion is independent of assumptions about optimality. We restrict the spilling choices as little as possible. Our criterion in particular supports *arbitrary live range splitting*, i.e., the choice whether a variable should reside in memory or registers is made per program point. This is mandatory to produce spillings with acceptable performance [4]. A value may also reside in a register and in memory simultaneously.

As a case study, we use the predicate to verify three spilling algorithms. The first is a trivial one which loads before instructions and spills afterwards. The second tries to minimize the number of loads and spills by loading as late and

---

[1] https://www.ps.uni-saarland.de/~sdschn/LVC.

as little as possible, and only spilling variables that are overwritten and live in the program continuation.

The third spilling algorithm is similar to a translation validator that takes spilling information from an untrusted source as input. Instead of only validating the spilling information, our algorithm corrects mistakes in the untrusted spilling. We formally show that our algorithm transforms any spilling information (correct or not) to an ultimately correct spilling, and that spilling information that already satisfies our criterion remains unchanged. Interestingly, the algorithm is not much more complicated than a translation validator. To our knowledge, this is the first algorithm of its kind. This approach unites the flexibility of translation validation with the guarantees of full verification.

Our results are formalized in Coq and part of the Linear Verified Compiler (LVC). The development is available online[2]. In summary, this paper makes the following contributions:

- A modular framework for correctness of spilling for term-based SSA
- Verification of two simple spilling algorithms for term-based SSA
- A translation validator for spilling that not only accepts valid spillings, but also repairs incorrect spilling from an external untrusted source.

Outline. The paper is organized as follows. Section 3 contains the semantics of the language IL, and Sect. 4 discusses liveness information. Section 5 discusses the representation of spilling information and the generation of the spilled program. In Sect. 6, we define the correctness criterion for spilling and in Sect. 7 we prove its soundness. Section 8 contains two case studies. Section 9 describes our translation validator with repair. Section 10 concludes.

2   Related Work

Global register allocation was pioneered by Chaitin [5]. Since Chaitin's initial work, there have been several improvements to graph coloring that mostly concentrated on coalescing, i.e. the removal of copy instructions. Most graph coloring approaches decide for every variable globally whether it resides in a register (and if so, in which) or a spill slot. Especially, graph coloring allocators do not attempt to split live ranges sophistically but rather transfer spilled variables from/to memory upon each access. This gives a simple spilling scheme that is also amenable to formal verification (see below). However, in practice the spilling quality of these algorithms is not sufficient to achieve acceptable performance [4].

Linear Scan by Poletto and Sarkar [9] is the basis for many practically popular approaches to register allocation. Linear scan splits live ranges, i.e. it allows a variable to be in a register at one program point and in memory at another. For performance reasons, linear scan over-approximates the live ranges of variables

---

[2] https://www.ps.uni-saarland.de/~rosemann/lvc-spill.

by linearizing control flow, hence the name. Linear scan intertwines spilling and register assignment.

Static Single Assignment (SSA) allows to decouple spilling and register assignment. In SSA, the number of simultaneously live variables equals the register pressure [7]. SSA-based spilling algorithms can hence effectively determine how many variables must be spilled at each program point without knowing the register assignment. Braun and Hack [4] provide an SSA-based spilling algorithm that is very sensitive to the underlying program structure.

Computational Complexity. Chaitin proves NP-completeness of global register allocation [5]. Bouchez et al. show that minimizing spills and loads is NP-complete in SSA [3]. Bouchez also shows NP-completeness of different coalescing problems, i.e. minimizing the number of copies/swaps required to implement SSA's ϕ-functions after the register allocation phase.

CompCert. Register allocation in the first version of CompCert used a translation validated graph coloring algorithm implemented in OCaml [8]. Spilling is verified and very simple: Variables not in a register are loaded before use and spilled after redefinition. Later Blazy et al. [2] fully verified Appel's [6] iterated register coalescing (IRC) approach, which includes spilling. Being a graph coloring technique, this algorithm suffers from the same drawbacks concerning spilling that we discussed above. Hence, especially for machines with few registers (such as IA32), the code quality is hardly acceptable. Instead of changing the fully verified spiller, which would have been a tremendous effort, Rideau and Leroy [10] developed a new translation validated algorithm for register allocation and spilling. The new spilling algorithm tracks recently spilled and loaded variables and thus avoids loading if the variable is still in a temporary register.

In contrast to the verified register allocation by Blazy et al., the second spilling algorithm we verify as case study splits live ranges. The algorithm follows a strategy similar to the translation validated algorithm of Rideau et al., is verified, but does not support overlapping registers yet. There is a project that aims to bring SSA to CompCert [1], but SSA-based register allocation for CompCert has not been explored yet.

CakeML. The compiler for CakeML [13] is verified in HOL4. The compiler represents loops as recursive functions and forces all variables a function uses to be parameters through closure conversion. This breaks all live ranges at loop headers. The CakeML compiler assumes all function parameters are live, hence register pressure may increase if closure conversion introduces dead parameters. CakeML does not use SSA with ϕ-functions and delegates register allocation to a non-SSA-based, verified IRC algorithm [6] that performs spilling and register assignment together. In contrast to the CakeML approach, our approach is SSA-based, separates spilling from register assignment, and allows fine-grained control over live range splitting. Our approach does not require closure conversion, but allows functions to refer to variables that are not parameters.

3   Syntax and Semantics of IL

The formal development in LVC uses the intermediate language IL with mutually recursive function definitions and external events (system calls) [12]. For
the presentation of spilling in this paper, we omit mutually recursive function
definitions and system calls for the sake of simplicity. IL as used in LVC has
a functional and an imperative semantic interpretation [11]. We verify spilling
with respect to the imperative semantics, as it simplifies the treatment of the
new definitions introduced by spills and loads.

3.1   Expressions

Let V be the type of values and **exp** be the type of expressions. By convention, v
ranges over values and e over expressions. The type of variables V is isomorphic
to the natural numbers N. An environment has the type $V \rightarrow V_\perp$ where $V_\perp$
includes V and $\perp$ in case there is no assignment available. Expression evaluation
is a function $\llbracket \cdot \rrbracket : \mathbf{exp} \rightarrow (V \rightarrow V_\perp) \rightarrow V_\perp$ that takes an expression and an
environment and returns a value or $\perp$ if the evaluation fails. For lists, we use
the notation $\overline{x}$ and we lift $\llbracket \cdot \rrbracket$ accordingly: $\llbracket \overline{e} \rrbracket$ yields $\perp$ if at least one of the
expressions in $\overline{e}$ failed to evaluate and the list of the evaluated values otherwise.
We use the usual function fv : $\mathbf{exp} \rightarrow$ set V that yields the *free variables* of an
expression. If V and V' agree on fv e, then $\llbracket e \rrbracket V = \llbracket e \rrbracket V'$. There is a function
$\beta : V \rightarrow \{\mathbf{t}, \mathbf{f}\}$ that simplifies the definition of the semantics of the conditional.

3.2   Syntax

IL is a first-order language with a tail-call restriction, which ensures that every
IL program can be implemented without a call stack. The syntax of IL is given in
Table 1. We use a separate alphabet $\mathcal{F}$ for function names to enforce a first-order
discipline. By convention, f ranges over $\mathcal{F}$.

Table 1. Syntax of IL

stmt $\ni$ s,t ::=	**let** x := e **in** s	let statement
	if e **then** s **else** t	conditional
	e	return statement
	fun f $\overline{x}$:= s **in** t	recursive function definition
	f $\overline{e}$	application

3.3   Semantics

A **context** is a list of named definitions. By convention, L ranges over contexts.
A definition in a context may refer to previous definitions and itself. Notationally,
we use contexts like functions and write L_f to access the first element with

name f. We have $L_f = \bot$ if no such element exists. We write L^{-f} for the context obtained from L by dropping all definitions before the first definition of f. We write ; for context concatenation and $\emptyset$ for the empty context.

Figure 1 shows the small-step transition relation $\longrightarrow$ of IL. The relation is defined on *configurations* (L, V, s) where L is a context containing tuples of type $\overline{\mathcal{V}} * \mathbf{stmt}$, V is an environment and s is an IL term. Often we write the configuration tuple $L \mid V \mid S$ to have the comma available as another separator. Since only tail recursion is syntactically allowed in IL, no call stack is required. Function application in IL is hence similar to a "goto" with a parallel copy on the variables resulting from parameter passing, and very different from a function call in a language with a call stack.

$$\frac{[\![e]\!]V = v}{L \mid V \mid \mathtt{let}\ x\ :=\ e\ \mathtt{in}\ s \longrightarrow L \mid V[x \mapsto v] \mid s} \quad \textsc{SemLet}$$

$$\frac{[\![e]\!]V = v \quad \beta(v) = b}{L \mid V \mid \mathtt{if}\ e\ \mathtt{then}\ s_t\ \mathtt{else}\ s_f \longrightarrow L \mid V \mid s_b} \quad \textsc{SemIf}$$

$$\frac{}{L \mid V \mid \mathtt{fun}\ f\ \overline{x}\ :=\ s\ \mathtt{in}\ t \longrightarrow f : (\overline{x}, s)\, ; L \mid V \mid t} \quad \textsc{SemFun}$$

$$\frac{[\![\overline{e}]\!]V = \overline{v} \quad L_f = (\overline{x}, s)}{L \mid V \mid f\ \overline{e} \longrightarrow L^{-f} \mid V[\overline{x} \mapsto \overline{v}] \mid s} \quad \textsc{SemApp}$$

Fig. 1. Semantics of IL

3.4   Renaming Apart

Many results in this paper require the input programs to be **renamed apart**, that is, every variable must be assigned at most once and defined before used. In general, renaming apart an imperative program requires SSA with ϕ-functions. As presented in our previous work [11], IL realizes SSA by interpreting variables as binders and emulates ϕ-functions through function applications. We previously established that renamed-apart IL programs are coherent, i.e. they behave equivalently under a semantic interpretation with binders and a semantics with imperative assignables. For this reason, our theorems require programs to be renamed apart, and at the same time rely on the imperative interpretation.

4   Liveness

Liveness over-approximates the semantic (and hence undecidable) notion that a variable is still used *later on*. The notion of liveness used in a register allocation approach greatly impacts the algorithm and its effectiveness. Consider, for example, the different notions of liveness used by graph-coloring register allocation [5] and linear scan [9].

We inductively define a soundness predicate $Z \mid \Lambda \vdash$ **live** $s : X$ that associates a set of variables X called **live set** with a program s. The **parameter context** Z maps every defined function to its parameters. The **live-in context** Λ maps every function to a set of variables that contains the variables live in the function body and the parameters, which we call the **live-in set** of the function. We embed the live-in sets in the IL syntax of function definitions, which from now on take the syntactic form $\text{fun } f \ \bar{x} := s_1 \ \{X_1\} \text{ in } s_2$, in which the function body s_1 is syntactically annotated with its live-in set X_1. We call these sets embedded in the syntax at function bodies **live-in annotations**. In contrast to these annotations, the live set X that appears in the judgment $Z \mid \Lambda \vdash$ **live** $s : X$ is not part of the syntax of IL. The inductive definition of liveness is given in Fig. 2 and similar to our previous definition [11]. The liveness predicate allows X to over-approximate the live set, that is, X may contain variables that are not used later on.

LIVELET
$$\frac{\text{fv}\,e \subseteq X \qquad X_s \setminus \{x\} \subseteq X \qquad x \in X_s \qquad Z \mid \Lambda \vdash \text{live } s : X_s}{Z \mid \Lambda \vdash \text{live } (\text{let } x := e \text{ in } s) : X}$$

LIVEAPP
$$\frac{\text{fv}\,\bar{e} \subseteq X \qquad \Lambda_f \setminus Z_f \subseteq X}{Z \mid \Lambda \vdash \text{live } f \ \bar{e} : X}$$

LIVERETURN
$$\frac{\text{fv}\,e \subseteq X}{Z \mid \Lambda \vdash \text{live } e : X}$$

LIVEIF
$$\frac{\text{fv}\,e \cup X_1 \cup X_2 \subseteq X \qquad Z \mid \Lambda \vdash \text{live } s_1 : X_1 \qquad Z \mid \Lambda \vdash \text{live } s_2 : X_2}{Z \mid \Lambda \vdash \text{live } (\text{if } e \text{ then } s_1 \text{ else } s_2) : X}$$

LIVEFUN
$$\frac{\bar{x} \subseteq X_1 \qquad f : \bar{x}, Z \mid f : X_1, \Lambda \vdash \text{live } s_1 : X_1 \qquad X_2 \subseteq X \qquad f : \bar{x}, Z \mid f : X_1, \Lambda \vdash \text{live } s_2 : X_2}{Z \mid \Lambda \vdash \text{live } (\text{fun } f \ \bar{x} := s_1 \ \{X_1\} \text{ in } s_2) : X}$$

Fig. 2. Inductive definition of liveness

4.1 Description of the Rules of the Inductive Predicate

LIVELET requires the live set X of the let statement to contain the free variables of the expression e, and the variables live in the continuation s, except the newly defined variable x. We also require x to be in the live set X_s of the continuation. This reflects that x must be considered live during the let-statement even if x is not used afterwards, because x is overwritten and hence cannot hold a value that is still used later on. LIVERETURN requires the free variables of the expression to be live. LIVEAPP requires the live-ins of the function that are not parameters to be live. LIVEIF requires the live variables of the consequence, the alternative, and the free variables of the condition to be live. LIVEFUN requires that variables live in continuation s_2 are live. The parameters are recorded in the context Z, and the live-ins X_1 are recorded in the context Λ. The live-ins X_1 contain all variables live in the function body and all parameters, regardless of whether a parameter is used: $\bar{x} \subseteq X_1$. This reflects that unused parameters are overwritten during function application, and hence occupy a register or a spill slot.

4.2   Minimal Live Sets and Live Set Annotations

Live-in annotations uniquely determine the *minimal* live set for every program point; those live sets can be computed by a bottom-up traversal. Since liveness annotations are part of the syntax, every algorithm or judgment formulated on the syntax can easily refer to and compute with the live-ins at function definitions. This allows us to concisely describe how the live sets change during spilling, we can explain changes to live sets by explaining them just for the live-ins at function definitions. The effect on the other live sets in the program is then uniquely determined. We write $Z \mid \Lambda \vdash \mathbf{live}\ s$ for $\exists X.\ Z \mid \Lambda \vdash \mathbf{live}\ s : X$ and use this notation whenever we want to hide the precise form of the live set.

5   Spilling

Spilling transforms a program into an equivalent program by inserting spills and loads such that the number of registers in the maximal live set is afterwards bounded by a given integer k. In our framework, spilling consists of two steps: First, a spilling algorithm inserts spilling annotations into the program that describe where spills and loads should be placed. Second, the spills and loads are inserted into the program as prescribed by these spilling annotations, which yields the **spilled program**. The spilled program also contains live set annotations at function definitions, and we describe in Sect. 7.2 how those are recomputed according to the spilling annotations.

Spilling annotations are three-tuples embedded in the syntax at every sub-term. A statement with spilling annotation has the form $s : (S, L, \_)$, where S is the set of variables to be spilled (**spill set**) and L is the set of variables to be loaded (**load set**). The third component is only required if s is a function applications or a function definition, and we discuss its purpose below. We call a statement that contains such annotations a **spill statement**.

A spill statement can be turned into a spilled program via the recursive function doSpill, which we now informally describe. We assume that the variables are partitioned into two countably-infinite sets $\mathcal{V} = \mathcal{V}_R \uplus \mathcal{V}_M$, and require that the

$$
\text{doSpillLocal}(s : (\underbrace{\{x_1, ..., x_n\}}_{\text{spills}}, \underbrace{\{y_1, ..., y_m\}}_{\text{loads}}, \_)) \quad = \quad
\begin{array}{l}
\texttt{let slot } \texttt{x}_1 = \texttt{x}_1 \texttt{ in } \ldots \\
\texttt{let slot } \texttt{x}_n = \texttt{x}_n \texttt{ in} \\
\texttt{let } \texttt{y}_1 = \texttt{slot } \texttt{y}_1 \texttt{ in } \ldots \\
\texttt{let } \texttt{y}_m = \texttt{slot } \texttt{y}_m \texttt{ in } s
\end{array}
$$

Listing 2: Definition of doSpillLocal

```
 1  fun f x y z :=              R_f = {y,z}, M_f = {c,x,z}    fun f X y z Z :=
 2  if y > 0 then                                            if y > 0 then
 3    let a := y+z in                                          let a := y+z in
 4      f x a z               R_app = {a,z}, M_app = {x,z}       f X a z Z
 5  else if y = 0 then                                       else if y = 0 then
 6                                                             let x := X in
 7                                                             let c := C in
 8    x + c                   L = {c,x}                        x + c
 9  else                                                     else
10    let w := y*y in                                          let w := y*y in
11    let a := y+w in                                          let a := y+w in
12      f x a z               R_app = {a}, M_app = {x,z}        f X a z Z
```

Listing 3: A spill statement on the left (non-empty sets in spilling annotations are indicated by equations) and the resulting spilled program on the right. The live-ins of f are $\{x,y,z,c\}$. The variable c is free in f. Lowercase variables denote registers, uppercase variables denote spill slots. In line 4, z is passed in register and memory to avoid loading z in line 3. The application in line 12 implicitly loads z (3rd parameter).

spill statement only contains variables from $\mathcal{V}_R$. We further assume an injection slot : $\mathcal{V}_R \to \mathcal{V}_M$ which we use to generate names for spill slots (cf. CompCert [8]).

To generate the spilled program for $s : (S, L, \_)$, doSpill first prepends the statement s with spills for each variable in S, followed by the loads for each variable in L as depicted in Listing 2. For let statements, conditionals, and return statements this is all that needs to be done. Function definitions and applications require some additional work, which we describe next.

Function definitions take a pair of sets (R_f, M_f) as third component of the spilling annotation: fun f $x_1, \ldots, x_n := s_1$ in s_2 : $(S, L, (R_f, M_f))$. We call the pair (R_f, M_f) the **live-in cover** and require it to cover the live-ins X_f of f, i.e. $R_f \cup M_f = X_f$. The set R_f specifies the variables the function expects to reside in registers, and the set M_f specifies the variables the function expects to reside in memory. The sets R_f and M_f are not necessarily disjoint, as a function may want a variable to reside both in register and in memory when it is applied (see Listing 3). Besides inserting spills and loads according to S and L as already described, the function parameters must be modified to account for parameters that are passed in spill slots. For this purpose, every parameter $x_i \in M_f \setminus R_f$ is replaced by the name slot x_i in $\overline{x}$. Furthermore, for any parameter $x_i \in M_f \cap R_f$ an additional parameter with name slot x_i is inserted directly after x_i.

Function applications have a pair of sets (R_{app}, M_{app}) as third component of spilling information and take the form f $y_1, \ldots, y_n$: $(S, L, (R_{app}, M_{app}))$. We require all function arguments y_i to be variables, and that $R_{app} \cup M_{app} = \{y_1, \ldots, y_n\}$. The sets R_{app} and M_{app} indicate the availability of argument variables at the function application. If an argument variable y_i is available in a register, then the spilling algorithm sets $y_i \in R_{app}$, if it is in memory, then $y_i \in M_{app}$. Besides inserting spills and loads according to S and L as already described, doSpill modifies the argument vector $y_1, \ldots, y_n$. For every parameter $x_i \in R_f$ such that the corresponding argument variable y_i is not in R_{app} (i.e. not available in a register), the variable y_i is replaced by the name slot y_i in the argument vector. For every parameter $x_i \in M_f \setminus R_f$ such that the corresponding argument variable y_i is in M_{app} (i.e. available in memory), the variable y_i is replaced by the name slot y_i in the argument vector. Furthermore, for every parameter $x_i \in M_f \cap R_f$ an additional argument is inserted directly after the corresponding argument variable (y_i or slot y_i) in $y_1, \ldots, y_n$, and the name of the additional argument is slot y_i if $y_i \in M_{app}$ and y_i otherwise. In this way, R_f and M_f are used to avoid implicit loads and stores at function application if availability, as indicated in R_{app} and M_{app}, permits. Since spill slots are just a partition of the variables, parameter passing can copy between spill slots and registers if the argument variable y_i for a register parameter x_i is only available in memory, or vice versa. This fits nicely in our setting, as we handle the generation of these implicit spills and loads later on, when parameter passing is lowered to parallel moves. In line 12 of Listing 3, for example, the application implicitly loads z. In contrast, availability of z in both register and memory at the application in line 4 allows avoiding any implicit loads and stores. Assuming $y > 0$ holds for most executions, this is beneficial for performance.

6   A Correctness Criterion for Spilling

We define a correctness predicate for spilling on spill statements of the form $Z \mid \Sigma \mid R \mid M \vdash \mathsf{spill}_k \; s : (S, L, \_)$. Note that as described in Sect. 5, the spilling annotation $(S, L, \_)$ is embedded in the syntax. The correctness predicate is defined relative to sets R and M, which contain the variables currently in registers, and in memory, respectively. Additionally, the *parameter context* Z maps function names to their parameter list, and the *live-in cover context* Σ maps functions to their live-in cover. The parameter k is the register bound. The rules defining the predicate are given in Fig. 3.

6.1   Description of the Rules of the Inductive Predicate

The predicate consists of two generic rules that handle spilling and loading, and one rule for each statement. Rules for statements only apply once spills and loads have been handled. This is achieved by requiring empty spill and load sets in statement rules, and requiring an empty spill set in the load rule. SPILLSPILL requires $S \subseteq R$ to ensure only variables currently in registers are spilled. The new memory state is $M \cup S$. SPILLLOAD requires the spill set to be empty. Its second premise ensures there are enough free registers to load all values. The *kill set* K represents the variables that may be overwritten because they are not used anymore or are already spilled. $R \setminus K \cup L$ is the new register state after loading. Clearly, K is most useful if $K \subseteq R$, but our proofs do not require this restriction. We also do not include K in the spilling annotation, as the spilling algorithm would have to compute liveness information to provide it. Simple

SPILLSPILL

$$\dfrac{S \subseteq R}{Z\,|\,\Sigma\,|\,R\,|\,M \cup S \vdash \mathbf{spill_k}\ s : (\emptyset, L, \_)}{Z\,|\,\Sigma\,|\,R\,|\,M \vdash \mathbf{spill_k}\ s : (S, L, \_)}$$

SPILLLOAD

$$\dfrac{L \subseteq M \qquad |R \setminus K \cup L| \le k}{Z\,|\,\Sigma\,|\,R \setminus K \cup L\,|\,M \vdash \mathbf{spill_k}\ s : (\emptyset, \emptyset, \_)}{Z\,|\,\Sigma\,|\,R\,|\,M \vdash \mathbf{spill_k}\ s : (\emptyset, L, \_)}$$

SPILLRETURN

$$\dfrac{\mathsf{fv}\,e \subseteq R}{Z\,|\,\Sigma\,|\,R\,|\,M \vdash \mathbf{spill_k}\ e : (\emptyset, \emptyset)}$$

SPILLAPP

$$\dfrac{\begin{array}{cc} \Sigma_f = (R_f, M_f) & R_f \setminus Z_f \subseteq R \\ M_f \setminus Z_f \subseteq M & \bar{y} = R_{app} \cup M_{app} \\ M_{app} \subseteq M & R_{app} \subseteq R \end{array}}{Z\,|\,\Sigma\,|\,R\,|\,M \vdash \mathbf{spill_k}\ (f\ \bar{y}) : (\emptyset, \emptyset, (R_{app}, M_{app}))}$$

SPILLIF

$$\dfrac{\mathsf{fv}\,e \subseteq R \qquad Z\,|\,\Sigma\,|\,R\,|\,M \vdash \mathbf{spill_k}\ s_1 \qquad Z\,|\,\Sigma\,|\,R\,|\,M \vdash \mathbf{spill_k}\ s_2}{Z\,|\,\Sigma\,|\,R\,|\,M \vdash \mathbf{spill_k}\ (\texttt{if}\ e\ \texttt{then}\ s_1\ \texttt{else}\ s_2) : (\emptyset, \emptyset)}$$

SPILLLET

$$\dfrac{\mathsf{fv}\,e \subseteq R \qquad |R \setminus K \cup \{x\}| \le k \qquad Z\,|\,\Sigma\,|\,R \setminus K \cup \{x\}\,|\,M \vdash \mathbf{spill_k}\ s}{Z\,|\,\Sigma\,|\,R\,|\,M \vdash \mathbf{spill_k}\ (\texttt{let}\ x := e\ \texttt{in}\ s) : (\emptyset, \emptyset)}$$

SPILLFUN

$$\dfrac{\begin{array}{cc} |R_f| \le k & f : \bar{x}; Z\,|\,f : (R_f, M_f); \Sigma\,|\,R_f\,|\,M_f \vdash \mathbf{spill_k}\ s_1 \\ R_f \cup M_f = X_f & f : \bar{x}; Z\,|\,f : (R_f, M_f); \Sigma\,|\,R\ \,|\,M\ \vdash \mathbf{spill_k}\ s_2 \end{array}}{Z\,|\,\Sigma\,|\,R\,|\,M \vdash \mathbf{spill_k}\ (\texttt{fun}\ f\ \bar{x} := s_1\ \{X_f\}\ \texttt{in}\ s_2) : (\emptyset, \emptyset, (R_f, M_f))}$$

Fig. 3. Inductive correctness predicate $\mathbf{spill_k}$

spilling algorithms, such as the one we verify in Sect. 8.1, never need to compute liveness information.

SPILLRETURN requires that the free variables are in the registers. SPILLIF requires the consequence and the alternative to fulfill the predicate on the same configuration, and that the variables used in the condition are in registers. SPILL-LET deals with the new variable x, which needs a register. The resulting register state is $R \setminus K \cup \{x\}$, the size of which must be bounded by the register bound k. This imposes a lower bound on k. The kill set K reflects that there might be a variable y holding the value of a variable required to evaluate the expression e, that is then overwritten to store the value of x. In this case $K = \{y\}$.

SPILLAPP uses the sets R_f and M_f from the corresponding function definition. The premises $R_f \setminus Z_f \subseteq R$ and $M_f \setminus Z_f \subseteq M$ require that all live-ins of the function except parameters are available in registers and memory at the application. The remaining premises require that all argument variables are available either in the registers (R_{app}) or in the memory (M_{app}), as discussed in Sect. 5. Note that the argument vector $\bar{y}$ is variables only, i.e. applications can only have variables as arguments. SPILLFUN refers to the live-in set X_f embedded in the syntax to require that the live-in cover (R_f, M_f) covers the live-ins X_f of the program: $R_f \cup M_f = X_f$. The rule also requires the function to expect at most k variables in registers: $|R_f| \le k$. The parameters and the live-in cover are recorded in the context. The condition for the function body s_1 uses R_f and M_f as register and memory sets, respectively.

6.2 Formalization of the Spill Predicate in Coq

The predicate $\textbf{spill}_k$ is realized with five rules in the Coq development instead of the seven rules presented here. Each of the five rules corresponds to a consecutive application of SPILLSPILL, SPILLLOAD and one of the statement-specific rules. The five-rule system behaves better under inversion and induction in Coq, but we think the formulation with seven rules provides more insight. The Coq development contains a formal proof of the equivalence of the two systems.

7 Soundness of the Correctness Predicate

In this section we show that our spilling predicate is sound. We show that if s is renamed apart and all variables in s are in $\mathcal{V}_R$, and the spilling and live-in annotations in s are sound, the following holds for the spilled program s':

(Section 7.1) all variables in s' are in a register when used
(Section 7.2) at most k registers are used in s'
(Section 7.3) s and s' have the same behavior.

7.1 Variables in Registers

Figure 4 defines a predicate that ensures every variable is in a register when used. The inference rules are straightforward. The predicate also ensures that let-statements that assign to memory have a single register on the right-hand side. We define $\mathsf{merge}\,(R, M) = R \cup M$ and $\mathsf{slotMerge}\,(R, M) = R \cup \mathsf{slot}\,M$ and analogously their pointwise liftings.

$$\frac{\text{ViRLoad}}{x \in \mathcal{V}_R \qquad y \in \mathcal{V}_M \qquad \textbf{vir}\,s}{\textbf{vir let } x := y \textbf{ in } s} \qquad \frac{\text{ViRLet}}{\mathsf{fv}\,e \subseteq \mathcal{V}_R \qquad \textbf{vir}\,s}{\textbf{vir let } x := e \textbf{ in } s} \qquad \frac{\text{ViRReturn}}{\mathsf{fv}\,e \subseteq \mathcal{V}_R}{\textbf{vir } e}$$

$$\frac{\text{ViRIf}}{\mathsf{fv}\,e \subseteq \mathcal{V}_R \qquad \textbf{vir}\,s \qquad \textbf{vir}\,t}{\textbf{vir if } e \textbf{ then } s \textbf{ else } t} \qquad \frac{\text{ViRApp}}{\textbf{vir } f \,\overline{y}} \qquad \frac{\text{ViRFun}}{\textbf{vir}\,s \qquad \textbf{vir}\,t}{\textbf{vir fun } f\,\overline{x} := s \textbf{ in } t}$$

Fig. 4. Predicate vir

Lemma 1. *Let* $Z\,|\,\Sigma\,|\,R\,|\,M \vdash \textbf{spill}_k\ s$ *and* $Z\,|\,\Lambda \vdash \textbf{live}\ s$ *and let* s *be renamed apart and let all variables in* s *be in* $\mathcal{V}_R$. *If* $R \cup M \cup \bigcup Z \subseteq \mathcal{V}_R$ *then* $\textbf{vir}\,(\mathsf{doSpill}\,Z\,\Lambda\,s)$.

Proof. The conditions follow directly by induction on $\textbf{spill}_k\ s$.

7.2   Register Bound

After the spilling phase, the liveness information in the program changed tremendously. Spills and loads introduce new live ranges, and shorten live ranges of already defined variables. To prove correctness of the spilling predicate, we must show that after spilling the register pressure is lowered to k. To formally establish the bound, we show that the number of variables from $\mathcal{V}_R$ in each live set in the spilled program is bounded by k. The following observation is key to this proof: The live-ins of a function after spilling can be obtained from the live-ins of the function before spilling by keeping the variables passed in registers, and adding the slots of the variables passed in memory. This property can be seen in the rule SPILLFUN, where we require $R_f \cup M_f = X_f$.

In the Coq development, the statements of the following lemmas involve the algorithm that reconstructs minimal liveness information we informally described in Sect. 4.2, but omitted in this presentation for the sake of simplicity.

Lemma 2. *Let $Z \mid \Sigma \mid R \mid M \vdash$ spill$_k$ s and $Z \mid$ merge $\Sigma \vdash$ live s and let s be renamed apart and let all variables in s be in $\mathcal{V}_R$. If $R \cup M \cup \bigcup Z \subseteq \mathcal{V}_R$ then $Z \mid$ slotMerge $\Sigma \vdash$ live doSpill $Z \Sigma s$.*

Proof. By induction on spill$_k$ s; mostly simple but tedious set constraints.

Lemma 3. *Let $Z \mid \Sigma \mid R \mid M \vdash$ spill$_k$ s and let s be renamed apart and let all variables in s be in $\mathcal{V}_R$. If $|R| \le k$ and $R \cup M \cup \bigcup Z \subseteq \mathcal{V}_R$ then for live set X in the minimal liveness derivation $Z \mid$ slotMerge $\Sigma \vdash$ live doSpill $Z \Sigma s$ the bound $|\mathcal{V}_R \cap X| \le k$ holds.*

Proof. By induction on s. The proof uses a technical lemma about the way the liveness reconstruction deals with forward-propagation that was difficult to find.

7.3   Semantic Equivalence

In this section we show that the spilled program is semantically equivalent to the original program. Semantic equivalence means trace-equivalence à la CompCert. As proof tool we use a co-inductively defined simulation relation. See our previous work [11,12] for details on simulation and proof technique. The verification is done with respect to the imperative semantics of IL. This allows for a simple treatment of the new variables that each spill and each load introduces. A typical spill and load looks as follows:

```
let x = 5 in            let x = 5 in
fun f () = x in         fun f () = x in
                        let X = x in        // spill
...                     ...
                        let x = X in        // load
f()                     f()
```

Note that in a semantics with binding, serious effort would be required to introduce additional function parameters after spilling and loading. In the above example, f would need to take x as a parameter. We postpone the introduction of additional parameters to a phase after spilling, where we switch to the functional semantics again to do register allocation. Changing the semantics from imperative to functional corresponds to SSA construction and is in line with practical implementations of SSA-based register allocation [4] that break the SSA invariant during spilling, and then perform some form of SSA (re-)construction.

Lemma 4. *Let s be a spill statement where all variables are renamed apart and in $\mathcal{V}_R$. Let $Z \mid \Sigma \mid R \mid M \vdash$ spill$_k$ s and $Z \mid$ merge $\Sigma \vdash$ live $s : X$ and $V =_R V'$ and $V =_M (\lambda x.V'(\text{slot } x))$. If V' is defined on $R \cup \text{slot } M$ and $R \cup M \subseteq \mathcal{V}_R$ and L and L' are suitably related then (L, V, s) and $(L', V', \text{doSpill } Z \Sigma s)$ are in simulation.*

8 Case Study: Verified Spilling Algorithms

A spilling algorithm translates a statement with live-in annotations to a spill statement, that is, it inserts spilling annotations. The following algorithms are implemented in Coq and verified using the correctness predicate.

8.1 SimpleSpill

The naive spilling algorithm simpleSpill loads the required values before each statement, without considering that some values might still be available in a register. After a variable is assigned, the algorithm immediately spills the variable. This is a very simple algorithm, and it corresponds to the spilling strategy used in the very first version of CompCert [8].

Theorem 1. *Let $Z \mid$ merge $\Sigma \vdash$ live s and let s be renamed apart and let all variables in s be in $\mathcal{V}_R$ and let every expression in s contain at most k different variables. If every live set X in s is bounded by $R \cup M$ and the first component in Σ_f is empty for every f then $Z \mid \Sigma \mid R \mid M \vdash$ spill$_k$ simpleSpill s.*

Proof. By induction on s in less than 100 lines.

8.2 SplitSpill

The spilling algorithm splitSpill follows three key ideas: Variables are loaded as late as possible, but in contrast to simpleSpill, only values not already available in registers. If a register must be freed for a load, the algorithm lets an oracle choose the variable to be spilled from the list of variables live and currently in a register. The correctness requirement for the oracle is trivial. The oracle enables live range splitting based on an external heuristic, similar to the approach of Braun and Hack [4]. In contrast to Braun's algorithm, splitSpill cannot hoist loads from their uses.

Theorem 2. *Let $Z \,|\, \text{merge } \Sigma \vdash$ live s and let s be renamed apart and let all variables in s be in $\mathcal{V}_R$ and let every expression in s contain at most k different variables. If every live set X in s is bounded by $R \cup M$ and and for every f such that $(R_f, M_f) = \Sigma_f$ we have $|R_f| \leq k$ then $\Sigma \,|\, Z \,|\, R \,|\, M \vdash \text{spill}_k$ splitSpill s.*

Proof. By induction on s in less than 500 lines.

9   Translation Validation with Repair

We devise a translation validator repairSpill for our correctness predicate. The translation validator repairSpill operates on a statement with liveness and spilling annotations, and assumes the liveness to be sound. Besides deciding whether the spilling annotation is sound, repairSpill also repairs the spilling annotation if necessary. The output of repairSpill always contains sound spilling annotations. Furthermore, we show that repairSpill leaves the spilling annotations unchanged if they are already sound with respect to the provided live-in annotation and our correctness predicate.

To explain the principle behind repairSpill, it is instructive to understand how the algorithm recomputes a live-in cover (R_f, M_f) from the (possibly unsound) spilling annotation using the corresponding live-ins X_f from the sound live-in annotation. Let take $k\,X$ be a function that yields a k-sized subset from X or X if $|X| \leq k$. The new live-in cover (R'_f, M'_f) is obtained as follows:

$$R'_f = \text{take } k\,(R_f \cap X_f)$$
$$M'_f = (X_f \setminus R'_f) \cup (M_f \cap X_f)$$

These equations have two important properties: First, it holds $R'_f \cup M'_f = X_f$, so the equations produce a correct live-in cover independent of the input sets R_f and M_f. Second, if $R_f \cup M_f = X_f$ and $|R_f| \leq k$ then $R'_f = R_f$ and hence $M'_f = M_f$, i.e. the original live-in cover is retained, if it is valid. repairSpill transforms every spill and load set in the spilling annotation in a similar way such that these two properties hold.

The kill sets K appearing in the derivation of spill_k are not recorded in the spilling annotation, because we did not want to require the spilling algorithm to compute them. To check whether a spilling annotation is correct, repairSpill must reconstruct kill sets. Maximal kill sets can be reconstructed in a backwards fashion from spilling annotation similar to how minimal liveness information can be reconstructed (Sect. 4.2). A maximal kill set upper-bounds the variables that can be soundly killed. The correctness of the kill sets repairSpill reconstructs depends on the correctness of the spilling annotation. For this reason, we designed repairSpill in such a way that the correctness of its output does not depend on the correctness of the kill sets. This is similar to the fact that the correctness of R'_f and M'_f in the equations above does not depend on the correctness of R_f and M_f. If the spilling annotation is correct, however, the kill sets are correct and ensure that the algorithm does not change the spilling annotation.

Consider, for example, a conditional if e then s else $t : (S, L, \_)$ where S and L are the potentially unsound spill and load sets. From the memory state (R, M) and the assumption fv $e \subseteq R \cup M$ the algorithm produces sound spill and load sets S' and L' that agree with S and L if those are already sound. For correctness, we only require fv $e \subseteq R$. We use the following definitions:

$$\mathsf{pick}\, k\, s\, t = s \cup \mathsf{take}\, (k - |s|)\, (t \setminus s)$$
$$\mathsf{pickload}\, k\, R\, M\, S\, L\, e = (\mathsf{fv}\, e \cap R \cap Q) \cup P$$
$$\text{where } Q = L \cap ((S \cap R) \cup M)$$
$$\text{and where } P = \mathsf{pick}\, (k - |\mathsf{fv}\, e \cap R|)\, (\mathsf{fv}\, e \setminus R)\, (Q \setminus (\mathsf{fv}\, e \cap R))$$

We can now pick the new load set $L' = \mathsf{pickload}\, k\, R\, M\, S\, L\, e$. Lemmas 5 and 6 establish that L' satisfies the register bound and loads the variables necessary to evaluate e. Lemma 7 shows that $L' = L$ if L was already correct.

Lemma 5. *If* $|\mathsf{fv}\, e| \leq k$ *then* $|\mathsf{pickload}\, k\, R\, M\, S\, L\, e| \leq k - |\mathsf{fv}\, e \cap R \setminus (L \cap (S \cup M))|$.

Lemma 6. $(\mathsf{fv}\, e \cap R \setminus (L \cap (S \cup M))) \cup (\mathsf{fv}\, e \setminus R) \subseteq \mathsf{pickload}\, k\, R\, M\, S\, L\, e$.

Lemma 7. *Let* $S \subseteq R$ *and* $L \subseteq S \cup M$ *and* fv $e \setminus R \subseteq L$ *and* $|\mathsf{fv}\, e \cup L| \leq k$ *then* $\mathsf{pickload}\, k\, R\, M\, S\, L\, e = L$.

We developed definitions similar to pickload that allow repairSpill to transform every spill and load set in the spilling annotation. Correct sets are retained, and incorrect sets are repaired. NP-completeness of the spilling problem makes it unlikely that quality guarantees hold for a polynomial-time repair algorithm.

Theorem 3 (Correctness). *Let* $Z \,|\, \mathsf{merge}\, \Sigma \vdash$ **live** $s : X$ *and let* R, M *be sets of variables such that* $X \subseteq R \cup M$ *and let every expression in* s *contain at most* k *different variables. If for every* f *such that* $(R_f, M_f) = \Sigma_f$ *we have* $|R_f| \leq k$ *then* $Z \,|\, \Sigma \,|\, R \,|\, M \vdash \mathbf{spill_k}$ (repairSpill $k\, Z\, \Sigma\, R\, M\, s$).

Theorem 4 (Idempotence). *Let* s *be renamed apart and let* $Z \,|\, \mathsf{merge}\, \Sigma \vdash$ **live** $s : X$ *and let* $Z \,|\, \Sigma \,|\, R \,|\, M \vdash \mathbf{spill_k}\, s$. *If for every* f *such that* $(R_f, M_f) = \Sigma_f$ *we have* $|R_f| \leq k$ *then* repairSpill $k\, Z\, \Sigma\, R\, M\, s = s$.

10   Conclusion

We presented a correctness predicate for spilling algorithms that permits arbitrary live range splitting. To our knowledge, it is the first formally proven correctness predicate for spilling on term-based SSA and the first to support arbitrary live range splitting. The conditions of our correctness predicate are mainly set constraints, and our case studies show that the predicate simplifies correctness proofs of spilling algorithms.

Based on the correctness predicate, we defined a translation validator for spilling algorithms with repair. The algorithm takes any spilling annotation and repairs it if necessary. Our algorithm combines the flexibility of translation validation with the correctness guarantees of verification.

This work is part of the verified compiler LVC. LVC has about 50k LoC and extracts to an executable verified compiler. The spilling framework presented in this paper is about 8k LoC. A considerable difference between the paper and the formal proofs is the presentation of liveness information: In the formal development, liveness reconstruction must, of course, be handled by a Coq function, and we must prove that is function is correct and yields minimal live sets.

References

1. Barthe, G., Demange, D., Pichardie, D.: A formally verified SSA-based middle-end. In: Seidl, H. (ed.) ESOP 2012. LNCS, vol. 7211, pp. 47–66. Springer, Heidelberg (2012). doi:10.1007/978-3-642-28869-2_3
2. Blazy, S., Robillard, B., Appel, A.W.: Formal verification of coalescing graph-coloring register allocation. In: Gordon, A.D. (ed.) ESOP 2010. LNCS, vol. 6012, pp. 145–164. Springer, Heidelberg (2010). doi:10.1007/978-3-642-11957-6_9
3. Bouchez, F., Darte, A., Rastello, F.: On the complexity of spill everywhere under SSA form. In: LCTES, San Diego, California, USA, 13–15 June 2007
4. Braun, M., Hack, S.: Register spilling and live-range splitting for SSA-form programs. In: de Moor, O., Schwartzbach, M.I. (eds.) CC 2009. LNCS, vol. 5501, pp. 174–189. Springer, Heidelberg (2009). doi:10.1007/978-3-642-00722-4_13
5. Chaitin, G.J.: Register allocation & spilling via graph coloring. In: PLDI, Boston, Massachusetts, USA, 23–25 June 1982
6. George, L., Appel, A.W.: Iterated register coalescing. ACM Trans. Program. Lang. Syst. **18**(3), 300–324 (1996)
7. Hack, S., Grund, D., Goos, G.: Register allocation for programs in SSA-form. In: Mycroft, A., Zeller, A. (eds.) CC 2006. LNCS, vol. 3923, pp. 247–262. Springer, Heidelberg (2006). doi:10.1007/11688839_20
8. Leroy, X.: A formally verified compiler back-end. JAR **43**(4), 363–446 (2009)
9. Poletto, M., Sarkar, V.: Linear scan register allocation. TOPLAS **21**(5), 895–913 (1999)
10. Rideau, S., Leroy, X.: Validating register allocation and spilling. In: Gupta, R. (ed.) CC 2010. LNCS, vol. 6011, pp. 224–243. Springer, Heidelberg (2010). doi:10.1007/978-3-642-11970-5_13
11. Schneider, S., Smolka, G., Hack, S.: A linear first-order functional intermediate language for verified compilers. In: Urban, C., Zhang, X. (eds.) ITP 2015. LNCS, vol. 9236, pp. 344–358. Springer, Cham (2015). doi:10.1007/978-3-319-22102-1_23
12. Schneider, S., Smolka, G., Hack, S.: An inductive proof method for simulation-based compiler correctness (2016). CoRR abs/1611.09606
13. Tan, Y.K., et al.: A new verified compiler backend for CakeML. In: ICFP, Nara, Japan, 18–22 September 2016

A Verified Generational Garbage Collector for CakeML

Adam Sandberg Ericsson, Magnus O. Myreen[✉], and Johannes Åman Pohjola

Chalmers University of Technology, Gothenburg, Sweden
myreen@chalmers.se

Abstract. This paper presents the verification of a generational copying garbage collector for the CakeML runtime system. The proof is split into an algorithm proof and an implementation proof. The algorithm proof follows the structure of the informal intuition for the generational collector's correctness, namely, a partial collection cycle in a generational collector is the same as running a full collection on part of the heap, if one views pointers to old data as non-pointers. We present a pragmatic way of dealing with ML-style mutable state, such as references and arrays, in the proofs. The development has been fully integrated into the in-logic bootstrapped CakeML compiler, which now includes command-line arguments that allow configuration of the generational collector. All proofs were carried out in the HOL4 theorem prover.

1 Introduction

High-level programming languages such as ML, Haskell, Java, Javascript and Python provide an abstraction of memory which removes the burden of memory management from the application programmer. The most common way to implement this memory abstraction is to use garbage collectors in the language runtimes. The garbage collector is a routine which is invoked when the memory allocator finds that there is not enough free space to perform allocation. The collector's purpose is to produce new free space. It does so by traversing the data in memory and deleting data that is unreachable from the running application. There are two classic algorithms: mark-and-sweep collectors mark all live objects and delete the others; copying collectors copy all live objects to a new heap and then discard the old heap and its dead objects.

Since garbage collectors are an integral part of programming language implementations, their performance is essential to make the memory abstraction seem worthwhile. As a result, there have been numerous improvements to the classic algorithms mentioned above. There are variants of the classic algorithms that make them incremental (do a bit of garbage collection often), generational (run the collector only on recent data in the heap), or concurrent (run the collector as a separate thread alongside the program).

This paper's topic is the verification of a generational copying collector for the CakeML compiler and runtime system [15]. The CakeML project has produced a formally verified compiler for an ML-like language called CakeML. The

© Springer International Publishing AG 2017
M. Ayala-Rincón and C.A. Muñoz (Eds.): ITP 2017, LNCS 10499, pp. 444–461, 2017.
DOI: 10.1007/978-3-319-66107-0_28

compiler produces binaries that include a verified language runtime, with supporting routines such as an arbitrary precision arithmetic library and a garbage collector. One of the main aims of the CakeML compiler project is to produce a verified system that is as realistic as possible. This is why we want the garbage collector to be more than just an implementation of one of the basic algorithms.

Contributions.

- To the best of our knowledge, this paper presents the first completed formal verification of a generational garbage collector. However, it seems that the CertiCoq project [1] is in the process of verifying a generational garbage collector.
- We present a pragmatic approach to dealing with mutable state, such as ML-style references and arrays, in the context of implementation and verification of a generational garbage collector. Mutable state adds a layer of complexity since generational collectors need to treat pointers from old data to new data with special care. The CertiCoq project does not include mutable data, i.e. their setting is simpler than ours in this respect.
- We describe how the generational algorithm can be verified separately from the concrete implementation. Furthermore, we show how the proof can be structured so that it follows the intuition of informal explanations of the form: a partial collection cycle in a generational collector is the same as running a full collection on part of the heap if one views pointers to old data as non-pointers.
- This paper provides more detail than any previous CakeML publication on how algorithm-level proofs can be used to write and verify concrete implementations of garbage collectors for CakeML, and how these are integrated into the full CakeML compiler and runtime. The updated in-logic bootstrapped compiler comes with new command-line arguments that allow configuration of the generational garbage collector.

2   Approach

In this section, we give a high-level overview of the work and our approach to it. Subsequent sections will cover some—but for lack of space, not all—of these topics in more detail.

Algorithm-Level Modelling and Verification:

- The intuition behind the copying garbage collection is important in order to understand this paper. Section 3.1 provides an explanation of the basic Cheney copying collector algorithm. Section 3.2 continues with how the basic algorithm can be modified to run as a generational collector. It also describes how we deal with mutable state such as ML-style references and arrays.

- Section 3.3 describes how the algorithm has been modelled as HOL functions. These algorithm-level HOL functions model memory abstractly, in particular we use HOL lists to represent heap segments. This representation neatly allows us to avoid awkward reasoning about potential overlap between memory segments. It also works well with the separation logic we use later to map the abstract heaps to their concrete memory representations, in Sect. 4.2.
- Section 3.4 defines the main correctness property, gc_related, that any garbage collector must satisfy: for every pointer traversal that exists in the original heap from some root, there must be a similar pointer traversal possible in the new heap.
- A generational collector can run either a partial collection, which collects only some part of the heap, or a full collection of the entire heap. We show that the full collection satisfies gc_related. To show that a run of the partial collector also satisfies gc_related, we exploit a simulation argument that allows us to reuse the proofs for the full collector. Intuitively, a run of the partial collector on a heap segment h simulates a run of the full collector on a heap containing only h. Section 3.4 provides some details on this.

Implementation and Integration into the CakeML Compiler:

- The CakeML compiler goes through several intermediate languages on the way from source syntax to machine code. The garbage collector is introduced gradually in the intermediate languages DATALANG (abstract data), WORD-LANG (machine words, concrete memory, but abstract stack) and STACKLANG (more concrete stack).
- The verification of the compiler phase from DATALANG to WORDLANG specifies how abstract values of DATALANG are mapped to instantiations of the heap types that the algorithm-level garbage collection operates over, Sect. 4.1. We prove that gc_related implies that from DATALANG's point of view, nothing changes when a garbage collector is run.
- For the verification of the DATALANG to WORDLANG compiler, we also specify how each instantiation of the algorithm-level heap types maps into WORD-LANG's concrete machine words and memory, Sect. 4.2. Here we implement and verify a *shallow embedding* of the garbage collection algorithm. This shallow embedding is used as a primitive by the semantics of WORDLANG.
- Further down in the compiler, the garbage collection primitive needs to be implemented by a *deep embedding* that can be compiled with the rest of the code. This happens in STACKLANG, where a compiler phase attaches an implementation of the garbage collector to the currently compiled program and replaces all occurrences of Alloc by a call to the new routine. Implementing the collector in STACKLANG is tedious because STACKLANG is very low- level—it comes after instruction selection and register allocation. However, the verification proof is relatively straight-forward since one only has to show that the STACKLANG deep embedding computes the same function as the shallow embedding mentioned above.
- Finally, the CakeML compiler's in-logic bootstrap needs updating to work with the new garbage collection algorithm. The bootstrap process itself does

not need much updating, illustrating the resilience of the bootstrapping procedure to such changes. We extend the bootstrapped compiler to recognise command-line options specifying which garbage collector is to be generated: `--gc=none` for no garbage collector; `--gc=simple` for the previous non-generational copying collector; and `--gc=gen`*size* for the generational collector described in the present paper. Here *size* is the size of the nursery generation in number of machine words. With these command-line options, users can generate a binary with a specific instance of the garbage collector installed.

Mechanised Proofs. The development was carried out in HOL4. The sources are available at http://code.cakeml.org/. The algorithm and its proofs are under `compiler/backend/gc`; the shallow embedding and its verification proof is under `compiler/backend/proofs/data_to_word_gcProofScript.sml`; the STACKLANG deep embedding is in `compiler/backend/stack_allocScript.sml`; its verification is in `compiler/backend/proofs/stack_allocProofScript.sml`.

Terminology. The *heap* is the region of memory where heap elements are allocated and which is to be garbage collected. A *heap element* is the unit of memory allocation. A heap element can contain pointers to other heap elements. The collection of all program visible variables is called the *roots*.

3   Algorithm Modelling and Verification

Garbage collectors are complicated pieces of code. As such, it makes sense to separate the reasoning about algorithm correctness from the reasoning about the details of its more concrete implementations. Such a split also makes the algorithm proofs more reusable than proofs that depend on implementation details. This section focuses on the algorithm level.

3.1   Intuition for Basic Algorithm

Intuitively, a Cheney copying garbage collector copies the live elements from the current heap into a new heap. We will call the heaps old and new. In its simplest form, the algorithm keeps track of two boundaries inside the new heap. These split the new heap into three parts, which we will call h1, h2, and unused space.

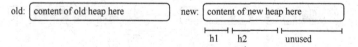

Throughout execution, the heap segment h1 will only contain pointers to the new heap, and heap segment h2 will only contain pointers to the old heap, i.e. pointers that are yet to be processed.

The algorithm's most primitive operation is to move a pointer ptr, and the data element d that ptr points at, from the old heap to the new one. The move primitive's behaviour depends on whether d is a forward pointer or not. A forward pointer is a heap element with a special tag to distinguish it from other heap elements. Forward pointers will only ever occur in the heap if the garbage collector puts them there; between collection cycles, they are never present nor created.

If d is not a forward pointer, then d will be copied to the end of heap segment h2, consuming some of the unused space, and ptr is updated to be the address of the new location of d. A forward pointer to the new location is inserted at the old location of d, namely at the original value of ptr. We draw forward pointers as hollow boxes with dashed arrows illustrating where they point. Solid arrows that are irrelevant for the example are omitted in these diagrams.

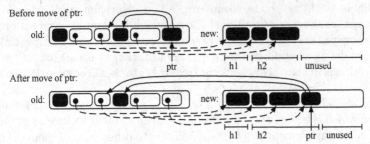

If d is already a forward pointer, the move primitive knows that this element has been moved previously; it reads the new pointer value from the forward pointer, and leaves the memory unchanged.

The algorithm starts from a state where the new heap consists of only free space. It then runs the move primitive on each pointer in the list of roots. This processing of the roots populates h2.

Once the roots have been processed, the main loop starts. The main loop picks the first heap element from h2 and applies the move primitive to each of the pointers that that heap element contains. Once the pointers have been updated, the boundary between h1 and h2 can be moved, so that the recently processed element becomes part of h1.

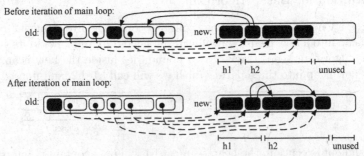

This process is repeated until h2 becomes empty, and the new heap contains no pointers to the old heap. The old heap can then be discarded, since it only contains data that is unreachable from the roots. The next time the garbage collector runs, the previous old heap is used as the new heap.

3.2 Intuition for Generational Algorithm

Generational garbage collectors attempt to run the collector only on part of the heap. The motivation is that new data tends to be short-lived while old data tends to stay live. By running the collector on new data only, one avoids copying around old data unnecessarily.

The intuition is that a partial collection focuses on a small segment of the full heap and ignores the rest, but operates as a normal full collection on this small segment.

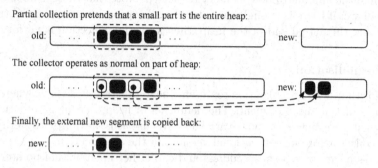

For the partial collection to work we need:

(a) the partial algorithm to treat all pointers to the outside (old data) as non-pointers, in order to avoid copying old data into its new memory region.
(b) that outside data does not point into the currently collected segment of the heap, because the partial collector should be free to move around and delete elements in the segment it is working on without looking at the heap outside.

In ML programs, most data is immutable, which means that old data cannot point at new data. However, ML programs also use references and arrays (henceforth both will be called references) that are mutable. References are usually used sparingly, but are dangerous for a generational garbage collector because they can point into the new data from old data.

Our pragmatic solution is to make sure immutable data is allocated from the bottom of the heap upwards, and references are allocated from the top downwards, i.e. the memory layout is as follows. This diagram also shows that we use a GC trigger pointer, which causes a GC invocation whenever one attempts to allocate past the GC trigger pointer.

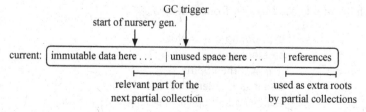

We modify the simple garbage collection algorithm described above to maintain this layout, and we make each run of the partial collection algorithm treat the references as roots that are not part of the heap. This way we can meet the two requirements (a) and (b) from above.

Our approach means that references will never be collected by a partial collection. However, they will be collected when the full collection is run.

Full collections happen if there is a possibility that the partial collector might fail to free up enough space, i.e. if the amount of unused space prior to collection is less than the amount of new memory requested. Note that there is no heuristic involved here: if there is enough space for the allocation between the GC trigger pointer and the actual end of the heap, then a partial collection is performed.

3.3   Formalisation

The algorithm-level formalisation represents heaps abstractly as lists, where each element is of type heap_element. The definition of heap_element is intentionally somwewhat abstract with type variables. We use this flexiblity to verify the partial collector for our generational version, in the next section.

Addresses are of type heap_address and can either be an actual pointer with some data attached, or a non-pointer Data. A heap element can be unused space, a forward pointer, or actual data.

$$\alpha \text{ heap\_address } = \text{ Pointer num } \alpha \mid \text{Data } \alpha$$

$$(\alpha, \beta) \text{ heap\_element } =$$
$$\quad \text{Unused num}$$
$$\quad \mid \text{ForwardPointer num } \alpha \text{ num}$$
$$\quad \mid \text{DataElement } (\alpha \text{ heap\_address list}) \text{ num } \beta$$

Each heap element carries its concrete length, i.e. how many machine words the eventual memory representation will hold. The length function, el_length, returns l plus one because we do not allow heap elements of length zero.

$$\text{el\_length (Unused } l) = l + 1$$
$$\text{el\_length (ForwardPointer } n \ d \ l) = l + 1$$
$$\text{el\_length (DataElement } xs \ l \ data) = l + 1$$

The natural number (type num in HOL) in Pointer values is an offset from the start of the relevant heap. We define a lookup function heap_lookup that fetches the content of address a from a heap xs:

$$\text{heap\_lookup } a \ [] = \text{None}$$
$$\text{heap\_lookup } a \ (x :: xs) =$$
$$\quad \text{if } a = 0 \text{ then Some } x$$
$$\quad \text{else if } a < \text{el\_length } x \text{ then None}$$
$$\quad \text{else heap\_lookup } (a - \text{el\_length } x) \ xs$$

The generational garbage collector has two main routines: gen_gc_full which runs a collection on the entire heap including the references, and gen_gc_partial which runs only on part of the heap, treating the references as extra roots. Both use the record type gc_state to represent the heaps. In a state s, the old heap is in s.heap, and the new heap comprises the following fields: s.h1 and s.h2 are the heap segments h1 and h2 from before, s.n is the length of the unused space, and s.r2, s.r1 are for references what s.h1 and s.h2 are for immutable data; s.ok is a boolean representing whether s is a well-formed state that has been arrived at through a well-behaved execution. It has no impact on the behaviour of the garbage collector; its only use is in proofs, where it serves as a convenient trick to propagate invariants downwards in refinement proofs.

Figure 1 shows the HOL function implementing the move primitive for the partial generational algorithm. It follows what was described informally in the section above: it does nothing when applied to a non-pointer, or to a pointer that points outside the current generation. When applied to a pointer to a forward pointer, it follows the forward pointer but leaves the heap unchanged. When applied to a pointer to some data element d, it inserts d at the end of h2, decrements the amount of unused space by the length of d, and inserts at the old location of d a forward pointer to its new location. When applied to an invalid pointer (i.e. to an invalid heap location, or to a location containing unused space) it does nothing except set the ok field of the resultant state to false; we prove later that this never happens.

```
gen_gc_partial_move conf state (Data d) = (Data d,state)
gen_gc_partial_move conf state (Pointer ptr d) =
  let ok = state.ok ∧ ptr < heap_length state.heap in
    if ptr < conf.gen_start ∨ conf.refs_start ≤ ptr then
      (Pointer ptr d,state with ok := ok)
    else
      case heap_lookup ptr state.heap of
        None ⇒ (Pointer ptr d,state with ok := F)
      | Some (Unused v₉) ⇒ (Pointer ptr d,state with ok := F)
      | Some (ForwardPointer ptr' v₁₁ l') ⇒ (Pointer ptr' d,state)
      | Some (DataElement xs l dd) ⇒
        let ok = ok ∧ l + 1 ≤ state.n ∧ ¬conf.isRef dd;
          n = state.n − (l + 1);
          h₂ = state.h2 ++ [DataElement xs l dd];
          (heap,ok) = write_forward_pointer ptr state.heap state.a d ok;
          a = state.a + l + 1 in
        (Pointer state.a d,
          state with ⟨h2 := h₂; n := n; a := a; heap := heap; ok := ok⟩)
```

Fig. 1. The algorithm implementation of the move primitive for gen_gc_partial.

The HOL function gen_gc_full_move implements the move primitive for the full generational collection; its definition is elided for space reasons. It is similar to gen_gc_partial_move, but differs in two main ways: first, it does not consider generation boundaries. Second, in order to maintain the memory layout it must distinguish between pointers to references and pointers to immutable data, allocating references at the end of the new heap's unused space and immutable data at the beginning. Note that gen_gc_partial_move does not need to consider pointers to references, since generations are entirely contained in the immutable part of the heap.

The algorithms for an entire collection cycle consist of several HOL functions in a similar style; the functions implementing the move primitive are the most interesting of these. The main responsibility of the others is to apply the move primitive to relevant roots and heap elements, following the informal explanations in previous sections.

3.4   Verification

For each collector (gen_gc_full and gen_gc_partial), we prove that they do not lose any live elements. We formalise this notion with the gc_related predicate shown below. If a collector can produce $heap_2$ from $heap_1$, there must be a map f such that gc_related f $heap_1$ $heap_2$. The intuition is that if there was a heap element at address a in $heap_1$ that was retained by the collector, the same heap element resides at address f a in $heap_2$.

The conjuncts of the following definition state, respectively: that f must be an injective map into the set of valid addresses in $heap_2$; that its domain must be a subset of the valid addresses into $heap_2$; and that for every data element d at address $a \in$ domain f, every address reachable from d is also in the domain of f, and f a points to a data element that is exactly d with all its pointers updated according to f. Separately, we require that the roots are in domain f.

> gc_related f $heap_1$ $heap_2$ $\iff$
>   injective (apply f) (domain f)
>   { a | isSomeDataElement (heap_lookup a $heap_2$) } $\land$
>   ($\forall i.\ i \in$ domain $f \Rightarrow$ isSomeDataElement (heap_lookup i $heap_1$)) $\land$
> $\forall i\ xs\ l\ d.$
> $i \in$ domain $f \land$ heap_lookup i $heap_1$ = Some (DataElement xs l d) $\Rightarrow$
>   heap_lookup (apply f i) $heap_2$ =
>   Some (DataElement (addr_map (apply f) xs) l d) $\land$
> $\forall ptr\ u.$ mem (Pointer ptr u) $xs \Rightarrow ptr \in$ domain f

Proving a gc_related-correctness result for gen_gc_full, as below, is a substantial task that requires a non-trivial invariant, similar to the one we presented in earlier work [10]. The main correctness theorem is as follows; we will not give further details of its proofs in this paper; for such proofs see [10].

⊢ roots_ok *roots heap* ∧ heap_ok *heap conf*.limit ⇒
  ∃ *state f*.
  gen_gc_full *conf* (*roots,heap*) = (addr_map (apply *f*) *roots,state*) ∧
  (∀ *ptr u*. mem (Pointer *ptr u*) *roots* ⇒ *ptr* ∈ domain *f*) ∧
  gc_related *f heap* (*state*.h1 ++ heap_expand *state*.n ++ *state*.r1)

The theorem above can be read as saying: if all roots are pointers to data elements in the heap (abbreviated roots_ok), if the heap has length *conf*.limit, and if all pointers in the heap are valid non-forward pointers back into the heap (abbreviated heap_ok), then a call to gen_gc_full results in a state that is gc_related via a mapping *f* whose domain includes the roots (and hence, by definition of gc_related, all live elements).

The more interesting part is the verification of gen_gc_partial, which we conduct by drawing a formal analogy between how gen_gc_full operates and how gen_gc_partial operates on a small piece of the heap. The proof is structured in two steps:

1. we first prove a simulation result: running gen_gc_partial is the same as running gen_gc_full on a state that has been modified to pretend that part of the heap is not there and the references are extra roots.
2. we then show a gc_related result for gen_gc_partial by carrying over the same result for gen_gc_full via the simulation result.

For the simulation result, we instantiate the type variables in the gen_gc_full algorithm so that we can embed pointers into Data blocks. The idea is that encoding pointers to locations outside the current generation as Data causes gen_gc_full to treat them as non-pointers, mimicking the fact that gen_gc_partial does not collect there.

The type we use for this purpose is defined as follows:

$$(\alpha, \beta) \text{ data\_sort} = \text{Protected } \alpha \mid \text{Real } \beta$$

and the translation from gen_gc_partial's pointers to pointers on the pretend-heap used by gen_gc_full in the simulation argument is:

to_gen_heap_address *conf* (Data *a*) = Data (Real *a*)
to_gen_heap_address *conf* (Pointer *ptr a*) =
  if *ptr* < *conf*.gen_start then Data (Protected (Pointer *ptr a*))
  else if *conf*.refs_start ≤ *ptr* then Data (Protected (Pointer *ptr a*))
  else Pointer (*ptr* − *conf*.gen_start) (Real *a*)

Similar to_gen functions, elided here, encode the roots, heap, state and configuration for a run of gen_gc_partial into those for a run of gen_gc_full. We prove that for every execution of gen_gc_partial starting from an ok state, and the corresponding execution of gen_gc_full starting from the encoding of the same state through the to_gen functions, encoding the results of the former with to_gen yields precisely the results of the latter.

Initially, we made an attempt to do the gc_related proof for gen_gc_partial using the obvious route of manually adapting all loop invariants and proofs for gen_gc_full into invariants and proofs for gen_gc_partial. This soon turned out to overly cumbersome; hence we switched to the current approach because it seemed more expedient and more interesting. As a result, the proofs for gen_gc_partial are more concerned with syntactic properties of the encoding than with semantic properties of the collector as such. The syntactic arguments are occasionally quite tedious, but we believe this approach still leads to more understandable and less repetitive proofs.

Finally, note that gc_related is the same correctness property that we use for the previous copying collector; this makes it straightforward to prove that the top-level correctness theorem of the CakeML compiler remains true if we swap out the garbage collector.

3.5   Combining the Partial and Full Collectors

An implementation that uses the generational collector will mostly run the partial collector and occasionally the full one. At the algorithm level, we define a combined collector and leave it up to the implementation to decide when a partial collection is to be run. The choice is made visible to the implementation by having a boolean input do_partial to the combined function. The combined function will produce a valid heap regardless of the value of do_partial.

Our CakeML implementation (next section) runs a partial collection if the allocation will succeed even if the collector does not manage to free up any space, i.e., if there is already enough space on the other side of the GC trigger pointer before the GC starts (Sect. 3.2).

4   Implementation and Integration into CakeML Compiler

The concept of garbage collection is introduced in the CakeML compiler at the point where a language with unbounded memory (DATALANG) is compiled into a language with a concrete finite memory (WORDLANG). Here the garbage collector's role is to automate memory deallocation and to implement the illusion of an unbounded memory.

This section sketches how the collector algorithm's types get instantiated, how the data refinement is specified, and how an implementation of the garbage collector algorithm is verified.

4.1   Instantiating the Algorithm's Types

The language which comes immediately prior to the introduction of the garbage collector, DATALANG, stores values of type v in its variables.

$$v \ = \ \text{Number int} \mid \text{Word64 (64 word)} \mid \text{Block num (v list)}$$
$$\mid \text{CodePtr num} \mid \text{RefPtr num}$$

DATALANG gets compiled into a language called WORDLANG where memory is finite and variables are of type word_loc. A word_loc is either a machine word Word w, or a code location Loc l_1 l_2.

$$\alpha \ \text{word\_loc} \ = \ \text{Word} \ (\alpha \ \text{word}) \mid \text{Loc num num}$$

In what follows we will show through an example how an instance of v is represented. We would have liked to provide more detail, but the definitions involved are simply too verbose to be included here. We will use the following DATALANG value as our running example.

Block 3 [Number 5; Number 80000000000000]

The relation v_inv specifies how values of type v relate to the heap_addresses and heaps that the garbage collection algorithms operate on. Below is the Number case from the definition of v_inv. If integer i is small enough to fit into a tagged machine word, then the head address x must be Data that carries the value of the small integer, and there is no requirement on the heap. If integer i is too large to fit into a machine word, then the heap address must be a Pointer to a heap location containing the data for the bignum representing integer i.

v_inv $conf$ (Number i) $(x,f,heap)$ $\iff$
  if small_int $(: \alpha)$ i then $x = $ Data (Word (Smallnum i))
  else
   $\exists\, ptr.$
    $x = $ Pointer ptr (Word $0w$) $\wedge$
    heap_lookup ptr $heap$ $-$ Some (Bignum i)

Bignum $i = $
  let $(sign,payload)$ $= $ sign_and_words_of_integer i
  in
   DataElement [] (length $payload$) (NumTag $sign$,map Word $payload$)

In the definition of v_inv, f is a finite map that specifies how semantic location values for reference pointers (RefPtr) are to be represented as addresses.

v_inv $conf$ (RefPtr n) $(x,f,heap)$ $\iff$
  $x = $ Pointer (apply f n) (Word $0w$) $\wedge$ $n \in $ domain f

The Block case below shows how constructors and tuples, Blocks, are represented.

$\text{v\_inv } conf \text{ (Block } n \text{ } vs) \text{ } (x, f, heap) \iff$
 $\quad \text{if } vs = [] \text{ then}$
 $\qquad x = \text{Data (Word (BlockNil } n)) \land n < \text{dimword } (: \alpha) \text{ div } 16$
 $\quad \text{else}$
 $\qquad \exists \, ptr \text{ } xs.$
 $\qquad \text{list\_rel } (\lambda \text{ } v \text{ } x'. \text{ v\_inv } conf \text{ } v \text{ } (x', f, heap)) \text{ } vs \text{ } xs \land$
 $\qquad x = \text{Pointer } ptr \text{ (Word (ptr\_bits } conf \text{ } n \text{ (length } xs))) \land$
 $\qquad \text{heap\_lookup } ptr \text{ } heap = \text{Some (BlockRep } n \text{ } xs)$

When v_inv is expanded for the case of our running example, we get the following constraint on the heap. The address x must be a pointer to a DataElement which contains Data representing integer 5, and a pointer to some memory location which contains the machine words representing bignum 80000000000000. Here we assume that the architecture has 32-bit machine words. Below one can see that the first Pointer is given information, ptr_bits $conf$ 3 2, about the length, 2, and tag, 3, of the Block that it points to. Such information is used to speed up pattern matching. If the information fits into the lower bits of the pointer, then the pattern matcher does not need to follow the pointer to know whether there is a match.

$\vdash \text{v\_inv } conf \text{ (Block 3 [Number 5; Number 80000000000000]) } (x, f, heap) \iff$
 $\quad \exists \, ptr_1 \text{ } ptr_2.$
 $\quad x = \text{Pointer } ptr_1 \text{ (Word (ptr\_bits } conf \text{ 3 2))} \land$
 $\quad \text{heap\_lookup } ptr_1 \text{ } heap =$
 $\quad \text{Some}$
 $\qquad \text{(DataElement [Data (Word (Smallnum 5)); Pointer } ptr_2 \text{ (Word } 0w)] \text{ 2}$
 $\qquad \text{(BlockTag 3, [])) } \land$
 $\quad \text{heap\_lookup } ptr_2 \text{ } heap = \text{Some (Bignum 80000000000000)}$

The following is an instantiation of *heap* that satisfies the constraint set out by v_inv for representing our running example.

$\vdash \text{v\_inv } conf \text{ (Block 3 [Number 5; Number 80000000000000])}$
 $\quad \text{(Pointer 0 (Word (ptr\_bits } conf \text{ 3 2)), } f,$
 $\quad \text{[DataElement [Data (Word (Smallnum 5)); Pointer 3 (Word } 0w)] \text{ 2}$
 $\quad \text{(BlockTag 3, []); Bignum 80000000000000])}$

As we know, the garbage collector moves heap elements and changes the addresses. However, it will only transform heaps in a way that respects gc_related. We prove that v_inv properties can be transported from one heap to another if they are gc_related. In other words, execution of a garbage collector does not interfere with this data representation.

$\vdash \text{gc\_related } g \text{ } heap_1 \text{ } heap_2 \land (\forall \, ptr \text{ } u. \text{ } x = \text{Pointer } ptr \text{ } u \Rightarrow ptr \in \text{domain } g) \land$
 $\quad \text{v\_inv } conf \text{ } w \text{ } (x, f, heap_1) \Rightarrow$
 $\quad \text{v\_inv } conf \text{ } w \text{ (addr\_apply (apply } g) \text{ } x, g \circ f, heap_2)$

Here addr_apply f (Pointer x d) = Pointer (f x) d.

4.2   Data Refinement down to Concrete Memory

The relation provided by v_inv only gets us halfway down to WORDLANG's memory representation. In WORDLANG, values are of type word_loc, and memory is modelled as a function, α word $\rightarrow \alpha$ word_loc, and an address domain set.

We use separation-logic formulas to specify how lists of heap_elements are represented in memory. We define separating conjunction ∗, and use fun2set to turn the memory function m and its domain set dm into something we can write separation logic assertions about. The relevant definitions are:

$$\vdash \text{split } s \ (u,v) \iff u \cup v = s \land u \cap v = \emptyset$$
$$\vdash p \ast q = (\lambda s. \ \exists u \ v. \ \text{split } s \ (u,v) \land p \ u \land q \ v)$$
$$\vdash a \mapsto x = (\lambda s. \ s = \{ \ (a,x) \ \})$$
$$\vdash \text{fun2set } (m,dm) = \{ \ (a, m \ a) \mid a \in dm \ \}$$

Using these, we define word_heap a *heaf conf* to assert that a heap_element list *heap* is in memory, starting at address a, and word_el asserts the same thing about individual heap_elements. Figure 2 shows an expansion of the word_heap assertion applied to our running example.

word_heap a
  [DataElement [Data (Word (Smallnum 5)); Pointer 3 (Word 0w)] 2
    (BlockTag 3,[]); Bignum 80000000000000] *conf* (fun2set (m,dm)))
⟺
(word_el a
    (DataElement [Data (Word (Smallnum 5)); Pointer 3 (Word 0w)] 2
      (BlockTag 3,[])) *conf* ∗
  word_el $(a + 12w)$ (Bignum 80000000000000) *conf*) (fun2set (m,dm)))
⟺
$(a \mapsto$ (Word (make_header *conf* 12w 2)) ∗
$(a + 4w) \mapsto$ (word_addr *conf* (Data (Word (Smallnum 5)))) ∗
$(a + 8w) \mapsto$ (word_addr *conf* (Pointer 3 (Word 0w))) ∗
$(a + 12w) \mapsto$ (Word (make_header *conf* 3w 2)) ∗
$(a + 16w) \mapsto$ (Word 1939144704w) ∗ $(a + 20w) \mapsto$ (Word 18626w))
(fun2set (m,dm)))
⟺
$(a \mapsto$ (Word (make_header *conf* 12w 2)) ∗ $(a + 4w) \mapsto$ (Word 20w) ∗
$(a + 8w) \mapsto$ (Word (get_addr *conf* 3 (Word 0w))) ∗
$(a + 12w) \mapsto$ (Word (make_header *conf* 3w 2)) ∗
$(a + 16w) \mapsto$ (Word 1939144704w) ∗ $(a + 20w) \mapsto$ (Word 18626w))
⟺
$m \ a =$ Word (make_header *conf* 12w 2) $\land m \ (a + 4w) =$ Word 20w $\land$
$m \ (a + 8w) =$ Word (get_addr *conf* 3 (Word 0w)) $\land$
$m \ (a + 12w) =$ Word (make_header *conf* 3w 2) $\land$
$m \ (a + 16w) =$ Word 1939144704w $\land m \ (a + 20w) =$ Word 18626w $\land$
$dm = \{ \ a; \ a + 4w; \ a + 8w; \ a + 12w; \ a + 16w; \ a + 20w \ \} \land$
all_distinct $[a; \ a + 4w; \ a + 8w; \ a + 12w; \ a + 16w; \ a + 20w]$

Fig. 2. Running example expanded to concrete memory assertion

4.3 Implementing the Garbage Collector

The garbage collector is used in the WORDLANG semantics as a function that the semantics of Alloc applies to memory when the allocation primitive runs out of memory. At this level, the garbage collector is essentially a function from a list of roots and a concrete memory to a new list of roots and concrete memory.

To implement the new garbage collector, we define a HOL function at the level of a concrete memory, and prove that it correctly mimics the operations performed by the algorithm-level implementation from Sect. 3. The following is an excerpt of the theorem relating gen_gc_partial_move with its refinement word_gen_gc_partial_move. This states that the concrete memory is kept faithful to the algorithm's operations over the heaps. We prove similar theorems about the other components of the garbage collectors.

$\vdash$ gen_gc_partial_move gc_conf s $x = (x_1, s_1)$ $\land$
     word_gen_gc_partial_move $conf$ (word_addr $conf$ $x, \ldots$) $= (w, \ldots)$ $\land \ldots \land$
     (word_heap a s.heap $conf$ $*$ word_heap p s.h2 $conf$ $* \ldots$) (fun2set (m, dm)) $\Rightarrow$
          $w =$ word_addr $conf$ x_1 $\land \ldots \land$
          (word_heap a s_1.heap $conf$ $*$ word_heap p_1 s_1.h2 $conf$ $* \ldots$) (fun2set (m_1, dm))

5 Discussion of Related Work

Anand et al. [1] reports that the CertiCoq project has a "high-performance generational garbage collector" and a project is underway to verify this using Verifiable C in Coq. Their setting is simpler than ours in that their programs are purely functional, i.e. they can avoid dealing with the added complexity of mutable state. The text also suggests that their garbage collector is specific to a fixed data representation. In contrast, the CakeML compiler allows a highly configurable data representation, which is likely to become more configurable in the future. The CakeML compiler generates a new garbage collector implementation for each configuration of the data representation.

CakeML's original non-generational copying collector has its origin in the verified collector described in Myreen [10]. The same verified algorithm was used for a verified Lisp implementation [11] which in turn was used underneath the proved-to-be-sound Milawa prover [2]. These Lisp and ML implementations are amongst the very few systems that use verified garbage collectors as mere components of much larger verified implementations. Verve OS [16] and Ironclad Apps [7] are verified stacks that use verified garbage collectors internally.

Numerous abstract garbage collector algorithms have been mechanically verified before. However, most of these only verify the correctness at the algorithm-level implementation and only consider mark-and-sweep algorithms. Noteworthy exceptions include Hawblitzel and Petrank [8] and McCreight [9]; recent work by Gammie et al. [4] is also particularly impressive.

Hawblitzel and Petrank [8] show that performant verified x86 code for simple mark-and-sweep and Cheney copying collectors can be developed using the Boogie verification condition generator and the Z3 automated theorem prover.

Their method requires the user to write extensive annotations in the code to be verified. These annotations are automatically checked by the tools. Their collector implementations are realistic enough to show good results on off-the-shelf C# benchmarks. This required them to support complicated features such as interior pointers, which CakeML's collector does not support. We decided to not support interior pointers in CakeML because they are not strictly needed and they would make the inner loop of the collector a bit more complicated, which would probably cause the inner loop to run a little slower.

McCreight [9] verifies copying and incremental collectors implemented in MIPS-like assembly. The development is done in Coq, and casts his verification efforts in a common framework based on ADTs that all the collectors refine.

Gammie et al. [4] verify a detailed model of a state-of-the-art concurrent collector in Isabelle/HOL, with respect to an x86-TSO memory model.

Pavlovic et al. [13] focus on an earlier step, namely the synthesis of concurrent collection algorithms from abstract specifications. The algorithms thus obtained are at a similar level of abstraction to the algorithm-level implementation we start from. The specifications are cast in lattice-theoretic terms, so e.g. computing the set of live nodes is fixpoint iteration over a function that follows pointers from an element. A main contribution is an adaptation of the classic fixpoint theorems to a setting where the monotone function under consideration may change, which can be thought of as representing interference by mutators.

This paper started by listing incremental, generational, and concurrent as variations on the basic garbage collection algorithms. There have been prior verifications of incremental algorithms (e.g. [6,9,12,14]) and concurrent ones (e.g. [3–5,13]), but we believe that this paper is the first to report on a successful verification of a generational garbage collector.

6   Summary

This paper describes how a generational copying garbage collector has been proved correct and integrated into the verified CakeML compiler. The algorithm-level part of the proof is structured to follow the usual informal argument for a generational collector's correctness: a partial collection is the same as running a full collection on part of the heap if pointers to old data are treated as non-pointers. To the best of our knowledge, this paper is the first to report on a completed formal verification of a generational garbage collector.

What We Did Not Do. The current implementation lacks support for (a) nested nursery generations, and (b) the ability to switch garbage collector mode (e.g. from non-generational to generational, or adjust the size of the nursery) midway through execution of the application program. We expect both extensions to fit within the approach taken in this paper and neither to require modification of the algorithm-level proofs. For (a), one would keep track of multiple nursery starting points in the immutable part of the heap. These parts are left untouched by collections of the inner nursery generations. For (b), one could run a full generational collection to introduce the special heap layout when necessary. This is possible since the correctness theorem for gen_gc_full does not assume that the references are at the top end of the heap when it starts.

Acknowledgements. We thank Ramana Kumar for comments on drafts of this text. This work was partly supported by the Swedish Research Council and the Swedish Foundation for Strategic Research.

References

1. Anand, A., Appel, A., Morrisett, G., Paraskevopoulou, Z., Pollack, R., Belanger, O.S., Sozeau, M., Weaver, M.: CertiCoq: a verified compiler for Coq. In: Coq for Programming Languages (CoqPL) (2017)
2. Davis, J., Myreen, M.O.: The reflective Milawa theorem prover is sound (down to the machine code that runs it). J. Autom. Reason. **55**(2), 117–183 (2015)
3. Dijkstra, E.W., Lamport, L., Martin, A.J., Scholten, C.S., Steffens, E.F.M.: On-the-fly garbage collection: an exercise in cooperation. Commun. ACM **21**(11), 966–975 (1978)
4. Gammie, P., Hosking, A.L., Engelhardt, K.: Relaxing safely: verified on-the-fly garbage collection for x86-TSO. In: Grove, D., Blackburn, S. (eds.) Programming Language Design and Implementation (PLDI). ACM (2015)
5. Gonthier, G.: Verifying the safety of a practical concurrent garbage collector. In: Alur, R., Henzinger, T.A. (eds.) CAV 1996. LNCS, vol. 1102, pp. 462–465. Springer, Heidelberg (1996). doi:10.1007/3-540-61474-5_103
6. Havelund, K.: Mechanical verification of a garbage collector. In: Rolim, J., et al. (eds.) IPPS 1999. LNCS, vol. 1586, pp. 1258–1283. Springer, Heidelberg (1999). doi:10.1007/BFb0098007
7. Hawblitzel, C., Howell, J., Lorch, J.R., Narayan, A., Parno, B., Zhang, D., Zill, B.: Ironclad apps: end-to-end security via automated full-system verification. In: Operating Systems Design and Implementation (OSDI), pp. 165–181. USENIX Association, Broomfield (2014)
8. Hawblitzel, C., Petrank, E.: Automated verification of practical garbage collectors. In: ACM SIGPLAN Notices, vol. 44, no. 1, pp. 441–453 (2009). http://dl.acm.org/citation.cfm?id=1480935
9. McCreight, A.: The Mechanized Verification of Garbage Collector Implementations. Ph.D. thesis, Yale University, December 2008
10. Myreen, M.O.: Reusable verification of a copying collector. In: Leavens, G.T., O'Hearn, P., Rajamani, S.K. (eds.) VSTTE 2010. LNCS, vol. 6217, pp. 142–156. Springer, Heidelberg (2010). doi:10.1007/978-3-642-15057-9_10
11. Myreen, M.O., Davis, J.: A verified runtime for a verified theorem prover. In: Eekelen, M., Geuvers, H., Schmaltz, J., Wiedijk, F. (eds.) ITP 2011. LNCS, vol. 6898, pp. 265–280. Springer, Heidelberg (2011). doi:10.1007/978-3-642-22863-6_20
12. Nieto, L.P., Esparza, J.: Verifying single and multi-mutator garbage collectors with Owicki-Gries in Isabelle/HOL. In: Nielsen, M., Rovan, B. (eds.) MFCS 2000. LNCS, vol. 1893, pp. 619–628. Springer, Heidelberg (2000). doi:10.1007/3-540-44612-5_57
13. Pavlovic, D., Pepper, P., Smith, D.R.: Formal derivation of concurrent garbage collectors. In: Bolduc, C., Desharnais, J., Ktari, B. (eds.) MPC 2010. LNCS, vol. 6120, pp. 353–376. Springer, Heidelberg (2010). doi:10.1007/978-3-642-13321-3_20

14. Russinoff, D.M.: A mechanically verified incremental garbage collector. Formal Aspects Comput. **6**(4), 359–390 (1994)
15. Tan, Y.K., Myreen, M.O., Kumar, R., Fox, A., Owens, S., Norrish, M.: A new verified compiler backend for CakeML. In: Garrigue, J., Keller, G., Sumii, E. (eds.) International Conference on Functional Programming (ICFP). ACM (2016)
16. Yang, J., Hawblitzel, C.: Safe to the last instruction: automated verification of a type-safe operating system. In: Programming Language Design and Implementation (PLDI), pp. 99–110. ACM, New York (2010)

A Formalisation of Consistent Consequence for Boolean Equation Systems

Myrthe van Delft[1], Herman Geuvers[2,3], and Tim A.C. Willemse[3(✉)]

[1] Fortiss, München, Germany
mecvandelft@gmail.com
[2] Radboud University Nijmegen, Nijmegen, The Netherlands
herman@cs.ru.nl
[3] Eindhoven University of Technology, Eindhoven, The Netherlands
T.A.C.Willemse@TUe.nl

Abstract. Boolean equation systems are sequences of least and greatest fixpoint equations interpreted over the Boolean lattice. Such equation systems arise naturally in verification problems such as the modal μ-calculus model checking problem. Solving a Boolean equation system is a computationally challenging problem, and for this reason, *abstraction* techniques for Boolean equation systems have been developed. The notion of *consistent consequence* on Boolean equation systems was introduced to more effectively reason about such abstraction techniques. Prior work on consistent consequence claimed that this notion can be fully characterised by a sound and complete derivation system, building on rules for logical consequence. Our formalisation of the theory of consistent consequence and the derivation system in the proof assistant Coq reveals that the system is, nonetheless, unsound. We propose a fix for the derivation system and show that the resulting system (system **CC**) is indeed sound and complete for consistent consequence. Our formalisation of the consistent consequence theory furthermore points at a subtle mistake in the phrasing of its main theorem, and how to correct this.

1 Introduction

The model checking problem for the modal μ-calculus, a typical problem in software and hardware verification, is polynomial-time equivalent to solving a Boolean equation system [10]. Indeed, several state-of-the-art tool sets, such as mCRL2 [3] and CADP [6], solve their verification problems by transforming these to solving such equation systems. A Boolean equation system is essentially a sequence of equations of the form $\sigma X = f$, where f is a propositional formula, and where one is interested in either the least (if $\sigma = \mu$) or greatest (if $\sigma = \nu$) solution to X that is logically equivalent to f.

M. van Delft—Partially funded by the European Union's Horizon 2020 Framework Programme for Research and Innovation under grant agreement no. 674875.

© Springer International Publishing AG 2017
M. Ayala-Rincón and C.A. Muñoz (Eds.): ITP 2017, LNCS 10499, pp. 462–478, 2017.
DOI: 10.1007/978-3-319-66107-0_29

The complexity of computing the solution to a Boolean equation system depends on the alternations of the fixpoint symbols μ and ν, the mutual dependency of equations and the size of the equation system. Boolean equation systems resulting from model checking problems often have reasonably low degrees of alternations between fixpoint symbols so, in practice, the major factor is the size of the equation system. These observations have led to the development of several forms of *abstraction* that help reduce the size of a given Boolean equation system while preserving knowledge about the solution of the original equation system. The latter can be quite involved due to the intricacies in the semantics of a Boolean equation system. For this reason [7] introduces the notion of *consistent consequence*, generalising the notion of consistent correlation [18]; consistent consequence is further studied in [2].

A key observation in [2, 7] is that abstraction can often be understood as a way to build an argument for a statement that is stronger than needed for what one needs to prove. While in a purely propositional logic setting this is somehow captured by *modus ponens*, in an equation system setting, this is captured by consistent consequence. In this context, if one can show that a propositional variable Y is a consistent consequence of propositional variable X, and X has solution tt, then so has Y. In [2, 8, 18] it is shown that various forms of abstraction yield, by construction, a consistent consequence between a 'concrete' Boolean equation system and its (smaller) abstraction.

To better understand consistent consequence and illustrate that it is a natural generalisation of logical consequence to the setting of Boolean equation systems, in [7] a derivation system for logical consequence for propositional formulae in positive form was extended by two derivation rules. The resulting derivation system was claimed to soundly and completely characterise consistent consequence. Unfortunately, as we show in this paper (see our Example 28), the soundness claim of the derivation system is not valid due to an unsound derivation rule. Moreover, one of the main theorems of [7] turns out to be ill-phrased as we show using a small example (see our Example 23) that contradicts the theorem.

Contributions. First, we show that by removing the unsound rule we still obtain a complete derivation system, which we call system **CC**. This essentially restores the claim of correctness of the derivation system postulated in [7]. Second, we adjust the definition of consistent consequence in such a way that the main theorem of [7] holds. Third, we provide a formalisation of the theory of Boolean equation systems, consistent consequence and its main theorem, the derivation system and the proofs of soundness and completeness in the proof assistant Coq [14].

Related Work. Boolean equation systems are instances of the more general *fixpoint equation systems*, in which equations are interpreted over arbitrary complete lattices rather than the Boolean lattice. A formalisation in PVS of various theorems and lemmata for fixpoint equation systems, including those found in [10], is described in [15]. We note that this formalisation does not include the notion of consistent consequence, nor its proof system. Boolean equation systems are also intimately tied to *parity games*. These are two-player games played on coloured, directed graphs. One of the main results for such games is that they are positionally determined; a formalised proof thereof, in Isabelle/HOL, can be found in [4]. In [12], an algorithm for the model checking problem for the modal μ-calculus is formalised in Coq. More general considerations concerning the formalisation of fixpoint theorems in Coq include, *e.g.* [1].

Structure. In Sect. 2 we briefly reiterate standard mathematical results about fixpoints and complete lattices. Section 3 recalls Boolean equation systems and Sect. 4 introduces the notion of consistent consequence. The (corrected) derivation system for consistent consequence, along with its formalisation in Coq and the proofs of soundness and completeness, is discussed in Sect. 5. We finish with a brief outlook on future work in Sect. 6. The Coq code accompanying this paper was developed in Coq version `CoqIDE 8.5pl2`. For more detailed descriptions of the proofs we refer to [16] and the formalisation in Coq, see [17].

2   Preliminaries

A *poset* $(A, \leq)$, is a set A paired with a binary relation $\leq \subseteq A \times A$ that is reflexive, antisymmetric and transitive. If all subsets of a given poset $(A, \leq)$ have both a *supremum* (least upper bound) and an *infimum* (greatest lower bound), then $(A, \leq)$ is a *complete lattice*.

For the remainder of this section, we fix some arbitrary complete lattice $(A, \leq)$. We denote the supremum of a set $A' \subseteq A$ by $\bigsqcup A'$; its infimum is denoted $\bigsqcap A'$. The element $A_\perp$ is an abbreviation of $\bigsqcup \emptyset$ whereas $A_\top$ abbreviates $\bigsqcap \emptyset$. Let $f : A \to A$ be an arbitrary endofunction on A. A *fixpoint* of f is an element $a \in A$ such that $f(a) = a$. The least fixpoint of f, denoted by μf, is the fixpoint of f such that, for all other fixpoints a of f, $\mu f \leq a$. Dually, the *greatest fixpoint* of f, written νf, is the fixpoint of f such that, for all other fixpoints a of f, $a \leq \nu f$. Furthermore, we say that f is *monotone* iff for all $a \leq b$, also $f(a) \leq f(b)$. By Tarski's Theorem [13], monotone endofunctions on a complete lattice are guaranteed to have least and greatest fixpoints; these are given by $\mu f = \bigsqcap \{a \in A \mid f(a) \leq a\}$ and $\nu f = \bigsqcup \{a \in A \mid f(a) \geq a\}$. Furthermore, in a complete lattice, the least and greatest fixpoints of f can be obtained using a transfinite approximation, see also, *e.g.* [11].

Definition 1. *For ordinal α, limit ordinal λ and $\sigma \in \{\mu, \nu\}$, the approximant $\sigma^\alpha f$ of f is defined by induction:*

$$\sigma^0 f = A_\sigma \qquad\qquad \mu^\lambda f = f(\bigsqcup_{\alpha < \lambda} \sigma^\alpha f)$$

$$\sigma^{\alpha+1} f = f(\sigma^\alpha f) \qquad\qquad \nu^\lambda f = f(\bigsqcap_{\alpha < \lambda} \sigma^\alpha f)$$

Lemma 2. *If $(A, \leq)$ is a complete lattice then there is some α such that $\mu f = \mu^\alpha f$; likewise, there is some α such that $\nu f = \nu^\alpha f$.*

If $(A, \leq)$ is a complete lattice and B is an arbitrary non-empty set, then the set of functions from B to A, denoted by A^B, together with the ordering $f \sqsubseteq g$, defined as $f(b) \leq g(b)$ for all $b \in B$, is a complete lattice.

3   Boolean Equation Systems

Boolean equation systems (BESs) are essentially finite sequences of least and greatest fixpoint equations of the form $\sigma X = f$. Each right-hand side of an equation is a propositional formula in positive form and each left-hand side consists of a fixpoint sign $\sigma \in \{\mu, \nu\}$ and a propositional variable X taken from a countable set $\mathcal{X}$ of propositional variables. For a thorough exposition on the theory of Boolean equation systems, we refer to [10]; we here recall only the concepts and results needed for understanding the core results of the paper.

Definition 3. *In Coq:* `propForm`.
The set of propositional formulae is given through the following grammar:

$$f ::= \top \mid \bot \mid X \in \mathcal{X} \mid f \vee f \mid f \wedge f$$

The semantics of a propositional formula is given in the context of an environment. An *environment* is a mapping from the set of propositional variables $\mathcal{X}$ to the set of Booleans $\mathbb{B} = \{ff, tt\}$. Note that the set $(\mathbb{B}, \leq)$, with $ff \leq tt$, is a complete lattice. We typically use symbols η, θ to denote environments. We write $\eta \leq \theta$ iff for all $X \in \mathcal{X}$ we have $\eta(X) \leq \theta(X)$. The supremum of η and θ, denoted $\eta + \theta$ is then the pointwise disjunction of η and θ, *viz.* $(\eta + \theta)(X) = \eta(X) \sqcup \theta(X)$. We define the least environment $\theta_\bot$ as $\theta_\bot(X) = ff$ for all X and, dually, the largest environment $\theta_\top$ as $\theta_\top(X) = tt$ for all X.

Definition 4. *In Coq:* `propForm_solution`.
Let $\theta : \mathcal{X} \to \mathbb{B}$ be an arbitrary environment. The *semantics* of a propositional formula f is defined inductively as follows:

$$[\![\top]\!]\theta = tt \qquad\qquad [\![X]\!]\theta = \theta(X) \qquad\qquad [\![\bot]\!]\theta = ff$$
$$[\![f \wedge g]\!]\theta = [\![f]\!]\theta \sqcap [\![g]\!]\theta \qquad\qquad\qquad [\![f \vee g]\!]\theta = [\![f]\!]\theta \sqcup [\![g]\!]\theta$$

Note that the set of operators on propositional formulae excludes implication and negation for reasons that will become apparent when we introduce the semantics of Boolean equation systems. Nonetheless, the theory of consistent consequence put forward in [7] revolves around a notion of logical consequence among variables in a Boolean equation system. Logical consequence and equivalence between propositional formulae is defined as follows.

Definition 5. *In Coq:* `propForm_cons`.
Let f, g be arbitrary propositional formulae. We say that f implies g, written $f \Rightarrow g$, iff for all environments θ we have $[\![f]\!]\theta \leq [\![g]\!]\theta$. We say that f is equivalent to g, denoted $f \Leftrightarrow g$, if both $f \Rightarrow g$ and $g \Rightarrow f$.

We here follow [5] in our treatment of Boolean equation systems; *i.e.* a Boolean equation system is represented as a sequence of blocks (which are sequences of Boolean equations), where each block is paired with a fixpoint symbol $\sigma \in \{\mu, \nu\}$. Formally, a *Boolean equation* is an equation of the form $(X = f)$, where X is a propositional variable and f is a propositional formula; a *block* is then a non-empty sequence $\langle X_i = f_{X_i} \rangle_{i=1}^n$ of Boolean equations.

Definition 6. *In Coq:* BES.
A Boolean equation system $\mathcal{E}$ is a sequence of fixpoint symbol and block pairs generated by the following grammar:

$$\mathcal{E} ::= \mathcal{E}_\mu \mid \mathcal{E}_\nu \qquad \mathcal{E}_\mu ::= \epsilon \mid (\mu B)\, \mathcal{E}_\nu \qquad \mathcal{E}_\nu ::= \epsilon \mid (\nu B)\, \mathcal{E}_\mu$$

where ϵ denotes the empty list and B is a block.

Given a block B, we write $\mathbf{bnd}_B$ for the set of propositional variables occurring at the left-hand sides in the equations of B; if the context is clear, we simply write $\mathbf{bnd}$. If for each $X \in \mathbf{bnd}_B$ there is exactly one equation in B with X as its left-hand side, we say that B is *well-formed* and we write f_X to refer to the right-hand side propositional formula belonging to the equation for X. We lift the definitions of $\mathbf{bnd}_B$, f_X and well-formed from blocks to Boolean equation systems in the natural way. Throughout this paper, we only consider well-formed blocks and BESs. A BES $\mathcal{E}$ is *closed* if all right-hand sides of its equations refer only to constants or variables taken from $\mathbf{bnd}_\mathcal{E}$.

Definition 7. *In Coq:* rank.
Let $\mathcal{E}$ be a BES. Variables X, Y have equal *rank*, written $X \sim_\mathcal{E} Y$, iff either $X, Y \notin \mathbf{bnd}_\mathcal{E}$, or there exists a block B in $\mathcal{E}$ such that $X, Y \in \mathbf{bnd}_B$.

Example 8. *An example of a closed Boolean equation system $\mathcal{F}$, consisting of two blocks, is given below:*

$$\nu \langle (X_0 = X_1 \vee X_4)\ (X_1 = X_2 \vee X_5)\ (X_2 = X_0 \vee X_6)\ (X_3 = X_7) \rangle$$
$$\mu \langle (X_4 = X_0)\ (X_5 = X_0 \wedge X_2)\ (X_6 = X_1)\ (X_7 = X_7 \wedge X_3) \rangle$$

Note that we have $X_0 \sim X_1 \sim X_2 \sim X_3$ and $X_4 \sim X_5 \sim X_6 \sim X_7$.   □

Each block B in a BES induces a monotone operator on the set of environments; the semantics of a BES is then defined inductively on the structure of the BES using the monotone operator induced by the blocks of the BES.

Definition 9. *In Coq:* unfold_block.
Let $B = \langle X_i = f_{X_i} \rangle_{i=1}^n$ be a block of Boolean equations. Block B induces an operator $\|B\|$ on environments θ as follows:

$$\|B\|\theta = \theta[\langle X_i := [\![f_{X_i}]\!]\theta \rangle_{i=1}^n]$$

where $\theta[\langle X_i := b_i \rangle_{i=1}^n]$ is the environment that assigns b_i to X_i for $i \in [1, \ldots, n]$ and $\theta(Y)$ to all $Y \notin \mathbf{bnd}_B$.

We define the operator $\|B\|^n$ inductively as $(\|B\|^0)\theta = \theta$ and $(\|B\|^{n+1})(\theta) = \|B\|((\|B\|^n)\theta)$; *i.e.* $\|B\|^n$ denotes applying operator $\|B\|$ n-times to an environment.

Lemma 10. *In Coq:* `unfold_block_monotone`.
Let B be a block of Boolean equations. Operator $\|B\|$ is a monotone operator on the complete lattice $(\mathbb{B}^{\mathcal{X}}, \leq)$ of environments.

Consequently, operator $\|B\|$ has a least and a greatest fixpoint. Observe that the lattice $(\mathbb{B}^{\mathcal{X}}, \leq)$ is infinite, so it is not obvious that the least and greatest fixpoint of this operator can be computed. However, it follows from some simple observations that we can compute least and greatest fixpoints of this operator using a finite approximation. This result essentially follows from the fact that $\|B\|$ is monotone and the value of each of the (fixed number of) propositional variables bound in B can change at most once. More formally, for a block B consisting of n Boolean equations, we have $(\|B\|^n)\theta = (\|B\|^{n+1})\theta$ for all environments θ satisfying $\|B\|\theta \leq \theta$ or $\theta \leq \|B\|\theta$. We next define the semantics of a Boolean equation system.

Definition 11. *In Coq:* `BES_solution`.
The semantics of a BES $\mathcal{E}$ in the context of an environment θ, denoted $[\![\mathcal{E}]\!]\theta$ is defined inductively as follows:

$$[\![\epsilon]\!]\theta = \theta$$
$$[\![(\sigma B)\mathcal{E}]\!]\theta = [\![\mathcal{E}]\!](\theta[\langle X_i := (\sigma F(\mathcal{E}, B, \theta))(X_i)\rangle_{i=1}^n])$$

where B is a block $\langle X_i = f_{X_i}\rangle_{i=1}^n$ and F is an operator on environments defined as follows:

$$F(\mathcal{E}, B, \theta) = \lambda \eta \in \mathbb{B}^{\mathcal{X}}.\|B\|([\![\mathcal{E}]\!](\theta[\langle X_i := \eta(X_i)\rangle_{i=1}^n]))$$

We remark that for a closed equation system $\mathcal{E}$, the semantics of $\mathcal{E}$ assigns a truth value to a bound variable that is independent of the environment θ; that is, we have $[\![\mathcal{E}]\!]\theta(X) = [\![\mathcal{E}]\!]\eta(X)$ for all $X \in \mathbf{bnd}_{\mathcal{E}}$ in case $\mathcal{E}$ is closed.

Example 12. *A simple algorithm for 'solving' a Boolean equation system is Gauß elimination [9, 10]. This algorithm is based on the standard laws for Boolean simplification and the following three rules: (1) local fixpoint elimination, (2) left substitution and (3) right substitution. Rule (1), local fixpoint elimination, replaces an equation $X = f$ by $X = f[X := \bot]$ in a least fixpoint block and by $X = f[X := \top]$ in a greatest fixpoint block. Rule (2), left substitution, allows for replacing an equation $Y = g$ by $Y = g[X := f]$ whenever there is an equation $X = f$ that is either in the same block as Y's equation or some block following Y's. Rule (3), right substitution, allows for replacing an equation $Y = g$ by $Y = g[X := f]$ whenever there is an equation $X = f$ that is either in the same block as Y's equation or some block preceding Y's; right substitution is only permitted when f contains no propositional variables. The Gauß elimination algorithm solves the first equation of a Boolean equation system by* *alternatingly applying rule (1), local fixpoint elimination, and, subsequently, rule (2), left substitution (exhaustively), processing the equations last-to-first. As a last step, the algorithm exhaustively applies rule (3), right substitution, replacing every 'solved' propositional variable with its solution. Repeating the Gauß elimination algorithm, possibly several times, will solve the entire system. This requires at most N runs of the Gauß elimination algorithm, where N is the number of equations in a system.*

Reconsider the Boolean equation system $\mathcal{F}$ of Example 8. For $\mathcal{F}$, Gauß elimination starts by replacing the equation $X_7 = X_7 \wedge X_3$ by $X_7 = \bot \wedge X_3$, which we can further simplify to $X_7 = \bot$. Next, the algorithm replaces the equation $X_3 = X_7$ by $X_3 = \bot$. Note that this effectively removes all references to X_7 in all equations preceding that of X_7. Next, the algorithm will apply local fixpoint elimination for $X_6 = X_1$ (which is 'ineffective') and replace the equation $X_2 = X_0 \vee X_6$ by $X_2 = X_0 \vee X_1$; and so forth. By repeating this process one eventually solves X_0. Using right substitution, every propositional variable in $\mathcal{F}$ is solved; we find that for all environments θ we have:

$$[\![\mathcal{F}]\!]\theta(X_3) = [\![\mathcal{F}]\!]\theta(X_7) = \mathit{ff}, \text{ and}$$
$$[\![\mathcal{F}]\!]\theta(X_0) = [\![\mathcal{F}]\!]\theta(X_1) = [\![\mathcal{F}]\!]\theta(X_2) = [\![\mathcal{F}]\!]\theta(X_4) = [\![\mathcal{F}]\!]\theta(X_5) = [\![\mathcal{F}]\!]\theta(X_6) = \mathit{tt}$$

We remark that the solution to X_0 is 'easy' to compute since X_0 effectively only depends on X_4 and X_4 only depends on X_0; the dependence of X_0 on X_1 turns out to be inessential for the solution to X_0. The solution to X_1, on the other hand, is 'harder' to compute since it depends on many more equations.   □

4   Consistent Consequences

An important technique often used in model checking is to apply methods of abstraction to approximate solutions, to avoid the state space explosion problem. Since Boolean equation systems can encode model checking problems, it is natural to consider abstraction techniques on such equation systems. However, proving soundness of such techniques is often quite tedious. The notion of *consistent consequence* [7] facilitates the reasoning involved in these proofs of soundness, reducing the problem of showing soundness to showing that the abstraction technique induces a consistent consequence relation. Showing that a given relation is a consistent consequence is often easier than reasoning directly with the semantics of Boolean equation systems. We here briefly review the notion of consistent consequence and how it relates to the semantics of a Boolean equation system.

Let $R \subseteq \mathcal{X} \times \mathcal{X}$ be an arbitrary relation on propositional variables. We denote the reflexive, transitive closure of R by R^*. A relation on propositional variables induces a set of environments that are *consistent* with the relation.

Definition 13. *In Coq:* `consistent_environment`.
Let θ be an environment. We say that θ is *consistent* with R iff for all X, Y for which $(X, Y) \in R$, we have $\theta(X) \leq \theta(Y)$. The set of all environments consistent with R is denoted Θ_R.

Note that in particular the environments $\theta_\perp$ and $\theta_\top$ are consistent with every relation R. We generalise logical consequence to logical consequence *relative* to a given relation R.

Definition 14. *In Coq:* `rel_cons`.
Let f, g be arbitrary propositional formulae. We say that g is a consequence of f relative to R, denoted $f \overset{R}{\Longrightarrow} g$ iff for all $\theta \in \Theta_R$ we have $[\![f]\!]\theta \le [\![g]\!]\theta$.

We next lift the notion of relative consequence to Boolean equation systems.

Definition 15. *In Coq:* `consistent_consequence`.
Let $\mathcal{E}$ be a BES. A relation $R \subseteq \mathcal{X} \times \mathcal{X}$ is a *consistent consequence* on $\mathcal{E}$ iff, for all $(X, Y) \in R$, we have:

1. $X \sim_\mathcal{E} Y$,
2. If $X, Y \in \mathbf{bnd}_\mathcal{E}$ then also $f_X \overset{R}{\Longrightarrow} f_Y$

We say that Y is a consistent consequence of X, written $X \lessdot_\mathcal{E} Y$, if there exists a consistent consequence relation $R \subseteq \mathbf{bnd}_\mathcal{E} \times \mathbf{bnd}_\mathcal{E}$ on $\mathcal{E}$ for which $(X, Y) \in R$. If $\mathcal{E}$ is clear from the context we simply write $X \lessdot Y$. In Coq: `cc_max`.

In fact, $\lessdot_\mathcal{E}$ is again a consistent consequence relation on $\mathcal{E}$ and it is the largest consistent consequence relation contained in $\mathbf{bnd}_\mathcal{E} \times \mathbf{bnd}_\mathcal{E}$. This follows from the following lemma.

Lemma 16. *In Coq:* `union_maintains_cc`.
For BES $\mathcal{E}$ and relations R, S which are consistent consequence relations on $\mathcal{E}$, also $R \cup S$ is a consistent consequence relation on $\mathcal{E}$.

The notion of consistent consequence and the semantics of a Boolean equation system are tightly linked. Informally, a consistent consequence underapproximates the semantics of a Boolean equation system. We make this statement more precise in the following two lemmata and Theorem 20, and illustrate it in the example below.

Example 17. *Reconsider the Boolean equation system $\mathcal{F}$ of Example 8 once more. We obtain, among others, that $X_3 \lessdot X_0 \lessdot X_1 \lessdot X_2 \lessdot X_0$ and $X_7 \lessdot X_4 \lessdot X_5 \lessdot X_6 \lessdot X_4$. Proving that this is the case, one can take the transitive closure of the relation R defined as:*

$$\{(X_3, X_0), (X_0, X_1), (X_1, X_2), (X_2, X_0), (X_7, X_4), (X_4, X_5), (X_5, X_6), (X_6, X_4)\}.$$

One can check that R^* is such that for all environments $\theta \in \Theta_{R^*}$ the second condition of Definition 15 is met for all pairs in R^*.   □

The first lemma claims that each operator $\|B\|$ induced by a block B transforms environments consistent with R into environments again consistent with R.

Lemma 18. *In Coq:* `unfold_block_maintains_consistency`.
Let B be a block of Boolean equations, and let R be a consistent consequence on σB for some $\sigma \in \{\mu, \nu\}$. Then for all $\theta \in \Theta_R$ we have $(\|B\|\theta) \in \Theta_R$.

Furthermore, given two environments consistent with a consistent consequence relation on a BES consisting of a single block, the result of replacing in one of the environments the interpretation of all variables in the BES with their interpretation in the other environment results in an environment consistent with the consistent consequence relation.

Lemma 19. *In Coq:* `redef_bnd_consistent`.
Let $\mathcal{E}$ be a Boolean equation system and let $\langle X_i = f_{X_i} \rangle_{i=1}^n$ be a block in $\mathcal{E}$. If R is a consistent consequence on $\mathcal{E}$ and we have $\eta, \theta \in \Theta_R$ then $(\theta[\langle X_i := \eta(X_i) \rangle_{i=1}^n]) \in \Theta_R$.

The theorem below firmly links the notion of consistent consequence and the semantics of a Boolean equation system; it is one of the main theorems in [7]. Informally, it states that the solution to a Boolean equation system is consistent with every consistent consequence relation.

Theorem 20. *In Coq:* `cc_BES_semantics`.
Let $\mathcal{E}$ be a BES and R a consistent consequence relation on $\mathcal{E}$. Then for all $\theta \in \Theta_R$ we have $[\![\mathcal{E}]\!]\theta \in \Theta_R$.

As a corollary of the Theorem above we find the following result, which substantiates the informal claim made earlier, stating that a consistent consequence underapproximates the solution to a Boolean equation system.

Corollary 21. *Let $\mathcal{E}$ be a closed BES. Then, for all environments θ and all variables X, Y such that $X <_{\mathcal{E}} Y$ we have $[\![\mathcal{E}]\!]\theta(X)$ implies $[\![\mathcal{E}]\!]\theta(Y)$.*

Example 22. *From Example 17 we know that $X_3 < X_0$. Since $[\![\mathcal{F}]\!]\theta(X_3) = f\!f$, we cannot deduce the truth value of $[\![\mathcal{F}]\!]\theta(X_0)$. However, from $X_0 < X_1$ and the fact that $[\![\mathcal{F}]\!]\theta(X_0) = t\!t$, we can conclude $[\![\mathcal{F}]\!]\theta(X_1) = t\!t$.* □

In practice, this means that if one can 'cheaply' compute that $[\![\mathcal{E}]\!]\theta(X) = t\!t$ and prove that $X < Y$, then we have a way of proving that $[\![\mathcal{E}]\!]\theta(Y) = t\!t$ without explicitly computing $[\![\mathcal{E}]\!]\theta(Y)$. The abstraction techniques explored in, *e.g.* [2] are based on exactly this concept, guaranteeing that a (finite representation of a potentially infinite) BES is reduced to a smaller BES whose semantics approximates the semantics of the original BES.

Variations on Consistent Consequence. In [7] the definition of consistent consequence only places restrictions on related *bound* variables; that is, in [7] a relation R is a consistent consequence relation if for all *bound* variables X, Y such that $(X, Y) \in R$, properties 1 and 2 hold. However, using this definition, Theorem 20 (Theorem 1 in [7]) fails, as the following example illustrates.

Example 23. *Consider the Boolean equation system $\mathcal{E}$ given by $\mu\langle (X = Z)(Y = Y) \rangle$. The relation $R = \{(X, Y), (Z, Y)\}$ is clearly a consistent consequence in the sense of [7]. However, for $\theta_\top \in \Theta_R$ we have $[\![\mathcal{E}]\!]\theta_\top(X) = [\![\mathcal{E}]\!]\theta_\top(Z) = t\!t$ whereas $[\![\mathcal{E}]\!]\theta_\top(Y) = f\!f$. But then $[\![\mathcal{E}]\!]\theta_\top \notin \Theta_R$, contradicting Theorem 20.* □

The main problem with the definition from [7] is that unbound variables can be related to bound variables; we forbid this explicitly in our definition.

In [2] the consistent consequence relation is defined for *parameterised* Boolean equation systems. Like in [7], the notion of consistent consequence allows for relating bound and unbound variables in [2]. The counterpart of our Theorem 20 in [2] again reasons about arbitrary consistent consequence relations for bound variables, falling into the pitfall of the example above. However, contrary to [7], but similar to [18], the relation $\prec$ in [2] coincides with $\prec$ in our Definition 15 since it requires the existence of a consistent consequence relation R that is restricted to bound variables *only*. As a result, for that relation, Theorem 20 does hold.

5   A Derivation System for Consistent Consequence

The two-stage approach of defining $\prec$ as the union of all consistent consequence relations can make the notion hard to understand. An additional complication is that checking that an arbitrary relation R is a consistent consequence relation requires reasoning about all environments θ taken from the infinite set Θ_R. In [7], a derivation system for consistent consequence has been presented, showing that a consistent consequence can indeed be understood as a form of logical consequence on propositional formulae lifted to Boolean equation systems.

While in [7] soundness and completeness of the derivation system are claimed, it turns out that there is a subtle mistake in the derivation system rendering the entire system unsound, see also Example 28. We show that by removing the dubious rule, one arrives at a sound and complete derivation system for consistent consequence.

5.1   The Derivation System CC

We first present the corrected derivation system, which we dub *system* **CC**, and which we base upon the derivation system of [7].

Definition 24. *In Coq:* `prv_tree`.
Given a BES $\mathcal{E}$, we define the derivation system for consistent consequence for $\mathcal{E}$, which we call **CC**. (The dependency on $\mathcal{E}$ is left implicit.) **CC** derives judgments of the form $\Gamma \vdash_{cc} \alpha \sqsubset \beta$ using rules from Table 1. Here, $\alpha, \beta, \ldots$ are propositional formulae and Γ is a context consisting of a sequence of relations on propositional variables of the form $X \sqsubset Y$, and $\alpha \sqsubset \beta$ is a relation between propositional formulae. For $X \sqsubset Y \in \Gamma$ we require that $X, Y \in \mathbf{bnd}_{\mathcal{E}}$ and $X \sim_{\mathcal{E}} Y$. The formulas f_X and f_Y in the rule **CC** are the formulas bound to X, resp. Y, in $\mathcal{E}$.

The rules for logical consequence on negation-free propositional formulae axiomatise associativity (**AS**), distributivity (**DS**), absorption (**AB**), idempotence (**ID**), supremum (**SUP**) and infimum (**INF**) and top (**TOP**) and bottom (**BOT**). These axioms, together with the reflexivity axiom (**REF**), the transitivity rule (**TRA**) and the context rule (**CTX**) form the basis of our derivation systems. From the two lemmata below it follows that these derivation rules exactly characterise logical consequence.

Table 1. Derivation system for logical consequence on negation-free propositional formulae and consistent consequence on an equation system $\mathcal{E}$.

Axioms for negation-free propositional logic

rules of the form $\dfrac{}{\Gamma \vdash_{cc} A}$, where A ranges over the following laws:

AS1	$\alpha \wedge (\beta \wedge \gamma) \subset (\alpha \wedge \beta) \wedge \gamma$	DS1	$\alpha \vee (\beta \wedge \gamma) \subset (\alpha \vee \beta) \wedge (\alpha \vee \gamma)$
AS2	$(\alpha \wedge \beta) \wedge \gamma \subset \alpha \wedge (\beta \wedge \gamma)$	DS2	$(\alpha \vee \beta) \wedge (\alpha \vee \gamma) \subset \alpha \vee (\beta \wedge \gamma)$
AS3	$\alpha \vee (\beta \vee \gamma) \subset (\alpha \vee \beta) \vee \gamma$	DS3	$\alpha \wedge (\beta \vee \gamma) \subset (\alpha \wedge \beta) \vee (\alpha \wedge \gamma)$
AS4	$(\alpha \vee \beta) \vee \gamma \subset \alpha \vee (\beta \vee \gamma)$	DS4	$(\alpha \wedge \beta) \vee (\alpha \wedge \gamma) \subset \alpha \wedge (\beta \vee \gamma)$
COM1	$\alpha \wedge \beta \subset \beta \wedge \alpha$	AB1	$\alpha \vee (\alpha \wedge \beta) \subset \alpha$
COM2	$\alpha \vee \beta \subset \beta \vee \alpha$	AB2	$\alpha \subset \alpha \wedge (\alpha \vee \beta)$
ID1	$\alpha \subset \alpha \wedge \alpha$	ID2	$\alpha \vee \alpha \subset \alpha$
SUP	$\alpha \subset \alpha \vee \beta$	INF	$\alpha \wedge \beta \subset \alpha$
TOP	$\alpha \subset \alpha \wedge tt$	BOT	$\alpha \vee f\!f \subset \alpha$

Inequality logic rules

REF $\quad \dfrac{}{\Gamma \vdash_{cc} \alpha \subset \alpha}$

TRA $\quad \dfrac{\Gamma \vdash_{cc} \alpha \subset \beta \quad \Gamma \vdash_{cc} \beta \subset \gamma}{\Gamma \vdash_{cc} \alpha \subset \gamma}$

CTX $\quad \dfrac{\Gamma \vdash_{cc} \alpha \subset \beta}{\Gamma \vdash_{cc} \gamma[X := \alpha] \subset \gamma[X := \beta]}$

Consistent consequence rules

CC $\quad \dfrac{\Gamma, X \subset Y \vdash_{cc} f_X \subset f_Y}{\Gamma \vdash_{cc} X \subset Y}\; X \sim Y$

CNT $\quad \dfrac{}{\Gamma \vdash_{cc} X \subset Y}\;(X \subset Y) \in \Gamma$

Lemma 25. *In Coq:* complete_cons.
For all propositional formulae f, g, if $f \Rightarrow g$ then we can derive $\Gamma \vdash_{cc} f \subset g$ without rules CC and CNT.

Lemma 26. *For propositional formulae f, g, if we can derive $\Gamma \vdash_{cc} f \subset g$ without rules CC and CNT, then $f \Rightarrow g$.*

It turns out that for a sound and complete derivation system for consistent consequence it suffices to add two rules to the axiomatisation of logical consequence. The first rule, rule CNT allows one to conclude 'facts' from the context Γ. The second rule, rule CC states that we can conclude (in the context of some Γ) that $X \subset Y$ holds, representing that Y is a consistent consequence of X, if we can derive that the right-hand side of Y is a consequence of X assuming $X \subset Y$ (in addition to Γ). Note that there is an inkling of circular reasoning in this rule since the right-hand side of Y and X might themselves consist of Y and X again.

Example 27. *Reconsider Example 17. There, we claimed that we have $X_3 < X_0$. Using the derivation system we are able to derive the very same using, among others, the CC and CNT rules:*

$$\dfrac{\dfrac{\Gamma_2 \vdash_{cc} X_7 \wedge X_3 \subset X_3 \;\; INF \;\; \dfrac{\Gamma_2 \vdash_{cc} X_3 \subset X_0}{\Gamma_2 \vdash_{cc} X_7 \wedge X_3 \subset X_0} \; CNT}{\Gamma_1 \vdash_{cc} X_7 \subset X_4} \; CC \qquad \dfrac{\Gamma_1 \vdash_{cc} X_4 \subset X_1 \vee X_4}{\Gamma_1 \vdash_{cc} X_7 \subset X_1 \vee X_4} \; SUP}{\vdash_{cc} X_3 \subset X_0} \; CC$$

where context Γ_1 contains $X_3 \subset X_0$ and context Γ_2 contains both $X_3 \subset X_0$ and $X_7 \subset X_4$. ☐

The derivation system of [7] includes, in addition to the rules of Table 1, the following substitution rule:

$$SUB \quad \dfrac{\Gamma \vdash_{cc} \alpha \subset \beta}{\Gamma \vdash_{cc} \alpha\varsigma \subset \beta\varsigma}$$

where ς is a substitution mapping propositional variables to propositional formulae. While substitution is in itself sound for logical consequence for negation-free propositional logic (axiomatised by the first block of axioms in Table 1), it causes problems in the setting of consistent consequence as illustrated by the following example.

Example 28. *Consider the following Boolean equation system:*

$$\mu\langle(X = A)(Y = B)\rangle \; \nu\langle(A = A)\rangle \; \mu\langle(B = B)\rangle$$

We can deduce that X has solution tt and Y has solution ff. By Corollary 21, we cannot have $X < Y$. Combining rule SUB and CC we are nonetheless capable of deriving $X \subset Y$:

$$\dfrac{\dfrac{X \subset Y \vdash_{cc} X \subset Y \;\; CNT}{X \subset Y \vdash_{cc} A \subset B} \; SUB \;\; using \; substitution \; \varsigma(X) = A, \varsigma(Y) = B}{\vdash_{cc} X \subset Y} \; CC$$

Consequently, the derivation system for $\subset$ of [7] is not sound for $<$. ☐

*Formalising **CC** in Coq.* In Coq, the derivation system is represented as the inductive type `prv_tree : BES→statement→Prop`, which has a constructor for every derivation rule in Table 1. A term of type `statement` is a triple of a relation and two propositional formulas: $\Gamma \vdash_{cc} f \subset g$; the type `statement` is defined as the Inductive type with one constructor `stmt : (relation propVar)→propForm→propForm→statement`. Constructing a derivation tree is done by combining constructors of the type `prv_tree`.

Reasoning on the properties of the derivation system in Coq, often amounts to showing that certain properties are sufficient for knowing that it is possible to create a derivation tree or *vice versa*. For example, (as we will show), if we have $X <_\varepsilon Y$ for a pair of variables, then we can create a proof tree with $\emptyset \vdash_{cc} X \subset Y$ as the root. *Vice versa*, if we can derive $\emptyset \vdash_{cc} X \subset Y$ for bound X, Y, then $X \overset{\leq_\varepsilon}{\Longrightarrow} Y$, which (as we will later see) tells us that $X <_\varepsilon Y$.

5.2   Soundness and Completeness

We recall that **CC** without the rules CC and CNT is sound and complete for negation-free propositional logic (Lemmas 25 and 26). For soundness and completeness of the full **CC** system we want to prove that for X, Y bound in $\mathcal{E}$, $\emptyset \vdash_{cc} X \subset Y$ is derivable if and only if $X \lessdot_{\mathcal{E}} Y$. In the derivation system we use a Γ, so we will have to prove something stronger: a property about $\Gamma \vdash_{cc} X \subset Y$. This implies that we also have to strengthen the notion $X \lessdot_{\mathcal{E}} Y$ by adding Γ.

Definition 29. *In Coq:* `relative_cc`.
Let $\Gamma \subseteq \mathcal{X} \times \mathcal{X}$ be a relation on propositional variables. A relation $R \subseteq \mathcal{X} \times \mathcal{X}$ is a *consistent consequence* on $\mathcal{E}$ *relative to* Γ if, for all $(X, Y) \in R$, we have:

1. $X \sim_{\mathcal{E}} Y$,
2. If $X, Y \in \mathbf{bnd}$, then $f_X \xrightarrow{R \cup \Gamma} f_Y$

We say that Y is a consistent consequence of X relative to Γ, written $\lessdot_{\mathcal{E}}^{\Gamma}$ if there exists a consistent consequence relation $R \subseteq \mathbf{bnd}_{\mathcal{E}} \times \mathbf{bnd}_{\mathcal{E}}$ on $\mathcal{E}$ relative to Γ such that $(X, Y) \in \Gamma \cup R$. In Coq: `rel_cc_max`.

The link between consistent consequence and consistent consequence relative to a context Γ is given by the following Lemma.

Lemma 30. *In Coq:* `cc_relative_empty`, `empty_relative_cc`.
For any relation R, R is a consistent consequence on $\mathcal{E}$ iff R is a consistent consequence on $\mathcal{E}$ relative to $\emptyset$.

We will prove that for variables X, Y bound in $\mathcal{E}$, $\Gamma \vdash_{cc} X \subset Y$ is derivable if and only if $X \lessdot_{\mathcal{E}}^{\Gamma} Y$; thus, the derivation system is sound and complete for consistent consequence when considering the bound variables in a BES. We first consider soundness. The following Proposition is proved by induction on the derivation. (In Coq terminology: by induction on `prv_tree E (stmt G f g)`.)

Proposition 31. *In Coq:* `soundness`.
Let f, g be propositional formulae and let Γ be an arbitrary context. If $\Gamma \vdash_{cc} f \subset g$ is derivable, then $f \xrightarrow{\lessdot_{\mathcal{E}}^{\Gamma}} g$.

This proposition underlies the proof of soundness, which is a direct consequence by taking $\Gamma = \emptyset$.

Theorem 32. *In Coq:* `prv_system_sound_bnd`.
For propositional variables $X, Y \in \mathbf{bnd}$, if $\emptyset \vdash_{cc} X \subset Y$ is derivable, then $X \lessdot Y$.

Similar to [7], our proof of completeness is built on the fact that **CC** is complete for logical consequence for negation-free propositional formula. In [7], the (faulty) proof of completeness was built on the unsound SUB rule. Nonetheless, the main idea for proving completeness that is exploited in [7] remains valid. The idea is to decompose a goal $f \subset g$ into a set of simple goals (relating only propositional variables) of the form $X \subset Y$. This is made precise by the following Lemma.

Lemma 33. *In Coq:* `derivable_set_derivable_rel_cons`.
For all contexts $\Gamma \subseteq \mathcal{X} \times \mathcal{X}$, relations $R \subseteq \mathbf{bnd} \times \mathbf{bnd}$, and propositional formulae f, g satisfying $f \overset{R}{\Longrightarrow} g$, the following rule is derivable:

$$\frac{\{\Gamma \vdash_{cc} X \subset Y \mid (X,Y) \in R\}}{\Gamma \vdash_{cc} f \subset g}$$

The proof uses a number of properties of the system **CC**. First of all, the 'standard' logical rules are derivable in **CC**: For example, if $\Gamma \vdash_{cc} f_1 \subset g$ and $\Gamma \vdash_{cc} f_2 \subset g$, then $\Gamma \vdash_{cc} f_1 \vee f_2 \subset g$, as the following derivation illustrates:

$$\frac{\dfrac{\Gamma \vdash_{cc} f_2 \subset g}{\Gamma \vdash_{cc} f_1 \vee f_2 \subset f_1 \vee g} \, \text{CTX} \quad \dfrac{\dfrac{\Gamma \vdash_{cc} f_1 \subset g}{\Gamma \vdash_{cc} f_1 \vee g \subset g \vee g} \, \text{CTX} \quad \dfrac{\Gamma \vdash_{cc} g \vee g \subset g}{} \, \text{ID2}}{\Gamma \vdash_{cc} f_1 \vee g \subset g} \, \text{TRA}}{\Gamma \vdash_{cc} f_1 \vee f_2 \subset g} \, \text{TRA}$$

Using these, one can prove, inside **CC**, that every formula f is equivalent to its *disjunctive normal form* $\mathrm{DNF}(f)$. A disjunctive normal form is a disjunction of conjunctions of literals, where the literals are $\bot$, $\top$ or a variable X. These conjunctions are called *clauses* and in our case of negation-free propositional logic, there are only positive literals (*i.e.* we only have X as literal, not $\neg X$).

Furthermore, the fact that we are in negation-free propositional logic means that our logic is constructive in the sense of (a) below. We also have (b).

(a) If $c_1 \vee \ldots \vee c_n \overset{R}{\Longrightarrow} d_1 \vee \ldots \vee d_m$, where all c_i and d_j are clauses, then $\forall i \exists j (c_i \overset{R}{\Longrightarrow} d_j)$.

(b) If $X_1 \wedge \ldots \wedge X_p \overset{R}{\Longrightarrow} Y_1 \wedge \ldots \wedge Y_\ell$, where all X_i and Y_j are variables, then $\forall j \exists i (X_i \overset{R}{\Longrightarrow} Y_j)$.

These properties are the crucial steps for proving Lemma 33. We write $c_f \in f$ to denote that c_f is a clause in f and $X \in c_f$ to denote that X is a variable in the clause c_f.

Proof (Lemma 33). Consider the propositional formulas f, g such that $f \overset{R}{\Rightarrow} g$ for some relation R. We may assume that f and g are in disjunctive normal form, and that $f \not\Leftrightarrow \bot$. Then, for every clause c_f in f, there is a clause c_g in g, such that $c_f \overset{R}{\Rightarrow} c_g$, by (a) above. Therefore, by (b) above,

$$\forall c_f \in f \; \exists d_g \in g \; \forall Y_g \in d_g \; \exists X_f \in c_f (X_f \overset{R}{\Rightarrow} X_g) \tag{1}$$

It can be shown (in Coq: `min_env_conj`) that $X_f \overset{R}{\Rightarrow} X_g$ implies $(X_f, Y_g) \in R^*$ (the transitive reflexive closure of R).

The proof now follows from the fact that the system is sound for logical consequence: From $\{\Gamma \vdash_{cc} X \subset Y \mid (X,Y) \in R\}$ one can derive $\{\Gamma \vdash_{cc} X \subset Y \mid (X,Y) \in R^*\}$ and it can be shown that if (1) holds, then $f \overset{R}{\Rightarrow} g$. □

Now the proof of completeness of the derivation system **CC** proceeds roughly as follows. Suppose $X \lessdot Y$. We use rule CC to conclude $\emptyset \vdash_{cc} X \subset Y$ from $X \subset Y \vdash_{cc} f_X \subset f_Y$. The latter goal can be decomposed using the lemma above, so we need to prove $\{\Gamma \vdash_{cc} X \subset Y \mid (X,Y) \in R\}$ for some Γ and R. Repeating the CC rule to prove the next subgoals we eventually must arrive at conclusions that can be drawn from the context using the CNT rule. Note that the number of applications of CC is limited since in every application of CC, Γ grows by one and we can have at most $|\mathbf{bnd}|^2$ different pairs of propositional variables to which rule CC is applicable before 'revisiting' a subgoal.

Lemma 34. *In Coq:* `complete_propVar`.
For any context Γ and for all X, Y such that $X \lessdot_{\mathcal{E}}^{\Gamma} Y$ we can derive $\Gamma \vdash_{cc} X \subset Y$.

Proof. Take a relation $\Gamma \subseteq (\mathbf{bnd}_{\mathcal{E}})^2$, and variables X, Y s.t. $X \lessdot_{\mathcal{E}}^{\Gamma} Y$. We proceed by induction on the size of $(\mathbf{bnd}_{\mathcal{E}})^2 \setminus \Gamma$.

- Case $(\mathbf{bnd}_{\mathcal{E}})^2 \setminus \Gamma = 0$. Then $\lessdot_{\mathcal{E}}^{\Gamma} \subseteq \Gamma$, thus $X \subset Y \in \Gamma$. We can complete the derivation using CNT.
- Case $(\mathbf{bnd}_{\mathcal{E}})^2 \setminus \Gamma = n + 1$. If $X \subset Y \in \Gamma$ we can complete the derivation using CNT. If $X = Y$ then we can complete the derivation using REF. Otherwise, there exists a consistent consequence relation R relative to Γ, such that $(X, Y) \in R$. We derive $\Gamma, X \subset Y \vdash_{cc} f_X \subset f_Y$ (and then $\Gamma, X \subset Y$ by the rule CC). We have $f \xRightarrow{R} g$, so, by Lemma 33, we are done if we derive $\{\Gamma, X \subset Y \vdash_{cc} X' \subset Y' \mid (X', Y') \in R\}$. This is possible according to the induction hypothesis: we know $X' \lessdot_{\mathcal{E}}^{\Gamma, X \subset Y} Y'$, so $\Gamma, X \subset Y \vdash_{cc} X' \subset Y'$. $\square$

From the previous lemma and Lemma 30 we can conclude the completeness of the system.

Theorem 35. *In Coq:* `complete`.
For all propositional variables X, Y satisfying $X \lessdot Y$, we can derive $\emptyset \vdash_{cc} X \subset Y$.

Proof. Follows from Lemmas 30 and 34. $\square$

6   Conclusions

We have formalised the notion of *consistent consequence* [7] in the Coq proof assistant. Consistent consequence underlies various forms of abstraction in the setting of Boolean equation systems (BES) [10], which have applications in software and hardware verification. We have shown that the derivation system of [7] for consistent consequence, which is claimed to be sound and complete, contains a serious flaw caused by an unsound rule. We have proved that soundness can be recovered by removing this rule. For our proof of completeness of the modified derivation system we have had to deviate substantially from the original proof (which essentially relied on the unsound rule, and was thus faulty). Moreover, our formalisation revealed an inaccuracy in the phrasing of the main correctness theorem for consistent consequence in [7], and which we could trace back to an omission in the definition of consistent consequence. We have shown that by fixing the latter, we can restore validity of the main theorem.

The Coq code, see [17], consists of over 5500 lines of Coq, comprising 178 Lemmas and 79 Definitions; of these Definitions, 2 are Records, 7 are Inductive definitions and 22 are Fixpoint definitions. Many of the notions in the field of BES are defined inductively, and therefore can be encoded into Coq quite directly. Most functions are defined by structural recursion (Fixpoint in Coq) and proofs proceed by induction, which are both features that are readily available in Coq. Similarly, the derivation system **CC** is straightforwardly defined in Coq by defining the inductive type family of derivation trees, indexed by a BES and a triple $\Gamma \vdash_{cc} f \subset g$, the conclusion of the derivation tree. This inductive type facilitates the soundness proof, which proceeds by induction on the derivation. The completeness proof is more interesting, as it uses a transformation to disjunctive normal forms, and it also essentially uses the fact that all propositional formulas are negation-free. A technical complication in the proof is that the inductive argument that proves completeness is not on the size of Γ but on the number of variables that are *not bound in Γ*.

Future Work. The main point for future work is to extend the system **CC** to reason about consistent consequence for *parametrised Boolean equation systems* (PBES) [8,18]. Then variables range over predicates over data, for example we would have $X(n)$ where n ranges over $\mathbb{N}$, and we would look for the least (or greatest) $X(n)$ satisfying certain equations. In the case of PBES, the least solution to an equation may require a transfinite iteration (as summarised in Sect. 2). This would also require a serious modification of our formalisation in Coq, which now relies on the fact that we can compute the least solution in a finite number of steps.

References

1. Bertot, Y., Komendantsky, V.: Fixed point semantics and partial recursion in Coq. In: PPDP, pp. 89–96. ACM (2008)
2. Cranen, S., Gazda, M., Wesselink, W., Willemse, T.A.C.: Abstraction in fixpoint logic. ACM Trans. Comput. Log. **16**(4/29), 29:1–29:39 (2015)
3. Cranen, S., Groote, J.F., Keiren, J.J.A., Stappers, F.P.M., Vink, E.P., Wesselink, W., Willemse, T.A.C.: An overview of the mCRL2 toolset and its recent advances. In: Piterman, N., Smolka, S.A. (eds.) TACAS 2013. LNCS, vol. 7795, pp. 199–213. Springer, Heidelberg (2013). doi:10.1007/978-3-642-36742-7_15
4. Dittmann, C.: Positional determinacy of parity games. In: Archive of Formal Proofs (2015)
5. Garavel, H., Lang, F., Mateescu, R.: Compositional verification of asynchronous concurrent systems using CADP. Acta Informatica **52**(4), 337–392 (2015)
6. Garavel, H., Mateescu, R., Lang, F., Serwe, W.: CADP 2006: a toolbox for the construction and analysis of distributed processes. In: Damm, W., Hermanns, H. (eds.) CAV 2007. LNCS, vol. 4590, pp. 158–163. Springer, Heidelberg (2007). doi:10.1007/978-3-540-73368-3_18

7. Gazda, M.W., Willemse, T.A.C.: Consistent consequence for boolean equation systems. In: Bieliková, M., Friedrich, G., Gottlob, G., Katzenbeisser, S., Turán, G. (eds.) SOFSEM 2012. LNCS, vol. 7147, pp. 277–288. Springer, Heidelberg (2012). doi:10.1007/978-3-642-27660-6_23

8. Keiren, J.J.A., Wesselink, W., Willemse, T.A.C.: Liveness analysis for parameterised boolean equation systems. In: Cassez, F., Raskin, J.-F. (eds.) ATVA 2014. LNCS, vol. 8837, pp. 219–234. Springer, Cham (2014). doi:10.1007/978-3-319-11936-6_16

9. Mader, A.: Modal μ-calculus, model checking and Gauß elimination. In: Brinksma, E., Cleaveland, W.R., Larsen, K.G., Margaria, T., Steffen, B. (eds.) TACAS 1995. LNCS, vol. 1019, pp. 72–88. Springer, Heidelberg (1995). doi:10.1007/3-540-60630-0_4

10. Mader, A.: Verification of modal properties using boolean equation systems. Ph.D. thesis, Technische Universität München (1997)

11. Sangiorgi, D.: Introduction to Bisimulation and Coinduction. Cambridge University Press, New York (2011)

12. Sprenger, C.: A verified model checker for the modal μ-calculus in Coq. In: Steffen, B. (ed.) TACAS 1998. LNCS, vol. 1384, pp. 167–183. Springer, Heidelberg (1998). doi:10.1007/BFb0054171

13. Tarski, A.: A lattice-theoretical fixpoint theorem and its applications. Pacific J. Math. 5(2), 285–309 (1955)

14. The Coq Development Team. http://coq.inria.fr

15. van de Pol, J.C.: Operations on fixpoint equation systems. Unpublished note; available from the author upon request

16. van Delft, M.E.C.: Consistent consequences formalized. Master's thesis, Eindhoven University of Technology (2016)

17. van Delft, M.E.C., Geuvers, H., Willemse, T.A.C. http://doi.org/10.4121/uuid:a06e90c7-9ca1-45df-ad37-e99bdbf75b78

18. Willemse, T.A.C.: Consistent correlations for parameterised boolean equation systems with applications in correctness proofs for manipulations. In: Gastin, P., Laroussinie, F. (eds.) CONCUR 2010. LNCS, vol. 6269, pp. 584–598. Springer, Heidelberg (2010). doi:10.1007/978-3-642-15375-4_40

Homotopy Type Theory in Lean

Floris van Doorn[1], Jakob von Raumer[2], and Ulrik Buchholtz[3]([envelope])

[1] Carnegie Mellon University, Pittsburgh, USA
fpvdoorn@gmail.com
[2] University of Nottingham, Nottingham, UK
jakob@von-raumer.de
[3] TU Darmstadt, Darmstadt, Germany
ulrikbuchholtz@gmail.com

Abstract. We discuss the homotopy type theory library in the Lean proof assistant. The library is especially geared toward synthetic homotopy theory. Of particular interest is the use of just a few primitive notions of higher inductive types, namely quotients and truncations, and the use of cubical methods.

Keywords: Homotopy type theory · Formalized mathematics · Lean · Proof assistants

1 Introduction

Homotopy type theory (HoTT) refers to the homotopical interpretation of Martin-Löf's dependent type theory [3,22], which grew out of the groupoid model of [13]. In the standard interpretation, every type-theoretical construct corresponds to a homotopy-invariant construction on spaces. An important example is the identity type, which corresponds to the path space construction.

Just like extensional type theory can be interpreted in a variety of categories, for instance elementary toposes, it is expected that homotopy type theory has homotopy-coherent interpretations in higher toposes. Conversely, the interpretation has inspired new type-theoretic ideas such as higher inductive types (HITs) and Voevodsky's univalence axiom. (See the HoTT book [21] for more about HoTT.)

Most previous formalizations of HoTT used proof assistants that were not originally designed with the homotopy interpretation in mind. In Coq we have both Voevodsky et al.'s *UniMath* project [23] and the HoTT library [4]. In Agda, there is another substantial HoTT library [5]. The former library eschews the use of HITs by instead using Voevodsky's resizing axiom. Common for all of these libraries is that certain tricks are used to accommodate HoTT: resizing is implemented bluntly in UniMath using the inconsistent principle *type-in-type*, while HITs are implemented in the other libraries using "Licata's trick" [15]. There is also an impressive experimental proof assistant implementing cubical type theory [7] which is designed with the homotopy interpretation in mind, but it lacks many features that make a proof assistant convenient to use, and the library is so far rudimentary.

© Springer International Publishing AG 2017
M. Ayala-Rincón and C.A. Muñoz (Eds.): ITP 2017, LNCS 10499, pp. 479–495, 2017.
DOI: 10.1007/978-3-319-66107-0_30

Contributions. In this paper, we report on a new library[1] for HoTT in the proof assistant Lean [18]. Lean is open source and implements dependent type theory. It is designed to have a small kernel, with many features built outside the kernel. We describe Lean in greater detail in Sect. 2. The `cloc` tool[2] reports the library as having 30 400 lines of specification and proof and 3 600 lines of comments. Thus, our library is roughly the same size as the Coq HoTT library, which has 29 800 lines of specification and proof. Our library includes many theorems from synthetic homotopy theory and a large algebraic hierarchy. We describe the library in more detail in Sect. 3. In the library we heavily use cubical techniques for higher path algebra, see Sect. 4. We also have a novel approach to implement HITs, which amounts to having two simple built-in HITs and reducing everything else to those, as described in Sect. 5.

2 The Lean Proof Assistant

Lean [18] is an interactive theorem prover which is mainly developed at Microsoft Research and Carnegie Mellon University. The project was started in 2013 by Leonardo de Moura and has since gained the attention of academics as well as hands-on users. Lean is an open-source program released under the Apache License 2.0 and welcomes additions to its code and mathematical libraries.

In its short history, Lean has undergone several major changes. The second version (Lean 2) supports two kernel modes. The standard mode is for proof irrelevant reasoning, in which Prop, the bottom universe, contains types whose objects are considered to be judgmentally equal. Since this is incompatible with homotopy type theory, a second HoTT mode was added, where proof irrelevance is not present. In 2016, the third major version of Lean (Lean 3) was released [17]. In this version, many components of Lean have been rewritten. Of note, the unification procedure has been restricted, since the full higher-order unification which is available in Lean 2 can lead to timeouts and error messages that are unrelated to the actual mistakes. Due to certain design decisions, such as proof erasure in the virtual machine and a function definition package which requires axiom K [11], the homotopy type theory mode is currently not supported in Lean 3. This has led to the situation that the homotopy type theory library is kept in the still maintained but not further developed Lean 2. In the future we hope that we will find a way to support a version of homotopy type theory in Lean 3 or a fork thereof.

The HoTT kernel of Lean 2 provides the following primitive notions:

- *Type universes* Type.{u} : Type.{u + 1} for each universe level $u \in \mathbb{N}$. In Lean, this chain of universes is non-cumulative, and all universes are predicative.

---

[1] Available as part of: https://github.com/leanprover/lean2.

[2] https://github.com/AlDanial/cloc.

- *Function types* A → B : Type.{max u v} for types A : Type.{u} and B : Type.{v} as well as *dependent function types* Πa, B a :Type.{max u v} for each type A : Type.{u} and type family B : A → Type.{v}. These come with the usual β and η rules.
- *inductive types* and *inductive type families*, as proposed by Peter Dybjer [10]. Every inductive definition adds its constructors and dependent recursors to the environment. Pattern matching is *not* part of the kernel
- two kinds of *higher inductive types*: n-truncation and (typal) quotients (cf. Sect. 5).

Outside the kernel, Lean's elaborator uses backtracking search to infer implicit information. It does the following simultaneously.

- The elaborator fills in *implicit arguments*, which can be inferred from the context, such as the type of the term to be constructed and the given explicit arguments. Users mark implicit arguments with curly braces. For example, the type of equality is eq : Π{A : Type}, {A → A → Type}, which allows the user to write eq a_1 a_2 or a_1 = a_2 instead of @eq A a_1 a_2. The symbol @ allows the user to fill in implicit arguments explicitly. The elaborator supports both first-order unification and higher-order unification.
- We can mark functions as *coercions*, which are then "silently" applied when needed. For example, we have equivalences f : A ≃ B, which is a structure consisting of a function A → B with a proof that the function is an equivalence. The map (A ≃ B) → (A → B) is marked as a coercion. This means that we can write f a for f : A ≃ B and a : A, and the coercion is inserted automatically.
- Lean was designed with *type classes* in mind, which can provide canonical inhabitants of certain types. This is especially useful for algebraic structures (see Sect. 3.3) and for type properties like truncatedness and connectedness. Type class instances can refer to other type classes, so that we can chain them together. This makes it possible for Lean to automatically infer why types are n-truncated if our reasoning requires this, for example when we are eliminating out of a truncated type. For example we show that the type of functors between categories C and D is equivalent to an iterated sigma type.

$$(\Sigma (F_0 : C \to D) (F_1 : \Pi \{a\ b\}, \text{hom}\ a\ b \to \text{hom}\ (F_0\ a)\ (F_0\ b)),$$
$$(\Pi (a), F_1 (ID\ a) = ID\ (F_0\ a)) \times$$
$$(\Pi \{a\ b\ c\} (g : \text{hom}\ b\ c) (f : \text{hom}\ a\ b),$$
$$F_1 (g \circ f) = F_1\ g \circ F_1\ f)) \simeq \text{functor}\ C\ D$$

Note the use of coercions here: F_0 : C → D really means a function from the objects of C to the objects of D. From this equivalence, Lean's type class inference can automatically infer that functor C D is a set if the objects of D form a set. Type class inference will repeatedly apply the rules when sigma-types and pi-types are sets, and use the facts that hom-sets are sets and that equalities in sets are sets (in total 20 rules are applied for this example).

- Instead of giving constructions by explicit terms, we can also make use of Lean's *tactics*, which give us an alternative way to construct terms step by step. This is especially useful if the proof term is large, or if the elaboration relies heavily on higher-order unification.
- We can define custom syntax, including syntax with binding. In the following example we declare two custom notations.

```
infix ·  := concat
notation `Σ` binders `, `r:(scoped P, sigma P) := r
```

The first line allows us to write p · q for path concatenation concat p q. The second line allows us to write Σ x, P x instead of sigma P. This notation can also be chained: Σ (A : Type) (a : A), a = a means sigma (λ(A : Type), sigma (λ(a : A), a = a)).

2.1   Consistency of HoTT Lean

Voevodsky's model of univalence in simplicial sets [14] covers the type theory with empty, unit, disjoint sums, pi, sigma, identity, and W-types and one univalent universe à la Tarski closed under the these type formers. The model validates the β and η rules for function types.

The cubical type theory of [8] interprets Martin-Löf type theory using Andrew Swan's construction of the identity types. (The cubical path types of this model do not satisfy the computation rule for identity types.) It has been checked that the corresponding model in cubical sets based on de Morgan algebras models two HITs, namely suspension and propositional truncation. (The model even satisfies the computation rules for the path constructors.) The technique used also covers pushouts, so by the reduction of n-truncation to pushouts [19], the models covers all n-truncations. We believe this model also covers all the ordinary inductive families supported by Lean, but this has not been checked in detail.

Mark Bickford's formalization of the cubical model[3] covers a whole hierarchy of universes like we have in the Lean kernel. It additionally verifies some novel type constructors such as a higher dimensional intersection type.

These models provide us with high confidence that the logic implemented by the Lean HoTT kernel is consistent. Furthermore, the kernel is very small compared to other kernels implementing dependent type theory. The kernel does not contain pattern matching, a termination checker, fixpoint operations or module management. This increases the confidence that the kernel implements the logic correctly. Furthermore, the only thing we do outside the kernel to extend the logic is to posit the univalence axiom; we do not use type-in-type or Licata's trick or anything else which might introduce inconsistencies.

---

[3] http://www.nuprl.org/wip/Mathematics/cubical!type!theory/.

3   The Structure of the Library

In this section we describe the overall structure of the homotopy type theory library and we highlight some examples.

The library contains a markdown file in each folder to describe the contents of the files in that folder. For readers familiar with [21], the library includes a file[4] `book.md` that describes where in the library the various parts of the book are formalized.

Figures 1, 2, 3 and 4 contain graphs of the files in various parts of the library; the edges denote the dependencies of the files. Each folder contains a file `default` which only contains imports of various files in the folder and which is imported if the user imports the folder. There are also three additional folders: `types` (see Subsect. 3.2), `cubical`, related to the cubical methods discussed in Sect. 4; and `hit`, related to higher inductive types as discussed in Sect. 5. There are also some files in the root folder which we do not describe here.

There is a separate `Spectral` repository,[5] the goal of which is to formalize the Serre Spectral Sequence, and which will be merged into the Lean-HoTT library in the future. Some examples below are located in this repository.

3.1   The Initial Part of the Library

Figure 1 illustrates the files of the initial part of the library. These files are imported by default when opening a Lean file. The very first file, `datatypes`, defines the basic datatypes, such as `unit`, `empty`, `eq`, `prod`, `sum`, `sigma`, `bool`, `nat`. Higher up, the `path` file develops the basic properties of the identity type (also called equality or identification type) in HoTT. This includes the basic properties of homotopies, transport and the low-dimensional ∞-groupoid structure of types.

In the rest of the files we define equivalences, posit the univalence axiom and derive function extensionality from univalence (in `equiv`, `ua` and `funext`, respectively). However, in order to be able to track which definitions only depend on function extensionality and not univalence, via the print axioms command, we also add function extensionality directly as an axiom.

Lastly, we develop n-truncated types, initialize the primitive HITs, prove that types with decidable equality are sets [12] and define the basic notions of pointed types (in `trunc`, `hit`, `hedberg` and `pointed`, respectively).

3.2   Facts About Types

The files in subdirectory `types` develop in more detail the properties and constructions related to individual types and type formers. For types like `sum`, `sigma` and `pi` we characterize the equality in that type, define the functorial action

---

[4] https://github.com/leanprover/lean2/blob/master/hott/book.md.

[5] It has 7 700 lines of code and 1 400 lines of comments. It is available at https://github.com/cmu-phil/Spectral.

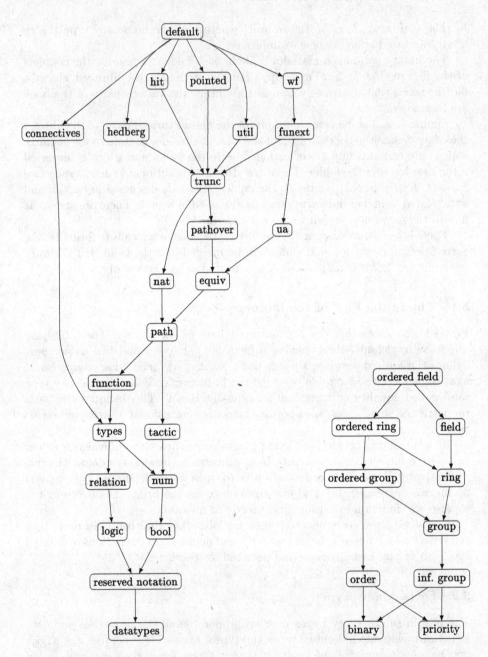

Fig. 1. The initial part of the library   **Fig. 2.** The algebraic hierarchy

and show that the functorial action preserves equivalences. In univ we prove properties of type universes, such as the object classifier property. Of particular importance is the file pointed, which contains properties of pointed types, maps, equivalences and homotopies, which contains over 2 000 lines (also counting the corresponding file in the Spectral repository).

3.3   The Algebraic Hierarchy

The algebraic hierarchy, all in the algebra subdirectory, is structured as seen in Fig. 2. That figure does not contain files that depend on the category theory sublibrary. The algebraic hierarchy defines common algebraic structures, starting with small structures, like semigroups and partial orders, and extending them to groups, rings, all the way up to discrete linear ordered fields. (Discrete means that the order is decidable.)

We combine the "partially bundled" approach with the "fully bundled" app-roach in the algebraic hierarchy, similar to how algebraic structres in the Coq library are defined [20]. The partially bundled approach means that given a type A we define what it means that A has a group-structure or ring-structure. This is used for concrete structures, and we use type classes to infer these inhabi-tants. For example, we prove that $\mathbb{N}$ forms a decidable linear ordered semiring, and mark this as an instance. If we want to show that for n m k : $\mathbb{N}$ we have (n * m) * k = n * (m * k), we can use mul.assoc, the theorem that multi-plication in any semigroup is associative. Then type class inference will try to show that $\mathbb{N}$ is a semigroup, and it will use the instance that every decidable linear ordered semiring is a semigroup. We use Lean's extend syntax to easily define new algebraic structures. For example, the following code defines a struc-ture ab_group of abelian groups, which consists of the fields of both group and comm_monoid. Also, the instances ab_group A → group A and ab_group A → comm_monoid A are automatically generated.

```
structure ab_group [class] (A : Type)
  extends group A, comm_monoid A
```

We use the fully bundled approach when doing group theory and other algebra. A bundled structure is a type together with a structure on that type. For example, this is the definition of a bundled group:

```
structure Group := (carrier : Type) (struct : group carrier)
```

We define Group.carrier to be a coercion. We make Group.struct an instance, which means that if we have to synthesize a term of type group (Group.carrier G), Lean will automatically find this instance. We use the bundled structures for group theory. For example, if G H : Group then we define the product group G ×g H. We use ×g for the product of two groups to disambiguate it from other products, like the product of two types, two pointed types or two truncated types (type class inference does not work well to disambiguate here, since all these structures coerce to types).

If we go back to the example (n * m) * k = n * (m * k) on N, we also interpret the multiplication symbol on N using type class inference. In this case, Lean will try to find an instance of has_mul N, where has_mul is a type class stating that the type has a multiplication. Lean can find this instance since we have a general instance semigroup A → has_mul A. However, since we want to also have additive semigroups, we have a different notion of additive semigroups, add_semigroup, with corresponding instance add_semigroup A → has_add A.

To minimize overhead, we can define additive structures as the multiplicative counterpart, and then prove theorems about additive structures by using the corresponding theorem for multiplicative structures. We do have to manually define the instances for additive structures. Here is an example for semigroups:

```
definition add_semigroup [class] : Type →  Type := semigroup
definition has_add_of_add_semigroup [instance] (A : Type)
  [s : add_semigroup A] : has_add A :=
has_add.mk (@semigroup.mul A s)
definition add.assoc {A : Type} [s : add_semigroup A] (a b c :
  A)
  : (a + b) + c = a + (b + c) :=
@mul.assoc A s a b c
```

This approach has advantages and disadvantages. An advantage is that theorem names are different for additive structures and multiplicative structures, so we can write add.assoc for associativity of addition and mul.assoc for associativity of multiplication. Furthermore, we can easily define a ring by extending an additive abelian group and a multiplicative monoid (plus distributivity).

A disadvantage is that operations that are traditionally not written using + or *, such as concatenating two lists, do not fall in either category. Also, in our formalization we make a distinction between additive and multiplicative groups. Since we define additive groups as multiplicative groups, we can still apply theorems about multiplicative groups to additive groups, but some care is needed when doing this: if one applies a theorem about multiplicative groups with assumption n * k = 1 to an additive group, the new subgoal becomes n * k = 1, even though in an additive group this really means n + k = 0.

All the algebraic structures we mentioned so far (not including has_mul and has_add) are assumed to be sets, i.e., 0-truncated. We also have variants of some of these structures which are not assumed to be sets. For example, we have inf_group and inf_ab_group, which are (abelian) groups without the assumption that they are sets, but without higher coherences. This is useful for, e.g., loop spaces or pointed maps into loop spaces, since those types are not groups, but will become groups (the homotopy and cohomology groups) after applying set-truncation.

3.4   Homotopy Theory

The homotopy theory part of the library is organized as shown in Fig. 3. Almost all results in Chap. 8 of the HoTT book have been formalized in Lean. In particular it contains various results about connectedness, a version of the Freudenthal suspension theorem, the complex and quaternionic Hopf fibration [6] and the long exact sequence of homotopy groups. Together these results show:

```
definition πnSn (n : ℕ) : πg[n+1] (S* (n+1)) ≃g gℤ
definition π3S2 : πg[3] (S* 2) ≃g gℤ
```

This is to say that the n-th homotopy group of the n-sphere (for n ≥1) and the 3rd homotopy group of the 2-sphere are group isomorphic to the integers. Of note here is the notation πg[n] A which denotes the n-th homotopy group of A, as a group. In contrast, we also have the operation π[n] A which is the n-th homotopy group of A as a pointed type, which is also defined for n = 0. Originally, we defined ghomotopy_group : ℕ → Type → Group where ghomotopy_group n A is the (n+1)-st homotopy group of A and we had notation πg[n+1] A for this. However, this requires the user to write the third homotopy group as πg[2+1]. To remedy this, we changed the definition of ghomotopy_group to have type Π(n : ℕ) [H : is_succ n], Type → Group, where H is a proof that n is a successor of a natural number, and which is synthesized using type class inference.

We also prove Whitehead's principle for truncated types and the Seifert-van Kampen theorem, and we define the Eilenberg-Maclane spaces and show that they are unique. Furthermore, we define operations on types of homotopy theoretic significance, such as cofibers, joins, and wedge and smash products, and prove various properties about them, such as the associativity of the join and smash products and the fact that the suspension and smash product have right adjoints, respectively loop spaces and pointed maps.

3.5 Category Theory

It seems a constant across many libraries of formalized mathematics that the development of category theory takes up a substantial fraction of the files, and our library is the same way, as can be seen in Fig. 4. Highlights include the Yoneda lemma and the Rezk completion [1].

As an example from this part of the library, consider this excerpt which formalizes the fact that the Yoneda embedding preserves existing limits:

```
definition yoneda_embedding (C : Precategory) : C ⇒ cset ^c
    Cᵒᵖ

variables {C D : Precategory}
definition preserves_existing_limits [class] (G : C ⇒ D) :=
Π(I : Precategory) (F : I ⇒ C)
    [H : has_terminal_object (cone F)],
    is_terminal (cone_obj_compose G (terminal_object (cone F)))

theorem preserves_existing_limits_yoneda_embedding
    (C : Precategory)
    : preserves_existing_limits (yoneda_embedding C)
```

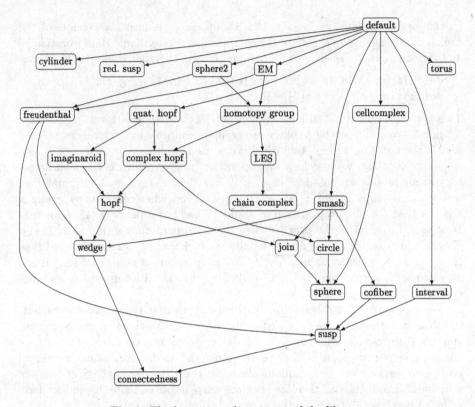

Fig. 3. The homotopy theory part of the library

4   Path Algebra and Cubical Methods

The core innovation in homotopy type theory is its new interpretation of equality. In contrast to proof irrelevant Martin-Löf type theory, we need to be careful about choosing well-behaved equality proofs in the library since we might need to prove lemmas about these proof objects themselves. We want to maintain brevity using tactics and equational rewriting while making sure that the generated proofs do not become unwieldy.

After defining equality on a type A in the library's prelude as an inductive type family over two objects of A which is generated by the reflexivity witness refl : Π(x : A), x = x, we can provide operations and proofs for the basic *higher groupoid structure* of these "equality paths": Concatenation p · q and inversion p^{-1} of paths as well as proofs about associativity and cancellation. These are constructed using the dependent recursor of equality which we call *path induction* and which, for each a : A, provides a function Π(b : A) (p : a = b), P b p given the reflexivity case P a (refl a). Likewise, we can prove the functoriality of functions with respect to equality: For a function f : A → B and p : a = a' we define ap f p : f a = f a' by induction on p. Using an equality p : a = a' in a type A to compare elements of two

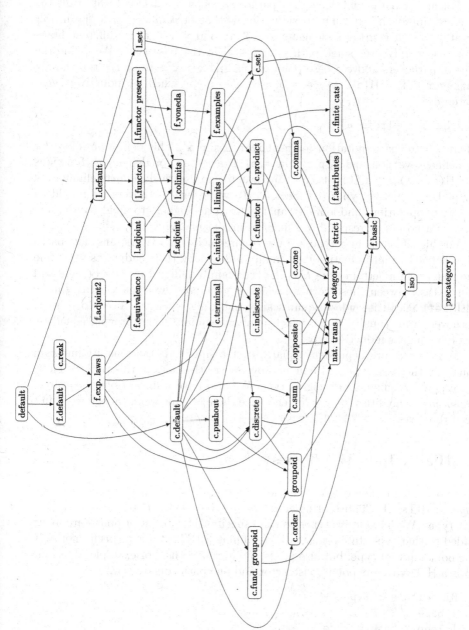

Fig. 4. The category theory part of the library (c = constructions, f = functor, l = limits)

fibers in a type family C over A, we define the *transport* of an element x : C a along p as p ▸ x : C a'.

For higher paths and dependent paths, we follow what Dan Licata calls the "cubical approach" [16]. The basic notion is that of *pathover*, or a "path over a path", which compares elements x : C a and y : C a' in different fibers of a type family over some path p : a = a' in the base type. We define the type of pathovers above a base point a : A and x : C a to be the type family pathover C x : Π{a': A}, a = a'→ C a'→ Type which is inductively generated by

```
idpo : x =[refl a] x
```

where x =[p] x' is notation for pathover ₋ x p x'. This definition allows us to define a version apd f p : f a =[p] f a' of ap for dependent functions f : Π(a : A), C a. It is also used by Lean to express the dependent eliminators for higher inductive types (c.f. Sect. 5). To work with pathovers we provide a variety of operations and lemmas, analogous to the higher groupoid structure of paths. Pathovers correspond to equalities in a sigma type.

For higher paths in a type, we use squares and squareovers. Just like paths were defined as an inductive type family indexed by their endpoints we define the squares in a certain type A as the type family indexed by four corners and four paths between those corners, which is generated by some identity square with refl on all its sides. Squares arise naturally when you need to prove a pathover in an equality type, which is often required when proving equalities involving higher inductive types.

Squareovers are dependent squares over a square. It takes as arguments a square in the base type and four pathovers over the sides of this square. These correspond to squares in a sigma type. We also have a library of cubes three-dimensional equalities. We could generalize these to cubeovers, though we didn't need those yet.

5   Higher Inductive Types

One novel idea in homotopy type theory is the introduction of *higher inductive types* or HITs [21, Chap. 6]. Higher inductive types are a generalization of inductive types. With inductive types you can specify which terms or points are freely added to that type. In contrast, when defining a HIT, you can specify not only the points in that type, but also paths and higher paths. For example, the circle S^1 is a HIT with one point constructor and one path constructor:[6]

```
HIT circle : Type :=
| base : S¹
| loop : base = base
```

---

[6] Although we use syntax inspired by the Lean syntax for inductive types, this is not valid syntax in Lean.

This means that the circle is generated by one point and one path
`loop : base = base`. There will be more loops in the circle, such as `refl base`
and `loop · loop` and `loop`$^{-1}$, which are all different. Higher inductive types have
elimination principles analogous to those of ordinary inductive types.

The most commonly used proof assistants which have HoTT support (such as
Coq and Agda) do not support HITs natively. Just adding HITs as constants is
not satisfactory, because then the computation rules are not judgmental equalities. Instead, users of Coq use "Dan Licata's trick" [15]. The idea is that to define
a higher inductive type, one first defines a private inductive type inside a module
with only the point constructors, and then adds the path constructors as axioms.
One then defines the desired induction principle using the induction principle
of the private inductive type and adds the computation rules of this induction
principle on paths as additional axioms. Then the user closes the module, and
the result is that only the data of the higher inductive type are accessible, while
the induction principle of the private inductive type is hidden. This ensures that
the computation rules are judgmentally true for point constructors (but not for
path constructors), but a disadvantage is that inside the module inconsistent
axioms were assumed, and one needs to trust that the code in these modules
does not introduce an inconsistency in the system. In Agda the rewriting feature
is used so that users can extend the kernel with judgmental rewrite rules, though
there are no checks for any rewrite rule declared in this way.

In Lean we follow an approach similar to Agda rewriting feature, by building in judgmental rewrite ryles. However, we only extend the kernel with the
rewrite rules for two "trusted" higher inductive types, namely the n-truncation
and the typal quotient (quotient for short). The quotient is parameterized by a
type A and a family of types R : A $\rightarrow$ A $\rightarrow$ Type. So "typal" (the adjective of
"type") means that we quotient by a family of types and not a family of mere
propositions. The quotient is the following HIT:

```
HIT quotient (A : Type) (R : A →  A →  Type) : Type :=
| i : A →  quotient A R
| e : Π{x y : A}, R x y →  i x = i y
```

For the n-truncation and the quotient, we add the type formation rule, point and
path constructors, and induction principle as constants/axioms.[7] Then we add
the judgmental computation rules for the point constructors to the Lean kernel;
the Lean kernel is extensible in such a way that certain new computation rules
can be added to it. After that, we add the computation rules on paths as axioms.
As remarked in Subsect. 2.1, we know that the resulting type theory is consistent,
because n-truncations and typal quotients can be reduced to pushouts, and type
theory with univalent universes closed under pushouts is modeled by [8].

Given these two HITs, we define all other HITs in the Lean HoTT
library using just these two. Some reductions are simple, for example the

---

[7] For the n-truncation we treat the fact that the new type is n-truncated as a "path-
constructor." In [21, Sect. 7.3] it is explained that the fact that a type is n-truncated
can be reduced to (recursive) path constructors.

homotopy pushout of $f : A \to B$ and $g : A \to C$ is the quotient on type $B + C$ with the edges R defined as an inductive family with constructor $\Pi(a : A), R \ (\text{inl} \ (f \ a)) \ (\text{inr} \ (g \ a))$. Proving the usual induction principle for the pushout is then trivial. Given the pushout, we have defined the other usual HITs: the suspension, circle, join, smash, wedge, cofiber, mapping cylinder and spheres. In particular, we define circle as sphere 1, which is susp (susp empty). We can then *prove* the usual induction principle for this type, and it satisfies the computation rules on the point constructors judgmentally.

We can also define HITs with 2-path constructors using quotients. This uses a method similar to the hubs-and-spokes method described in [21, Sect. 6.7]. From the elimination principle of the circle it follows that for any path $p : x = x$ in type A we can define a map $f : S^1 \to A$ with ap f loop = p by circle induction. Then we can prove the equivalence

$$(p = \text{refl} \ x) \simeq \Sigma(x_0 : A), \Pi(z : S^1), f \ z = x_0$$

This equivalence informally states that filling in a loop is the same as adding a new point x_0, the *hub*, and *spokes* $f \ z = x_0$ for every $z : S^1$, similar to the spokes in a wheel. This means that in a higher inductive type, we can replace a 2-path constructor $p = \text{refl} \ x$ by a new point constructor $x_0 : A$ and a family of 1-path constructors $\Pi(z : S^1), f \ z = x_0$.

However, this does not quite define 2-HITs in terms of the quotient, since this family of path constructors refers to other path constructors (via the definition of f), which is not allowed in quotients. For this reason, we construct 2-HITs using two nested quotients. We first define a quotient with only the 1-paths and the hubs, and then use another quotient to add the spokes.

For a formal treatment of this, we need the following inductive family, which are the paths in a graph:

```
inductive path {A} (R : A →  A →  Type) : A →  A →  Type :=
| of_rel  : Π{a a' : A}, R a a' →  path R a a'
| of_path : Π{a a' : A}, a = a' →  path R a a'
| symm  : Π{a a' : A}, path R a a' →  path R a' a
| trans : Π{a a' a''}, path R a a' →  path R a' a'' →  path R a a''
```

A *specification for a (nonrecursive) 2-HIT* consists of a type A and two families $R : A \to A \to \text{Type}$ and $Q : \Pi\{a \ a': A\}, \text{path} \ R \ a \ a' \to \text{path} \ R \ a \ a' \to \text{Type}$. Using this, we define the 2-HIT two_quotient A R Q with constructors

```
HIT two_quotient A R Q : Type :=
| i₀ : A →  two_quotient A R Q
| i₁ : Π{a a' : A}, R a a' →  i₀ a = i₀ a'
| i₂ : Π{a a' : A} {r r' : path R a a'}, Q r r' →
        extend i₁ r = extend i₁ r'
```

where `extend` i_1 `r` is the action of i_1 on paths in R, e.g. `extend` i_1(`trans` r_1 r_2) := `extend` i_1 $r_1 \cdot$ `extend` i_1 r_2. We first define a special case where the 2-path constructor has only reflexivities on the right hand side. We call this `simple_two_quotient A R Q'`, where Q' has type Π(a : A), `path R a a` $\rightarrow$ Type and where

$$i_2' : \Pi\{a\} \{r : \text{path R a } a'\}, Q\ r \rightarrow \quad \text{extend } i_1\ r = \text{refl } (i_0\ a)$$

As mentioned before, we define `simple_two_quotient A R Q` in two steps. We first define a type X with only the 1-path constructors and the hubs:

$$X := \text{quotient A R} + \Sigma(a : A)\ (r : \text{path R a a}),\ Q'\ r$$

We then define `simple_two_quotient A R Q'`:= `quotient X R'` where

```
inductive R' : X → X → Type :=
| mk : Π{a : A} (r : path R a a) (q : Q' r) (x : S¹),
        R' (f q x) (inr (a,q))
```

with `f q` : $S^1 \rightarrow$ X defined by induction so that `ap (f q) (loop)` = `extend (inl oe) r` for `q : Q'r`.

We now prove the expected (dependent) induction principle, (nondependent) recursion principle, and computation rules for this two-quotient. The only computation rule which we did not manage to prove is the computation rule of the induction principle on 2-paths. However, this rule is not necessary to determine the type up to equivalence.

We then define the general version, `two_quotient A R Q`, to be equal to `simple_two_quotient A R Q'` where:

```
inductive Q' : Π{a : A}, path R a a →  Type :=
| q₀ : Π{a a' : A} {r r' : path R a a'},
        Q r r' →  Q' (trans r (symm r'))
```

We then show that `two_quotient A R Q` and `trunc n (two_quotient A R Q)` have the right elimination principles and computation rules. It (perhaps surprisingly) requires quite some work to show that the correct computation rules of the truncated version follow from the untruncated version.

This allows us to define all nonrecursive HITs with point, 1-path and 2-path constructors. For example, we define the torus T^2:= `two_quotient unit R Q`. Here R $\star\star$ = `bool`, which gives two path constructors p and q from the basepoint to itself. Q is generated by the constructor q_0 : Q (`trans [ff] [tt]`) (`trans [ff] [tt]`) where `[b]` is notation for `of_rel b`. This gives a path p $\cdot$ q = q $\cdot$ p. We also define the *groupoid quotient*: For a groupoid G we define its quotient as `trunc 1 (two_quotient G (@hom G) Q)` where:

```
inductive Q :=
| q₀ : Π(a b c) (g : hom b c) (f : hom a b),
        Q (g ∘ f) (trans f g)
```

If G is just a group (considered as a groupoid with a single object), then the groupoid quotient of G is exactly the Eilenberg-MacLane space K G 1.

We have also defined the propositional truncation just using quotients in Lean [9]. An extension of this construction to n-truncations has been given on paper [19]. If we formalize this generalization in Lean, it is possible to remove n-truncations as a primitive HIT in Lean.

6   Conclusion

We have described the HoTT library for the Lean proof assistant, which formalizes many results in HoTT, including higher inductive types, synthetic homotopy theory and category theory. It has a large library of pointed types, and uses cubical methods for reasoning about higher paths. In the future, we hope to make a HoTT mode for Lean 3, possibly using a version of cubical type theory [2,8].

Acknowledgments. We wish to thank the members of the HoTT group at Carnegie Mellon University for many fruitful discussions and Lean hacking sessions, and in particular Steve Awodey and Jeremy Avigad who have been very supportive of our work. Additionally, we deeply appreciate all the times Leonardo de Moura fixed an issue in the Lean kernel to accommodate our library. Lastly, we want to thank all contributors to the HoTT library and the Spectral repository, most notably Egbert Rijke and Mike Shulman.

The first and second authors gratefully acknowledge the support of the Air Force Office of Scientific Research through MURI grant FA9550-15-1-0053. Any opinions, findings and conclusions or recommendations expressed in this material are those of the authors and do not necessarily reflect the views of the AFOSR.

References

1. Ahrens, B., Kapulkin, K., Shulman, M.: Univalent categories and the Rezk completion. Mathe. Struct. Comput. Sci. **25**(5), 1010–1039 (2015)
2. Angiuli, C., Harper, R., Wilson, T.: Computational higher-dimensional type theory. In: Proceedings of the 44th Annual ACM SIGPLAN-SIGACT Symposium on Principles of Programming Languages, POPL 2017. ACM (2017)
3. Awodey, S., Warren, M.A.: Homotopy theoretic models of identity types. In: Mathematical Proceedings of the Cambridge Philosophical Society, vol. 146, no. 1, pp. 45–55 (2009)
4. Bauer, A., Gross, J., LeFanu Lumsdaine, P., Shulman, M., Sozeau, M., Spitters, B.: The HoTT library: a formalization of homotopy type theory in Coq. ArXiv e-prints, October 2016
5. Brunerie, G., Hou (Favonia), K.B., Cavallo, E., Finster, E., Cockx, J., Sattler, C., Jeris, C., Shulman, M., et al.: Homotopy Type Theory in Agda (2017). Code library. https://github.com/HoTT/HoTT-Agda
6. Buchholtz, U., Rijke, E.: The cayley-dickson construction in homotopy type theory. ArXiv e-prints, October 2016
7. Cohen, C., Coquand, T., Huber, S., Mörtberg, A.: Cubical type theory. Code library. https://github.com/mortberg/cubicaltt

8. Cohen, C., Coquand, T., Huber, S., Mörtberg, A.: Cubical type theory: a constructive interpretation of the univalence axiom. In: 21st International Conference on Types for Proofs and Programs (TYPES 2015). LIPIcs. Leibniz International Proceedings in Informatics, Schloss Dagstuhl. Leibniz-Zent. Inform., Wadern (2016, to appear)

9. van Doorn, F.: Constructing the propositional truncation using non-recursive hits. In: Proceedings of the 5th ACM SIGPLAN Conference on Certified Programs and Proofs, pp. 122–129. ACM (2016)

10. Dybjer, P.: Inductive families. Formal Aspects Comput. **6**(4), 440–465 (1994)

11. Goguen, H., McBride, C., McKinna, J.: Eliminating dependent pattern matching. In: Futatsugi, K., Jouannaud, J.-P., Meseguer, J. (eds.) Algebra, Meaning, and Computation. LNCS, vol. 4060, pp. 521–540. Springer, Heidelberg (2006). doi:10.1007/11780274_27

12. Hedberg, M.: A coherence theorem for Martin-Löf's type theory. J. Funct. Program. **8**(4), 413–436 (1998)

13. Hofmann, M., Streicher, T.: The groupoid interpretation of type theory. In: Twentyfive Years of Constructive Type Theory (Venice, 1995). Oxford Logic Guides, vol. 36, pp. 83–111. Oxford University Press, New York (1998)

14. Kapulkin, C., Lumsdaine, P.L.: The simplicial model of univalent foundations (after voevodsky) (2012, preprint)

15. Licata, D.: Running circles around (in) your proof assistant; or, quotients that compute. blog post, April 2011. http://homotopytypetheory.org/2011/04/23/running-circles-around-in-your-proof-assistant/

16. Licata, D., Brunerie, G.: A cubical approach to synthetic homotopy theory. In: Proceedings of the 2015 30th Annual ACM/IEEE Symposium on Logic in Computer Science (LICS), LICS 2015, pp. 92–103. IEEE Computer Society, Washington, DC (2015)

17. de Moura, L., Ebner, G., Roesch, J., Ullrich, S.: The Lean theorem prover. Slides, January 2017. https://leanprover.github.io/presentations/20170116_POPL

18. de Moura, L., Kong, S., Avigad, J., van Doorn, F., van Raumer, J.: The lean theorem prover (system description). In: Felty, A.P., Middeldorp, A. (eds.) CADE 2015. LNCS, vol. 9195, pp. 378–388. Springer, Cham (2015). doi:10.1007/978-3-319-21401-6_26

19. Rijke, E.: The join construction. ArXiv e-prints, January 2017

20. Spitters, B., van der Weegen, E.: Developing the algebraic hierarchy with type classes in Coq. In: Kaufmann, M., Paulson, L.C. (eds.) ITP 2010. LNCS, vol. 6172, pp. 490–493. Springer, Heidelberg (2010). doi:10.1007/978-3-642-14052-5_35

21. The Univalent Foundations Program: Homotopy Type Theory: Univalent Foundations of Mathematics. Institute for Advanced Study (2013). http://homotopytypetheory.org/book

22. Voevodsky, V.: A very short note on the homotopy λ-calculus (2006). http://www.math.ias.edu/vladimir/Site3/Univalent_Foundations_files/Hlambda_short_current.pdf

23. Voevodsky, V., Mörtberg, A., Ahrens, B., Lelay, C., Pannila, T., Matthes, R.: UniMath: Univalent Mathematics (2017). Code library. https://github.com/UniMath

Verifying a Concurrent Garbage Collector Using a Rely-Guarantee Methodology

Yannick Zakowski[1]([⊠]), David Cachera[1], Delphine Demange[2], Gustavo Petri[3], David Pichardie[1], Suresh Jagannathan[4], and Jan Vitek[5]

[1] ENS Rennes – IRISA – Inria, Rennes, France
yannick.zakowski@irisa.fr
[2] Université Rennes 1 – IRISA – Inria, Rennes, France
[3] IRIF – Université Paris Diderot, Paris, France
[4] Purdue University, West Lafayette, USA
[5] Northeastern University, Boston, USA

Abstract. Concurrent garbage collection algorithms are an emblematic challenge in the area of concurrent program verification. In this paper, we address this problem by proposing a mechanized proof methodology based on the popular Rely-Guarantee (RG) proof technique. We design a specific compiler intermediate representation (IR) with strong type guarantees, dedicated support for abstract concurrent data structures, and high-level iterators on runtime internals. In addition, we define an RG program logic supporting an incremental proof methodology where annotations and invariants can be progressively enriched.

We formalize the IR, the proof system, and prove the soundness of the methodology in the Coq proof assistant. Equipped with this IR, we prove a fully concurrent garbage collector where mutators never have to wait for the collector.

1 Introduction

Modern programming languages like ML, Java, and C# rely on garbage collection (GC) for the automatic reclamation of memory no longer used by the application. The GC is considered to be one of the most subtle parts of modern runtime systems, carefully engineered to minimize runtime overheads of the applications it supports. A family of garbage collection algorithms, named *on-the-fly* garbage collectors [2], allows the detection of garbage and its reclamation to occur concurrently with an application's threads. Such algorithms are notably difficult to implement, test, and prove, and constitute a significant challenge for mechanized verification. Many on-the-fly algorithms are inherently racy, and some algorithms never require application threads (called *mutators*) to wait for the *collector* thread, which detects and frees unused memory. This paper focuses on an emblematic algorithm in this landscape [3–5], where no locks are required – *i.e.* it is *lock-free*.

This material is based upon work supported by grants ANR 14-CE28-0004, NSF CCF-1318227, CCF-1544542, SHF-1518844 and ONR N00014-15-1-2332.

M. Ayala-Rincón and C.A. Muñoz (Eds.): ITP 2017, LNCS 10499, pp. 496–513, 2017.
DOI: 10.1007/978-3-319-66107-0_31

This challenge has been identified and addressed in various settings [8,9,11,12]. This paper provides an independent proof, and it explores a different proof method in the design space. First, the backbone of the formalization is a new compiler intermediate representation, named RtIR, which we use to implement the garbage collector. Our experience implementing on-the-fly garbage collectors [20] indicates that the choice of programming abstractions is of paramount importance in reasoning and optimizing this kind of algorithm. This concern necessitates a representation that makes the expression and proof of invariants tractable. Moreover, in this work, we strive to make our proof well suited to the context of a larger project, described in [1,14], aiming at the formal verification of a compiler for concurrent, managed languages. Our intermediate representation has special support for the implementation of efficient runtime mechanisms: (i) strong type guarantees, (ii) abstract concurrent data structures, (iii) high-level iterators for reflective inspection of objects, used to implement low-level services, *e.g.* ensuring the garbage collector visits every live object (iv) native support for threads, and (v) native support for the root management of a concurrent garbage collector (each thread must be able to iterate over the set of memory references it can access directly).

Another important characteristic of our approach is the dedicated rely-guarantee program logic that accompanies our intermediate representation. While previous approaches [8,9,12] attack the proof by means of an abstract state transition system requiring a monolithic global invariant to be established, we follow the well established rely-guarantee [15] (RG) methodology. RG is a major technique for proving the correctness of concurrent programs that provides explicit thread-modular reasoning. In this setting, interferences between threads are described using binary relations: *relies* and *guarantees*. Each thread is proved correct under the assumption it is interleaved with threads fulfilling a *rely* relation. The effect of the thread itself on the shared memory must respect its *guarantee* relation. This guarantee must also be coherent with respect to the relies that the other threads assume. Being able to reason in a thread modular way is key to realize a tractable correctness proof because it avoids the need to explicitly consider all possible interleavings. We prove the soundness of our RG logic, and develop a set of tactics that reduce the proof effort required to discharge the invariants.

Finally, we report on an original *incremental* proof technique that we put in place to carry out this massive endeavour. Starting from the full GC implementation, we progressively annotate the program in order to prove stronger and stronger invariants. At each level, dedicated specification annotations and tactics allow us to refine and reuse what has been proved at the previous levels.

Using the Coq proof assistant, we achieved the following formalizations: (i) the syntax, semantics and the soundness of an RG program logic for our intermediate representation, (ii) a number of tactics and structural lemmas to facilitate the so-called *stability proofs* required by the RG methodology, (iii) a realistic implementation of Domani *et al.*'s GC algorithm [5] in our intermediate representation and (iv) an RG proof ensuring the correctness of the GC: the collector never frees references accessible by the running threads. Our formal development is available online [7].

2   The RTIR ITNTERMEDIATE REPRESENTATION

2.1   Syntax

Figure 1 shows the syntax of RTIR (RunTime IR). It provides two kinds of variables: *global* or *shared* variables that can be accessed by all threads, and *local variables* used for thread-local computations. Expressions (e) are built from constants and local variables with the usual arithmetic and boolean operators. Commands include standard instructions, such as skip, assume e, local variable update $x = e$, and classic combinators: sequencing, non-deterministic choice ($c_1 \oplus c_2$), and loops. The usual conditional (if e then c_1 else c_2) can be defined as (assume e; c_1) $\oplus$ (assume $!e$; c_2), where we write $!e$ for the boolean negation of e. While loops and repeat-until loops can be encoded similarly. RTIR also provides atomic blocks (atomic c). In our GC, we use atomic blocks only to add ghost-code – code only used for the proof, not taking part in the computation – and to model linearizable data structures. These atomic constructs can be refined into low-level, fine-grained implementations using techniques like [14, 26].

$$X, Y \in \text{gvar} \qquad x, y \in \text{lvar} \qquad t, m, C \in \text{tid} \qquad f \in \text{fid} \qquad rn \in \text{list fid}$$

$$
\begin{array}{llll}
\text{cmd} \ni c := \text{skip} & |\ \text{assume } e & |\ x = e & \\
|\ c_1 \,;\, c_2 & |\ c_1 \oplus c_2 & |\ \text{loop}(c) & |\ \text{atomic } c \\
|\ x = \text{alloc}(rn) & |\ \text{free}(x) & |\ \text{isFree?}(x) & \\
|\ x = Y & |\ X = e & |\ x = y.f & |\ x.f = e \\
|\ x.\text{push}(y) & |\ x = y.\text{empty?}() & |\ x = y.\text{top}() & |\ x.\text{pop}() \quad |\ X = y.\text{copy}() \\
|\ \text{foreach } (x \text{ in } l) \text{ do } c \text{ od} & & |\ \text{foreachField } (f \text{ of } x) \text{ do } c \text{ od} & \\
|\ \text{foreachObject } x \text{ do } c \text{ od} & & |\ \text{foreachRoot } (x \text{ of } t) \text{ do } c \text{ od} &
\end{array}
$$

Fig. 1. Simplified syntax of RTIR. Proof annotations elided.

Instruction alloc(rn) allocates a new object in the heap by extracting a fresh reference from the freelist – a pool of unused references – and initializing all of its fields in the record name rn to their default value. Conversely, free puts a reference back into the freelist. Instruction isFree? looks up the freelist to test whether a reference is in it. We use these memory management primitives to implement the GC.

In RTIR, basic instructions related to shared-memory accesses are fine-grained, *i.e.* they perform exactly one global operation (either read or write). These include loads and stores to global variables and field loads and updates. This allows us, when conducting the proofs, to consider each possible interleaving of memory operations arising from different threads, while keeping the semantics reasonably simple. Apart from these basic memory accesses, RTIR provides abstract concurrent queues which implement the *mark buffers* of [5], accessible through standard operations $y = x.\text{top}()$, $x.\text{pop}()$, $x.\text{push}(y)$, $x = y.\text{empty?}()$. The use of these buffers, necessary for the implementation of the GC, will be made clear in Sect. 4. While we could implement these data structures directly in RTIR, we argue that to carry out the proof of the GC, it is better to reason about them at a higher level, and hence to assume that they behave atomically. Implementing these data structures in a correct and linearizable [13] fashion is an orthogonal problem, that we address separately [26]. Mark buffers also provide an operation $X = y.\text{copy}()$, to perform a deep copy, only used in ghost code.

A salient ingredient of RTIR is its native support for *iterators*, allowing to easily express many bookkeeping tasks of the GC. The iterator foreach (x *in* 1) do c od, where the variable x can be free in command c, iterates c through all elements x of the static list 1. Some more sophisticated bookkeeping tasks include the visiting of all the fields of a given object, the marking of each of the *roots* – references bound to local variables – of mutators, or the visiting of every object in the heap (performed during the *sweeping* phase). In those cases, the lists of elements to be iterated upon is not known statically, so we provide dedicated iterators. The iterator foreachField (f of x) do c od iterates c on all the fields f of the object stored in x. Command foreachRoot (r of t) do c od iterates over the roots of mutator thread t, while foreachObject x do c od iterates over all objects. We stress the fact that iterators have a fine-grained behavior: the body command c executes in a small-step fashion.

2.2 Operational Semantics

The operational semantics of RTIR is mostly standard. We provide two kinds of operational semantics: (i) a *big-step* semantics, used to define the semantic validity of Hoare-like tuples for basic instructions (see Sect. 3), as well as commands in atomic blocks; (ii) a small-step interleaving semantics used to prove our final soundness results. We only present here the description of execution states, and refer the interested reader to the Coq development [7] for the formal semantics.

Typing Information. The semantics of RTIR is enriched with typing information. Basic types in typ include TNum for numeric constants, TRef for references to regular objects (see below), and TRefSet for non-null references to abstract mark-buffers. Local variables, global variables, and field identifiers are declared to have exactly one of these types, respectively accessible through functions lvar_typ, gvar_typ and fid_typ. RTIR manipulates two kinds of values: numeric values in the Coq type Z and references in ref. Types are mapped to values with the function value of type typ $\rightarrow$ Type.

```
typ ≜ { TNum, TRef, TRefSet }        Definition value (t:typ):Type :=
lvar ≜ varId × typ                      match t with
gvar ≜ varId × typ                      | TNum ⇒ Z
fid ≜ fieldId × typ                     | TRef | TRefSet ⇒ ref end.
```

Execution States. Local (resp. global) environments map local (resp. global) variables to values of their declared type. Environments are hence dependent functions of type:

```
Definition lenv := ∀ x:lvar, value (lvar_typ x).
Definition genv := ∀ X:gvar, value (gvar_typ X).
```

A thread-local state is defined by a local environment and a command to execute. A global state includes a global environment ge and a heap hp – a partial map from references to objects. We consider two distinct kinds of objects: regular objects, mapping fields to values, and abstract mark-buffers.

```
Definition thread_state := (cmd * lenv).
Record gstate := {  ge: genv;                    freelist: ref → bool;
                    hp: ref → option object;   roots: tid → ref → nat }.
```

Global states also include two components essential to the implementation of a GC: roots and a freelist. The freelist is indeed a shared data structure, while roots are considered to be thread-local – mutators are responsible for handling their own roots with thread-local counters. Here, we model roots as part of the global state only to ease proof annotations – our final theorem is an invariant of the program global state.

Finally, execution states include the states of all threads and a global state.

```
Definition state := ((tid → option thread_state) * gstate).
```

Well-Typedness Invariants. A number of invariants are guaranteed by typing: (i) each variable in the local or global environment contains a value of the appropriate type, (ii) any reference of type TRef is either null, in the domain of the heap, or in the freelist, and (iii) each abstract mark-buffer is accessible from a unique global variable, indexed by a thread identifier. This mechanism enforces separation of mark-buffers by typing.

3   RtIR Proof System

On top of RtIR, we design a program logic, based on a variation of rely-guarantee (RG). In a nutshell, RG [15] extends Hoare-logic to handle concurrency in a thread modular fashion. In addition to the standard Hoare-tuples, side conditions ensure that program annotations take into account the possible interferences of other threads. When thinking about a particular thread's code, we shall refer to the actions of the other concurrent threads as its *context*. This context is formally encoded as a *rely* relation stating its possible execution steps. Thus, each annotation in the code of a thread must be proved to be *stable w.r.t.* its rely condition, meaning that its validity is not affected by possible state changes induced by any number of rely steps. We follow a similar approach to encode guarantees (cf. Sect. 1). In fact, throughout our development we only ever need to define guarantees, and we synthesize the relies of other threads from guarantees.

High-Level Design Choices of Proof Rules. In our approach, we firstly annotate a program, as is usually done on paper, and then prove the annotated program using syntax-directed proof rules. We thus extend the syntax of commands to include *annotations*. Syntax-directed proof rules were capital for proof automation.

The proof system decouples sequential and concurrent reasoning. Its first layer is a Hoare-like system, with no use of relies or guarantees. A second layer handles interference: proof obligations about relies, guarantees and stability checks of annotations.

Finally, to avoid polluting programs with routine annotations, typically the global invariants, the first layer of the system *assumes* that such invariants hold, and the second layer requires to separately prove their invariance as a stability check.

Annotations. We use a shallow embedding into Coq, with annotations of type either $\text{pred} \triangleq \text{gstate} \rightarrow \text{lenv} \rightarrow \text{Prop}$, or $\text{gpred} \triangleq \text{gstate} \rightarrow \text{Prop}$ when they deal with the global state only. Typically, the global invariant of the GC is of type gpred. We also define the usual logical connectives on pred and gpred with the expected meaning. Conjunction is written $A \wedge B$ and implication is written $A \longrightarrow B$. Annotations of type gpred are automatically cast into pred when needed.

The syntax presented in Sect. 2 is extended to take annotations into account. While elementary commands that do not utilize the global state do not need to be extended, basic commands accessing memory (*e.g.* field loads and updates, global loads and stores, and mark-buffer operations) have to take an extra argument of type pred, representing the pre-condition of the command. This is also the case for loops, annotated with a loop-invariant, and atomic blocks, whose body may affect the global state. The semantics of RTIR completely ignores annotations which are only relevant for proofs.

In the sequel, we use the informal notation $P @ c$ for a command c annotated with P.

Sequential Layer. We start by defining the following predicate, $I \vDash t: \langle P \rangle\, c\, \langle Q \rangle$ that corresponds to the validity of a sequential Hoare tuple, with respect to the big-step operational semantics of commands. This semantic judgment asserts that, for thread t, if command c runs in a state satisfying precondition P, and if the execution terminates, the final state must satisfy post-condition Q under the assumption that the global predicate I is an invariant. Proving that I is indeed invariant is done separately.

First-layer logic judgments for commands are of the form $I \vdash t: \langle P \rangle\, c\, \langle Q \rangle$. For basic commands which do not require annotations and simple command compositions (sequence, non-deterministic choice and loops), proof rules follow the traditional weakest-precondition style. This can be seen in the following rules:

$$\frac{}{I \mid t: \langle P \rangle\, \text{skip}\, \langle P \rangle} \qquad \frac{I \vdash t: \langle P \rangle\, c_1\, \langle R \rangle \quad I \vdash t: \langle R \rangle\, c_2\, \langle Q \rangle}{I \vdash t: \langle P \rangle\, c_1; c_2\, \langle Q \rangle} \qquad \frac{I \vdash t: \langle P1 \rangle\, c_1 \langle Q \rangle \quad I \vdash t: \langle P2 \rangle\, c_2 \langle Q \rangle}{I \vdash t: \langle P1 \wedge P2 \rangle\, c_1 \oplus c_2\, \langle Q \rangle}$$

On the other hand, commands that require annotations directly embed the semantic judgment $I \vDash t: \langle P \rangle\, c\, \langle Q \rangle$ as a proof obligation. For instance:

$$\frac{I \vDash t: \langle P \rangle\, P @ X = e\, \langle Q \rangle}{I \vdash t: \langle P \rangle\, P @ X = e\, \langle Q \rangle} \qquad \frac{I \vDash t: \langle P \rangle\, c\, \langle Q \rangle}{I \vdash t: \langle P \rangle\, P @ \text{atomic}\, c\, \langle Q \rangle}$$

Interference Layer. This layer takes into account threads interference with a given command, handling the validity of guarantees and the stability of program annotations *w.r.t.* the context. This can be seen in the definition of a valid RG tuple:

```
Record RGt (t:tid) (R:rg) (G:list rg) (I:gpred) (P Q:pred) (c:cmd) := {
  RGt_hoare:     I ⊢ t: ⟨P⟩ c ⟨Q⟩
; RGt_stable:    stable I P R ∧ stable I Q R ∧ AllStable I c R
; RGt_guarantee: AllRespectGuarantee t I c G }.
```

Here, the type $\text{rg} \triangleq \text{gstate} \rightarrow \text{gstate} \rightarrow \text{Prop}$ defines relies and guarantees as binary relations between global states. In our development, we build them from annotated commands. For a command P@c, the associated rg is defined by running the (big-step) operational semantics of c from a pre-state satisfying P to a post-state (in Sect. 5, we explain how our proof methodology benefits from this definition).

Predicate stable defines the stability of a pred *w.r.t.* a rely, given some invariant:

```
Definition stable (I:gpred) (H:pred) (R:rg) : Prop := ∀ gs1 gs2 l,
  I gs1 ∧ H gs1 l ∧ R gs1 gs2 ∧ I gs2 → H gs2 l.
```

The predicate AllStable builds the conjunction of the stability conditions for all assertions syntactically appearing therein. We omit its formal definition here.

The validity of the guarantee of a command (predicate AllRespectGuarantee) follows the same principle, this time accumulating proof obligations that all elementary effects of the command are reflected by an elementary guarantee in the list G.

Program RG Specification. The RG specification of a program p is defined as a record considering guarantees G and pre and post-conditions P and Q for all threads. Formally:

```
Record RGt_prog (G:tid → rg) (I:gpred) (P Q:tid → pred) (p:program) := {
  RGp_t:∀ t ∈ (threads p), RGt t (Rely G t) (G t) I (P t) (Q t) (cmd t p)
; RGp_I:∀ t, stable TTrue I (G t) }.
```

Obligation RGp_t requires that each thread's command is proved valid. It is worth noting that only guarantees need to be considered: for each thread, we build its rely from other threads' guarantees (Rely G t). This significantly reduces redundancies in specifications. Second, obligation RGp_I requires that I is invariant. We encode this as a stability condition under the union of all threads' guarantees, assuming the trivial invariant TTrue ≜ (fun _ _ ⇒ True). Indeed, as all threads' code satisfy their guarantees, this is enough to prove that the global invariant I is preserved by any number of program steps.

Reasoning About Iterators. As expected, the case of iterators is more involved. We illustrate their treatment on foreach. Though more technically involved, others iterators are similar. Recall that foreach iterates on a list of data of type A, morally representing a loop. Hence, its proof involves a loop invariant, predicated over the visited elements of the list. Predicates annotating foreach are thus indexed by a list of visited elements. And, as the loop body may include annotations about visited elements, we also index it by a list of visited elements and a current element. Summing up, the syntax of foreach, extended with annotations is P@foreach (*x in* l) do c od where annotation P has type list A → pred, and c has type list A → A → cmd. The associated proof rule is:

$$\frac{\begin{array}{c} \forall\ a\ seen,\ prefix\ (seen\!+\!+\![a])\ l \to \\ I \vdash t: \langle P\ seen \rangle\ (c\ seen\ a)\ \langle P\ (seen\!+\!+\![a]) \rangle \\ P\ l \wedge\!\!\wedge\ I \longrightarrow Q \end{array}}{I \vdash t: \langle P\ nil \rangle\ P@\texttt{foreach}\ (x\ in\ l)\ \texttt{do}\ c\ \texttt{od}\ \langle Q \rangle}$$

The first premise amounts to proving a valid tuple whose pre- and post-conditions are adjusted to the list of already visited elements. The second premise requires precondition P applied to the whole list of elements to entail the post-condition of the iterator itself. We define a more general rule in Coq, to get an induction principle usable to prove the soundness of the logic.

Soundness of the Logic. Soundness states that invariant I holds in every state reachable from a well-formed initial state – which must satisfy I by construction – through the small-step semantics mentionned in Sect. 2. Formally:

```
Hypothesis init_wf : ∀ tsi gsi, init_state p (tsi,gsi) →
     RGt_prog G I P p Q                          (* program RG spec *)
   ∧ (∀ t c le, tsi(t) = Some(c, le) → P t gsi le) (* pre-conds. hold *)
   ∧ I gsi.                                        (* I holds initially *)
Theorem soundness : ∀ ts gs, reachable init_state p (ts,gs) → I gs.
```

The proof of this theorem relies on an auxiliary proof system, proved equivalent to the one presented earlier. The auxiliary system reuses the same basic components, but proof rules now require to prove everything in situ: the invariant, the pre- and postconditions, the stability of annotations, and the validity of guarantees. For instance, compare the rule for instruction $X = e$ in the previous system (left) with the proof rule of the auxiliary system (right):

$$
\frac{I \vDash t: \langle P \rangle\ P@X = e\ \langle Q \rangle}{I \vdash t: \langle P \rangle\ P@X = e\ \langle Q \rangle}
$$

$$
\frac{TTrue \vDash t: \langle P \wedge I \rangle\ P@X = e\ \langle Q \wedge I \rangle \quad stable\ TTrue\ (P \wedge I)\ G \quad stable\ TTrue\ (Q \wedge I)\ G \quad RespectGuarantee\ t\ I\ G\ (P@X - e)}{R,\ G,\ I \vdash t: \langle P \rangle\ P@X = e\ \langle Q \rangle}
$$

This auxiliary system is very close to the classic RG [15,25]. Its verbosity makes it easier to reason about the soundness proof.

The soundness proof itself consists in a subject-reduction lemma *w.r.t.* the following property: in the current execution state, every possible thread currently running is in fact running a piece of code that conforms to RGt (the pre-conditions map P is hence updated at each step), and the global invariant I holds. Invariance of I follows from the fact that, in each rule of the auxiliary system, the invariant is part of the pre- and post-conditions, which are stable against any step of the rely and the guarantee of the stepping thread.

4   The Concurrent Garbage Collector

We now describe our implementation in RTIR of the concurrent GC, and its associated correctness theorem. The algorithm is based on [5], a variant of the well known concurrent *mark-and-sweep* algorithm due to Doligez et al. [3,4].

Main Theorem. Intuitively, we want to show that the collector thread never reclaims memory that could potentially be used by mutators. To do this, we program the collector and the mutators in RTIR, and prove that their parallel composition preserves an invariant on global execution states, using the soundness theorem of our program logic.

The particularity of mutators is that they participate to the bookkeeping required for the collection to be correct. In practice, bookkeeping code is injected in client code by the compiler. Here, we consider a *Most General Client* (MGC) representing a collector thread composed with an arbitrary number of mutators with identifiers in Mut, each running relevant injected pieces of code.[1]

$$
\begin{aligned}
\texttt{mutator} \triangleq\ & \texttt{loop(\,update(x, f, v)} \\
& \oplus \texttt{load(x, f)} \oplus \texttt{alloc()} \\
& \oplus \texttt{cooperate()} \oplus \texttt{changeRoots())} \\
\texttt{mgc} \triangleq\ & \texttt{collector} \,\|\, \texttt{mutator} \,\|\, ... \,\|\, \texttt{mutator}
\end{aligned}
$$

Recall that the special global variable freelist a pool of unused references. Hence, upon allocation, a reference is fetched from the freelist. Symmetrically, to reclaim an unused object, the collector puts back its reference into the freelist.

Our main invariant establishes that in a given state gs, any reference r reachable from any mutator m is not in the freelist, and hence has not been collected.

```
Definition I_correct: gpred :=
    fun gs ⇒ ∀ m r, In m Mut ∧ Reachable_from m gs r ⟶ ¬ in_freelist gs r.
```

We can now formulate our main theorem. It uses the predicate reachable_mgc stating that a global state gs can be reached, from a predefined initial state, by the code of the mgc shown above.

```
Theorem gc_sound: ∀ gs, reachable_mgc gs → I_correct gs.
```

The initial state we consider is obtained by a startup phase of the runtime, that carefully initializes intrinsic features of the runtime, and establishes key invariants.

Evidently, this theorem would be impossible to prove without the aid of other intermediate invariants. In the sequel we explain the important aspects of the implementation, and a few salient auxiliary invariants. Describing the algorithm and our code in full details is out of the scope of this paper. We refer the reader to the explanations in [5] and to the formal proof [7] for details.

High-Level Principles of the Algorithm. Our GC is of the *mark and sweep* family: the heap is traversed, marking objects that are presumably alive, *i.e.* reachable from mutators local variables, henceforth called *roots*. Once the marking procedure finishes, the sweeping procedure reclaims objects detected as not reachable by putting them back in the freelist.

The marking conventions to denote the reachability of objects follows the tricolor convention [2]. Color WHITE is used for objects not yet visited. GREY is used for visited, hence presumably live objects, whose children (through fields accesses) have not yet been visited. BLACK is used for visited objects whose children have all been visited. In our implementation, colors WHITE and BLACK are implemented with numerical constants. We explain the encoding of GREY later. The heap traversal (marking) procedure is called *tracing*, and completes once no GREY objects remain.

---

[1] We present a simplified pseudo-code of the MGC, with variable x, field f, and value v assumed non-deterministically chosen from the thread environment. The actual definition in Coq is an operational characterization of this thread system.

```
// collector ::=
while (true) do
  atomic  // ghost
    stage[C] = CLEAR
    phantom_flipped = 0
  atomic  // linearizable[4]
    foreachObject o do
      if !(isFree?(o)) then
        o.color = WHITE
    od
    phantom_flipped = 1
  handshake()  // SYNCH1
  handshake()  // SYNCH2
  stage[C] = TRACING
  handshake()  // ASYNCH
  trace()
  stage[C] = SWEEPING
  sweep()
  stage[C] = RESTING
od
```

Listing 1. Collector

```
// handshake() ::=
phantom_hdsk = 1 phase[C] =
    phase[C] + 1 mod 3
  foreach (m in Mut) do
    repeat skip
    until phase[m]==phase[C]
od phantom_hdsk = 0
```

Listing 2. Handshake

```
// tid m : cooperate ::=
if phase[m] != phase[C] then
  if phase[C] == ASYNCH then
    foreachRoot (r of m) do
      markGrey(buffer[m], r)
    od
  phase[m] = phase[C]
```

Listing 3. Cooperate

```
// tid m : update(x,f,v) ::=
if (phase[m] != ASYNCH
    stage[C] == TRACING) then
  old = x.f
  markGrey(buffer[m],old)
  markGrey(buffer[m],v)
x.f = v
```

Listing 4. Write Barrier

```
// markGrey(buffer,x) ::=
if (x != NULL
    && x.color != BLACK) then
  buffer.push(x)
```

Listing 5. MarkGrey

Extra care is required to cope with the concurrent execution of mutators: they could modify the object graph at any point, and thus invalidate the properties of the coloring. In particular, mutators are responsible for publishing their own roots by marking them as GREY before tracing begins. This is the goal of the cooperate procedure. Similarly, object field update should not break color-related reachability invariants during tracing. This is the goal of the so-called *write-barriers*, implemented by the update procedure. Finally, the right color should be assigned to newly allocated objects. For space reasons, we elude the details of the alloc procedure that we have implemented, and refer to our formalization [7] and the descriptions in [5] for the details.

All these subtle procedures, run by the collector and the mutators, are orchestrated using the global variable stage[C], which encodes for the various stages of the collection cycle (including the tracing and sweeping), and the global variables phase[m] – one for each mutator – and phase[C] – one for the collector – to coordinate mutators with the collector. A diagrammatic representation of a collection cycle is shown in Fig. 2, gathering all previously mentioned ingredients. We will refer to it below in more detail.

RtIR Implementation and Main Invariants. Let us now describe the implementation. Code snippets in RtIR use a simplified syntax from the one we presented above for space and readability reasons.

Stage and Phase Protocol. The code of the collector is presented in Listing 1. For the moment, we concentrate only on the calls to the handshake() procedure (Listing 2), and its counterpart cooperate() (Listing 3) executed by the mutators. A collection cycle is structured using four stages: CLEAR, TRACING, SWEEPING and RESTING. The current stage is written by the collector to a global variable stage[C]. This global variable allows mutators to coordinate with the collector at a coarse level. At a finer level, a handshake mechanism is required, and the status of each thread, the mutators and the collector, is tracked with a phase variable, with values ranging over ASYNCH, SYNCH1 or SYNCH2. Each phase is encoded with a dedicated integer between 0 and 2. Instead of presenting

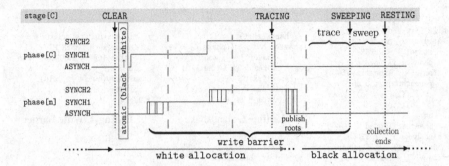

Fig. 2. Timeline of a collection cycle. All mutators are coalesced into the bottom line, and the collector is shown in the top line. Dotted lines represent the GC start of a new stage, and dashed lines represent the end of a phase change (handshake).

a detailed description to justify these phases, let us point out that the original algorithm of [4] used only two phases, which was later discovered to be incorrect. A new phase was added to correct it in [3].

We concentrate now on the horizontal lines of Fig. 2, showing the evolution of phase[C], as well as the aggregated representation of all the phase[m] variables of mutators. Each phase starts by the collector modifying the phase[C] variable (second line of Listing 2). Mutators query it (first line of Listing 3), to acknowledge possible changes, in which case mutators *respond* by updating their own phase[m] variable (the last line of Listing 3). When the collector acknowledges that all mutators have updated phase[m], the phase transition is completed (dashed line in Fig. 2). Importantly, phase[C] and phase[m] are subject to race conditions. We also point out that threads do never stop their execution while executing cooperate.

An important invariant relating the phases of the collector and the mutators is that any mutator's phase is at most one step behind the collector's phase.

```
Definition I_phases : gpred := fun gs ⇒
∀ m, In m Mut → phase[C]gs = phase[m]gs ∨ phase[C]gs=(phase[m]gs+1) mod 3.
```

Buffers and GREY. Objects are marked GREY with the markGrey procedure (Listing 5) when mutators publish their roots (Listing 3) and during the write barriers (Listing 4). Each mutator owns a buffer[m] abstract data structure, in which it adds references to be traced. Hence, buffer[m] serves as an interface between mutators and the collector to mark objects as GREY. In other words, an object is considered GREY if it is present in any buffer and its color field is WHITE. In this sense, GREY is a convention rather than a constant like BLACK or WHITE.

Write Barriers. Their code is shown in Listing 4. The barrier will conditionally either directly update the field f (fast-path) or markGrey two objects (slow-path).[2] Notice that the slow-path of the write barrier is only executed when the collector is ready to start tracing, and not after it starts sweeping (see Fig. 2). The code of write barriers is intrinsically racy since the client code itself might contain races at the field; moreover, the

---

[2] The write barrier in [5] avoids marking old in some cases. We drop this optimization.

```
 1  // trace() ::=                              22    while (!buffer[C].isEmpty()) do
 2  all_empty = false                           23      all_empty = false
 3  while (!all_empty) do                        24      ob = buffer[C].top()
 4    atomic // ghost code                       25      if (ob.color == WHITE) then
 5      foreach (m in Mut) do                    26        foreachField (f of ob) do
 6        phantom_buffer[m].copy(buffer[m])      27          if (ob.f!=NULL
 7      od                                       28            && ob.f.color==WHITE) then
 8    all_empty = true                           29            buffer[C].push(ob.f)
 9    foreach (m in Mut) do                      30        od
10      is_empty = buffer[m].isEmpty()           31        ob.color = BLACK
11      while (!is_empty) do                     32      buffer[C].pop()
12        all_empty = false                      33    od
13        x = buffer[m].top()                    34  od
14        if (x.color == WHITE) then             35
15          buffer[C].push(x)                    36
16          buffer[m].pop()                      37  // sweep() ::=
17        else buffer[m].pop()                   38  foreachObject o do
18        is_empty = buffer[m].isEmpty()         39    if (!isFree?(o) && o.color == WHITE) then
19      od                                       40      free(o)
20    od                                         41  od
21
```

Listing 6. Trace and Sweep (Collector)

accesses to the `buffer` data structures are not protected by synchronization between mutators and the collector. Finally, we emphasize that the order in which the `markGrey` operations are performed in the write barrier is critical to the GC correctness.

Trace. This is the most challenging code to verify, and its verification by means of program logics would be remarkably hard without some of the design choices of RTIR, and our proof methodology.

The `trace` procedure (Listing 6) traverses the object graph starting from GREY objects. More precisely, the collector visits each of the mutators `buffer[m]` in the `foreach` loop at Line 9, transferring their contents into its own `buffer[C]`. If the collector sees empty buffers for all mutators, tracing ends. Otherwise, it traverses the graph starting from objects in `buffer[C]`, and marking BLACK objects whose children have been seen.

Regarding the complexity of the code, we emphasize that it contains three nested loops, a number of `foreach` constructs, and heavily uses the buffer abstract data structures. Moreover, it exhibits races in all threads (through write barriers and buffer operations) since it traverses the object graph, while mutators concurrently modify it.

An important invariant establishes that during the tracing phase, any WHITE object that is alive must be reachable from a GREY object, signaling that it still has to be visited. Since another invariant, `I_black_to_white`, states that any path from a BLACK object to a WHITE object goes through a GREY object, this translates to the property that all objects reachable from the roots are either BLACK, or reachable from a GREY one.

Definition I_trace_grey_reach_white : gpred := fun gs ⇒ ∀ m r,
stage[C]gs ≠ CLEAR ∧ In m Mut ∧ phase[m]gs=ASYNCH ∧ Reachable_from m gs r→
Black gs r ∨ (∃ r0, Grey Mut gs r0 ∧ reachable gs r0 r).

When this code terminates, we are able to prove that: (i) there are no more GREY objects, (ii) all objects reachable from the mutators roots are BLACK, and consequently (iii) there are no WHITE objects reachable from any of the mutators roots.

Property (i), namely that all buffers are *simultaneously* empty at the end of tracing (Listing 6, Line 35), is particularly difficult to prove, given the write barriers executed concurrently by mutators. We prove that this property is established at Line 4 of the last iteration of the enclosing while loop. We proceed as follows. We first prove that, at Line 4, buffer[C] is always empty. As for mutators' buffers, we use ghost variables phantom_buffer[m] to take their snapshot at Line 4. Mutators can only push on their buffers, so, in a given iteration of the enclosing while loop, if a mutator buffer is empty, so was its ghost counterpart during the same iteration. In the last iteration of the while loop, all buffers are witnessed empty, one at a time. But this implies that all phantom buckets are simultaneously empty at Line 8. This, in turn, implies that all buffers are, this time *simultaneously*, empty at Line 4. This property remains true until Line 35: it is both stable under mutators' guarantees, and preserved by the while loop. Indeed, if all buffers are empty (there are no GREY objects), the above invariant I_trace_grey_reach_white implies that both the old and new objects that markGrey could push on a buffer are in fact BLACK, and thus not pushed on any buffer (Listing 5). As a consequence, no reference is pushed on the collector's buffer (Line 15).

Sweep. The sweep phase (Listing 6) recycles all the objects that remain WHITE after TRACING. This is the only place where instruction free is ever used. Note that this code is also non-blocking. A key property, whose proof we have sketched above, is that during sweeping, no GREY objects remains. Formally,

```
Definition I_sweep_no_grey : gpred := fun gs ⇒
  (stage[C]gs = SWEEPING ∨ stage[C]gs = RESTING) → ∀ r, ¬ Grey Mut gs r.
```

This invariant, with I_trace_grey_reach_white above, implies that no WHITE object is reachable from any thread-local variable.

5   Proof Methodology

Mechanizing such a sizable proof raises methodological concerns. While the proof system of Sect. 3 separates proof concerns between sequential reasoning and stability checks, we deal here with the intrinsic complexity of the proof and its scalability.

First, stating upfront the right set of invariants, guarantees, and assertions is unrealistic for such a proof. To tackle this issue, we group invariants related to distinct aspects, *e.g.* the phase protocol or coloring invariants. To reflect this structure in our proof, and avoid constant refactoring of proof scripts, we design an incremental workflow.

Second, we must deal with the quantity of proof obligations. For the GC code, proof obligation RGp_I involves 18 invariants, which must be proved stable under 17 guarantees, thus requiring 306 stability proof obligations. On top of this, proof obligation RGt_stable adds more than 60 annotated lines of code, each bearing several predicates, that must be proved stable under significant subsets of the 17 guarantees. This becomes quickly intractable without a disciplined methodology and automation.

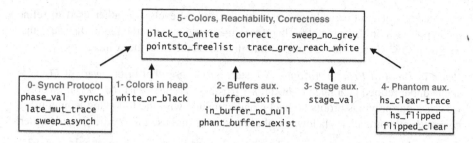

Fig. 3. Main Invariants of the GC. Numbers are timestamps in the incremental proof methodology. Dependencies are shown with boxes (inter-dependency) and arrows.

5.1 Workflow

Figure 3 shows the major invariants of the GC, organized in groups. In each boxed group, invariants are inter-dependent, while arrows indicate a dependency of the target group on the source group.

RG proofs are thread-modular, but RG does not solve the interdependency problem: invariants, guarantees and code annotations are all eventually connected to form the end-result. To maximize proof reuse, we use a simple mechanism: invariants I and guarantees G are indexed by a natural number – morally a timestamp of their introduction into the development (Fig. 3). When introducing a new increment to an invariant, all invariants with a lower timestamp are not modified. Nor are their proofs, resulting in an incremental, non-destructive methodology. More concretely, at each level:

1. we enrich the invariant, refine the guarantees and code annotations;
2. we prove the new stability proof obligations, for which we can reuse prior stability proofs, and we use automation to discharge as many obligations as possible;
3. we adapt sequential Hoare proofs, and prove that enriched guarantees are still valid.

This workflow proved robust during our development, allowing for an incremental and manageable proof effort. We detail below the first two items of this methodology.

5.2 Incremental Proofs

Let us focus on obligation RGp_I from Sect. 3, which requires establishing the invariant stability under all threads' guarantees. Let us fix a thread and index both the invariant and guarantee by n. The obligation is thus (stable TTrue (I n) (G n)). Let us now see how we establish (stable TTrue (I n+1) (G n+1)) by using the already proved (stable TTrue (I n) (G n)) obligation.

Monotonicity of I *and* G. We build (I n+1) as a conjunction of prior established invariant (I n), and the increment at the current level: (I n+1) ≜ (I n) ⋏ (Ic n+1). Hence, we have that ∀n, (I n+1) ⟶ (I n).

Recall that in our proof system, guarantees are expressed through the effect of a command, under certain hypotheses on the pre-state. At each level, the command will not change – it is effectively executed by the code. Levels are rather used to refine the hypotheses on the pre-state. Therefore, guarantees are monotonic in the sense that $\forall n, (G\ n+1) \subseteq (G\ n)$: they are made more precise as the level index increases.

Reuse of Proof of Prior Invariants. We start by proving that prior invariant (I n) is stable under refined guarantee (G n+1), *i.e.* (stable TTrue (I n) (G n+1)). To do so, we reuse our previous proofs at level n and conclude with the following lemma using guarantee monotonicity – below, we abuse notations and use _ as a valid Coq identifier.

```
Lemma stable_refineG: ∀_ I G1 G2, G2 ⊆ G1 ∧ stable _ I G1 → stable _ I G2.
```

New Invariant Stability. It remains to prove the stability of increment (Ic n+1) under refined guarantee (G n+1). In simple cases, (stable TTrue (Ic n+1) (G n+1)) is provable independently from prior invariants. In this case, we combine the stabilities of (Ic n+1) and (I n) into the one of (I n+1) with lemma stable_and:

```
Lemma stable_and: ∀ _ I1 I2 G,
  stable _ I1 G ∧ stable _ I2 G → stable _ (I1 ⋏ I2) G.
```

However, the situation is often more involved, requiring prior invariants to prove the stability of (Ic n+1). Formally, we have (stable (I n) (Ic n+1) (G n+1)). We can then combine the stability of (I n) and (Ic n+1) under (G n+1) using this lemma:

```
Lemma stable_with: ∀ _ I1 I2 G,
  stable _ I1 G ∧ stable I1 I2 G → stable _ (I1 ⋏ I2) G.
```

5.3 Proof Scalability

To tackle the blowup of stability checks alluded to earlier, we built a toolkit of structural stability lemmas, and develop some tactic-based partial automation. This allowed us to discharge automatically 186 obligations among the 306 obligations induced by RGp_I. The remaining obligations are also partially reduced by the automation.

Structural Lemmas. Structural lemmas serve three purposes. First, they are critical to enable the incremental methodology delineated above. Second, they allow for complex stability proof obligations to be simplified: both annotations, invariants, and interferences can be structurally split up. Thus, intrinsically complex arguments are isolated from trivial ones, that are automatically discharged. Finally, to reuse as much proofs as possible, we rely on a custom notion of stability under extra-hypotheses:

```
Definition stable_hyps (I: gpred) (H: pred) (R: rg): Prop := ∀ gs1 gs2 l,
  I gs1 ∧ H gs1 l ∧ P gs1 l ∧ R gs1 gs2 ∧ I gs2 ∧ H gs2 l → P gs2 l.
```

Typically, this notion allows to leverage stability results from previous levels, notably through the following lemmas:

```
Lemma stable_weakI: ∀ I1 I2 P G, I2 ⊆ I1 → stable I1 P G → stable I2 P G.
Lemma stable_weakH : ∀ I (H: pred) R,
    stable I H R → stable_hyps I H P R → stable I (H ⋏ P) R.
```

By decomposing annotations and relaxing interferences, we can factor out the proof of stability of annotations that reappear in the code.

Automation. We developed a set of tactics that simplify stability goals into elementary ones before attempting to solve them. This leads to clearer goals and more tractable proof contexts. The tactics combine our structural lemmas with two additional ideas: systematic inversion on guarantee actions – defined operationally using commands – and rewriting in predicates.

6   Related Work

Concurrent GC. The literature on garbage collection is vast. We refer the reader to [16] for a comprehensive and up-to-date presentation of garbage collection techniques. We use [5] as a starting point. It is a state-of-the-art non-blocking concurrent GC based on the earlier DLG algorithm [3,4]. Many of the invariants we prove are inspired by those of [3].

Mechanized GC Proofs. Many prior efforts have tackled the verification of sequential GCs [11,18]. Unfortunately, the addition of concurrency renders these approaches inadequate. Insofar our work could be subsequently integrated into a verified run-time, it is possible to reuse some methodological aspects of [19], such as the structuring in a multi-layer refinement of the garbage collection specification.

The first mechanized proof of a concurrent GC was presented by Gonthier [9]. Unlike ours, Gonthier's proof rests on an abstract encoding of the algorithm. Our development sidesteps this additional modelling step by proving the implementation in RTIR. A similar remark can be made of the approaches in [8,10], which formalize GCs in the PVS and Isabelle/HOL provers respectively.

Liang et al. [17] provide a proof of a mostly-concurrent GC based on the RGSim methodology. While the meta-theory of the logic is mechanized, the proof of the GC itself is not.

Mechanized Concurrent Program Logics. In [21] an RG logic for a simple imperative concurrent language is formalized and proved sound in Isabelle/HOL. In contrast, our program logic is customized for runtime system implementations, and therefore supports local and global environments, references, iterators, etc. Also, the proof rules of [21] mix sequential reasoning with side conditions about stability and guarantee checks. We decouple these aspects and avoid redundancies by extracting relies from the guarantees of the context.

Other approaches to the mechanized verification of concurrent code are [6,17,23, 24] to mention but a few. These works are mostly concerned with concurrent data structure correctness, whereas we are concerned with the implementation of a runtime system.

7   Conclusion

This paper presents the mechanized proof of an emblematic challenge in program verification: an on-the-fly concurrent garbage collector. Overcoming this challenge requires a number of methodological advances. We follow a programming language-based approach: a well-chosen intermediate representation, a companion program logic, and a dedicated proof workflow. RᴛIR strikes a balance between low-level features for the expression of efficient concurrent code, and high-level features which remove the burden of dealing with low-level details in the proofs. Our program logic is inspired by Rely-Guarantee, a milestone in concurrency proof techniques, but one that has heretofore not been used for the mechanized verification of garbage collectors. Our incremental proof workflow, combined with specific and efficient tool support via Coq tactics, is efficient and flexible enough for such a verification challenge.

There are two major avenues for future work. The first is pragmatic, and concerns the embedding of our work in a verified compiler tool chain. Using our theorem about the most-general client, we can build a refinement proof between an IR with implicit memory management and RᴛIR. We then need to have a fully executable version of the GC. This would require cleaning up ghost code, coding iterators as low-level macros, and implementing abstract concurrent data structures natively supported by RᴛIR. The two first tasks are essentially administrative. The third task is more challenging, requiring us to formally prove an atomicity refinement result for linearizable, fine-grained data-structures. To that end, we have developed the meta-theory in [26].

The second is methodological. Our proof is the first GC proof to be mechanized using Rely-Guarantee, but it does not take advantage of other tools like Separation Logic [22]. Methods combining RG and Separation Logic exist [25]. It remains to be seen how (or if) these techniques could improve our current proof.

Acknowledgement. We thank the anonymous reviewers and Peter Gammie for their thorough comments and suggestions on how to improve the final version of the paper. We also thank Vincent Laporte for his work earlier in this project, and his help on implementing parts of the garbage collector presented here.

References

1. Demange, D., Laporte, V., Zhao, L., Jagannathan, S., Pichardie, D., Vitek, J., Plan, B.: a buffered memory model for java. In: POPL 2013, pp. 329–342 (2013)
2. Dijkstra, E.W., Lamport, L., Martin, A.J., Scholten, C.S., Steffens, E.F.M.: On-the-fly garbage collection: an exercise in cooperation. Commun. ACM **21**(11), 966–975 (1978)
3. Doligez, D., Gonthier, G.: Portable, unobtrusive garbage collection for multiprocessor systems. In: Proceedings POPL 1994, pp. 70–83 (1994)
4. Doligez, D., Leroy, X.: A concurrent, generational garbage collector for a multithreaded implementation of ML. In: Proceedings of POPL 1993, pp. 113–123 (1993)
5. Domani, T., Kolodner, E.K., Lewis, E., Salant, E.E., Barabash, K., Lahan, I., Levanoni, Y., Petrank, E., Yanover, I.: Implementing an on-the-fly garbage collector for Java. In: Proceedings of ISMM 2000, pp. 155–166 (2000)
6. Elmas, T., Qadeer, S., Tasiran, S.: A calculus of atomic actions. In: Proceedings of POPL 2009, pp. 2–15 (2009)
7. Zakowski, Y., et al.: Verifying a concurrent garbage collector using an RG methodology (2017). http://www.irisa.fr/celtique/ext/cgc/
8. Gammie, P., Hosking, A.L., Engelhardt, K.: Relaxing safely: verified on-the-fly garbage collection for x86-TSO. In: Proceedings of PLDI 2015, pp. 99–109 (2015)
9. Gonthier, G.: Verifying the safety of a practical concurrent garbage collector. In: Alur, R., Henzinger, T.A. (eds.) CAV 1996. LNCS, vol. 1102, pp. 462–465. Springer, Heidelberg (1996). doi:10.1007/3-540-61474-5_103

10. Havelund, K.: Mechanical verification of a garbage collector. In: Rolim, J., et al. (eds.) IPPS 1999. LNCS, vol. 1586, pp. 1258–1283. Springer, Heidelberg (1999). doi:10.1007/BFb0098007

11. Hawblitzel, C., Petrank, E.: Automated verification of practical garbage collectors. In: Proceedings of POPL 2009, pp. 441–453 (2009)

12. Hawblitzel, C., Petrank, E., Qadeer, S., Tasiran, S.: Automated and modular refinement reasoning for concurrent programs. In: Kroening, D., Păsăreanu, C.S. (eds.) CAV 2015. LNCS, vol. 9207, pp. 449–465. Springer, Cham (2015). doi:10.1007/978-3-319-21668-3_26

13. Herlihy, M., Wing, J.M.: Linearizability: a correctness condition for concurrent objects. ACM Trans. Program. Lang. Syst. **12**(3), 463–492 (1990)

14. Jagannathan, S., Laporte, V., Petri, G., Pichardie, D., Vitek, J.: Atomicity refinement for verified compilation. ACM Trans. Program. Lang. Syst. **36**(2), 6:1–6:30 (2014)

15. Jones, C.B.: Tentative steps toward a development method for interfering programs. ACM Trans. Program. Lang. Syst. **5**(4), 596–619 (1983)

16. Jones, R., Hosking, A., Moss, E.: Handbook, The Garbage Collection: The Art of Automatic Memory Management, 1st edn. Chapman & Hall/CRC, Boca Raton (2011)

17. Liang, H., Feng, X., Fu, M.: Rely-guarantee-based simulation for compositional verification of concurrent program transformations. ACM Trans. Program. Lang. Syst. **36**, 3 (2014)

18. McCreight, A., Chevalier, T., Tolmach, A.P.: A certified framework for compiling and executing garbage-collected languages. In: Proceedings of ICFP 2010, pp. 273–284 (2010)

19. Myreen, M.O.: Reusable Verification of a Copying Collector. In: VSTTE 2010 (2010)

20. Pizlo, F., Ziarek, L., Maj, P., Hosking, A.L., Blanton, E., Vitek, J.: Schism: fragmentation-tolerant real-time garbage collection. In: Proceedings of PLDI (2010)

21. Nieto, L.P.: The rely-guarantee method in Isabelle/HOL. In: Degano, P. (ed.) ESOP 2003. LNCS, vol. 2618, pp. 348–362. Springer, Heidelberg (2003). doi:10.1007/3-540-36575-3_24

22. Reynolds, J.C.: Separation logic: a logic for shared mutable data structures. In: Proceedings of LICS 2002, pp. 55–74 (2002)

23. Sergey, I., Nanevski, A., Banerjee, A.: Mechanized verification of fine-grained concurrent programs. In: Proceedings of PLDI 2015, pp. 77–87. ACM (2015)

24. Vafeiadis, V.: Concurrent separation logic and operational semantics. Electron. Notes Theor. Comput. Sci. **276**, 335–351 (2011)

25. Vafeiadis, V., Parkinson, M.: A marriage of rely/guarantee and separation logic. In: Caires, L., Vasconcelos, V.T. (eds.) CONCUR 2007. LNCS, vol. 4703, pp. 256–271. Springer, Heidelberg (2007). doi:10.1007/978-3-540-74407-8_18

26. Zakowski, Y., Cachera, D., Demange, D., Pichardie, D.: Compilation of linearizable data structures - a mechanised RG logic for semantic refinement. Technical report (2017). https://hal.archives-ouvertes.fr/hal-01538128

Formalization of the Fundamental Group in Untyped Set Theory Using Auto2

Bohua Zhan[(✉)]

Massachusetts Institute of Technology, Cambridge, USA
bzhan@mit.edu

Abstract. We present a new framework for formalizing mathematics in untyped set theory using auto2. Using this framework, we formalize in Isabelle/FOL the entire chain of development from the axioms of set theory to the definition of the fundamental group for an arbitrary topological space. The auto2 prover is used as the sole automation tool, and enables succinct proof scripts throughout the project.

1 Introduction

Auto2, introduced by the author in [17], is a proof automation tool for the proof assistant Isabelle. It is designed to be a powerful, extensible prover that can consistently solve "routine" tasks encountered during a proof, thereby enabling a style of formalization using succinct proof scripts written in a custom, purely declarative language.

In this paper, we present an application of auto2 to formalization of mathematics in untyped set theory[1]. In particular, we discuss the formalization in Isabelle/FOL of the entire chain of development from the axioms of set theory to the definition of the fundamental group for an arbitrary topological space. Along the way, we discuss several improvements to auto2 as well as strategies of usage that allow us to work effectively with untyped set theory.

The contribution of this paper is two-fold. First, we demonstrate that the auto2 system is capable of independently supporting proof developments on a relatively large scale. In the previous paper, several case studies for auto2 were given in Isabelle/HOL. Each case study is at most several hundred lines long, and the use of auto2 is mixed with the use of other Isabelle tactics, as well as proof scripts provided by Sledgehammer. In contrast, the example we present in this paper is a unified development consisting of over 13,000 lines of theory files and 3,500 lines of ML code (not including the core auto2 program). The auto2 prover is used exclusively starting from basic set theory.

Second, we demonstrate one way to manage the additional complexity in proofs that arise when working with untyped set theory. For a number of reasons, untyped set theory is considered to be difficult to work with. For example, everything is represented as sets, including objects such as natural numbers that

---

[1] Code available at https://github.com/bzhan/auto2.

© Springer International Publishing AG 2017
M. Ayala-Rincón and C.A. Muñoz (Eds.): ITP 2017, LNCS 10499, pp. 514–530, 2017.
DOI: 10.1007/978-3-319-66107-0_32

we usually do not think of as sets. Moreover, statements of theorems tend to be longer in untyped set theory than in typed theories, since assumptions that would otherwise be included in type constraints must now be stated explicitly. In this paper, we show that with appropriate definitions of basic concepts and setup for automation, all these complexities can be managed, without sacrificing the inherent flexibility of the logic.

We now give an outline for the rest of the paper. In Sect. 2, we sketch our choice of definitions of basic concepts in axiomatic set theory. In particular, we describe how to use tuples to realize extensible records, and build up the hierarchy of algebraic structures. In Sect. 3, we review the main ideas of the auto2 system, and describe several additional features, as well as strategies of usage, that allow us to manage the additional complexities of untyped set theory.

In Sect. 4, we give two examples of proof scripts using auto2, taken from the proofs of the Schroeder-Bernstein theorem and a challenge problem in analysis from Lasse Rempe-Gillen. In Sect. 5, we describe our main example, the definition of the fundamental group, in detail. Given a topological space X and a base point x on X, the fundamental group $\pi_1(X, x)$ is defined on the quotient of the set of loops in X based at x, under the equivalence relation given by path homotopy. Multiplication on $\pi_1(X, x)$ comes from joining two loops end-to-end. Formalizing this definition requires reasoning about algebraic and topological structures, equivalence relations, as well as continuous functions on real numbers. We believe this is a sufficiently challenging task with which to test the maturity of our framework, although it has been achieved before in the Mizar system. HOL Light and Isabelle/HOL also formalized the essential ideas on path homotopy. We review these and other related works in Sect. 6, and conclude in Sect. 7.

2   Basic Constructions in Set Theory

We now discuss our choice of definitions of basic concepts, starting with the choice of logic. Our development is based on the FOL (first-order logic) instantiation of Isabelle. The initial parts are similar to those in Isabelle/ZF, and we refer to [13,14] for detailed explanations.

The only Isabelle types available are i for sets, o for propositions (booleans), and function types formed from them. We call objects with types other than i and o *meta-functions*, to distinguish them from functions defined within set theory (which have type i). It is possible to define higher-order meta-functions in FOL, and supply them with arguments in the form of lambda expressions. Theorems can be quantified over variables with functional type at the outermost level. These can be thought of as theorem-schemas in a first-order theory. However, one can only quantify over variables of type i inside the statement of a theorem, and the only equalities defined within FOL are those between types i (notation $\cdot = \cdot$) and o (notation $\cdot \longleftrightarrow \cdot$). In practice, these restrictions mean that any functions that we wish to consider as first-class objects must be defined as set-theoretic functions.

2.1   Axioms of Set Theory

For uniformity of presentation, we start our development from FOL rather than theories in Isabelle/ZF. However, the list of axioms we use is mostly the same. The only main addition is the axiom of global choice, which we use as an easier-to-apply version of the axiom of choice. Note that as in Isabelle/ZF, several of the axioms introduce new sets or meta-functions, and declare properties satisfied by them. The exact list of axioms is as follows:

```
extension:    "∀z. z ∈ x ⟷ z ∈ y ⟹ x = y"
empty_set:    "x ∉ ∅"
collect:      "x ∈ Collect(A,P) ⟷ (x ∈ A ∧ P(x))"
upair:        "x ∈ Upair(y,z) ⟷ (x = y ∨ x = z)"
union:        "x ∈ ⋃C ⟷ (∃A∈C. x∈A)"
power:        "x ∈ Pow(S) ⟷ x ⊆ S"
replacement:  "∨x∈A. ∀y z. P(x,y) ∧ P(x,z) ⟶ y = z ⟹
                  b ∈ Replace(A,P) ⟷ (∃x∈A. P(x,b))"
foundation:   "x ≠ ∅ ⟹ ∃y∈x. y ∩ x = ∅"
infinity:     "∅ ∈ Inf ∧ (∀y∈Inf. succ(y) ∈ Inf)"
choice:       "∃x. x∈S ⟹ Choice(S) ∈ S"
```

Next, we define several basic constructions in set theory. They are summarized in the following table. See [13] for more explanations.

Notation	Definition
THE x. P(x)	⋃ (Replace({∅}, λx y. P(y)))
{b(x). x∈A}	Replace(A, λx y. y = b(x))
SOME x∈A. P(x)	Choice({x∈A. P(x)})
⟨a,b⟩	{{a}, {a,b}}
fst(p)	THE a. ∃b. p = ⟨a,b⟩
snd(p)	THE b. ∃a. p = ⟨a,b⟩
⟨a₁,...,aₙ⟩	⟨a₁, ⟨a₂, ⟨···, aₙ⟩⟩⟩
if P then a else b	THE z. P ∧ z=a ∨ ¬P ∧ z=b
⋃a∈I. X	⋃{X(a). a∈I}
A × B	⋃x∈A. ⋃y∈B. {⟨x,y⟩}

2.2   Extensible Records as Tuples

We now consider the problem of representing records. In our framework, records are used to represent functions, algebraic and topological structures, as well as morphisms between structures. It is often advantageous for records of different types to share certain fields. For example, groups and rings should share the multiplication operator, rings and ordered rings should share both addition and multiplication operators, and so on.

It is well-known that when formalizing mathematics using set theory, records can be represented as tuples. To achieve sharing of fields, the key idea is to assign each shared field a fixed position in the tuple.

We begin with the example of functions. A function is a record consisting of a source set (domain), a target set (codomain), and the graph of the function. In particular, we consider two functions with the same graph but different target sets to be different functions (another structure called *family* is used to represent functions without specified target set). The three fields are assigned to the first three positions in the tuple:

definition *"source(F) = fst(F)"*
definition *"target(F) = fst(snd(F))"*
definition *"graph(F)   = fst(snd(snd(F)))"*

A function with source S, target T, and graph G is represented by the tuple $\langle S,T,G,\emptyset \rangle$ (we append an $\emptyset$ at the end so the definition of **graph** works properly). For G to actually represent a function, it must satisfy the conditions for a functional graph:

definition *func_graphs* :: *"i ⇒ i ⇒ i"* **where**
  "func_graphs(X,Y) = {G∈Pow(X×Y). (∀a∈X. ∃!y. ⟨a,y⟩∈G)}"

The set of all functions from S to T (denoted $S \rightarrow T$) is then given by:

definition *function_space* :: *"i ⇒ i ⇒ i"* (**infixr** *"→"* 60) **where**
  "A → B = {⟨A,B,G,∅⟩. G∈func_graphs(A,B)}"

Functions can be created using the following constructor. Note this is a higher-order meta-function. The argument b can be supplied by a lambda expression.

definition *Fun* :: *"[i, i, i ⇒ i] ⇒ i"* **where**
  "Fun(A,B,b) = ⟨A, B, {p∈A×B. snd(p) = b(fst(p))}, ∅⟩"

Evaluation of a function f at x (denoted $f\,{}^{\backslash}x$) is then defined as:

definition *feval* :: *"i ⇒ i ⇒ i"* (**infixl** *"\"* 90) **where**
  "f ` x = (THE y. ⟨x,y⟩∈graph(f))"

2.3   Algebraic Structures

The second major use of records is to represent algebraic structures. In our framework, we will define structures such as groups, abelian groups, rings, and ordered rings. The carrier set of a structure is assigned to the first position. The order relation, additive data, and multiplicative data are assigned to the third, fourth, and fifth position, respectively. This is expressed as follows:

definition *"carrier(S)   = fst(S)"*
definition *"order_graph(S) = fst(snd(snd(S)))"*
definition *"zero(S)   = fst(fst(snd(snd(snd(S)))))"*
definition *"plus_fun(S)   = snd(fst(snd(snd(snd(S)))))"*
definition *"one(S)   = fst(fst(snd(snd(snd(snd(S))))))"*
definition *"times_fun(S)   = snd(fst(snd(snd(snd(snd(S))))))"*

Here *order_graph* is a subset of $S×S$, and *plus_fun*, *times_fun* are elements of $S×S→S$. Hence, the operators $\leq, +,$ and $*$ can be defined as follows:

definition "le(R,x,y) ⟷ ⟨x,y⟩∈order_graph(R)"
definition "plus(R,x,y) = plus_fun(R)`⟨x,y⟩"
definition "times(R,x,y) = times_fun(R)`⟨x,y⟩"

These are abbreviated to $x \leq_R y$, $x +_R y$, and $x *_R y$, respectively (in both theory files and throughout this paper, we use $*$ to denote multiplication in groups and rings, and $\times$ to denote product on sets and other structures). We also abbreviate $x \in$ carrier(S) to $x \in S$.

The constructor for group-like structures is as follows:

definition Group :: "[i, i, i ⇒ i ⇒ i] ⇒ i" **where**
  "Group(S,u,f) = ⟨S,∅,∅,∅,⟨u,λp∈S×S. f(fst(p),snd(p))∈S⟩,∅⟩"

The following predicate asserts that a structure contains *at least* the fields of a group-like structure, with the right membership properties (1_G abbreviates one(G)):

definition is_group_raw :: "i ⇒ o" **where**
  "is_group_raw(G) ⟷
    $1_G \in$. G ∧ times_fun(G) ∈ carrier(G) × carrier(G) → carrier(G)

To check whether such a structure is in fact a monoid/group, we use the following predicates:

definition is_monoid :: "i ⇒ o" **where**
  "is_monoid(G) ⟷ is_group_raw(G) ∧
    (∀x∈.G. ∀y∈.G. ∀z∈.G. (x *_G y) *_G z = x *_G (y *_G z)) ∧
    (∀x∈.G. 1_G *_G x = x ∧ x *_G 1_G = x)"

definition units :: "i ⇒ i" **where**
  "units(G) = {x ∈. G. (∃y∈.G. y *_G x = 1_G ∧ x *_G y = 1_G)}"

definition is_group :: "i ⇒ o" **where**
  "is_group(G) ⟷ is_monoid(G) ∧ carrier(G) = units(G)"

Note these definitions are meaningful on any structure that has multiplicative data. Likewise, we can define a predicate is_abgroup for abelian groups, that is meaningful for any structure that has additive data. These can be combined with distributive properties to define the predicate for a ring:

definition is_ring :: "i ⇒ o" **where**
  "is_ring(R) ⟷ (is_ring_raw(R) ∧ is_abgroup(R) ∧ is_monoid(R) ∧
    is_left_distrib(R) ∧ is_right_distrib(R) ∧ $0_R \neq 1_R$)"

Likewise, we can define the predicate for ordered rings, and constructors for such structures. Structures are used to represent the hierarchy of numbers: we let *nat int, ra*, and *real* denote the *set* of natural numbers, integers, etc., while $\mathbb{N}, \mathbb{Z}, \mathbb{Q}$, and $\mathbb{R}$ denote the corresponding structures. Hence, addition on natural numbers is denoted by $x +_N y$, addition on real numbers by $x +_R y$, etc. We can also state and prove theorems such as is_ord_field($\mathbb{R}$), which contains all proof obligations for showing that the real numbers form an ordered field.

2.4   Morphism Between Structures

Finally, we discuss morphisms between structures. Morphisms can be considered as an *extension* of functions, with additional fields specifying structures on the source and target sets. The two additional fields are assigned to the fourth and fifth positions in the tuple:

```
definition "source_str(F) = fst(snd(snd(snd(F))))"
definition "target_str(F) = fst(snd(snd(snd(snd(F)))))"
```

The constructor for a morphism is as follows (here S and T are the source and target structures, while the source and target sets are automatically derived):

```
definition Mor :: "[i, i, i ⇒ i] ⇒ i" where
  "Mor(S,T,b) = (let A = carrier(S) in let B = carrier(T) in
   ⟨A, B, {p∈A×B. snd(p) = b(fst(p))}, S, T, ∅⟩)"
```

The space of morphisms (denoted $S \rightharpoonup T$) is given by:

```
definition mor_space :: "i ⇒ i ⇒ i" (infix "⇀" 60) where
  "mor_space(S,T) = (let A = carrier(S) in let B = carrier(T) in
   {⟨A,B,G,S,T,∅⟩. G∈func_graphs(A,B)})"
```

Note the notation $f`x$ for evaluation still works for morphisms. Several other concepts defined in terms of evaluation, such as image and inverse image, continue to be valid for morphisms as well, as are lemmas about these concepts. However, operations that construct new morphisms, such as inverse and composition, must be redefined. We will use $g \circ f$ to denote the composition of two functions, and $g \circ_m f$ to denote the composition of two morphisms.

Having morphisms store the source and target structures means we can define properties such as homomorphism on groups as a predicate:

```
definition is_group_hom :: "i ⇒ o" where
  "is_group_hom(f) ⟷ (let S = source_str(f) in let T = target_str(f) in

       is_morphism(f) ∧ is_group(S) ∧ is_group(T) ∧
       (∀x∈.S. ∀y∈.S. f`(x *ₛ y) = f`x *ₜ f`y))"
```

The following lemma then states that the composition of two homomorphisms is a homomorphism (this is proved automatically using auto2):

```
lemma group_hom_compose:
  "is_group_hom(f) ⟹ is_group_hom(g) ⟹
   target_str(f) = source_str(g) ⟹ is_group_hom(g ∘ₘ f)"
```

3   Auto2 in Untyped Set Theory

In this section, we describe several additional features of auto2, as well as general strategies of using it to manage the complexities of untyped set theory.

We begin with an overview of the auto2 system (see [17] for details). Auto2 is a theorem prover packaged as a tactic in Isabelle. It works with a collection of rules of reasoning called *proof steps*. New proof steps can be added at any time within an Isabelle theory. They can also be deleted at any time, although it is rarely necessary to add and delete the same proof step more than once. In general, when building an Isabelle theory, the user is responsible for specifying, by adding proof steps, how to use the results proved in that theory. In return, the user no longer needs to worry about invoking these results by name in future developments.

The overall algorithm of auto2 is as follows. First, the statement to be proved is converted into contradiction form, so the task is always to derive a contradiction from a list of assumptions. During the proof, auto2 maintains a list of *items*, the two most common types of which are propositions (that are derived from the assumptions) and terms (that have appeared so far in the proof). Each item resides in a *box*, which can be thought of as a subcase of the statement to be proved (the box corresponding to the original statement is called the *home box*). A proof step is a function that takes as input one or two items, and outputs either new items, new cases, or the action of shadowing one of the input items, or resolving a box by proving a contradiction in that box.

The main loop of the algorithm repeatedly applies the current collection of proof steps and adds any new items and cases in a best-first-search manner, until some proof step derives a contradiction in the home box. In addition to the list of items, auto2 also maintains several tables. The most important of which is the *rewrite table*, which keeps track of the list of currently known equalities (not containing arbitrary variables), and maintains the congruence closure of these equalities. There are two other tables: the property table and the well-form table, which we will discuss later in this section.

There are two broad categories of proof steps, which we call the *standard* and *special* proof steps in this paper. A standard proof step applies an existing theorem in a specific direction. It matches the input items to one or two patterns in the statement of the theorem, and applies the theorem to derive a new proposition. Here the matching is up to rewriting (*E-matching*) using the rewrite table. A special proof step can have more complex behavior, and is usually written as an ML function. The vast majority of proof steps in our example are standard, although special proof steps also play an important role.

The auto2 prover is not intended to be complete. For example, it may intentionally apply a theorem in only one of several possible directions, in order to narrow the search space. For more difficult theorems, auto2 provides a custom language of proof scripts, allowing the user to specify intermediate steps of the proof. Generally, when proving a result using auto2, the user will first try to prove it without any scripts, and in case of failure, successively add intermediate steps, perhaps by referring to an informal proof of the result. In case of failure, auto2 will indicate the first intermediate step that it is unable to prove, as well as what it is able to derive in the course of proving that step. We will show examples of proof scripts in Sect. 4.

The current version of auto2 can be set up to work with different logics in Isabelle. It contains a core program, for reasoning about predicate logic and equality, that is parametrized over the list of constants and theorems for the target logic. In particular, auto2 is now set up and tested to work with both HOL and FOL in Isabelle.

3.1 Encapsulation of Definitions

One commonly cited problem with untyped set theory is that every object is a set, including those that are not usually considered as sets. Common examples of the latter include ordered pairs, natural numbers, functions, etc. In informal treatments of mathematics, these definitions are only used to establish some basic properties of the objects concerned. Once these properties are proved, the definitions are never used again.

In formal developments, when automation is used to produce large parts of the proof, one potential problem is that the automation may needlessly expand the original definitions of objects, rather than focusing on their basic properties. This increases the search space and obscures the essential ideas of the proof. Using the ability to delete proof steps in auto2, this problem can be avoided entirely. For any definition that we wish to drop in the end, we use the following three-step procedure:

1. The definition is stated and added to auto2 as rewrite rules.
2. Basic properties of the object being defined are stated and proved. These properties are added to auto2 as appropriate proof steps.
3. The rewrite rules for the original definition are deleted.

For example, after the definitions concerning the representation of functions as tuples in Sect. 2.2, we prove the following lemmas, and add them as appropriate proof steps (as indicated by the attributes in brackets):

lemma lambda_is_function [backward]:
    "∀x∈A. f(x)∈B ⟹ Fun(A,B,f) ∈ A → B"

lemma beta [rewrite]:
    "F = Fun(A,B,f) ⟹ x ∈ source(F) ⟹ is_function(F) ⟹ F`x = f(x)"

lemma feval_in_range [typing]:
    "is_function(f) ⟹ x ∈ source(f) ⟹ f`x ∈ target(f)"

After proving these (and a few more) lemmas, the rewriting rules for the definitions of Fun, function_space, feval, etc., are removed. Note that all lemmas above are independent of the representation of functions as tuples. Hence, this representation is effectively hidden from the point of view of the prover. Some of the original definitions may be temporarily re-added in rare instances (for example when defining the concept of morphisms).

3.2 Property and Well-Form Tables

In this section, we discuss two additional tables maintained by auto2 during a proof. The property table is already present in the version introduced in [17], but not discussed in that paper. The well-form table is new.

The main motivation for both tables is that for many theorems, especially those stated in an untyped logic, some of its assumptions can be considered as "side conditions". To give a basic example, consider the following lemma:

lemma `unit_l_cancel:`
    "is_monoid(G) $\implies$ y $\in$. G $\implies$ z $\in$. G $\implies$ x $*_G$ y = x $*_G$ z $\implies$
    x $\in$ units(G) $\implies$ y = z"

In this lemma, the last two assumptions are the "main" assumptions, while the first three are side conditions asserting that the variables in the main assumptions are well-behaved in some sense. In Isabelle/HOL, these side conditions may be folded into type or type-class constraints.

We consider two kinds of side conditions. The first kind, like the first assumption above, checks that one of the variables in the main assumptions satisfy a certain predicate. In Isabelle/HOL, these may correspond to type-class constraints. In auto2, we call these *property assumptions*. More precisely, given any predicate (in FOL this means constant of type $i \Rightarrow o$), we can register it as a property. The *property table* records the list of properties satisfied by each term that has appeared so far in the proof. Properties propagate through equalities: if $P(a)$ is in the property table, and $a = b$ is known from the rewrite table, then $P(b)$ is automatically added to the property table. The user can also add theorems of certain forms as further propagation rules for the property table (we omit the details here).

The second kind of side conditions assert that certain terms occuring in the main assumptions are *well-formed*. We use the terminology of well-formedness to capture a familiar feature of mathematical language: that an expression may make implicit assumptions about its subterms. These conditions can be in the form of type constraints. For example, the expression $a +_R b$ implicitly assumes that a and b are elements in the carrier set of R. However, this concept is much more general. Some examples of well-formedness conditions are summarized in the following table:

Term	Conditions
$\bigcap A$	$A \neq \emptyset$
$f \, ` \, x$	$x \in source(f)$
$g \circ f$	$target(f) = source(g)$
$g \circ_m f$	$target\_str(f) = source\_str(g)$
$a +_R b$	$a \in. R,\ b \in. R$
$inv(R,a)$	$a \in units(R)$
$a /_R b$	$a \in. R,\ b \in units(R)$
$subgroup(G,H)$	$is\_subgroup\_set(G,H)$
$quotient\_group(G,H)$	$is\_normal\_subgroup\_set(G,H)$

In general, given any meta-function f, any propositional expression in terms of the arguments of f can be registered as a well-formedness condition of f. In particular, well-formedness conditions are not necessarily properties. For example, the condition $a \in R$ for $a +_R b$ involves two variables and hence is not a property. The *well-form table* records, for every term encountered so far in the proof, the list of its well-formedness conditions that are satisfied. Whenever a new fact is added, auto2 checks against every known term to see whether it verifies a well-formedness condition of that term.

The property and well-form tables are used in similar ways in standard proof steps. After the proof step matches one or two patterns in the "main" assumptions or conclusion of the theorem that it applies, it checks for the side conditions in the two tables, and proceed to apply the theorem only if all side conditions are found. Of course, this requires proof steps to be re-applied if new properties or well-formedness conditions of a term becomes known.

3.3   Well-Formed Conversions

Algebraic simplification is an important part of any automatic prover. For every kind of algebraic structure, e.g. monoids, groups, abelian groups, and rings, there is a concept of normal form of an expression, and two terms can be equated if they have the same normal form. In untyped set theory, such computation of normal forms is complicated by the fact that the relevant rewriting rules have extra assumptions. For example, the rule for associativity of addition is:

$$\texttt{is\_abgroup}(R) \implies x \in R \implies y \in R \implies z \in R \implies$$
$$x +_R (y +_R z) = (x +_R y) +_R z$$

The first assumption can be verified at the beginning of the normalization process. The remaining assumptions, however, are more cumbersome. In particular, they may require membership status of terms that arise only during the normalization. For example, when normalizing the term $a +_R (b +_R (c +_R d))$, we may first rewrite it to $a +_R ((b +_R c) +_R d)$. The next step, however, requires $b +_R c \in R$, where $b +_R c$ does not occur initially and may not have occured so far in the proof. In typed theories, this poses no problem, since $b + c$ will be automatically given the same type as b and c when the term is created.

In untyped set theory, such membership information must be kept track of and derived when necessary. The concept of well-formed terms provides a natural framework for doing this. Before performing algebraic normalization on a term, we first check for all relevant well-formedness conditions. If all conditions are present, we produce a data structure (of type *wfterm* in Isabelle/ML) containing the certified term as well as theorems asserting well-formedness conditions. A theorem is called a *well-formed rewrite rule* if its main conclusion is an equality, each of its assumptions is a well-formedness condition for terms on the left side of the equality, and it has additional conclusions that verify all well-formedness conditions for terms on the right side of the equality that are not already present in the assumptions. For example, the associativity rule stated above is not yet a well-formed rewrite rule: there is no justification for $x +_R y \in. R$, which is a well-formedness condition for the term $(x +_R y) +_R z$ on the right side of the equality. The full well-formed rewrite rule is:

$$\texttt{is\_abgroup}(R) \implies x \in. R \implies y \in. R \implies z \in. R \implies$$
$$x +_R (y +_R z) = (x +_R y) +_R z \wedge x +_R y \in. R$$

Given a well-formed rewrite rule, we can produce a *well-formed conversion* that acts on *wfterm* objects, in a way similar to how equalities produce regular conversions that act on *cterm* objects in Isabelle/ML. Like regular conversions, well-formed conversions can be composed in various ways, and full normalization procedures can be written using the language of well-formed conversions. These normalization procedures in turn form the basis of several special proof steps. We give two examples:

– Given two terms s and t that are non-atomic with respect to operations in R, where R is a monoid (group/abelian group/ring), normalize s and t using the rules for R. If the normalizations are equal, output $s = t$.
– Given two propositions $a \leq_R b$ and $\neg (c \leq_R d)$, where R is an ordered ring. Compare the normalizations of $b -_R a$ and $d -_R c$. If they are equal, output a contradiction.

These proof steps, when combined with proof scripts provided by the user, allow algebraic manipulations to be performed rapidly. They replace the handling of associative-commutative functions for HOL discussed in [17].

3.4   Discussion

We conclude this section with a discussion of our overall approach to untyped set theory, and compare it with other approaches. One feature of our approach is that we do not seek to re-institute a concept of types in our framework, but simply replace type constraints with set membership conditions (or predicates, for constraints that cannot be described by a set). The aim is to fully preserve the flexibility of set-membership as compared to types. Empirically, most of the extra assumptions that arise in the statement of theorems can be taken care of by classifying them as properties or well-formedness conditions. Our approach can be contrasted with that taken by Mizar, which defines a concept of soft types [16] within the core of the system.

Every framework for formalizing modern mathematics need a way to deal with structures. In Mizar, structures are defined in the core of the system as partial functions on selectors [9,15]. In both Isabelle/HOL and IsarMathLib's treatement of abstract algebra, structures are realized with extensive use of locales. For Coq, one notable approach is the use of Canonical Structures [10] in the formalization of the Odd Order Theorem. We chose a relatively simple scheme of realizing structures as tuples, which is sufficient for the present purposes. Representing them as partial functions on selectors, as in Mizar, is more complicated but may be beneficial in the long run.

Finally, we emphasize that we do not make any modification to Isabelle/FOL in our development. The concept of well-formed terms, for example, is meaningful only to the automation. The whole of auto2's design, including the ability for users to add new proof steps, follows the LCF architecture. To have confidence in the proofs, one only need to trust the existing Isabelle system, the ten axioms stated in Sect. 2.1, and the definitions involved in the statement of the results.

4   Examples of Proof Scripts

Using the techniques in the above two sections, we formalized enough mathematics in Isabelle/FOL to be able to define the fundamental group. In addition to work directly used for that purpose, we also formalized several interesting results on the side. These include the well-ordering theorem and Zorn's lemma, the first isomorphism theorem for groups, and the intermediate value theorem. Two more examples will be presented in the remainder of this section, to demonstrate the level of succinctness of proof scripts that can be achieved.

Throughout our work, we referred to various sources including both mathematical texts and other formalizations. We list these sources here:

- Axioms of set theory and basic operations on sets, construction of natural numbers using least fixed points: from Isabelle/ZF [13,14].
- Equivalence and order relations, arbitrary products on sets, well-ordering theorem and Zorn's lemma: from Bourbaki's *Theory of Sets* [2].
- Group theory and the construction of real numbers using Cauchy sequences: from my previous case studies [17], which in turn is based on corresponding theories in the Isabelle/HOL library.
- Point-set topology and construction of the fundamental group: from *Topology* by Munkres [12].

4.1   Schroeder-Bernstein Theorem

For our first example, we present the proof of the Schroeder-Bernstein theorem. See [14] for a presentation of the same proof in Isabelle/ZF. The bijection is constructed by gluing together two functions. Auto2 is able to prove automatically that under certain conditions, the gluing is a bijection (lemma `glue_function2_bij`). For the Schroeder-Bernstein theorem, a proof script (provided by the user) is needed. This is given immediately after the statement of the theorem.

```
definition glue_function2 :: "i ⇒ i ⇒ i" where
  "glue_function2(f,g) = Fun(source(f)∪source(g), target(f)∪target(g),
    λx. if x ∈ source(f) then f`x else g`x)"

lemma glue_function2_bij [backward]:
  "f ∈ A ≅ B ⟹ g ∈ C ≅ D ⟹ A ∩ C = ∅ ⟹ B ∩ D = ∅ ⟹
  glue_function2(f,g) ∈ (A ∪ C) ≅ (B ∪ D)"
```

```
theorem schroeder_bernstein:
  "injective(f) ⟹ injective(g) ⟹ f ∈ X → Y ⟹ g ∈ Y → X ⟹
   equipotent(X,Y)"
  LET "X_A = lfp(X, λW. X - g``(Y - f``W))" THEN
  LET "X_B = X - X_A, Y_A = f``X_A, Y_B = Y - Y_A" THEN
  HAVE "X - g``Y_B = X_A" THEN
  HAVE "g``Y_B = X_B" THEN
  LET "f' = func_restrict_image(func_restrict(f,X_A))" THEN
  LET "g' = func_restrict_image(func_restrict(g,Y_B))" THEN
  HAVE "glue_function2(f', inverse(g')) ∈ (X_A ∪ X_B) ≅ (Y_A ∪ Y_B)"
```

4.2   Rempe-Gillen's Challenge

For our second example, we present our solution to a challenge problem proposed by Lasse Rempe-Gillen in a mailing list discussion[2]. See [1] for proofs of the same result in several other systems. The statement to be proved is:

Lemma 1. *Let f be a continuous real-valued function on the real line, such that $f(x) > x$ for all x. Let x_0 be a real number, and define the sequence x_n recursively by $x_{n+1} := f(x_n)$. Then x_n diverges to infinity.*

Our solution is as follows. We make use of several previously proved results: any bounded increasing sequence in $\mathbb{R}$ converges (line 2), a continuous function f maps a sequence converging to x to a sequence converging to $f`x$ (line 4), and finally that the limit of a sequence in $\mathbb{R}$ is unique.

```
lemma rempe_gillen_challenge:
  "real_fun(f) ⟹ continuous(f) ⟹ incr_arg_fun(f) ⟹ x0 ∈. ℝ ⟹
   S = Seq(ℝ, λn. nfold(f,n,x0)) ⟹ ¬upper_bounded(S)"
  HAVE "seq_incr(S)" WITH HAVE "∀n∈.ℕ. S`(n +ᵣ 1) ≥ᵣ S`n" THEN
  CHOOSE "x, converges_to(S,x)" THEN
  LET "T = Seq(ℝ, λn. f`(S`n))" THEN
  HAVE "converges_to(T,f`x)" THEN
  HAVE "converges_to(T,x)" WITH (
    HAVE "∀r>ᵣ0ᵣ. ∃k∈.ℕ. ∀n≥ₙk. |T`n −ᵣx|ᵣ <ᵣ r" WITH (
      CHOOSE "k ∈. ℕ, ∀n≥ₙk. |S`n −ᵣ x|ᵣ <ᵣ r" THEN
      HAVE "∀n≥ₙk. |T`n −ᵣ x|ᵣ <ᵣ r" WITH HAVE "T`n = S`(n +ₙ 1)"))
```

5   Construction of the Fundamental Group

In this section, we describe our construction of the fundamental group. We will focus on stating the definitions and main results without proof, to demonstrate the expressiveness of untyped set theory under our framework. The entire formalization including proofs is 864 lines long.

Let I be the interval $[0,1]$, equipped with the subspace topology from the topology on $\mathbb{R}$. Given two continuous maps f and g from S to T, a *homotopy* between f and g is a continuous map from the product topology on $S \times I$ to T that restricts to f and g at $S \times \{0\}$ and $S \times \{1\}$, respectively:

---

[2] http://www.cs.nyu.edu/pipermail/fom/2014-October/018243.html.

definition is_homotopy :: "[i, i, i] ⇒ o" **where**
  "is_homotopy(f,g,F) ⟷
    (**let** S = source_str(f) **in** **let** T = target_str(f) **in**
    continuous(f) ∧ continuous(g) ∧
    S = source_str(g) ∧ T = target_str(g) ∧ F ∈ S ×$_\tau$ I →$_\tau$ T ∧
    (∀x∈.S. F'⟨x,0$_\mathbb{R}$⟩ = f'x ∧ F'⟨x,1$_\mathbb{R}$⟩ = g'x))"

A *path* is a continuous function from the interval. A homotopy between two paths is a *path homotopy* if it remains constant on {0} × I and {1} × I:

definition is_path :: "i ⇒ o" **where**
  "is_path(f) ⟷ (f ∈ I →$_\tau$ target_str(f))"

definition is_path_homotopy :: "[i, i, i] ⇒ o" **where**
  "is_path_homotopy(f,g,F) ⟷
    (is_path(f) ∧ is_path(g) ∧ is_homotopy(f,g,F) ∧
    (∀t∈.I. F'⟨0$_\mathbb{R}$,t⟩ = f'(0$_\mathbb{R}$) ∧ F'⟨1$_\mathbb{R}$,t⟩ = f'(1$_\mathbb{R}$)))"

Two paths are *path-homotopic* if there exists a path homotopy between them. This is an equivalence relation on paths.

definition path_homotopic :: "i ⇒ i ⇒ o" **where**
  "path_homotopic(f,g) ⟷ (∃F. is_path_homotopy(f,g,F))"

The path product is defined by gluing two morphisms. It is continuous by the pasting lemma:

definition I1 = subspace(ℝ, closed_interval(ℝ, 0$_\mathbb{R}$, 1$_\mathbb{R}$ /$_\mathbb{R}$ 2$_\mathbb{R}$))
definition I2 = subspace(ℝ, closed_interval(ℝ, 1$_\mathbb{R}$ /$_\mathbb{R}$ 2$_\mathbb{R}$, 1$_\mathbb{R}$))
definition interval_lower = Mor(I1,I,λt. 2$_\mathbb{R}$ *$_\mathbb{R}$ t)
definition interval_upper = Mor(I2,I,λt. 2$_\mathbb{R}$ *$_\mathbb{R}$ t −$_\mathbb{R}$ 1$_\mathbb{R}$)

definition path_product :: "i ⇒ i ⇒ i" (infixl "⋆" 70) **where**
  "f ⋆ g = glue_morphism(I, f ∘$_m$ interval_lower, g ∘$_m$ interval_upper)"

The loop space is a set of loops on *X*. Path homotopy gives an equivalence relation on the loop space, and we define loop_classes to be the quotient set:

definition loop_space :: "i ⇒ i ⇒ i" **where**
  "loop_space(X,x) = {f ∈ I →$_\tau$ X. f'(0$_\mathbb{R}$) = x ∧ f'(1$_\mathbb{R}$) = x}"

definition loop_space_rel :: "i ⇒ i ⇒ i" **where**
  "loop_space_rel(X,x) = Equiv(loop_space(X,x), λf g.
path_homotopic(f,g))"

definition loop_classes :: "i ⇒ i ⇒ i" **where**
  "loop_classes(X,x) = loop_space(X,x) // loop_space_rel(X,x)"

Finally, the fundamental group is defined as:

definition `fundamental_group :: "i ⇒ i ⇒ i" ("π₁") where`
  `"π₁(X,x) = (let R = loop_space_rel(X,x) in`
    `Group(loop_classes(X,x), equiv_class(R,const_mor(I,X,x)),`
      `λf g. equiv_class(R,rep(R,f) ⋆ rep(R,g))))"`

To show that the fundamental group is actually a group, we need to show that the path product respects the equivalence relation given by path homotopy, and is associative up to equivalence (along with properties about inverse and identity). The end result is:

lemma `fundamental_group_is_group:`
  `"is_top_space(X) ⟹ x ∈. X ⟹ is_group(π₁(X,x))"`

An important property of the fundamental group is that a continuous function between topological spaces induces a homomorphism between their fundamental groups. This is defined as follows:

definition `induced_mor :: "i ⇒ i ⇒ i" where`
  `"induced_mor(k,x) =`
    `(let X = source_str(k) in let Y = target_str(k) in`
    `let R = loop_space_rel(X,x) in let S = loop_space_rel(Y,k`x) in`
    `Mor(π₁(X,x), π₁(Y,k`x), λf. equiv_class(S, k ∘ₘ rep(R,f))))"`

The induced map is a homomorphism satisfying functorial properties:

lemma `induced_mor_is_homomorphism:`
  `"continuous(k) ⟹ X = source_str(k) ⟹ Y = target_str(k) ⟹`
  `x ∈ source(k) ⟹ induced_mor(k,x) ∈ π₁(X,x) →_g π₁(Y,k`x)"`

lemma `induced_mor_id:`
  `"is_top_space(X) ⟹ x ∈. X ⟹`
  `induced_mor(id_mor(X),x) = id_mor(π₁(X,x))"`

lemma `induced_mor_comp:`
  `"continuous(k) ⟹ continuous(h) ⟹`
  `target_str(k) = source_str(h) ⟹ x ∈ source(k) ⟹`
  `induced_mor(h ∘ₘ k, x) = induced_mor(h, k`x) ∘ₘ induced_mor(k, x)"`

6   Related Work

In Isabelle, the main library for formalized mathematics using FOL is Isabelle/ZF. The basics of Isabelle/ZF is described in [13,14]. We also point to [13] for a review of older work on set theory from automated deduction and artificial intelligence communities. Outside the official library, IsarMathLib [5] is a more recent project based on Isabelle/ZF. It formalized more results in abstract algebra and point-set topology, and also constructed the real numbers. The initial parts of our development closedly parallels that in Isabelle/ZF, but we go further in several directions including constructing the number system. The primary difference between our work and IsarMathLib is that we use auto2 for proofs, and develop our own system for handling structures, so that we do not make use of Isabelle tactics, Isar, or locales.

Outside Isabelle, the major formalization projects using set theory include Metamath [11] and Mizar [4], both of which have extensive mathematical libraries. There are some recent efforts to reproduce the Mizar environment in HOL-type systems [6,8]. While there are some similarities between our framework and Mizar's, we do not aim for an exact reproduction. In particular, we maintain the typical style of stating definitions and theorems in Isabelle. More comparisons between our approach and Mizar are discussed in Sect. 3.4.

Mizar formalized not just the definition of the fundamental group [7], but several of its properties, including the computation of the fundamental group of the circle. There is also a formalization of path homotopy in HOL Light which is then ported to Isabelle/HOL. This is used for the proof of the Brouwer fixed-point theorem and the Cauchy integral theorem, although the fundamental group itself does not appear to be constructed.

In homotopy type theory, one can work with fundamental groups (and higher-homotopy groups) using synthetic definitions. This has led to formalizations of results about homotopy groups that are well beyond what can be achieved today using standard definitions (see [3] for a more recent example). We emphasize that our definition of the fundamental group, as with Mizar's, follows the standard one in set theory.

7   Conclusion

We applied the auto2 prover to the formalization of mathematics using untyped set theory. Starting from the axioms of set theory, we formalized the definition of the fundamental group, as well as many other results in set theory, group theory, point-set topology, and real analysis. The entire development contains over 13,000 lines of theory files and 3,500 lines of ML code, taking the author about 5 months to complete. On a laptop with two 2.0 GHz cores, it can be compiled in about 24 min. Through this work, we demonstrated the ability of auto2 to scale to relatively large projects. We also hope this result can bring renewed interest to formalizing mathematics in untyped set theory in Isabelle.

Acknowledgements. The author would like to thank the anonymous referees for their comments. This research is completed while the author is supported by NSF Award No. 1400713.

References

1. Blanchette, J.C., Kaliszyk, C., Paulson, L.C., Urban, J.: Hammering towards QED. J. Formalized Reason. **9**(1), 101–148 (2016)
2. Bourbaki, N.: Theory of Sets. Springer, Heidelberg (2000)
3. Brunerie, G.: On the homotopy groups of spheres in homotopy type theory. Ph.D. thesis. https://arxiv.org/abs/1606.05916

4. Grabowski, A., Kornilowicz, A., Naumowicz, A.: Mizar in a nutshell. J. Formaliz. Reason. Spec. Issue: User Tutor. I **3**(2), 153–245 (2010)
5. IsarMathLib. http://www.nongnu.org/isarmathlib/
6. Kaliszyk, C., Pak, K., Urban, J.: Towards a Mizar environment for Isabelle: foundations and language. In: Proceedings of the 5th ACM SIGPLAN Conference on Certified Programs and Proofs (CPP 2016), New York, pp. 58–65 (2016)
7. Kornilowicz, A., Shidama, Y., Grabowski, A.: The fundamental group. Formalized Math. **12**(3), 261–268 (2004)
8. Kuncar, O.: Reconstruction of the Mizar type system in the HOL light system. In: Pavlu, J., Safrankova, J. (eds.) WDS Proceedings of Contributed Papers: Part I - Mathematics and Computer Sciences, pp. 7–12. Matfyzpress (2010)
9. Lee, G., Rudnici, P.: Alternative aggregates in Mizar. In: Kauers, M., Kerber, M., Miner, R., Windsteiger, W. (eds.) Calculemus/MKM 2007. LNCS (LNAI), vol. 4573, pp. 327–341. Springer, Heidelberg (2007). doi:10.1007/978-3-540-73086-6_26
10. Mahboubi, A., Tassi, E.: Canonical structures for the working Coq user. In: Blazy, S., Paulin-Mohring, C., Pichardie, D. (eds.) ITP 2013. LNCS, vol. 7998, pp. 19–34. Springer, Heidelberg (2013). doi:10.1007/978-3-642-39634-2_5
11. Megill, N.D.: Metamath: a computer language for pure mathematics. http://us.metamath.org/downloads/metamath.pdf
12. Munkres, J.R.: Topology. Prentice Hall, Upper Saddle River (2000)
13. Paulson, L.C.: Set theory for verification: I. From foundations to functions. J. Automated Reason. **11**(3), 353–389 (1993)
14. Paulson, L.C.: Set theory for verification: II. Induction and recursion. J. Automated Reason. **15**(2), 167–215 (1995)
15. Trybulec, A.: Some features of the Mizar language. In: ESPRIT Workshop (1993)
16. Wiedijk, F.: Mizar's soft type system. In: Schneider, K., Brandt, J. (eds.) TPHOLs 2007. LNCS, vol. 4732, pp. 383–399. Springer, Heidelberg (2007). doi:10.1007/978-3-540-74591-4_28
17. Zhan, B.: AUTO2, a saturation-based heuristic prover for higher-order logic. In: Blanchette, J.C., Merz, S. (eds.) ITP 2016. LNCS, vol. 9807, pp. 441–456. Springer, Cham (2016). doi:10.1007/978-3-319-43144-4_27

Author Index

Printed in the United States
by Baker & Taylor Publisher Services